AF386478

VEGETATION DYNAMICS

Understanding ecosystem structure and function requires familiarity with the techniques, knowledge and concepts of the three disciplines of plant physiology, remote sensing and modelling. This is the first textbook to provide the fundamentals of these three domains in a single volume. It applies cross-disciplinary insights to multiple case studies in vegetation and landscape science. A key feature of these case studies is an examination of relationships among climate, vegetation structure and vegetation function, to address fundamental research questions. This book is for advanced students and researchers who need to understand and apply knowledge from the disciplines of plant physiology, remote sensing and modelling. It allows readers to integrate and synthesise knowledge to produce a holistic understanding of the structure, function and behaviour of forests, woodlands and grasslands.

DEREK EAMUS is a professor at the University of Technology Sydney (UTS). He is an internationally recognised plant ecophysiologist and ecohydrologist, specialising in plant water relations, the carbon and water balances of native woodlands and forests and forging cross-disciplinary links between our understanding of processes at cellular, whole plant and canopy scales. He has published more than 185 research publications in diverse journals, including *Nature, Nature Climate Change, Remote Sensing of Environment, Oecologia, Global Change Biology* and *Agricultural and Forest Meteorology.*

ALFREDO HUETE is a professor and geospatial ecologist at UTS. He uses remote sensing to examine ecosystem functioning, phenology and vegetation health, with an emphasis on extreme climate events. He has twenty-five years' experience in satellite earth observation for NASA and is a member of the EOS MODIS Science Team. The satellite products he developed are among the most widely used by the scientific and resource management communities. He has published several high-impact papers in journals such as *Nature, Science* and the *Proceedings of the National Academy of Sciences.*

QIANG YU is a professor at UTS and formerly at the Chinese Academy of Science, where he was awarded a professorship in the prestigious "Hundred Talents Program". His principle research interests are modelling of stomatal function, leaf and canopy photosynthesis and transpiration, carbon and water fluxes, and climate impacts on agriculture. He is the lead author of the China Agricultural Ecosystem Model. He has published more than 100 research papers in environmental modelling, climatology, ecology, agronomy and water resources.

VEGETATION DYNAMICS
A Synthesis of Plant Ecophysiology, Remote Sensing and Modelling

DEREK EAMUS

University of Technology Sydney

ALFREDO HUETE

University of Technology Sydney

QIANG YU

University of Technology Sydney

CAMBRIDGE
UNIVERSITY PRESS

Shaftesbury Road, Cambridge CB2 8EA, United Kingdom

One Liberty Plaza, 20th Floor, New York, NY 10006, USA

477 Williamstown Road, Port Melbourne, VIC 3207, Australia

314–321, 3rd Floor, Plot 3, Splendor Forum, Jasola District Centre, New Delhi – 110025, India

103 Penang Road, #05–06/07, Visioncrest Commercial, Singapore 238467

Cambridge University Press is part of Cambridge University Press & Assessment,
a department of the University of Cambridge.

We share the University's mission to contribute to society through the pursuit of
education, learning and research at the highest international levels of excellence.

www.cambridge.org
Information on this title: www.cambridge.org/9781107054202

First published 2016

A catalogue record for this publication is available from the British Library

Library of Congress Cataloging-in-Publication data
Names: Eamus, Derek, author. | Huete, Alfredo, author. | Yu, Qiang, 1962– author.
Title: Vegetation dynamics : a synthesis of plant ecophysiology, remote
sensing and modelling / Derek Eamus, University of Technology Sydney, Alfredo Huete,
University of Technology Sydney, Qiang Yu, University of Technology Sydney.
Description: New York, NY : Cambridge University Press, [2015] |
Includes bibliographical references and index.
Identifiers: LCN 2015029272| ISBN 9781107054202 (hardback) |
ISBN 9781107656666 (pbk.)
Subjects: LCSH: Plant ecophysiology. | Vegetation dynamics –
Remote sensing. | Vegetation dynamics – Mathematical models.
Classification: LC QK717.E16 2015 | DDC 581.7–dc23
LC record available at http://lccn.loc.gov/2015029272

ISBN   978-1-107-05420-2   Hardback

Contents

Section Three   Modelling

Preface

"Classical" plant physiology is the study of physiological processes of individual plants of a single species growing in pots in glasshouses, growth cabinets and controlled-environment chambers. Single-factor experiments are frequently used to manipulate one variable (e.g. water supply, temperature) in order to establish the response of individual processes (e.g. transpiration rate, phloem loading) or whole plants (e.g. growth rate) to that variable. It has been an immensely powerful science, contributing to increased food productivity and crop genetic selection for many decades.

Ecophysiology takes knowledge gained from plant physiological studies and applies them to plants growing "in the wild", in real landscapes. This adds several layers of complexity arising from (a) large spatial and temporal variations in multiple variables (e.g. rainfall, temperature, solar radiation); (b) the interactions amongst multiple variables; and (c) complexities arising from the fact that landscapes are composed of multiple species. Although manipulative experiments can be undertaken in ecophysiology (e.g. rainfall exclusion, and rainfall redistribution troughs), the majority of ecophysiological studies do not manipulate environmental variables. Rather, they allow natural seasonal and inter-annual variation to impact on the structure and function of natural vegetation and measure the response of individual leaves, plants (trees, grasses, etc.) and canopies and use statistical inferences and models to analyse these responses.

Modelling of plant function can similarly be undertaken at small (leaves; xylem function), intermediate (trees, canopies) and large scales (stands, regions, sub-continental, global) across a range of temporal scales (typically hours to centuries). These models incorporate plant physiological and ecophysiological data (e.g. light response curves of leaves, eddy covariance tower flux data) to model the function (e.g. gross primary productivity [GPP], net primary productivity [NPP], evapotranspiration [ET]) of landscapes and biomes.

Remote sensing (RS) uses air-borne and satellite platforms for remote surveillance of land and vegetation surfaces (e.g. reflectance of solar radiation across multiple wavebands, land surface temperature). Using these remotely sensed data, plant structural attributes (e.g. LAI) and functional attributes (e.g. NPP, ET) can be calculated. As is the case for modelling, RS as a discipline is increasingly using physiological

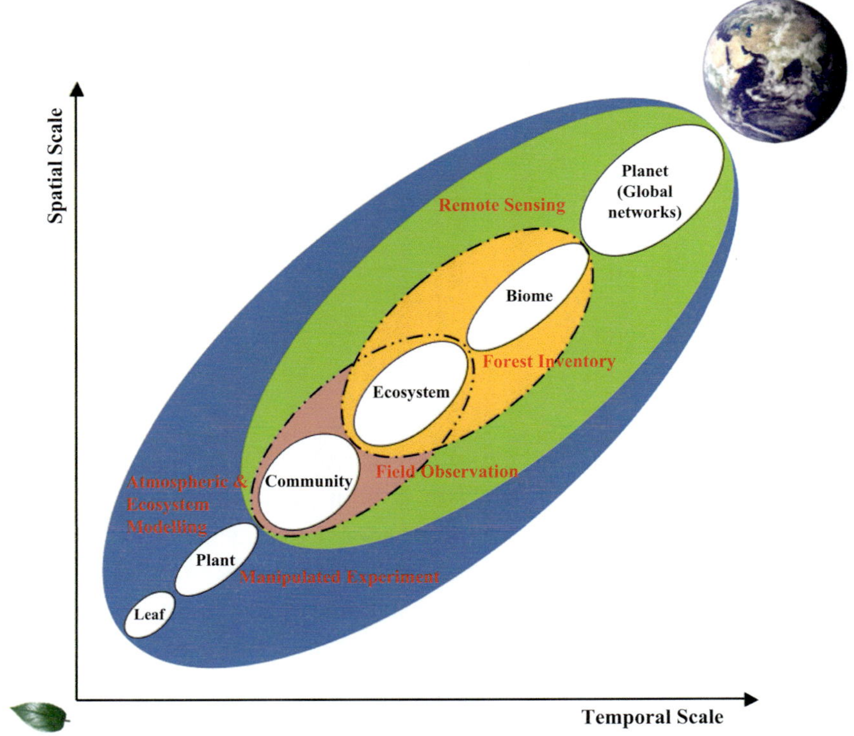

Figure I.1. A simplified conceptual representation of the different spatial and temporal scales of organisation (and study) encompassed in this textbook. Plant physiological and ecophysiological observations are mostly confined to cellular, leaf, whole plant, community and ecosystem spatial scales; most observations are made at sub-annual temporal scales and few observations have been made for longer than 15 years. In contrast, modelling can be undertaken from leaf-to-global-scales at almost any required temporal scale. Remote sensing is generally applied at community-to-global spatial scales and with weekly-to-decadal temporal scales.

and ecophysiological (e.g. canopy conductance, canopy gas fluxes, LAI) data to validate/test/compare remotely sensed estimates of landscape processes and vegetation structure. Figure I.1 provides a simplified representation of these three disciplines and their overlap.

Aims of This Book

The first aim of this book is to provide a relatively simple guide to some key aspects of plant physiology and plant ecophysiology, as they relate to the functional behaviour of natural landscapes, with particular emphasis on carbon (C) and water fluxes. This section should be of most value to those who are experienced modellers and remote sensing practitioners who need to increase their knowledge of plant physiology and ecophysiology. The focus on physiology pertaining to C and water fluxes is deliberate because these fluxes are two of the principal vegetation functions that drive all downstream aspects of landscape function (e.g. catchment water balance, productivity, biogeochemical cycling). The second and third aims are to provide similar introductions to the disciplines of remote sensing (RS) and modelling of vegetation structure and function. It is likely that ecophysiologists will, at some point in their career, need to include aspects of these two disciplines in their work; similarly, modellers will increasingly use RS data in conjunction with ecophysiological information whilst RS practitioners increasingly need the ability to develop and apply models and incorporate/understand ecophysiological data in relation to the insights generated through remote sensing.

Thus, the final aim of this book is to provide some level of integration of the three disciplines. It is our contention that a full understanding of landscape function requires integration across these disciplines. We hope that this text may facilitate that integration.

Structure of the Content of the Book

The book has four sections. Section One contains the basic plant physiology and ecophysiology required to examine landscape carbon and water fluxes. The second section provides an overview of the techniques available in remote sensing, including consideration of the physical principles of remote sensing and the different platforms available to examine landscape structure and function. The third section provides descriptions of the basic modelling of vegetation and landscape processes across multiple scales. The final section contains seven case studies where data from ecophysiological, modelling and RS studies are presented and combined to provide a richer and deeper understanding of landscape structure and function. These case studies include (1) Carbon and water fluxes of five contrasting biomes (boreal forests, arid and semi-arid grasslands, tropical montane forests, Amazonian forest, savannas); (2) groundwater-dependent ecosystems; (3) and global drought and forest mortality.

We hope you enjoy the read.

Section One

Plant Ecophysiology

1

An Introduction to Biogeography: Broad-Scale Relationships Amongst Climate, Vegetation Distribution and Vegetation Attributes

This chapter provides a broad-scale overview of the patterns of global distribution of rainfall, temperature and evaporative demand. In addition it provides an introduction to some of the causes of inter-annual variability (including the El Niño Southern Oscillation, the North Atlantic Oscillation and the Southern Annular Mode) in weather patterns. Examination of the impacts of inter-annual variability of rainfall and temperature is a recurrent theme in studies of vegetation ecophysiology, modelling of landscape function and the application of remote sensing to investigate landscape structure and function. This chapter describes how climate variables were originally used to classify vegetation assemblages into biomes but also discusses the concept of plant functional types and shows how these are increasingly used in the disciplines of modelling and remote sensing as a means of representing biomes. Finally, leaf and whole-plant attributes that are deemed important in land surface models and remote sensing are discussed and recent developments in the interpretation of plant functional types are presented.

1.1 Large-Scale Patterns in Climate

1.1.1 The Solar Constant and the Earth's Tilt

Directly above the equator, at noon, about 1367 watts of solar energy per square metre (1367 W m^{-2}) are received at the outer edge of the atmosphere. This is termed the solar constant, although the sun's output is not constant with time because of changes in solar activity, including the 11-year sunspot cycle. About 7 percent of solar radiation is ultraviolet, about 41 percent is in the visible range (0.4–0.7 µm) and 51 percent has a wavelength > 0.7 µm (near infrared). The global annual average receipt of solar radiation at the edge of the atmosphere is, however, about 342 W m^{-2}, that is, one quarter of the solar constant. This is because the surface area of a sphere (assuming the earth is a sphere) is four times larger than the surface area of a disc of the same radius and thus solar energy is spread across an area four times larger than if the earth were a simple flat disc.

The amount of solar radiation reaching the earth's surface can be calculated from Equation 1.1:

$$S_h = (S_c/r^2)\cos(Z) \tag{1.1}$$

where S_h is the amount of solar radiation received on a horizontal surface (in W m^{-2}) at the earth's surface (assuming the atmosphere doesn't attenuate the solar radiation at all), S_c is the solar constant, r is the radius vector, and Z is the zenith angle. The zenith angle is the angle between a line perpendicular to the earth's surface and the sun. A zenith angle of 10° means the sun is very low in the sky (morning or evening) while a zenith angle of 80° means the sun is close to solar noon. The radius vector is the ratio of the actual distance of the earth from the sun at any given date to the annual average distance of the sun-to-earth. Since the earth's orbit is not circular but elliptical, the sun-to-earth distance is largest in July (northern summer) and smallest in January (southern summer). However, the difference in solar receipt at the earth's surface because of this is relatively small (about 3.3% less in July than January). Seasonality in temperatures arises because of the tilt of the earth's axis (about 23.4° from the vertical). During the northern hemisphere summer, the northern hemisphere is tilted towards the sun and therefore is warmed more than the southern hemisphere, which is tilted away from the sun and is therefore in its southern winter. The lower angle of the sun in winter and the shorter day length both combine to make winters colder than summers in both hemispheres. The tilt of the earth also explains why the largest solar radiation receipt on the summer solstice (June 21/22) is not seen at the equator because the equator is south of the solar declination angle (23.4°) which means that the equator is not receiving solar radiation perpendicular to the earth's surface; solar radiation that is perpendicular to the earth's surface occurs at 23.4° N latitude in the northern summer.

Clouds, aerosols, dust and gases within the atmosphere interact with solar radiation by absorbing (water vapour and other diatomic gases absorb radiation), scattering or reflecting light. Two types of scattering occur: Rayleigh and Mie. The former occurs when airborne particles have a radius of about one tenth of the wavelength of light and explains why the sky is blue on a cloudless day (blue wavelengths are scattered more because they have a shorter wavelength than the other colours). Rayleigh scattering is responsible for the creation of the diffuse beam fraction of solar radiation. Mie scattering occurs when photons interact with larger particles (approximately the same size as the wavelength of light) as opposed to particles that are much smaller (= Rayleigh). Mie scattering is responsible for the grey/white colour of clouds, as the water droplets are a comparable size to the wavelengths of visible light.

1.1.2 Latitudinal Gradients in Temperature

At the North and South poles (90°N and 90°S, respectively), mean annual temperatures are sub-zero, with mean monthly temperatures ranging from about zero to about −27°C. The coldest temperature ever recorded on the earth's surface (−89.2°C)

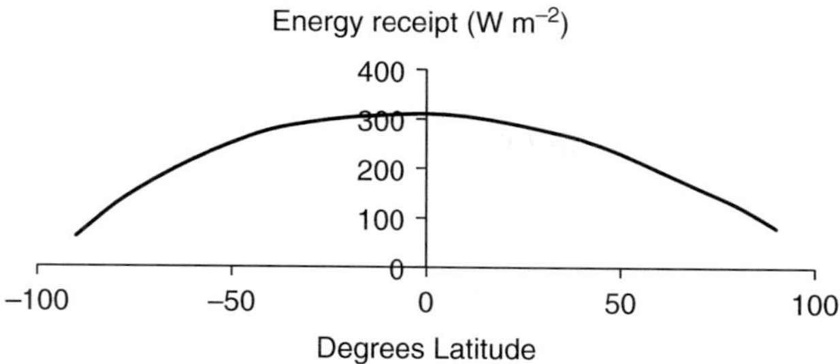

Figure 1.1. Energy receipt per square metre of land surface increases as latitude decreases (i.e., as we move towards the equator).

was recorded at Vostok Station, Antarctica. The three reasons for the poles being the coldest regions at sea-level are first, no solar radiation is received for six months of every year during winter and the ice-surface radiates heat to outer space (radiative cooling: the process of losing heat and hence cooling of a surface through loss of infra-red radiation to space); second, the sun's angle, is so oblique, even in summer, that the transfer of energy from solar radiation to the ice surface is much less efficient than in tropical regions where the solar angle is much closer to perpendicular at the ground surface; and, third, when the sun does irradiate the poles, the surface is covered with ice and snow which have a very high albedo (that is, reflect a large proportion of incoming solar radiation), thereby reducing the warming potential of the sun's rays. Note that for a zenith angle of 60° (30° from the horizontal) the area illuminated by a beam of light is double that (so the energy receipt per unit area is half) for a zenith angle of 0° (that is, perpendicular to the surface).

Total annual receipt of solar radiation (i.e., energy input) per square metre of land surface increases towards the equator (decreasing latitudes; Fig. 1.1). Consequently mean annual temperatures increase towards the equator. Annual solar radiation (energy) receipt increases because the sun's angle to the land surface increases towards the equator and therefore tends to the perpendicular.

Pole-to-equator temperature gradients are pronounced across all continents and reflect the gradient of annual solar radiation with latitude. The two following maps illustrate this (Fig. 1.2) using annual maximum temperatures (Australia) or annual mean temperatures (USA). This gradient in temperature is one of the three principle climatic determinants of the distributional patterns of vegetation. The other two climatic determinants are rainfall and potential evapotranspiration (the rate of evapotranspiration that would occur if water availability was unlimited; that is, the rate of evaporation that occurs at a site given the amount of solar radiation and atmospheric demand for water at that site, assuming water supply is not limiting to the rate of evaporation).

Coastal regions of the five continents tend to receive more rainfall than interior regions because of the close proximity to a source of water (and hence water vapour) on the coast. This is clearly seen in the rainfall maps for Australia and the United States (Fig. 1.3). Also apparent is the trend for east coasts of Australia and North

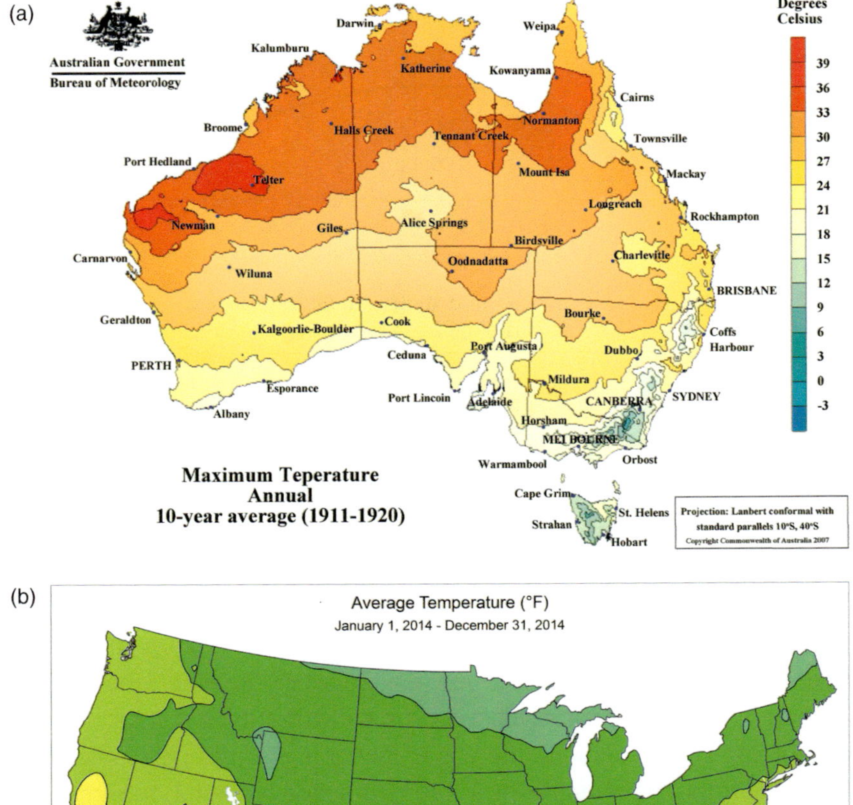

Figure 1.2. Maps of maximum annual temperature for Australia, and mean annual temperature for continental United States demonstrating the global trend of increasing temperatures towards the equator. Australian Bureau of Meteorology.

and South America to be wetter than west coasts because: (a) the prevailing wind directions bring moisture-laden wind off oceans onto the east coasts of these continents; and, (b) east coast water is warmer than land in summer but for the west coasts water is warmer than land in winter. In contrast, it is the western coast of Europe that receives the most rain because of the dominance of westerly winds bringing moisture laden air from the Atlantic Ocean, and Mediterranean, Adriatic, and Black Seas.

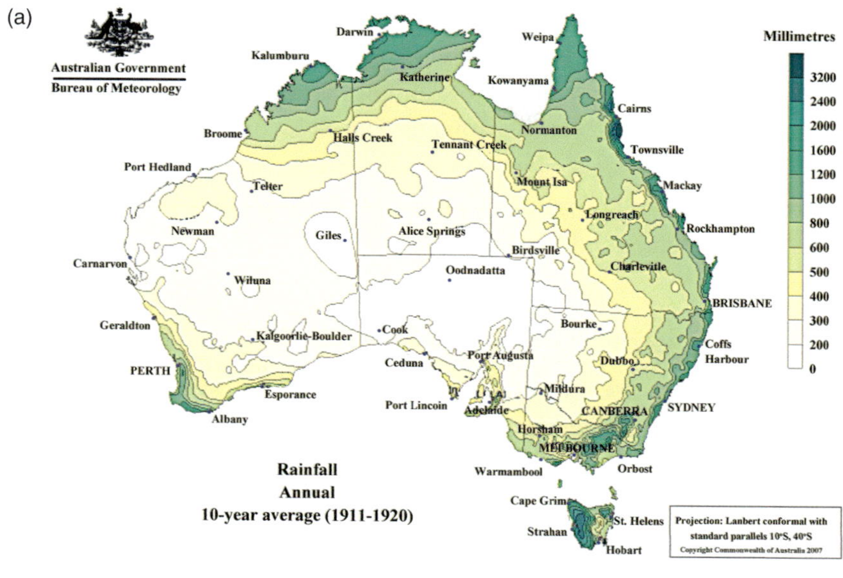

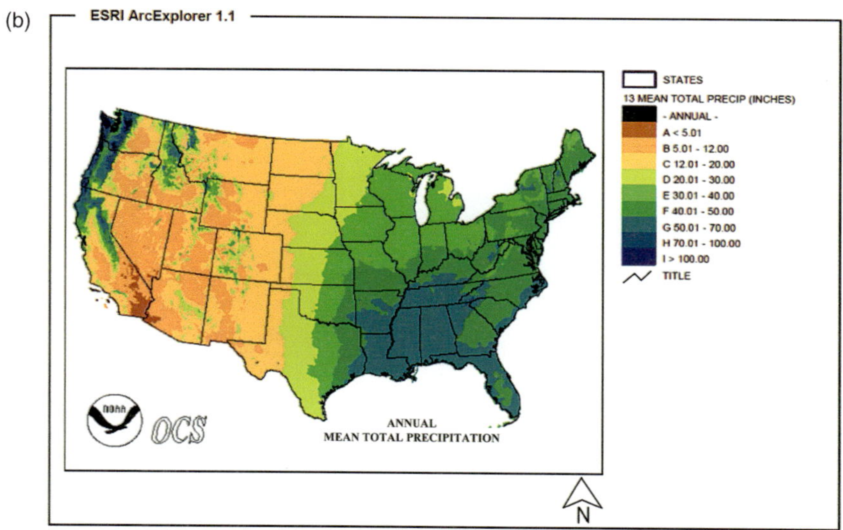

Figure 1.3. Maps of average annual rainfall for Australia and continental United States demonstrate the general trends of (a) coasts receive more rainfall (Australian Bureau of Meteorology); and (b) the east coast is wetter than the west coast (exceptions are discussed in the main text).

Notable exceptions to the trend for coastal regions to be wetter than the interiors of continents are the western coastal regions of the Sahara Desert in northern Africa and the western coastal edge of the Mojave Desert in the south western United States. These exceptions are caused by the influence of cold water off the coast which reduces evaporation and hence reduces the off-shore formation of moisture-filled clouds. Generally, however, there are gradients of rainfall from coastal to inland sites, although the presence of mountains can alter this broad pattern (for example, the Alps in south west Europe).

A key feature of rainfall with respect to vegetation distribution and function is seasonality. Rainfall that is received evenly across the whole year has a different effect on vegetation than the same amount of rainfall that falls only during a 6-month wet season (for example in the wet-dry tropics). The case studies chapters discussing the Amazon, savannas, tropical montane forests and arid-zone grasslands highlight the importance of the timing of rainfall in addition to the amount of rainfall, in influencing vegetation structure and function.

1.1.3 Evaporative Demand and the Water Balance Coefficient

In addition to consideration of the amount and timing of rainfall, it is important to consider the evaporative demand of the atmosphere. Rates of pan evaporation are the rates of evaporation that occur from an open water surface and reflect the influence of humidity, air temperature and prevailing wind speed. Annual pan evaporation rates can be larger or smaller than annual rainfall. This is clearly seen when comparing the map of rainfall for Australia (Fig. 1.3) or Europe (Fig. 1.5) with the map of pan evaporation rates for Australia or Europe (Figs. 1.4, 1.5). Pan evaporation rates are largest in tropical regions because temperatures are largest in these regions.

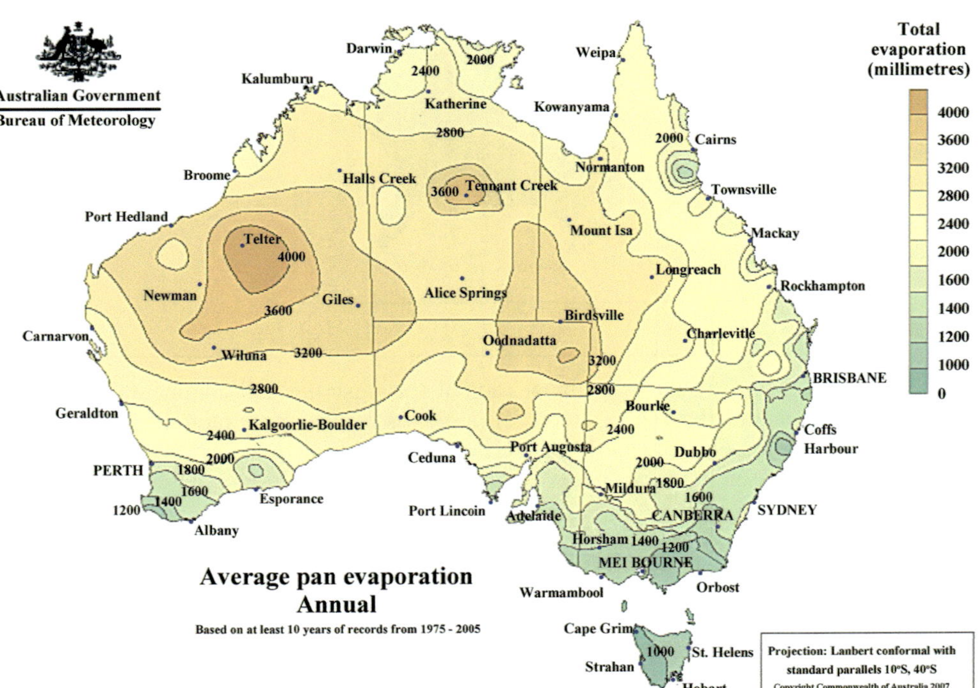

Figure 1.4. Average annual Australian pan evaporation rates. Northern, central and north-western regions display large rates; low rates occur in eastern and south-eastern regions. Annual pan evaporation rates exceed rainfall for most of Australia (refer to Fig. 1.3). Australian Bureau of Meteorology.

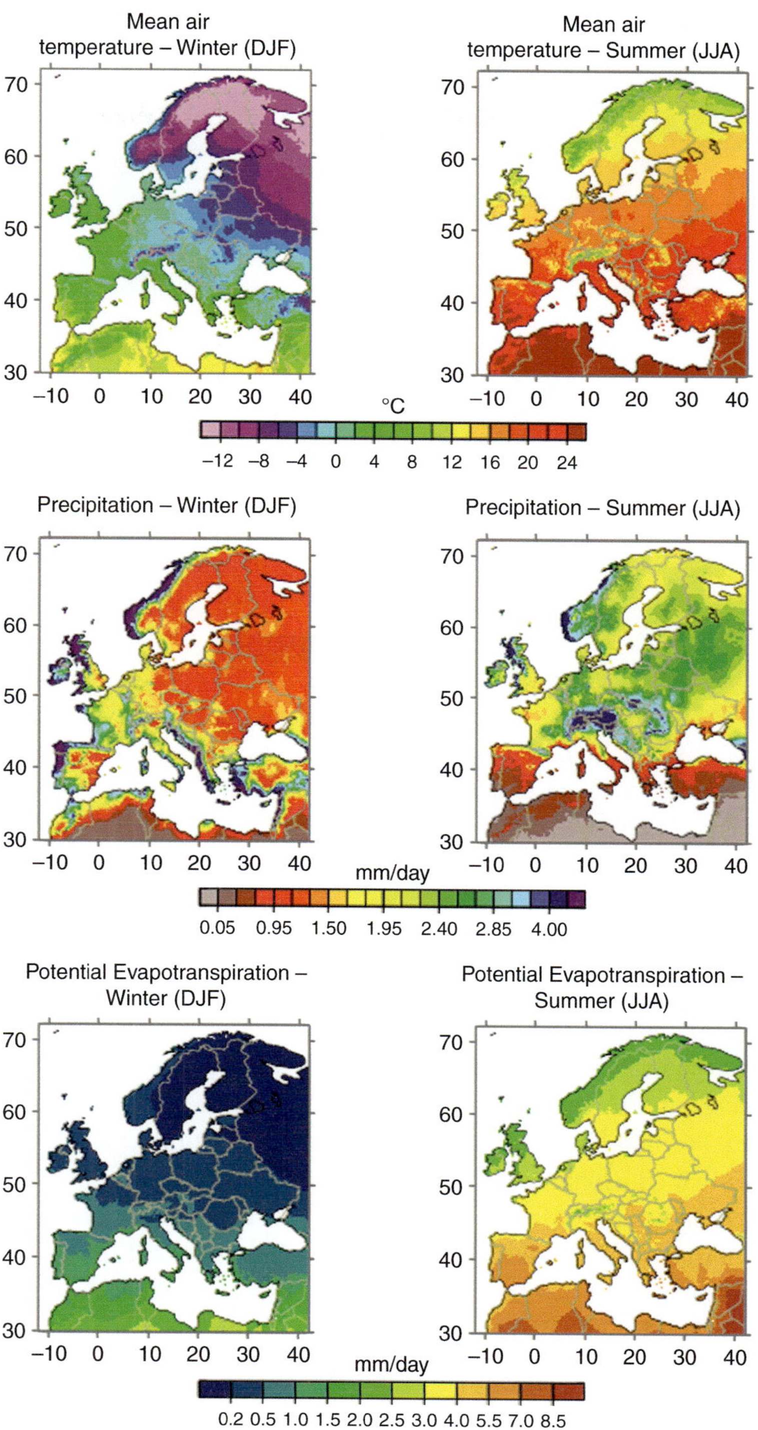

Figure 1.5. Mean air temperature, rainfall and potential evapotranspiration for summer and winter in western Europe. In summer, rates of potential evapotranspiration exceed rainfall for much of Central Europe. Note the different scale for the rainfall and potential evaporation. From the Climate Research Unit: http://www.cru.uea.ac.uk/projects/ecochange/climatedata/desc/

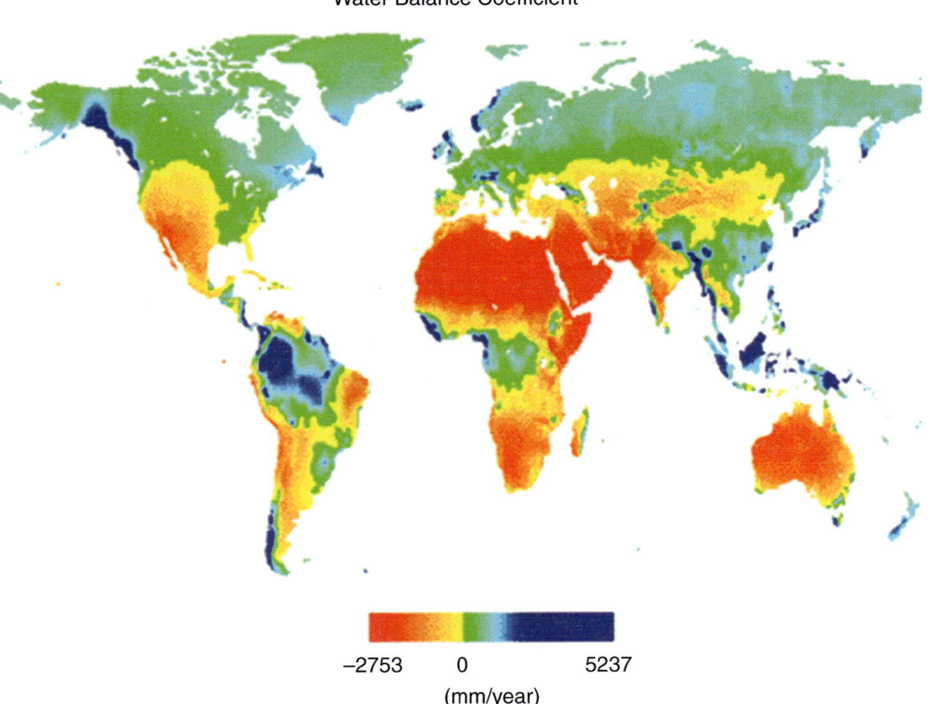

Figure 1.6. Water balance coefficient computed as the difference between annual precipitation and potential evapotranspiration. This coefficient was calculated at each 0.5° × 0.5° longitude/latitude grid cell. Potential evapotranspiration computed by the Priestley-Taylor method using the Cramer and Leemans (1991) global climate databases. From the IGBP Global Analysis, Integration and Modelling program of the IGBP: http://gaim.unh.edu/Products/Reports/Report_6/NPP.html

The water balance coefficient is the difference between annual rainfall and either pan evaporation or potential evapotranspiration rates. When annual rainfall exceeds annual potential evapotranspiration or pan evaporation rates the water balance coefficient is positive. Vegetation that has evolved in regions with a negative coefficient (rainfall smaller than evaporative demand) display very different ecophysiological attributes to that which has evolved in regions with a positive coefficient.

Figure 1.6 shows the global distribution of the water balance coefficient. Most of Australia, northern and southern Africa and the south west and central regions of the United States have strongly negative coefficients while northern Europe and Canada and northern South America have a strongly positive coefficient. Low temperatures limit evapotranspiration in northern Europe and Canada and this contributes to the positive balances in these regions.

1.2 Climate Classification Systems

Broad geographic patterns in rainfall, temperature and evaporative demand have been formalised and form the basis of the Koppen climate classification system, which has been revised into the Koppen-Geiger climate classification system. The classification system has five broad climate groups:

1. Tropical climates (Group A): where average monthly temperature exceeds 18°C for every month. There is little seasonal variation in temperature in tropical climates. Some texts call this the "humid tropical climate zone".
2. Dry climates (Group B): where annual potential evapotranspiration exceeds annual rainfall. Arid and semi-arid zones are dry climates.
3. Temperate climates (Group C): where the mean temperature in the warmest months exceeds 10°C and the mean temperature in the coldest months is between −3 and −18°C. Some authorities have a 0°C cut-off rather than the −3°C threshold for the cold months.
4. Continental climates (Group D): where mean temperatures in the warmest months is > 10°C and the mean temperature in the coldest month is < −3°C (or 0°C). These climates are commonly observed in the interior of continents in the northern hemisphere; they are uncommon in the southern hemisphere. Some texts call this the "moist continental climate".
5. Polar climates (Group E): where there are very cold winters and cold summers and mean temperatures for all months are below 10°C.

Within each Group (A–E) there are many sub-divisions, usually based upon rainfall (both the amount and the monthly distribution), but also often including temperature as a descriptor. For example Group A (Tropical Climates) can be sub-divided into: tropical rainforest climates (see case study chapter on tropical montane forests; Chapter 18) where rainfall is at least 60 mm for every month; tropical monsoonal climates, where rainfall is not evenly distributed across all months; and, tropical wet-dry climates, where a pronounced dry season occurs (see case study chapter on savannas; Chapter 16). Group B can be sub-divided into desert and steppe climates. Group C is sub-divided according to whether the winter or the summer is the driest (for example Mediterranean climates have wet cool winters and hot dry summers), and are commonly found on coastal western regions of the United States, Europe and Australia. Humid sub-tropical climates have wet summers and are common on eastern coasts or in the interior of continents. Group D can be sub-divided into hot summer, warm summer, continental sub-arctic and boreal climates (Chapter 14). Group E can be sub-divided into Tundra, which is found on the northern regions of Europe, Russia, China, and Canada and the Ice Cap climate zone, where all months have a mean temperature lower than 0°C.

A map of the Koppen-Geiger climate systems is presented in Figure 1.7, for which each colour represents a different climate. Note that the five climate zones are further sub-divided into six rainfall sub-groups and eight temperature zones. Therefore each of the 31 climates globally can be described by combinations of either two or three letters that represent the five climates, the six rainfall classes, and eight temperature classes (e.g., Af = the equatorial fully humid climate; Csc represents the "warm temperate summer with dry cool winter" climate zone). The Koppen-Geiger climate

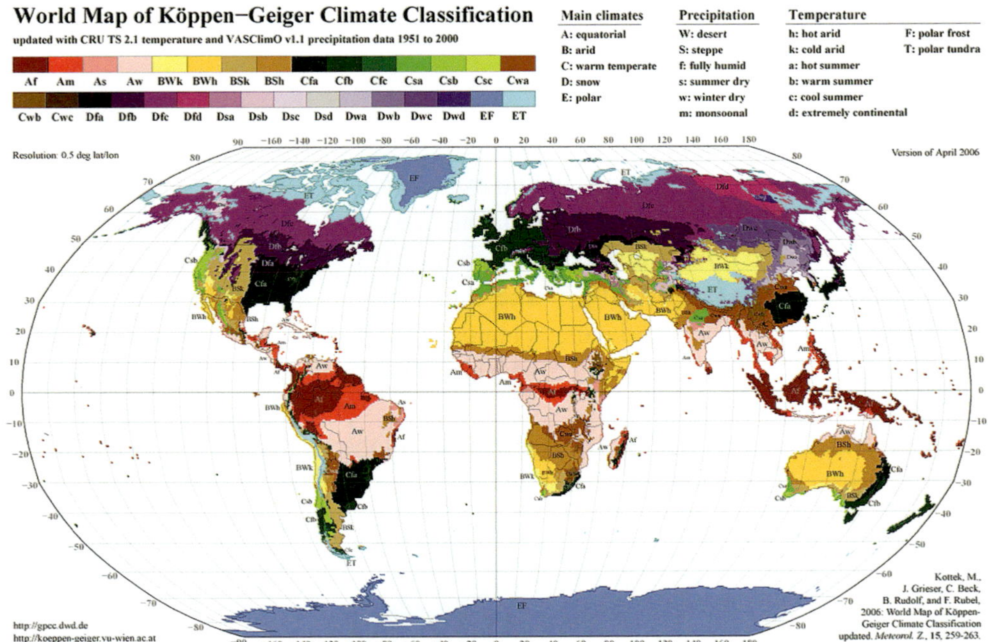

Figure 1.7. World map of the Koppen-Geiger climate classification system. From: http://koeppen-geiger.vu-wien.ac.at/pics/kottek_et_al_2006.gif. See Kottek *et al.* (2006).

classification system has practical applications. For example, mapping of the distribution of biting midges of the genus *Culicoides* spp. (which transmit several viral diseases) across Europe on the basis of this climate classification system has yielded improved ability to manage the present and future distribution of these species (Brugger and Rubel 2013).

1.2.1 Altitude and Aspect Modify Broad-Scale Patterns in Climate

Two additional topographical features modify these broad-scale patterns in climate: altitude and aspect. With increasing altitude, temperatures generally decline, typically, but not universally, at a rate of between 6.4°C and 10°C per 1000 m increase in altitude. The change in temperature with altitude (or elevation) is called the adiabatic lapse rate. Wetter air cools more slowly with altitude than dry air because wetter air is more dense and absorbs long wave radiation better than dry air, therefore a dry and a moist air adiabatic lapse rate of 6.4 and 10°C per I km, respectively, can be identified. The influences of this temperature gradient (and gradients of rainfall with altitude) are discussed further in the case study on Tropical Montane Forests (Chapter 18).

It is possible for temperatures to be higher at higher elevations than lower elevations if upward air movement occurs from warmer, down-slope locations (because

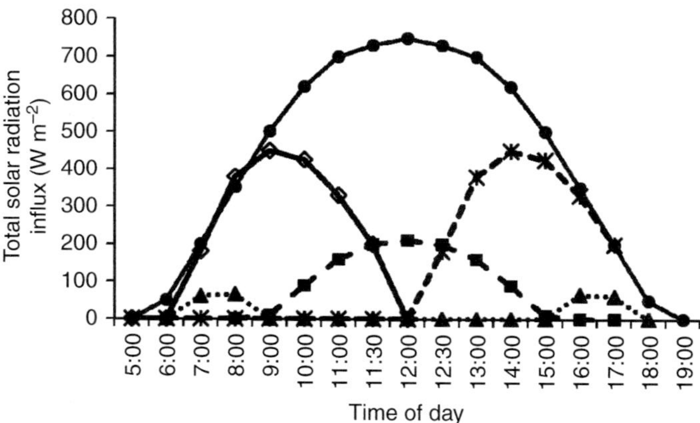

Figure 1.8. The amount of direct beam solar radiation varies with aspect. This hypothetical set of curves is for a cloud-free June 21 (summer solstice) for a set of surfaces located at 40°N. A horizontal surface (circles; solid line) receives the largest radiation input and peaks at solar noon. The north-facing vertical surface (triangles; dotted line) receives direct beam solar radiation only in the early morning and late afternoon when the sun is north of the east-west plane. For a south-facing vertical surface (squares; dashed line) a single peak occurs at solar noon but with a lower maximum than the horizontal surface because of the difference in solar angle for the two surfaces. The length of time during which direct beam radiation is received by the south-facing wall is less than that for the horizontal surface because the sun rises north of east in the morning and sets north of west in the afternoon. For a west-facing (stars) or east-facing (diamonds) wall, direct beam radiation is received only either in the afternoon (setting sun) or morning (rising sun). Redrawn from Bonan (2008).

warm air rises). Alternatively, cold air drainage (typically at night) from upslope locations to valley floors can occur with the result that the valley floor is cooler than higher, mid-slope locations. Cooling of warm, moist air as air moves up the side of high hills and mountains is often associated with condensation of water vapour to produce rain or mist or fog (Chapter 18). Rain produced from this effect is known as orographic rain (orography is the study of mountains). Consequently the leeward side of mountains receives less rain than the windward side. This difference can be very large, leading to major differences in vegetation composition and structure on the windward and leeward sides of mountains.

Aspect (direction that a slope is facing and the angle of the slope) also exerts a strong influence on temperature (Fig. 1.8). Outside of the tropics in the northern hemisphere, a south-facing slope receives more solar radiation than a north-facing slope because the sun is in the south. In the southern hemisphere, the reverse is true. Clearly the angle and the direction the slope is facing determine whether the slope receives solar radiation that is perpendicular to the land surface. The difference in temperature between north- and south-facing slopes can be large (ten degrees or more on any given day) and these temperature differences often result in large differences in species composition and vegetation structure for different sides of the same valley. In the Swiss Alps, for example, farming tends to be concentrated on south-facing slopes because of warmer conditions and larger availabilities of solar radiation to drive growth. In the Himalayas, southern slopes can be forested and relatively moist whilst the northern slopes are heavily glaciated, colder and with minimal vegetation.

As well as the direct beam radiation environment, the diffuse beam radiation component varies with slope and aspect. The sky, as viewed from a horizontal flat surface, can be represented by an inverted hemi-sphere. From any point on the flat horizontal surface, the entire hemisphere can be viewed and therefore 100 percent of the diffuse beam fraction can be received. In contrast, on a sloping surface a proportion of the sky is not visible (the proportion "behind" the slope) and consequently the fraction of the total of the diffuse beam component received by the slope is less than 100 percent.

1.3 Atmospheric and Oceanic Circulation Influence Regional Climates

1.3.1 Atmospheric Circulations

The El Niño Southern Oscillation is a natural cycle of weather that affects the Pacific Ocean and climate in Australia, Indonesia and the western coasts of South and North America. The Southern Oscillation Index (SOI) is a measure of difference in the atmospheric pressures in Darwin (north Australia) and Tahiti (an island approximately midway between the east coast of Australia and the west coast of South America; it is approximately due south of Honolulu). The SOI can be calculated as in Equation 1.2:

$$SOI = 10(\Delta P - P_{av})/sd\Delta P \qquad (1.2)$$

where ΔP is the difference in atmospheric pressure between Darwin and Tahiti at the time of interest, P_{av} is the long-term average pressure difference for the month of interest and $sd\Delta P$ is the long-term standard deviation of the pressure difference for the month of interest. Monthly values are usually calculated.

When the index is negative and lower than -8, an El Niño event is likely to occur or is occurring. A strong negative index was observed in late 2009/early 2010. In contrast, a strong positive index was recorded throughout most of 2010 and 2011 when conditions were in a La Niña phase.

During "normal" years, the seas of the western equatorial regions of the Pacific are warm (typically > 28°C) and the eastern Pacific waters off-shore of Peru and Ecuador are cool. During "normal" years, trade winds blow from east to west in the tropical Pacific Ocean bringing warm, moisture-laden air and warm surface water towards Indonesia and Australia. The warm pool of water around Indonesia and the east coast of Australia creates high rates of evaporation into the atmosphere and atmospheric warming. Consequently cloud cover is extensive and rainfall is "average". As the air rises and cools and rain falls, high altitude winds carry this drier air east towards the west coast of southern America, where it descends over the cooler regions of equatorial Pacific Ocean. This circulation of air, travelling west at low altitude and east at high altitudes, is called the Walker circulation. The strengthening and weakening of the trade winds is driven by changes in the gradient in atmospheric pressure over the tropical Pacific. Warming sea surface temperatures decrease atmospheric pressure because more heat enters the atmosphere, thereby making it more buoyant. Atmospheric pressure gradients influence sea surface temperatures, and *vice versa*.

During El Niño years, the central and eastern equatorial regions of the Pacific are warmer than usual because of a weakening of the westerly trade winds resulting in a weaker upwelling of cool, deep ocean water along the west coast of southern America. Consequently sea surface temperatures along the Peruvian and Ecuadorian coasts increase. In contrast, sea surface temperatures are lower than normal in the western Pacific around Indonesia and the east coast of Australia. High rates of evaporation from the warmer eastern and central regions of the equatorial Pacific occur and less evaporation occurs from the western equatorial regions of the Pacific. Rainfall and temperatures decline in the eastern seaboard of Australia and rainfall increases in the eastern Pacific regions. Thus Peru and Ecuador receive increased rainfall and warmer weather. During El Niño the southern oscillation is deemed to be in the 'negative' phase of the oscillation and the Walker circulation is weak or absent. El Niño is associated with reduced grain yields, reduced vegetation growth and increased frequency of bushfires in eastern and southern Australia. While generally viewed as having an approximately seven-year cycle, the frequency and duration of the positive and negative cycles is highly variable.

During La Niña years (the positive phase of the cycle), the Walker circulation is more intense and the warm pool of water moves west relative to the "normal" years. Convection and evaporation increase above the warm pool of water and the Australian monsoon strengthens. Rainfall in northern and eastern Australia increases, while the west coast of southern America is drier and cooler.

The North Atlantic Oscillation (NAO) is an oscillatory behaviour similar to that of the El Niño Southern Oscillation (ENSO). The reference points for the NAO are Lisbon, in Portugal (the southern reference point), and Reykjavik in Iceland (the northern reference point). Alternatively the Azores, a group of islands about 1400 km west of Portugal, is used as the southern reference point. During the summer there is a large high pressure sub-tropical system around the Azores. However, this high pressure system weakens as winter develops, and the low pressure system around Iceland becomes dominant over much of the North Atlantic.

Westerly winds bring moist air across the North Atlantic into Europe. When these westerlies are strong European summers are cool, winters are mild and rain is frequent. In contrast, when westerlies are absent European temperatures are more extreme with summer heat waves, freezing winters and reduced rainfall.

The low-pressure system over Iceland and the high-pressure system over the Azores control both the direction and strength of westerly winds into Europe. The NAO represents the change in relative strengths and locations of the Icelandic low and Azore high. Positive NAO years are associated with increased westerlies, cool

European summers and mild and wet winters. In contrast, a negative NAO is associated with reduced westerlies, freezing winters and increased rainfall in southern Europe and North Africa.

The NAO may also impact the weather of eastern North America. During the winter under a positive NAO, the Icelandic low draws a stronger south-westerly atmospheric circulation over eastern North America. This inhibits Arctic air from penetrating southward and, when this is combined with an El Niño year winters are much warmer than usual for north-eastern United States and south-eastern Canada. When the NAO index is negative the reverse is observed and the eastern seaboard and south-eastern United States experience freezing winters, even as far as Florida.

Together the ENSO and NAO account for a very large fraction of the inter-annual variation in tropical rainfall and inter-annual variation in northern hemisphere winter temperatures. The ecology of Europe, North America, Australia and Asia is therefore highly influenced by these large-scale patterns in climate. However, there are three additional weather patterns that are included in this brief overview of regional, continental and inter-annual variability in climate: the Pacific Decadal Oscillation, the Indian Ocean Dipole and the Southern Annular Mode.

The Pacific Decadal Oscillation (PDO) is a pattern of Pacific climate variability that can affect weather patterns along the entire western seaboard of the United States and Canada and the eastern seaboard of Australia. It is expressed physically as warmer-than-normal or cooler-than-normal sea surface temperatures over the north Pacific Ocean. A "pool" of warmer-than-normal sea surface temperatures (the "positive" or warm phase) occurs off the south coast of Alaska, the western coast of Canada and the United States as far south as the Tropic of Cancer, where it extends westward into the Pacific Ocean, almost reaching Indonesia, thereby producing a horseshoe shape around a region of cooler-than-normal Pacific Ocean. The cold, or negative, phase occurs when coastal waters are cooler-than-normal and the western Pacific is warmer than normal. The temperature difference between the cooler- and warmer-than-normal phases is typically 2–4°C. The PDO is confined to changes in the sea-surface temperatures of the north Pacific and has an approximate period of oscillation of two or three decades. During the positive phase, warm and moist air is advected north up the west coast of the United States and air temperatures are higher-than-normal for the Pacific Northwest to Alaska but cooler in Mexico and the South eastern seaboard of the United States. Droughts are more commonly observed for much of the northern United States if the Atlantic Multi-decadal Oscillation (AMO) is in its positive phase. The AMO is an ocean current within the North Atlantic Ocean with a cycle periodicity of 50 – 90 years. Droughts are more common in the south-western United States when the PDO is negative (cooler) and the AMO is positive.

The PDO interacts with and influences the ENSO and can either enhance or diminish the impacts of ENSO according to its phase. When both the PDO and ENSO are in phase, the impacts of El Niño/La Niña are increased. When the PDO and ENSO are out of phase the impacts of the ENSO are reduced. The location of the South Pacific Convergence Zone (SPCZ; a band of low-altitude atmospheric convergence, leading to increased cloud cover and rainfall in a narrow band that extends from the eastern coast of Indonesia in a south-easterly direction passing over Fiji and Samoa in the central Pacific) changes according to the ENSO and PDO phases. During El Niño the SPCZ moves northeast; the same occurs when the PDO is positive. Conversely, during La Niña years, and negative PDO years, the SPCZ moves southwest. It is possible to observe the PDO cycle within tree rings of North American and Asian forests. During the 1950s–mid 1970s, the PDO has been (mostly) strongly negative (cooler) but during the 1980s it was strongly positive. During the period 1000–1300 AD, the PDO was strongly negative. This period corresponds to the multi-century droughts of the south-western United States (Seager *et al.* 2007).

The cause of the PDO remains debated. Oceans integrate the rapid fluctuations in atmospheric temperature and winds, but "re-emergence mechanisms" (Alexander 2010) play a role. Re-emergence is the phenomenon whereby seasonal changes in the depth of the mixing layer of surface ocean water allows anomalies in sea surface temperatures to be stored at depth during the summer and return to the surface in the following winter. Unlike ENSO the PDO is not a single mode of the atmosphere and is responsive to a number of driving mechanisms, including stochastic (random) forcing by heat fluxes associated with random fluctuations in the Aleutian low pressure system, the "atmospheric bridge" plus re-emergence processes and wind driven changes in the North Pacific gyre (Alexander 2010; a gyre is a large system of rotating ocean currents). The Aleutian low pressure system is a semi-permanent low pressure system located in the Gulf of Alaska in winter and has a significant impact on atmospheric circulation of the Northern Hemisphere. The atmospheric bridge, or ENSO teleconnections (teleconnections mean causal links between two meteorological phenomena separated by large distances), link changes in equatorial Pacific to the North and South Pacific, the North Atlantic, and Indian Oceans so that ENSO-driven changes in temperatures and pressures can influence the near-surface air temperature, humidity, wind, and cloud cover many thousands of kilometres away from equatorial Pacific.

The Indian Ocean Dipole (IOD) is calculated from the difference in sea surface temperature between two reference points (or poles). The western pole is in the Arabian Sea in the western region of the Indian Ocean (the ocean that lies between the western coast of Australia, the eastern coast of Africa and India to the north). The eastern pole is located in the eastern Indian Ocean, south of Indonesia. The Indian Ocean Dipole (IOD) affects the climate of Australia and other countries that surround the Indian Ocean Basin. As seen for the El Niño / La Niña southern oscillation (ENSO), fluctuations in the IOD cause significant variability in the timing and amount of rainfall across this region.

Change in temperature gradients across the Indian Ocean alter spatial patterns of ascending and descending air and hence moisture. Importantly, the IOD is a coupled ocean-atmosphere phenomenon, similar to ENSO. The IOD is likely to have a link with ENSO events through an extension of the Walker Circulation to the west and associated Indonesian through-flow of warm tropical ocean water from the Pacific into the Indian Ocean. Thus, positive IOD events can be associated with El Niño and negative IOD events are linked with La Niña events. Occasionally the IOD and ENSO are in phase; when this occurs, the impacts of El Niño and La Niña events are generally more extreme over Australia. Conversely if the IOD and ENSO are out of phase the impacts of El Niño and La Niña events tend to be smaller.

Positive IODs occur when sea surface temperatures in the western Indian Ocean are warmer than the eastern part of the Indian Ocean. This results in reduced cloud cover in north Western Australia with concomitant reductions in rainfall across southern Australia and north western and the Top End of Australia. Negative IODs occur when sea surface temperatures in the western Indian Ocean are cooler. Cloudiness increases and hence rainfall increases in southern, western and north Western Australia.

The Southern Annular Mode (SAM) is also known as the Antarctic Oscillation (AAO) and is the name given to the north–south movement of the westerly wind belt that encircles Antarctica. This phenomenon dominates the middle to higher latitudes of the southern hemisphere. The changing north-south location of the westerly wind belt influences the strength and position of cold fronts and mid-latitude storm systems and contributes to variability in annual rainfall across southern Australia and southern South America.

In a positive SAM event the belt of westerly winds moves south towards Antarctica. Because of this there are weaker-than-usual westerly winds, but larger atmospheric pressures are recorded over southern Australia. Consequently the penetration of cold fronts from Antarctica to inland Australia is reduced. During autumn and winter, a positive SAM value can mean that cold fronts and storms are restricted to regions further south and southern Australia receives lower-than-average rainfall. In contrast, a strong positive SAM in spring and summer may mean that southern Australia is influenced by the northern half of high pressure systems, bringing moist, more easterly winds from the Tasman Sea, resulting in larger-than-average rainfall for southern Australia.

A negative SAM event occurs when there is a northerly expansion of the westerly winds. This results in more frequent and/or stronger storms and low pressure systems over southern Australia and this contributed to the Australian millennial drought of 1997–2010.

1.3.2 Oceanic Circulation

Oceans exert a strong influence on rainfall in coastal regions of continents. As previously noted, rainfall tends to be largest on coastal regions and declines as distance from the coast increases. On-shore winds pick up moisture from the ocean surface and this condenses as rain. The further inland the wind moves, the less moisture is contained in the atmosphere. However, oceans also affect coastal temperatures because the very large heat storage capacity of water (Chapter 3) gives the oceans a very large thermal mass or thermal inertia, which means oceans show smaller temperature fluctuations than the atmosphere. Heat absorbed and stored in oceanic water during the summer months is released in the winter, thereby moderating fluctuations in coastal air temperatures.

Oceans transport very large quantities of energy (including heat), from tropical to polar, regions. There are two causes of oceanic circulation: wind driven and density driven thermohaline circulations. Thermohaline derives from the words for temperature and salt, both of which determine the density of seawater. Another word for thermohaline circulation is "the ocean conveyor belt". A representation of the global conveyor belt is shown in Figure 1.9.

Wind driven ocean circulation is predominantly clockwise in the northern hemisphere and anti-clockwise in the southern hemisphere (Fig. 1.10). A major exception to this is the Antarctic circumpolar gyre which flows in a clockwise direction around

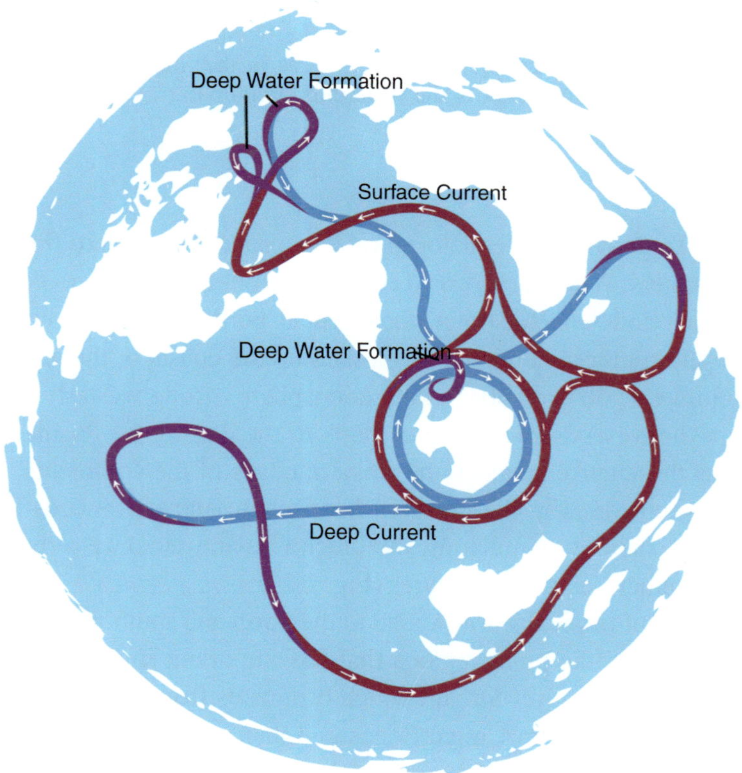

Figure 1.9. The global conveyor belt showing warmer surface flows in red and colder deeper flows in blue. Note that the continent offset slightly to the right of the centre of the image is Antarctica. The Pacific Ocean is the large Ocean dominating the lower left quadrant of the image. Image from: http://en.wikipedia.org/wiki/File:Conveyor_belt.svg

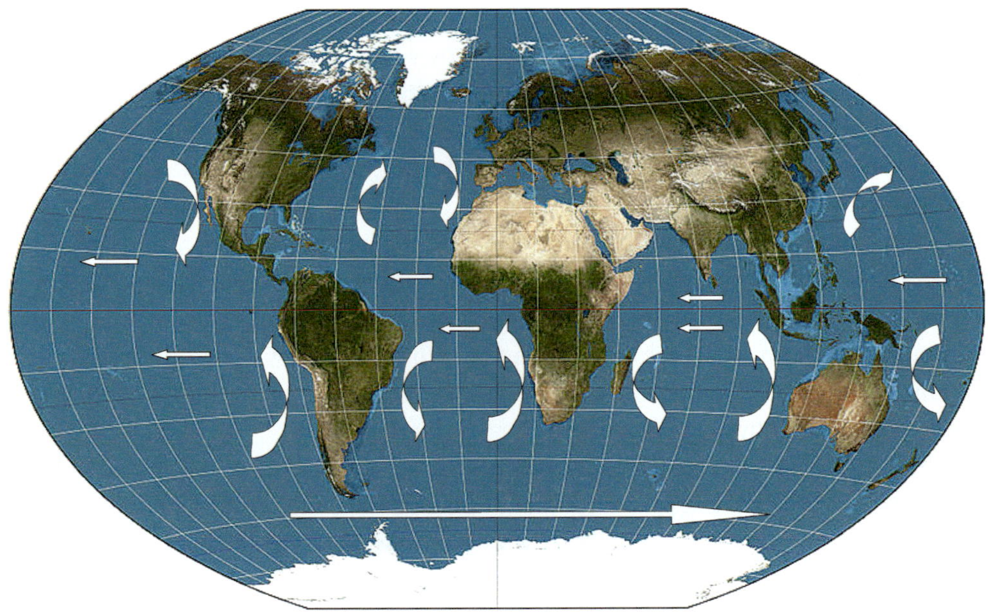

Figure 1.10. A simplified representation of some of the wind-driven ocean currents, showing the clockwise sub-tropical gyres in the North Atlantic and North Pacific oceans and the anti-clockwise sub-tropical gyres in the South Atlantic and South Pacific oceans. Also indicated is the Antarctic circumpolar gyre, flowing west-to-east.

the Antarctic. Trade winds in equatorial regions produce westward flowing currents in equatorial waters. When these warm currents encounter the eastern coasts of South America, Africa and Asia/Australia, they deflect north in the northern hemisphere and south in the southern hemisphere, moving up the coasts of the great continents. Between latitudes of 30° and 60° in both hemisphere, strong westerly winds (that is, coming from the west) drive surface currents to the east of the Pacific and Atlantic oceans and as these currents reach the western edges of the continents, the currents are deflected southwards in the northern hemisphere and northwards in the southern hemisphere, thereby completing the clockwise/anti-clockwise rotations of the northern and southern hemisphere gyres (Fig. 1.10).

The movement of tropical warm waters to the north and south, transport vast quantities of heat towards the polar regions that has important impacts on regional climates. For example, the Gulf Stream which originates from the tip of Florida flows north up the coast of eastern United States and then crosses the Atlantic. It divides into two and the northern flow crosses to Western Europe, making Western Europe much warmer than it otherwise would be. For example, La Coruna in northwest Spain has a winter average minimum/maximum temperature of 7°C to 13°C and Vladivostok in Russia has a winter average minimum/maximum temperature of −15°C to −8°C, yet both are located at 43°N latitude.

The Antarctic circumpolar current (ACC) essentially isolates the warmer waters of the Pacific and Atlantic oceans away from the Antarctic and this helps keep the Antarctic ice sheet intact. Fluctuations in the speed and intensity of the ACC occur as part of a periodic oscillation in the ACC and these changes have significant impacts on the climate of much of the southern hemisphere.

Thermohaline circulations arise from differences in the density of water in different parts of the oceans. Cold water is denser than warm water and seawater is denser than freshwater. Water in the polar regions is colder and more dense, and saltier, in the Antarctic Ocean than in tropical seawater. The Antarctic Ocean is saltier because the formation of sea-ice excludes salt from freezing water.

Cold water sinks to the bottom of the ocean and travels north from the Antarctic or south from the coastal area of Greenland. Cold, salty water from Antarctica flows north at depth, into the Indian Ocean or into the Pacific Ocean. In the Indian Ocean some of the cold salty water from the Atlantic is affected by the flow of warmer, fresher, upper ocean water from the tropical Pacific that results in a vertical exchange of cold, dense, sinking water with lighter warmer water above in a process called overturning. In the Pacific Ocean, the rest of the cold and salty water undergoes haline forcing, and becomes warmer and fresher more quickly. Haline forcing is where there is a net gain of freshwater in high latitudes (as rainfall) and high rates of oceanic evaporation in low latitudes leading to gradients in salinity between low and high latitude regions of the oceans.

Cold dense water in the North Atlantic Ocean creates sinking water that flows very slowly into the deep abyssal plains of the Atlantic. This high latitude cooling and low latitude heating drives the movement of the deep water in a polar southward flow from around Greenland. The movement of deep cold and salty water out of the Atlantic makes the sea level of the Atlantic slightly lower than that of the Pacific. Salinity of the Atlantic is slightly higher too, as recently established from the NASA Aquarius/SAC-D satellite which mapped Oceanic surface salinities. This generates a large but slow flow of warmer and fresher upper ocean water from the tropical Pacific to the Indian Ocean to replace the cold and salty Antarctic bottom water that has moved northwards at depth. Warmer, fresher water from the Pacific flows up through the South Atlantic to Greenland where it cools and sinks, thereby closing the thermohaline loop.

Thermohaline circulation (the warmer, surface layer part of the circulation) supplies large amounts of heat to the poles and can regulate the amount of sea ice that is produced and maintained. Changes in the thermohaline circulation may have large effects on the Earth's radiation budget by influencing the amount of ice and thus the Earth's albedo.

In 2005, it was discovered that the net flow of the northern Gulf Stream had decreased by about 30 percent since 1957. At the same time, it was discovered that the North Atlantic was getting fresher (less salty). It is possible that the Gulf stream may "fail" in the future because of climate change (changes in rainfall and temperature may change the thermohaline circulation too) and this is likely to cause significant cooling in western and northern Europe.

1.4 Biome Classification Systems

Having established a broad-brush understanding of climate zones and a climate classification system and seen how inter-annual variability in temperature and rainfall patterns can be generated, it is now possible to use this information to devise a classification system for the global distribution patterns of vegetation. This is now discussed.

1.4.1 Holdridge Life Zones

Holdridge was a botanist who devised a scheme that divided the world's biomes into thirty or more classes. He recognised the importance of temperature (he used 'biotemperature' which he defined as average temperatures for all months where temperature exceeds 0°C), annual rainfall and the evaporative demand of the atmosphere (expressed as the ratio of potential evapotranspiration to rainfall). In the following diagram note that annual rainfall, the potential evapotranspiration ratio and temperature are presented on a log scale (Fig. 1.11). Each biome is approximately bounded

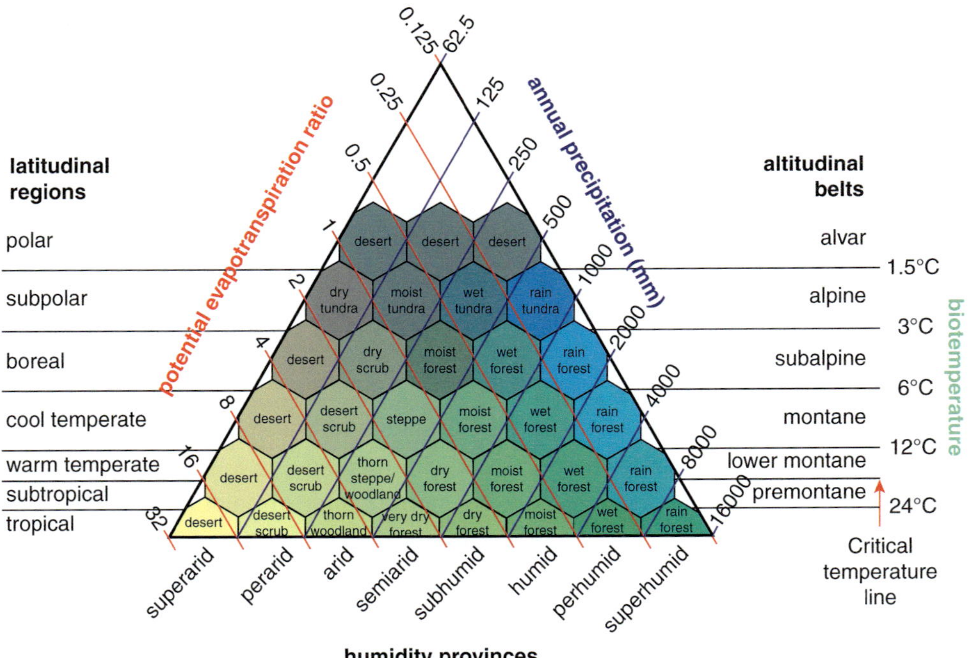

Figure 1.11. The Holdridge Life-Zone Classification Scheme. Image from Wikipedia: http://en.wikipedia.org/wiki/Holdridge_life_zones.

by an upper and lower temperature (e.g., 12–24°C for dry forests), an upper and lower annual rainfall (e.g., 100–500 mm for dry forests) and an upper and lower evapo-transpiration ratio (1–2 for dry forests). This classification system explicitly links the vegetation of the biome to the three dominant climate variables (temperature, rainfall and evaporative demand) that drive the distribution of vegetation. This scheme can be applied to latitudinal and altitudinal gradients of temperature. The nomenclature along the lower axis merely ascribes a name to a climate that is defined by the ratio of potential evaporation to rainfall (humidity provinces).

The Holdridge life-zone classification scheme can be applied globally (Fig. 1.12). Such biomes classes have been used extensively in remote sensing and modelling studies of landscape structure and function, as they provide a rapid and simple way to assign structural and functional attributes (for example LAI, root depth) to any ecosystem. Examples of such attributes and their application to modelling and remote sensing are discussed later in this chapter.

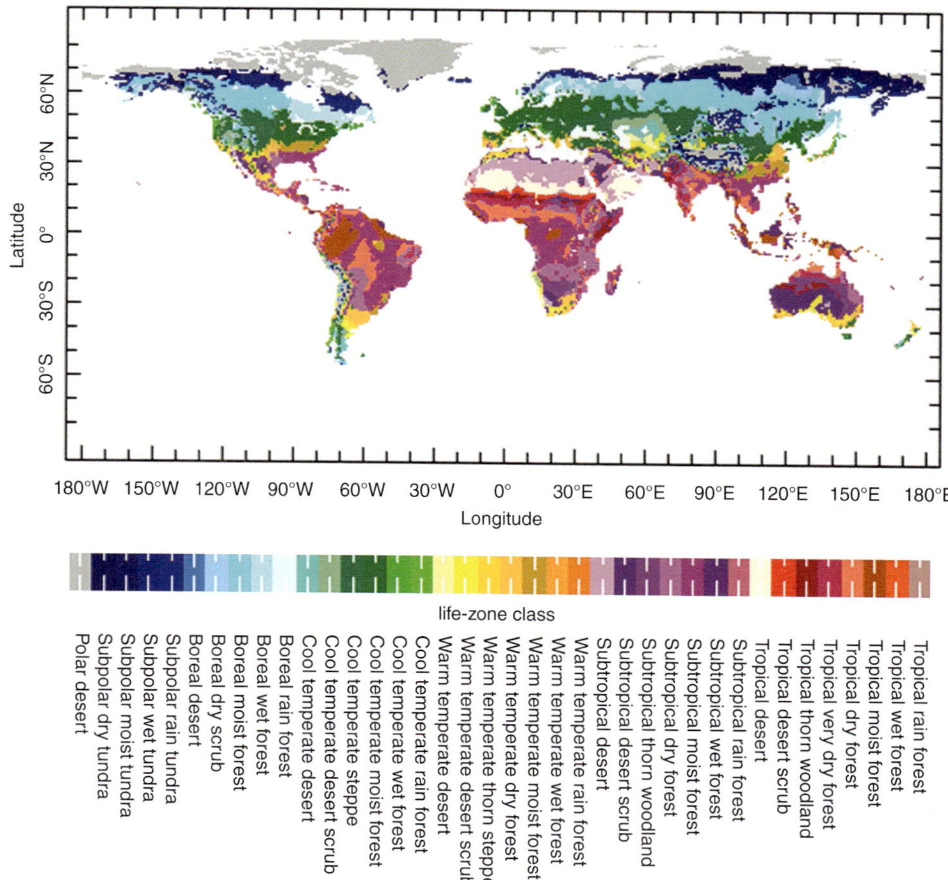

Figure 1.12. The Holdridge Life-Zone classification scheme. Thirty seven life-zones are identified in this classification. From Chapter 47 of the FAO Global Forestry Resources Assessment 2000.

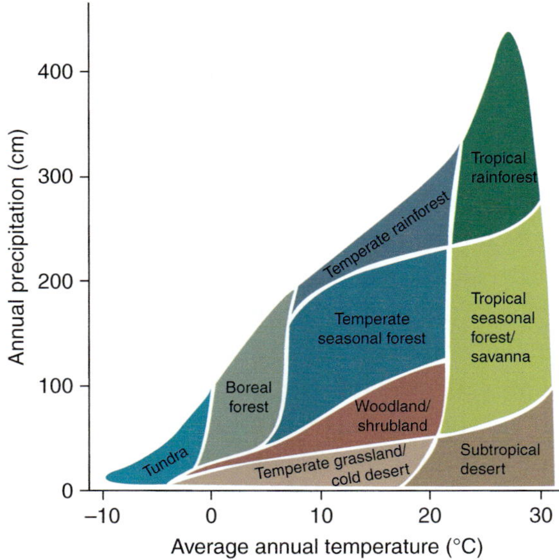

Figure 1.13. The Whittaker biome diagram. Nine biomes are identified cased upon annual rainfall and average temperature. From: http://en.wikipedia.org/wiki/Biome

1.4.2 The Whittaker Biome Classification

Whittaker simplified the Holdridge approach by recognising gradients in temperature and rainfall as the principle determinants of vegetation distribution (Whittaker 1970). He identified only eight broad vegetation types rather than the thirty-plus recognised by Holdridge. These are represented in the classic Whittaker biome diagram (Fig. 1.13). Lines separating biomes are only approximate and serve as guidelines rather than hard-and-fast delineations.

Alternatively, the World Wildlife Fund (WWF) has devised a scheme that recognises fourteen biomes or major habitat types:

1. Tropical and sub-tropical moist broadleaf forests (tropical and sub-tropical, humid)
2. Tropical and sub-tropical dry broadleaf forests (tropical and sub-tropical, semi-humid)
3. Tropical and sub-tropical coniferous forests (tropical and sub-tropical, semi-humid)
4. Temperate broadleaf and mixed forests (temperate, humid)
5. Temperate coniferous forests (temperate, humid to semi-humid)
6. Boreal forests/taiga (sub-arctic, humid)
7. Tropical and sub-tropical grasslands, savannas, and shrub lands (tropical and sub-tropical, semi-arid)
8. Temperate grasslands, savannas, and shrub lands (temperate, semi-arid)
9. Flooded grasslands and savannas (temperate to tropical, fresh or brackish water inundated)

10. Montane grasslands and shrub lands (alpine or montane climate)
11. Tundra (Arctic)
12. Mediterranean forests, woodlands, and scrub or Sclerophyll forests (temperate warm, semi-humid to semi-arid with winter rainfall)
13. Deserts and xeric shrub lands (temperate to tropical, arid)
14. Mangrove (sub-tropical).

These different classification schemes serve merely to present, order and identify putative markers along a continuum of environmental gradients (gradients in water availability, temperature and humidity, for example). Clearly defined boundaries are rarely observed, and ecotones exist where biomes come together. Ecotones have species of plants present from both adjacent biomes in addition to species that occur only in the ecotone.

Accurate representation of biomes within land-surface models and dynamic global vegetation models (Fig. 1.14) is a very difficult problem. Almost all biomes are composed of multiple species and multiple life forms (grass, herb, and tree) are usually present. It is not possible to parameterise land-surface models for every species and every life form that can be found in a single biome (that is, biomes are complex mixtures of species and life-forms). Consequently, classification schemes that simplify

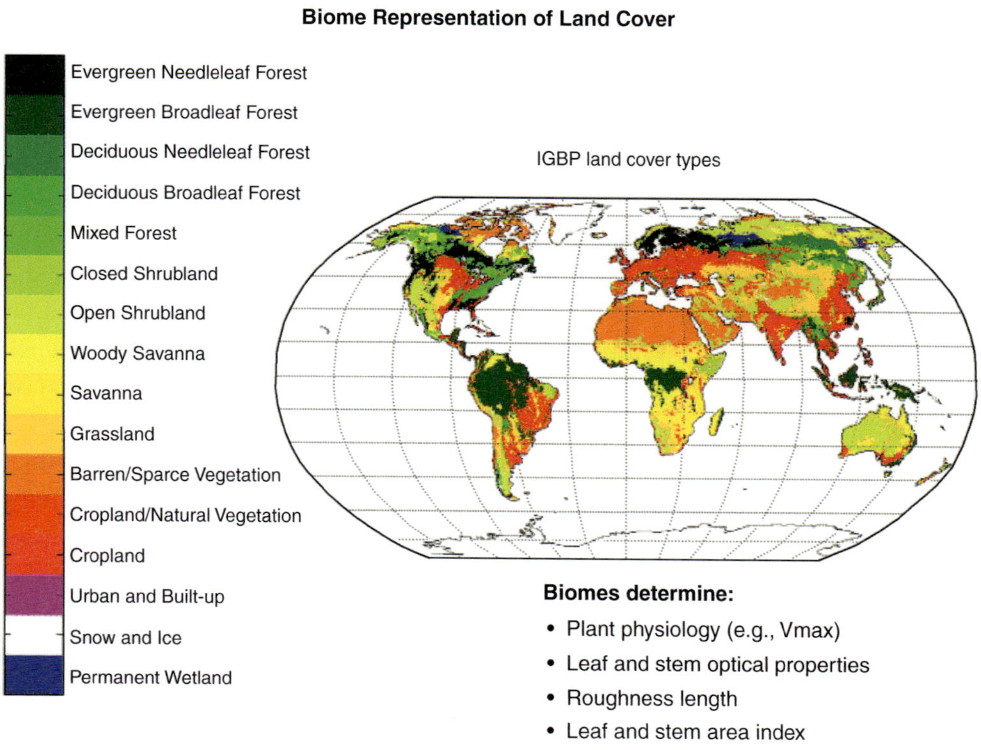

Figure 1.14. Land cover classes used by the International Geosphere-Biosphere Program. From: http://www.cgd.ucar.edu/tss/clm/pfts/index.html

the representation of biomes within land-surface models are being developed and these are now discussed.

1.5 Classifying Vegetation by Function and Form

An alternative to vegetation classification schemes based upon climate that has gained considerable support in the past decade, are schemes based upon functional and behavioural attributes. A very simple scheme, for example, classifies vegetation on the basis of patterns of water-use, known as ecohydrological or hydraulic functional types (Burgess 2006). Another scheme used the concept of plant functional types and this is discussed in Section 1.5.2.

1.5.1 Ecohydrological Types

Rainfall events can vary between the extremes of short and long duration with low or high intensity. Similarly the duration between rainfall events can be measured in days, weeks, months or years depending on the season and site. Different species exhibit different strategies in their response to variation in rainfall characteristics (Burgess 2006), and utilisation of these different strategies may explain the differential distribution of species at small spatial scales (hundreds of metres). Different rooting patterns may contribute to some of these different strategies. Extensive shallow roots are highly adaptive to low intensity and/or short duration rainfall events in that they explore a large volume of soil near the land surface. During brief or low-intensity rainfall events the plant roots rarely explore deep layers of the soil. In contrast dimorphic roots (proliferation of root mass at shallow and deep depths) are commonly observed in environments with highly seasonal rainfall such as monsoonal tropical Australia or Mediterranean climates. Such root systems may also favour the nocturnal redistribution of water from deeper wetter parts of the soil profile to shallower drier parts of the profile (hydraulic lift; see Chapter 3; Oliveira *et al.* 2005). Other attributes, including sapwood hydraulic conductivity (Chapter 3), branch angle, and leaf attributes, also contribute to the differentiation of water-use strategies among co-occurring species.

Burgess (2006) examined the response of sapflow in eucalypt species to a single unusual rainfall event in summer in a Mediterranean environment, where rainfall is winter dominant. The species growing on the lower slopes, where the soil was deep, showed no response in sapflow following the event. This was because their deep roots and deep stores of water resulted in a pattern of water-use that reflected daily patterns in solar radiation rather than changes in the upper soil moisture content. Such a response has been observed in deep rooted eucalypts growing at sites with access to deep stores of water (Zeppel *et al.* 2008). On the upper slopes (but only a few metres in terms of vertical height) the shallower soils, that is, those with smaller water storage capacity, resulted in trees having very low rates of transpiration in the summer, and some species displayed a very large capacity (approximately a five-fold increase) to rapidly increase their rate of transpiration in response to the rainfall event. In the mid-slope locations the responses tended to be smaller and a delay was introduced (of up to two weeks) in the response. It was also noted that small rainfall events (e.g., 4 mm) generally did not induce a response in any species and thus this size event was below the threshold required to trigger a response.

The preliminary findings of Burgess (2006) were recently expanded upon by Mitchell *et al.* (2008) who grouped twenty-one species (nineteen woody species) with similar strategies for water-use into 'hydraulic functional types (HFTs)'. Burgess examined a range of hydraulic features (see Chapter 3) that included wood density, xylem conduit diameter, sapwood hydraulic conductivity, minimum leaf water potential, water-use-efficiency, and specific leaf area, from which he identified five HFTs based on multiple trait associations. For example, Mitchell *et al.* (2008) identified the HFT "year-round active trees". Species within this group clustered together because of their convergence in sapwood specific hydraulic conductivity (K_s), xylem conduit diameter, maximum stomatal conductance and the maximum percentage embolism of the xylem in the summer. In contrast, the HFT "drought suppressed broad leaved shrubs" tended to have convergent values for water-use-efficiency, Huber value and leaf specific hydraulic conductivity (K_l). The HFT "drought suppressed narrow leaved shrubs" tended to have lower δC^{13} values, less negative minimum leaf water potentials and higher specific leaf areas (SLA). Burgess showed that SLA and the minimum leaf water potential were two key predictors of likely water-use strategies. He also noted that, on shallow soil ridge-top sites, a different suite of traits converge, especially traits associated with increased hydraulic safety (lower vulnerability to xylem embolism, lower sapwood hydraulic conductivity, lower maximum stomatal conductance values and lower SLA); while in the more mesic woodlands there was a much larger diversity in HFT.

1.5.2 Plant Functional Types

A more comprehensive species classification scheme that is being applied globally is that based on plant functional types (Bonan *et al.* 2002). Models of land surface behaviour or function, that is models that describe the exchange of CO_2, water and energy between the land surface and the atmosphere, need to represent the vegetation present in a uniform and relatively simple way. It is not possible to represent vegetation across the globe at a species level as there are far too many species to be able to describe their behaviour in a way that can be represented in a model. Consequently land-surface-exchange models (also called soil-vegetation-atmosphere-transfer models, SVATs, land-surface models (LSMs) and land-surface schemes (LSS)) have simplified the representation of vegetation by classifying biomes as collections of plant functional types. Examples of LSMs include JULES (Joint UK Land Environment

 An Introduction to Biogeography

Simulator), MOSES (Meteorological Office Surface Exchange Scheme), and CABLE (Community Atmosphere Biosphere Land Exchange). Land surface models are used to represent fluxes of energy, water and CO_2, and feed into larger-scale global circulation models. Dynamic global vegetation models (DGVMs) are land-surface models that model both the biogeochemical fluxes of landscapes (for example N cycling, water, carbon and energy fluxes), but also explicitly include vegetation dynamics (carbon allocation, growth, death, competition, seasonal changes in leaf area index) at intra- and inter-annual time-scales. Examples of DGVMs include ORCHIDEE, LPJ and TRIFFID. The key feature of LSMs and DGVMs is that they are process-based models rather than empirical models. Therefore, in theory, they are applicable to any vegetated landscape when parameterised appropriately. A brief comparison of some attributes of three LSMs is given in Table 1.1.

LSMs represent the terrestrial land surface as a set of "tiles" or "grid cells". A tile might be 1 km × 1 km, or 250 km × 250 km (or smaller or larger), depending on the resolution required to satisfy the aims of the model. Within each tile, models of the processes of photosynthesis, evapotranspiration and surface energy balances are run using a set of drivers (also called forcing inputs; e.g., meteorological variables, including rainfall, temperature, net radiation) to predict carbon, energy and water fluxes, growth and vegetation dynamics. Typically the model runs at a daily time-step for each tile. Because photosynthesis in C3 plants occurs identically in all C3 plants, and since transpiration is a physical process dependent on stomatal aperture, solar radiation and atmospheric evaporative demand, these processes are represented more-or-less identically across all LSMs. The key difference across terrestrial landscapes is therefore the behaviour of vegetation, which in turn is dependent on the type of vegetation present in each tile. The question is: how best to represent vegetation within each tile of an LSM?

For single-species biomes (e.g., crop fields) representing the ecophysiological attributes of a tile is relatively easy as there is only one species that dominates the behaviour of that land surface. But most biomes are complex mixes of species, with annuals and perennial species present; grasses and trees and herbs are often mixed in space and time. Both C3 and C4 photosynthetic processes can co-occur (for example in tropical savannas). These different types of plants differ in rooting depth (and hence the soil water storage volume differs), stomatal behaviour, light-use-efficiency, phenology (temporal pattern of leaf growth, expansion and senescence), and many other ecophysiological attributes (Chapters 2 and 3). To overcome this problem, the concept of plant functional types (PFTs) has been developed (Bonan *et al.* 2002). A biome then becomes a collection of different PFTs and each PFT is defined by a set of accepted "average" or representative attributes. The key assumption in the application of PFTs is that the underlying ecophysiology and hence behaviour is the same irrespective of the species. Thus, for example, all evergreen coniferous species are deemed to have the same ecophysiological behaviour, all broadleaf deciduous temperate species are deemed to have the same ecophysiological behaviour, irrespective

Table 1.1. A brief comparison of some of the attributes of three commonly used dynamic global vegetation models (DGVMs)

Attribute/ characteristic	DGVM		
	LPJ	ORCHIDEE	TRIFFID
Shortest time-step	1 day	0.5 h	0.5 h
Photosynthetic model	Farquhar model	Farquhar model	Farquhar model
Stomatal conductance model	Haxeltine and Prentice 1996	Ball *et al.* 1987	Cox *et al.* 1998
Evapotranspiration	Total ET via the Monteith equation	Components of ET using the Monteith equation	Penman-Monteith equation
Number of soil layers	Two	Two	Four
PFTs included	Tropical evergreen; Temperate broadleaf evergreen; Temperate needle leaf evergreen; Boreal evergreen needle leaf; Tropical rain-green; Temperate summer-green; C3 and C4 herbaceous	Tropical broadleaf evergreen; Temperate broadleaf evergreen; temperate needle leaf evergreen; boreal needle leaf evergreen; tropical broadleaf rain-green; temperate broadleaf summer-green; boreal broadleaf summer-green; boreal needle leaf summer-green; C3 and C4 herbaceous	Broadleaf; needle leaf; shrubs; C3 and C4 herbaceous
Competition	Area-based competition for light and water amongst all PFTs	Area-based competition for light and water amongst all PFTs	Lotka-Volterra model based on fractional cover
Mortality	Self-thinning based on C balance plus impacts on fire and extreme temperatures	Self-thinning based on C balance plus impacts on fire and extreme temperatures	Used-defined rate for each PFT

of their location on the earth's surface. Surface energy, water and carbon fluxes are calculated separately for each PFT in a tile.

Running *et al.* (1995) originally suggested six PFTs (1) needle leaf evergreen perennial; (2) broadleaf evergreen perennial; (3) needle leaf deciduous perennial; (4) broadleaf deciduous perennial; (5) broadleaf annual; and (6) grasses, using a classification based on permanence of aboveground biomass, leaf longevity, and leaf type. Subsequently Nemani and Running (1996) proposed that these six PFTs should be expanded by taking into account physiological variation in photosynthesis (C3 *versus* C4) and on the basis of simple climate rules. Generally LSMs divide the terrestrial biosphere into vegetation units based on the presence or absence of: trees, shrubs or grasses; evergreen or deciduous habit; C3 or C4 photosynthetic pathways; and, broad or needle leaves. Some of these characteristics are observable by remote sensing (Chapters 4–7) and all are deemed to be important determinants of ecophysiological behaviour because of consistent differences across these classes in key processes (e.g., stomatal conductance, photosynthesis, transpiration, phenology and carbon allocation).

There are several schemes in use today for classifying PFTs. Some teams recognise nine, rather than six, PFTs:

(a) Broadleaf evergreen tree
(b) Broadleaf deciduous tree
(c) Broadleaf and needle leaf trees
(d) Needle leaf evergreen tree
(e) Needle leaf deciduous tree
(f) Broadleaf shrub
(g) Dwarf trees and shrubs
(h) Agriculture/C3 grassland
(i) Short vegetation/C4 grassland

Other researchers recognise sixteen land-cover types which are mixtures of PFTs and land-surface types:

(a) broadleaf evergreen tree
(b) broadleaf deciduous tree
(c) mixed woodland
(d) needle leaf evergreen tree
(e) needle leaf deciduous tree
(f) evergreen shrub
(g) deciduous shrub
(h) tall grass (savanna)
(i) short grass
(j) tundra
(k) desert
(l) semi-desert
(m) cropland
(n) irrigated crop
(o) wetland
(p) glacier.

The International Geosphere-Biosphere Program is a research program examining the interactions amongst biological, physical and chemical processes globally. It uses a land cover scheme with the sixteen land covers listed earlier (Fig. 1.14).

Table 1.2. *Surface types and associated PFTs and fractional cover used in the standard NCAR LSM*

Surface Type	PFT 1	% cover	PFT2	%	PFT3	%
Glacier	B	1.00	..	0	..	..
Desert	B	1.00	..	0	..	..
Needle leaf evergreen forest, cool	NET	0.75	B	.25	..	..
Needle leaf deciduous forest, cool	NDT	0.50	B	.50	..	..
Broadleaf deciduous forest, cool	BDT	0.75	B	.25	..	..
Mixed forest, cool	NET	0.37	BDT	.37	B	0.26
Needle leaf evergreen forest, warm	NET	0.75	B	.25	..	..
Broadleaf deciduous forest, warm	BDT	0.75	B	.25	B	.26
Mixed forest, warm	NET	0.37	BDT	.37	..	..
Broadleaf evergreen forest, tropical	BET	0.95	B	.05	..	..
Broadleaf deciduous forest, tropical	TST	0.75	B	.25	..	..
Savanna	WG	0.70	TST	.3	B	.5
Forest tundra, evergreen	NET	0.25	AG	.25	B	.5
Forest tundra, deciduous	NDT	0.25	AG	.25	NET	.3
Forest crop, cool	C	0.40	BDT	.3	NET	.3
Forest crop, warm	C	0.40	BDT	.3	B	.2
Grassland, cool	CG	0.60	WG	.2	B	.2
Grassland, warm	WG	0.60	CG	.2	B	.4
Tundra	ADS	0.30	AG	.3	..	..
Shrub land, evergreen	ES	0.80	B	.2	..	..
Shrub land, deciduous	DS	0.80	B	.2	..	..
Semidesert	DS	0.10	B	.9	..	..
Irrigated crop, cool	C	0.85	B	.15	..	..
Crop, cool	C	0.85	B	.15	..	..
Irrigated crop, warm	C	0.85	B	.15	..	..
Crop, warm	C	0.85	B	.15	..	..
Wetland, forest	BET	0.80	B	.2	..	..
Wetland, nonforest	B	1.00	0	0	..	..

Abbreviations: NET, needle leaf evergreen tree; NDT, needle leaf deciduous tree; BET, broadleaf evergreen tree; BDT, broadleaf deciduous tree; TST, tropical seasonal tree; ES, evergreen shrub; DS, deciduous shrub; ADS, arctic deciduous shrub; CG, C3 grass; WG, C4 grass; AG, arctic grass; C, crop; B, bare.
Source: From Bonan *et al.* 2002.

Other land surface schemes recognise twenty-eight land surface covers, or biomes, (e.g., desert, glacier, cold broadleaf deciduous forest, tropical evergreen broadleaf forest, etc.) and have assigned up to three PFTs to each biome/surface type (Bonan *et al.* 2002), ranging from zero vegetation cover (glaciers, deserts) to a cover that is comprised of, for example, 20 percent bare soil plus 20 percent C3 grasses plus 60 percent C4 grasses (surface type "warm grassland" in Table 1.2). In this scheme, phenology of each biome has also been defined and is fixed across an annual cycle.

Combining physiological and morphological traits with climatic factors (C4 tropical grass as a PFT, or evergreen needle-leaf tree, for example), is a common practice in all PFT classifications. This is because these attributes are known to: (a) correlate with a number of important ecophysiological traits (e.g., leaf lifespan, SLA, A_{max}); (b) have a long history in ecological and ecophysiological studies; and, (c) can often be determined using remote sensing, thereby allowing large-scale classification of landscapes.

Using RS data and several climate rules (for example, the average temperature of the warmest and coolest months), Bonan *et al.* (2002, 2003) have produced a list of 15 PFTs, plus a set of climate rules that most closely reflects the factors determining the distribution of these PFTs across the globe (Table 1.3).

Once a grid-cell or tile has been identified (from remote sensing imagery; Chapters 4–7) as having one (or more) of the known land cover types, the PFTs associated with each land cover can be assigned along with the ecophysiological traits associated with each PFT (such as LAI, rooting depth and phenological pattern). The land surface model is then run for that tile. In most LSMs, the vegetation component of the land-surface is fixed through time. More recently, dynamic global vegetation models of the land surface include changes in the distribution of PFTs through time as the climate changes (e.g., gets warmer and drier) and as competition among PFTs occurs, such that one PFT can be replaced by another over time.

Application of PFTs within land surface models has had mixed success. In some applications, good agreement between model outputs and observed fluxes of C and water has been observed, while in others agreement has been poor. Grant *et al.* (2005) compared six land-surface models (Boreal ecosystem productivity simulator; Ecosys; Canadian land surface scheme; C-CLASSa; C-CLASSm; Ecological assimilation of climate and land observations; and the Canadian terrestrial ecosystem model) and compared observed fluxes with model simulations of energy, carbon and water fluxes of a Douglas-fir forest and a Jack Pine forest in Canada. They found very good agreement between model simulations and field observations.

1.6 Global Traits of Leaf Attributes and Leaf Function

Is there any evidence that it is reasonable to apply common rules of ecophysiological behaviour across a small number of plant functional types? A key research focus over the past 20 years has been determination of relationships among climatic variables, leaf traits and plant performance at the individual–and species-scales. In 2004 Wright and a large team of co-authors put together a data set covering 2548 species from 175 sites globally, covering a very wide range of rainfall (mean annual rainfall from 133 to 5300 mm) and temperature (mean annual temperature from −16.5°C to 27.5°C) regimes (Wright *et al.* 2004). They examined how plants invest carbon (C) in leaf structures to maximise C fixation over the lifetime of the leaf and present a globally

Table 1.3. *Plant functional types and their derivation from 1 km land cover data and climate rules*

Plant Functional Type	1 km Land Cover Data	Climate Rules
Needle evergreen tree, temperate	needle evergreen tree	$T_c > -19°C$; GDD > 1200
Needle evergreen tree, boreal	needle evergreen tree	$T_c < -19°C$ or GDD < 1200
Needle deciduous tree	needle deciduous tree	none
Broadleaf evergreen tree, tropical	broadleaf evergreen tree	$T_c > 15.5°C$
Broadleaf evergreen tree, temperate	broadleaf evergreen tree	$T_c > 15.5°C$
Broadleaf deciduous tree, tropical	broadleaf deciduous tree	$T_c > 15.5°C$
Broadleaf deciduous tree, temperate	broadleaf deciduous tree	$T_c > 15°C < Tc > 15.5\_C$ and GDD > 1200
Broadleaf deciduous tree, boreal	broadleaf deciduous tree	$T_c > 15°C$ or GDD _ 1200
Broadleaf evergreen shrub	temperate shrub	$T_c > 19°C$ and GDD > 1200 and $P_{ann} > 520$ mm and $P_{win} > 2/3\ P_{ann}$
Broadleaf deciduous shrub	temperate shrub	$T_c > 19°C$ and GDD > 120 and $P_{ann} > 520$ mm or $P_{win} > 2/3\ P_{ann})$
Broadleaf deciduous shrub	boreal shrub	$T_c > 19\_C$ or GDD > 1200
C3 grass	arctic grass	GDD < 1000
C3 grass	grass	GDD > 1000 and ($T_w < 22\_C$ or P_{mon} 25 mm and for months with $T > 22°C$)
C4 grass	grass	GDD > 1000 and $T_c > 22$ & driest month $P_{mon} > 25$ mm
Crop	crop	None

Abbreviations: T_c, temperature of coldest month; T_w, temperature of warmest month; GDD, growing-degree days above 5°C; P_{ann}, annual precipitation; P_{win}, winter precipitation (Northern Hemisphere, November through April; Southern Hemisphere, May through October); P_{mon}, monthly precipitation. A 1-km grid cell is assumed to be 50 percent C3 and C4 if GDD > 1000 and neither the C3 nor C4 criteria are met.
Source: From Bonan *et al.* (2002).

applicable spectrum of leaf traits that appears to operate independently of growth form (tree *versus* grass, for example), plant functional type (evergreen needle leaf *versus* broadleaf deciduous, for example) and biome (savanna versus rainforest, for example). They conclude that categories along the spectrum of traits and trait combinations would describe leaf economic variation better than plant functional types because different functional types display significant overlap in their leaf traits.

Wright and co-workers focused on six leaf traits:

1. Leaf mass per area (LMA) – a measure of leaf investment per unit light intercepting area; a high LMA means either a thicker leaf and/or denser leaves
2. Photosynthetic rate – under light saturated conditions with an adequate water supply
3. Foliar N content – leaves with a high rate of C fixation require large investments in N for Rubisco, electron transport proteins and light capture
4. Foliar P content – rapid cell growth requires a rapid influx of P to support membrane synthesis, nucleic acid synthesis and the production of ATP
5. Dark respiration rate – reflects the metabolic activity contained in synthesis and degradation of biomolecules
6. Leaf lifespan – the time available for a leaf to return the investment (of C, N and P) in its construction and maintenance, back to the plant.

The key finding of the analyses of data for 2548 species was that pairs and triples of mass-based leaf traits (for example, A_{mass} *versus* N_{mass} *versus* LMA) were tightly correlated across all species irrespective of growth form, plant functional group or biome (Fig. 1.15), thereby revealing a single globally applicable spectrum of variation across leaf traits. Eighty two per cent of all variation in A_{mass}, LMA and N_{mass} across species, for example, lay along the principal axis in three-trait space.

The spectrum of leaf traits runs from short-lived leaves with a high investment in foliar N and P, a high rate of photosynthesis and respiration and a low LMA. This is the so-called "live-fast-die-young" strategy. Trees with these traits are deciduous and the short-lived leaves must quickly return the investment of C made to them over a short growing season. In contrast, evergreen leaves have more time to return the investment so can have a lower maximum rate of photosynthesis. However, because they live longer they must be more durable and invest more resources into anti-herbivore chemical defences and mechanisms to tolerate less favourable conditions (freezing winters for conifers; prolonged dry seasons in savannas). They tend to have thicker, tougher leaves (larger LMA). This contrast between evergreen and deciduous leaves has been examined within the context of cost-benefit analyses (Section 1.7).

The relationships between LMA and leaf lifespan differ between climatic zones. For arid sites, the leaf lifespan was shorter at any given value for LMA, meaning the time available to return the investment was smaller in arid sites than mesic sites. Furthermore, the slope of the relationship between rainfall and lifespan was much lower than that for LMA and lifespan (Fig. 1.16); there is a shallower response of leaf lifespan to LMA with increasing aridity. A possible reason for this is the trade-off between leaf nitrogen, transpiration rate and photosynthesis, which is discussed in detail in Chapter 2.

A much larger global data set of plant traits (69,000 species; 52 groups of traits) has been recently established (Kattge *et al.* 2011). Importantly, the traits encompass a

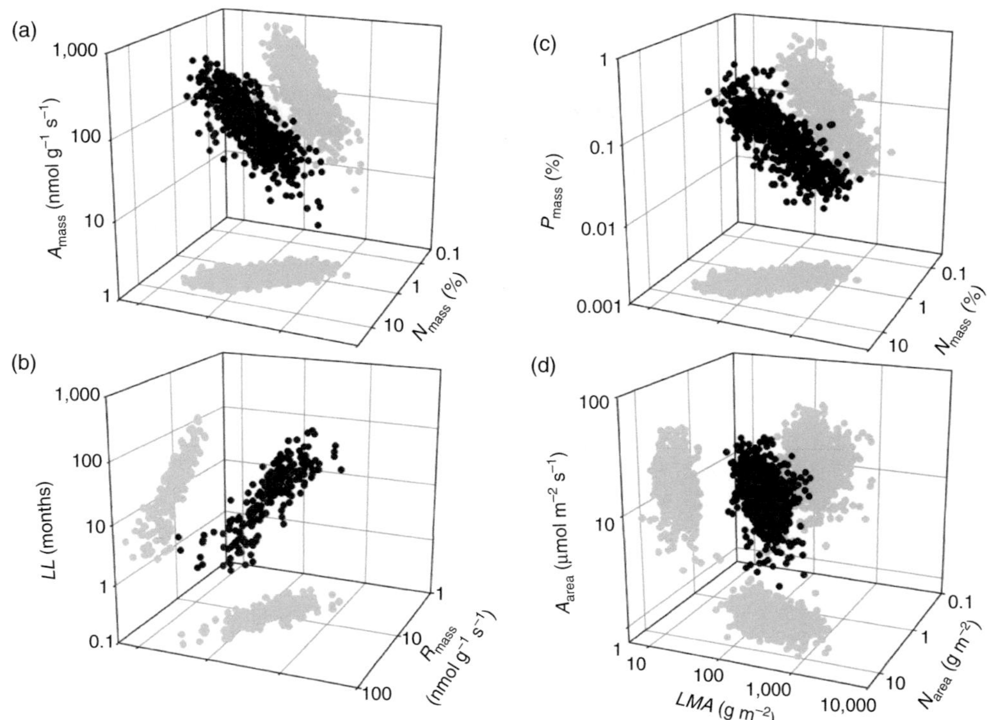

Figure 1.15. Correlations amongst eight key leaf traits. The direction of the data cloud in three-dimensional space can be ascertained from the shadows projected on the floor and walls of the three-dimensional space.
Reprinted from Wright et al. (2004). Used with permission from Nature Publishing Group.

much larger range of plant attributes, including traits pertaining to root function, plant hydraulics and plant height in addition to leaf attributes. In a comparison across species and across plant functional traits the following key conclusions were made for an analyses using 10 key traits (seed mass, plant height, leaf longevity, specific leaf area, N and P content per unit leaf dry mass, N content per unit leaf area, A_{max} per unit leaf area, per unit dry mass and per unit N content):

1. Intra-specific variation across ten key traits was substantial
2. Despite 1, above, variation within species was smaller than variation between species
3. Mean trait values are significantly different between species, with 60–98 percent of trait variance occurring between species (Fig. 1.17)
4. For all 10 key traits the mean values for each plant functional type differ significantly and for four traits (height, seed mass, leaf longevity and A_{max} per unit N) the variation within plant functional types (PFTs) is smaller than variation between PFTs
5. More than 60 percent of the variance in some traits occurs between PFTs

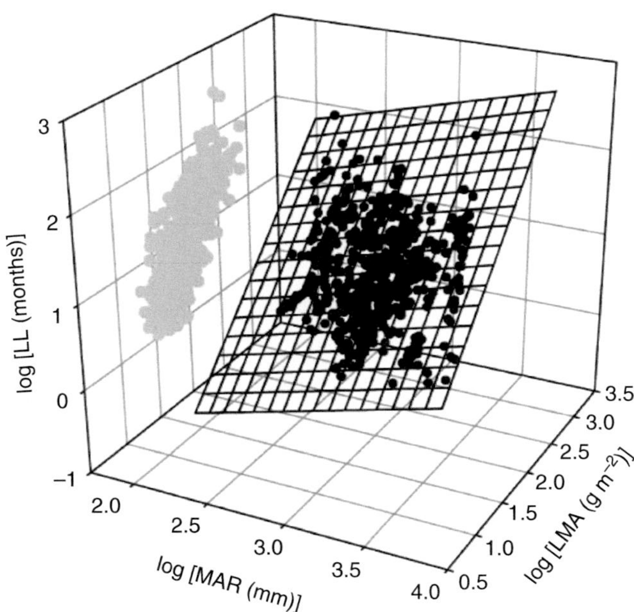

Figure 1.16. Relationships amongst mean annual rainfall (MAR), leaf lifespan (LL) and leaf mass per unit area (LMA) shows smaller slope for mean annual rainfall *versus* leaf lifespan and larger slope for LMA *versus* leaf lifespan. Is climate a strong determinant of leaf traits? Although it is broadly true that species growing in arid and semi-arid regions have high LMA (thick, tough) leaves, the global relationship between LMA and rainfall is very weak ($r^2 = 0.002$; $P = 0.032$). Taking into account mean annual temperature only improves the correlation slightly ($r^2 = 0.1$). Correlation of LMA with vapour pressure deficit or irradiance produced an r^2 of 0.15 and 0.18 respectively (Wright *et al.* 2004). While these were highly significant correlations, the explanatory power of these climate variables was small. Within site variation in traits was typically almost equal to or larger than between site variation in traits. Thus climate *per se* was not a strong predictor of leaf traits. Reprinted from Wright *et al.* (2004). Used with permission from Nature Publishing Group.

6. For many (but not all) traits, PFTs capture a large fraction of the variation in trait values (Fig. 1.17)
7. The parameter values assigned in 12 global vegetation models for many traits are considerably different to the global means of observed data for particular PFTs.

It is apparent, therefore, that the use of PFTs can be an acceptable and practical means of modelling land surface-atmosphere interactions across biomes, but the construction of ever-larger data sets (for example, compare Wright *et al.* 2004 with Kattge *et al.* 2011) should be used to update parameter values used in dynamic global vegetation maps to improve the representation of different biomes across the global landscape. It is also likely that additional, more refined, classifications of PFTs are required above the current classes commonly used.

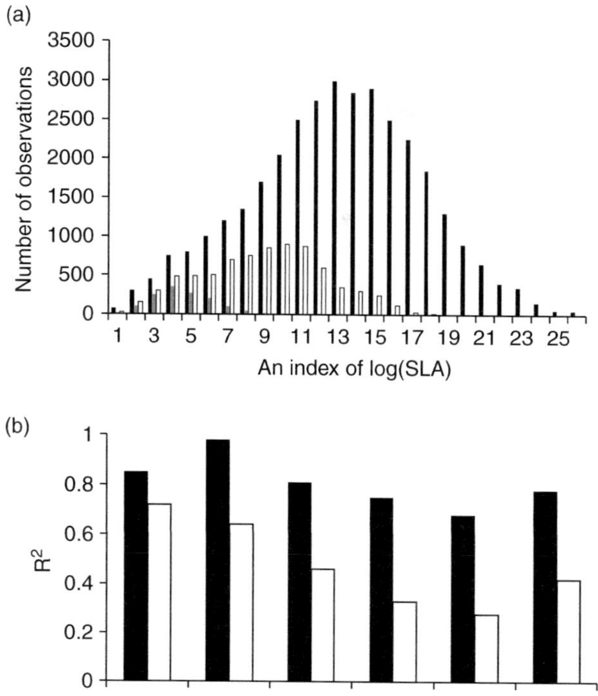

Figure 1.17. Upper panel: There is a log-normal frequency distribution for specific leaf area (SLA) values within the entire data set of Kattge *et al.* (2011) (solid black bars). The mean value of all species is larger than the mean for all tree data (hollow bars) and this is larger than the mean for the PFT needle-leaved evergreen trees (grey bars). Lower panel: the fraction of variance of explained by PFT (open bars) is smaller than the variance explained by species (black bars) but PFTs do capture an important fraction of variation in trait values.
Redrawn from Kattge *et al.* (2011).

1.7 Leaf Lifespan: "Live Fast Die Young" Interpreted Through Cost-Benefit Analysis

Cost-benefit analysis provides a conceptual framework for comparing the costs and benefits of different structures and strategies used by plants to survive in a resource-limiting environment (and all environments are limiting in one or more resources) (Sobrado 1991). The costs are usually valued in terms of C and N investment in growth and maintenance of, for example, leaves or sapwood of a plant. Costs can be divided into construction and maintenance costs. Benefits are usually calculated as the return in the investment in terms of C returns (for leaves) or ease of water transport and resistance to embolism (for investment in sapwood).

Construction costs can be calculated in one of four ways: (i) determining the biochemistry of construction of macromolecules (cellulose, proteins, lipid in membranes); (ii) elemental analyses of the major components of biomass; and (iii) from growth and CO_2 exchange analyses. These approaches are time, resource, and labour intensive. An alternative method, (iv) is to measure the heat of combustion of tissues to estimate the energy investment in that tissue (Williams *et al.* 1987, Eamus and Prichard 1998). This is faster, simpler and gives good agreement with the other methods (Williams *et al.* 1987). Construction costs are calculated from the following equation (Williams *et al.* 1987):

$$C = [(0.06968H_c - 0.065)(1 - A) + (kN/14.0067)(180.15/24)]/0.89 \qquad (1.3)$$

where C = construction costs (g glucose equivalent g_{dw}^{-1}), H_c = ash-free heat of combustion (kJ g_{dw}^{-1}), N = total N content of the leaf (g N g_{dw}^{-1}), and k = +5 when N is imported as nitrate and k = −3 when N is imported as ammonia.

Maintenance costs arise from all the metabolic processes that expend energy but that don't result in growth. Maintenance costs can be calculated from measurements of the ash, protein and lipid contents of leaves and from their maintenance coefficients (Merino *et al.* 1984). Much respiration occurs to maintain current standing biomass. Maintenance costs (g glucose equivalents g_{dw}^{-1} d^{-1}) can be calculated from determinations of lipid, protein and ash contents of tissues and the associated maintenance coefficients (0.0425 for lipids, 0.028 to 0.053 for protein and between 0.006 and 0.01 for ash).

Evergreen trees maintain leaves in their canopy all year. Leaf lifespan can be between thirteen months and many years. These leaves must be robust to survive unfavourable periods (e.g., evergreen conifers maintain leaves during the freezing snow-covered months of winter; Chapter 14) and they must also be resistant to herbivore attack. However, they have a long time (relative to deciduous trees) for their leaves to repay the investment made to them during their construction and maintenance. Leaves of deciduous trees, in contrast, may live for no more than six months. They must return all of the investment (in terms of C and N) made to them in a short burst of photosynthetic activity. This dichotomy of long-lived leaves versus short-lived leaves has been analysed in several ecosystems, including north Australian savannas (Eamus and Prichard 1998) where evergreen and deciduous species are present in approximately equal numbers (Williams *et al.* 1997). Key findings of Eamus and Prichard are:

1. The specific leaf area (SLA; leaf area/leaf dry weight) of deciduous species was almost 50 percent larger (0.0099 m^2 g^{-1}) than that of leaves of evergreen species (0.0069 m^2 g^{-1}). This is consistent with many other studies. Thus the leaves of evergreen species were thicker and more sclerophyllous (tougher) than those of deciduous trees. More C was invested in evergreen leaves and part of this was invested in thick cell walls (to make the leaves tougher and able to withstand drought better) and part of this was increased investment in tannins and other compounds to make the leaves more resistant to herbivore attack.
2. Light saturated rates of photosynthesis (expressed on a leaf DW basis) and foliar N content of deciduous species were 30 percent and 80 percent larger, respectively,

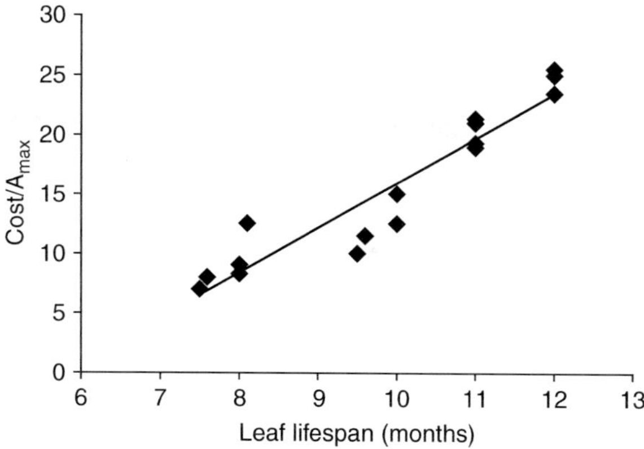

Figure 1.18. The relationship between cost-benefit ratio and leaf lifespan of four north Australian savanna tree species.
Redrawn from Eamus and Prichard (1998).

than for leaves from evergreen trees. The larger investment in N was required to support the higher rate of photosynthesis observed in leaves of deciduous species.

3. Construction costs of leaves of evergreen species were 10 percent larger than those of leaves of deciduous species while maintenance costs of leaves of evergreen species were 20 percent larger than that of leaves of evergreen species. The higher maintenance costs are associated with the larger A_{max} and larger foliar N content (indicative of larger protein content, which is expensive to maintain).

4. The cost to benefit ratio (construction cost: A_{max}) was linearly correlated with leaf lifespan, that is longer-lived leaves had a higher cost:A_{max} than short-lived leaves (Fig. 1.18).

5. It was concluded that shorter-lived leaves of deciduous species invest more N in order to support a larger A_{max} and this strategy has evolved in order to quickly return the C invested in them during the short (six month) wet season when soil moisture and atmospheric VPD were not limiting to photosynthesis.

1.8 Root Depth as a Function of PFTs

A key "look-up" feature (that is, a set of attributes defined by the user) for different biomes used in land-surface models is that of root depth. Each plant functional type is assigned a maximum rooting depth. This is done because the factors that determine the root depth of a particular plant at a particular site are poorly defined. Few modellers will measure root depth in the field. There have been several meta-analyses of root depth data. Canadell *et al.* (1996), Schenk and Jackson (2002) and Schenk and Jackson (2005) provide such analyses (Fig. 1.19). Even within a biome, variation in root depth is at least four-fold, often 10-fold or more.

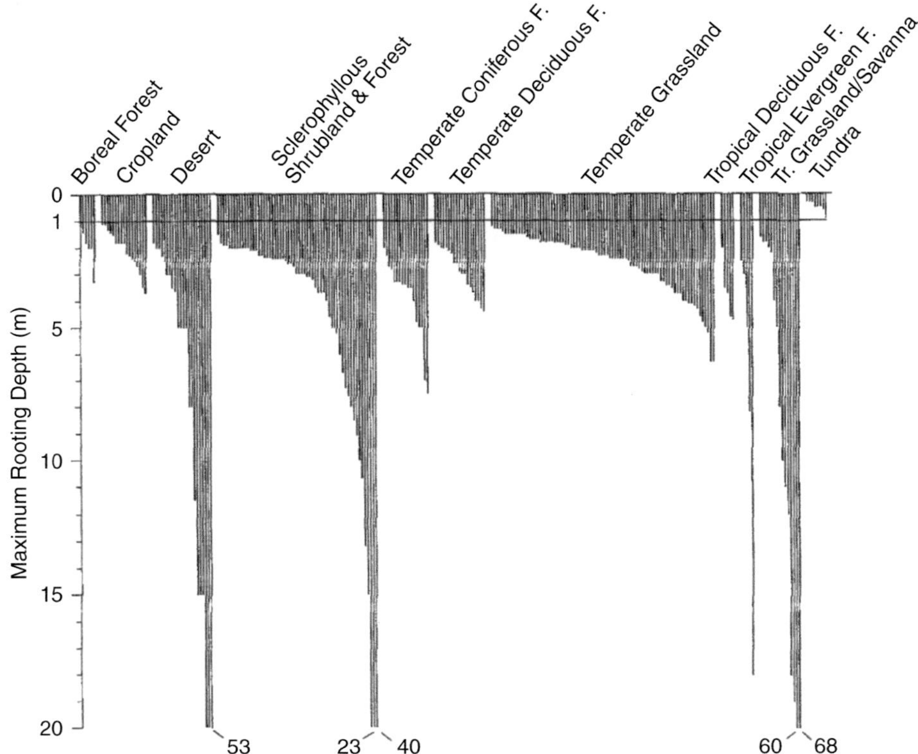

Figure 1.19. Meta-analyses of root depth data for 11 different biomes. From Canadell *et al.* (1996). Used with permission.

More recently Schenk (2008) developed a simple one-dimensional infiltration and extraction model (SWIEM) that simulated water infiltration to a depth of 6 m. Extraction of water by roots is determined by potential ET and the vertical distribution of water. The model was developed to test the hypothesis that root depth is the shallowest possible that still fulfils the evapotranspiration demands of the vegetation. Schenk was able to demonstrate that this hypothesis appeared to be confirmed both by the model and by comparisons of model outputs with field observations of drying profiles across nine different biomes (two biomes are shown in Figure 1.20).

In this chapter we have examined how the globe can be divided into regional climates and we have seen how atmospheric and oceanic circulation patterns occur globally. These patterns in climate distribution have a strong influence on the distribution of different biomes. Broad-scale patterns in several attributes of vegetation (for example, leaf longevity, leaf thickness, foliar N content) can be used to classify vegetation into plant functional types and this information is encapsulated in land surface and dynamic global vegetation models. In the next chapter, the fundamentals of plant ecophysiology, including processes governing carbon uptake and water fluxes, are described.

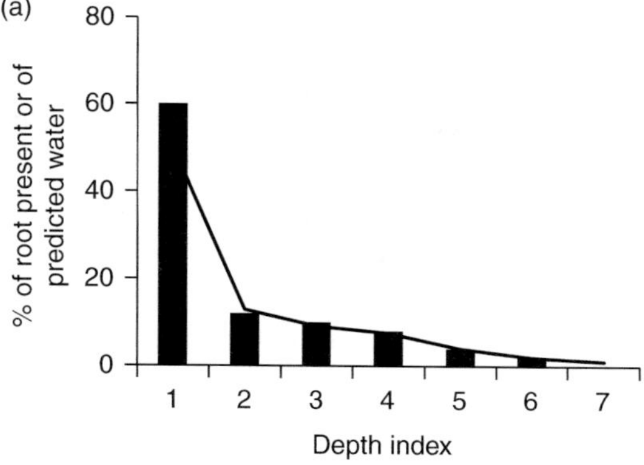

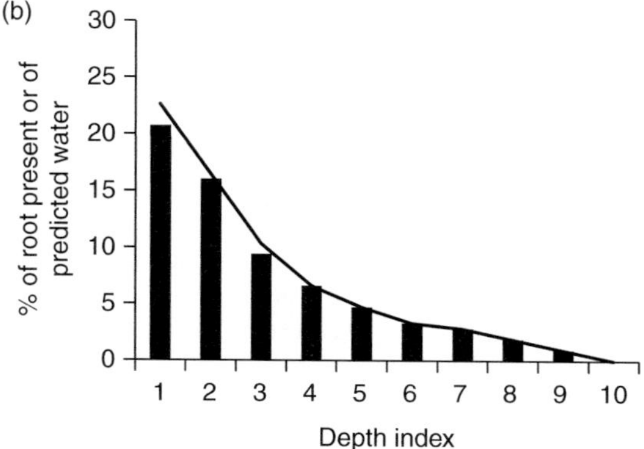

Figure 1.20. Comparisons of soil water extraction profiles predicted for two soil profiles using the SWIEM model compared with observed root distributions. The black bars are the observed root distributions as a function of depth and the line is the modelled water extraction profile. Rooting depth index is a non-linear index of depth, from 0.1 m to 1 m for the upper panel (Solling, UK) and from 0.1 m to 5 m in Paragominas (Brazil; lower panel). Very close agreement between observed root distributions and root water extraction profiles were obtained for the model. From Schenk (2008).

1.9 References

Alexander M, (2010). Extratropical air-sea interaction, SST variability and the Pacific Decadal Oscillation (PDO). In *Climate Dynamics: Why does Climate Vary*, D Sun and F Bryan (Eds.), AGU Monograph #189, Washington, DC, pp. 123–148.

Ball JT, IE Woodrow and JA Berry, (1987). A model predicting stomatal conductance and its contribution to the control of photosynthesis under different environmental conditions. In *Progress in Photosynthesis Research*, J Biggins (Ed.), Kluwer Academic Press, Dordrecht, pp. 221–224.

Bonan GB, S Levis, L Kergoat and KW Oleson, (2002). Landscapes as patches of plant functional types: an integrating concept for climate and ecosystem models. *Global Biogeochemical Cycles* 16 (2), 1021. DOI:1029/2000GF001360

Bonan GB, S Levis, S Sitch, M Vertenstein and KW Oleson, (2003). A dynamic global vegetation model for use with climate models: concepts and description of simulated vegetation dynamics. *Global Change Biology* 9, 1543–1566.

Brugger K and F Rubel, (2013). Characterizing the species composition of European *Culicoides* vectors by means of the Köppen-Geiger climate classification. *Parasites and Vectors* 6, 33.

Burgess SSO, (2006). Measuring transpiration responses to summer precipitation in a Mediterranean climate: a simple screening tool for identifying plant water-use strategies. *Physiologia Plantarum* 127, 339–342.

Canadell J, RB Jackson, JR Erlinger, HA Mooney, E Sala and ED Schulze, (1996). Maximum rooting depth of vegetation types at the global scale. *Oecologia* 108, 583–595.

Cox PM, C Huntingford and RJ Harding, (1998). A canopy conductance photosynthesis model for use in a GCM land surface scheme. *Journal of Hydrology* 212, 79–94.

Eamus D and H Prichard, (1998). A cost-benefit analysis of leaves of four Australian savanna species. *Tree Physiology* 18, 537–545.

Grant RF, A Arain, V Aroroa, A Barr, TA Black, J Chen, S Wang, F Yuan and Y Zhang, (2005). Inter-comparison of techniques to model high temperature effects on CO_2 and energy exchange in temperate and boreal coniferous forests. *Ecological Modelling* 188, 217–252.

Kattge J, *et al.*, (2011). TRY: a global database of plant traits. *Global Change Biology* 17(9), 2905–2935.

Kottek M, J Grieser, C Beck, B Rudolf and F Rubel, (2006). World map of Koppen-Geiger Climate Classification updated. *Meteorological Zeitschrift* 15, 259–263.

Merino J, C Field and HA Mooney, (1984). Construction and maintenance costs of Mediterranean-climate evergreen and deciduous leaves. II. Biochemical pathway analysis. *Acta Oecologica Oecologia Plantarum* 5, 211–229.

Mitchell PJ, EJ Veneklass, H Lambers, SSO Burgess, (2008). Using multiple trait associations to define hydraulic functional types in plant communities of south-western Australia. *Oecologia* 158, 385–397.

Nemani R and SW Running, (1996). Implementation of a hierarchical global vegetation classification in ecosystem function models. *Journal of Vegetation Science* 7, 337–346.

Oliveira RS, TE Dawson, SSO Burgess and DC Nepstad, (2005). Hydraulic redistribution in three Amazonian trees. *Oecologia* 145, 354–363.

Running SW, TR Lovel and and LL Pierce, (1995). A remote-sensing based vegetation classification logic for global land cover analyses. *Remote Sensing of Environment* 51, 39–48.

Schenk HJ, (2008). The shallowest possible water extraction profile: a null model for global root distributions. *Vadose Zone Journal* 7, 1119–1124.

Schenk HJ and RB Jackson, (2002). The global biogeography of roots. *Ecological Monographs* 72, 311–328.

Schenk HJ and RB Jackson, (2005). Mapping the global distribution of deep roots in relation to climate and soil characteristics. *Geoderma* 126, 129–140.

Seager R, N Graham, C Herweijer, AL Gordon, Y Kushnir and E Cook, (2007). Blueprints for Medieval hydroclimate. *Quaternary Science Reviews* 26 (19–21), 2322–2336.

Sobrado MA, (1991). Cost-benefit relationships in deciduous and evergreen leaves of tropical dry forest species. *Functional Ecology* 5, 608–616.

Whittaker RH, (1970). *Communities and ecosystems*. MacMillan, New York.

Williams K, F Percival, J Merino and HA Mooney, (1987). Estimation of tissue construction cost from heat of combustion and organic nitrogen content. *Plant Cell and Environment* 10, 725–734.

Wright IJ, PB Reich, M Westoby and DD Ackerly, *et al.*, 2004. The world-wide leaf economics spectrum. *Nature* 428, 821–827.

Zeppel M, CMO Macinnis-Ng, A Palmer, D Taylor, R Whitley, S Fuentes, I Yunusa, M Williams and D Eamus, (2008). An analysis of the sensitivity of sapflux to soil and plant variables assessed for an Australian woodland using a soil-plant-atmosphere model. *Functional Plant Biology* 35, 509–520.

2

An Introduction to Plant Structure and Ecophysiology

Plate 2.1. A narrow riparian forest of *Eucalyptus camaldulensis* in central Australia.

This chapter reviews the fundamentals of the plant structures and processes that govern ecophysiological functioning of vegetation at leaf, root, canopy, and stand-scales. That ecophysiologists need to understand both structure and function of vegetation across multiple scales is self-evident. However, modellers of vegetation and landscape functioning also need this understanding in order to improve their representation of vegetation in models. Similarly, an understanding of plant ecophysiology at cellular to regional scales is required for those who use remote sensing methods to understand temporal and spatial trends in landscape behaviour.

2.1 Leaf Anatomy and Leaf Attributes

Understanding the basics of leaf anatomy is central to understanding the processes of gas exchange (photosynthesis, transpiration), light absorption, reflection and transmission and hence energy balances (at leaf, canopy, and regional scales) of vegetated landscapes. This knowledge is then integral to modelling of vegetation-atmosphere

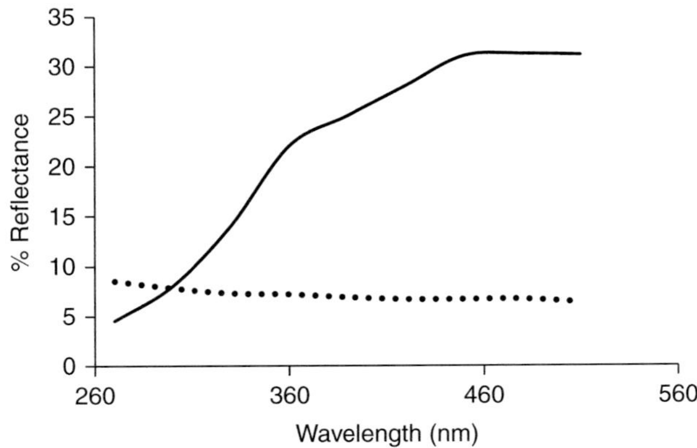

Figure 2.1. The reflectance of leaves of *Kalanchoe pumila* before (solid line) and after (dotted line) removal of the waxy cuticle.
Redrawn from Holmes and Keiller (2002).

exchanges of energy, heat, and momentum and interpretation of remote sensing data of terrestrial landscapes.

All leaves of terrestrial higher plants are covered with a waxy cuticle, the principle purpose of which is to limit the diffusion of water vapour from the underlying epidermis. The conductance to water vapour of a plant cuticle is much smaller (< 15 mmol m^{-2} s^{-1}) than that of open stomata, which typically ranges from 200 to 600 mmol m^{-2} s^{-1}. The cuticle contains cutin, a relatively inert set of polymers of large chain mono-carboxylic acid with a low permeability to water vapour. The structure and conductance to water vapour of the cuticle is responsive to many factors, including wind erosion, ozone pollution, UV environment, water availability, and temperature. For example, the conductance to water of the cuticle is much smaller in xerophytes (typically 0.5–4 mmol m^{-2} s^{-1}) than it is in annual crops (4–15 mmol m^{-2} s^{-1}). The cuticle of plants growing at altitude, where exposure to UV light is large, has a larger capacity to absorb UV, through accumulation of phenolics, than the cuticle of plants growing at sea-level. The cuticle can also increase the reflectance of leaves to incoming solar radiation in deserts, thereby reducing the heat-load imposed on the leaf and its biochemistry. Consequently, the albedo (reflectance) of leaves can be adjusted by changing cuticle thickness, composition and structure. The reflectance of leaves varies with wavelength (Fig. 2.1, Fig. 2.2) and reflectance is larger with a cuticle present than when the cuticle is removed experimentally (Fig. 2.1).

Directly beneath the cuticle lies the epidermis, a single layer of cells and within which stomata are located. Cells of the epidermis differ from most other leaf cells in not containing chlorophyll. The reason for this is to allow epidermal cells to act as lenses, thereby focusing incoming photosynthetically active radiation (PAR; that part of the spectrum [400–700 nm] of solar radiation that drives photosynthesis) into the underlying mesophyll cells and increasing the local intensity within the mesophyll by a factor of 1.5–6 times above ambient (Martin *et al.* 1991, Vogelmann *et al.* 1996).

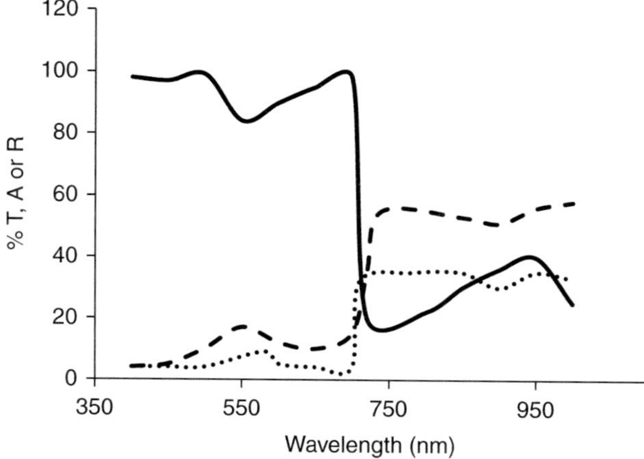

Figure 2.2. Percentage absorbance (A; solid line), transmission (T; dotted line) and reflectance (R; dashed line) for a generalised leaf, as a function of wavelength.

This is especially valuable in low light environments, for example, in the understorey of rainforests and closed canopy woodlands. It is especially effective for direct beam solar radiation rather than diffuse beam radiation. This might be an adaptation to make maximal use of sun-flecks that penetrate to the forest floor occasionally.

Stomata are the pores through which CO_2 (for photosynthesis) and water vapour (as transpiration) diffuse. Stomatal pores open and close through the inflation and deflation of two guard cells. Guard cells are specialised cells that are distinctive in being the only epidermal cells to contain chloroplasts and they are the only living plant cells to not have plasmodesmatal links to adjacent cells. Plasmodesmata are cytoplasmic bridges that allow diffusion of molecules between cells without the need to cross a membrane. Immediately beneath a pair of adjacent guard cells is an air-space – the sub-stomatal cavity. This cavity is in diffusional contact with both the air-spaces of the leaf and the ambient atmosphere.

The number of stomata on a leaf is highly variable (ranging between 50 mm^{-2} and 300 mm^{-2}) and is strongly influenced by environmental factors. Leaves taken from the upper sunlit part of a canopy have larger stomatal densities than leaves growing in the lower shaded part. Similarly, leaves collected in the nineteenth century have a larger stomatal density than leaves of the same species collected in the twenty-first century because atmospheric CO_2 levels have increased substantially over that time (Royer 2001). For the majority of annual crops and northern hemisphere trees, stomata are almost entirely located on the underside of leaves (abaxial location; hypostomatous leaves). In contrast, eucalypt trees have stomatal densities that are approximately equal for both surfaces (amphistomatous leaves). Hyperstomatous leaves have stomata only on the upper surface, for example, the floating leaves of some aquatic plants.

Stomatal pore width varies between zero (closed) and a maximum that generally is in the range 10–25 μm. As stomatal aperture increases, stomatal conductance increases (that is, stomatal resistance decreases) and therefore the ease with which

CO_2 and water vapour can diffuse into and out of the leaf increases. Stomatal opening is stimulated by light and a reduction in the concentration of CO_2 within the leaf (C_i) that results from the onset of photosynthesis. Stomatal opening is reduced in response to many factors, including declining soil moisture content, declining leaf water potential, increasing atmospheric water vapour pressure deficit (VPD) and when temperatures are supra-optimal. Stomata have a central and fundamental role in regulating photosynthesis and transpiration through regulation of gas diffusion. A full consideration of stomatal physiology is given later in this chapter. Mathematical modelling of stomatal conductance is presented in Chapter 11.

Leaves growing in hot and dry environments tend to be "hairy", that is, they are covered in leaf hairs which are highly reflective of solar radiation thereby increasing leaf albedo. Increased albedo can reduce leaf temperature by reducing the amount of solar radiation absorbed by the leaf. These hairs also increase the depth of the boundary layer of air around the leaf. Increasing the thickness of the boundary layer of air, that is, the layer of stationary air that is found adjacent to the surface of all objects on earth, reduces the conductance of the boundary layer to water thereby reducing transpirational water loss. Leaf hairs are part of the epidermis.

Immediately beneath the epidermis lie two to six layers of mesophyll cells. These are the cells that fix CO_2 during photosynthesis. In a typical broadleaf crop or tree, the upper one or two layers of mesophyll (the palisade layer) have relatively few airspaces, have cells that are longer than they are wide (columnar) and have a high density of chloroplasts. The cell walls of these cells are very effective light guides, channelling light to the lower layers of spongy mesophyll. Scattering of light also occurs in the palisade layer so that light is distributed more evenly through the palisade than would otherwise happen. The spongy mesophyll is characterised by cells that are round rather than columnar and consequently large air spaces occur between adjacent cells. This facilitates diffusion of CO_2 to the numerous chloroplasts. When ambient light levels are too high, the chloroplasts of the mesophyll are able to migrate to the side of the cells (anti-clinal walls), thereby reducing light absorption and minimising damage to the chloroplasts arising from photo-oxidation. Low light levels result in chloroplast movement to peri-clinal walls. Movement of chloroplasts is the result of the activity of actin filaments in the cytoskeleton and is induced by blue light in angiosperms and red light in algae, moss and ferns. Phototropin photoreceptors initiate chloroplast movement. Phototropins are flavoproteins that are responsible for phototropic responses in plants and are activated (become phosphorylated) in response to blue light.

The anatomy of eucalypt leaves and needles of conifers differ from the majority of broadleaf higher plant species. Most broadleaf species have leaves that are held approximately parallel to the ground surface, that is, they are held horizontally so that they expose the largest surface area possible to incoming solar radiation. These leaves are also typically hypostomatous and bilateral, having a layer of palisade cells immediately beneath the upper epidermis but a layer of spongy mesophyll immediately above the lower epidermis. Therefore the leaves are not symmetrical top-to-bottom (they are dorsiventral) through the leaf. In contrast, mature leaves of eucalypts are pendulous (hang approximately vertically) thereby minimising the surface area of leaf exposed to the midday sun. They are also isobilateral (symmetrical through the depth of the leaf) with palisade cells beneath both layers of epidermal cells and amphistomatous (stomata present in approximately equal numbers on both side of the leaf). Many eucalypt species also show very distinct morphological differences in leaves of seedlings and saplings compared to leaves of mature trees. Juvenile leaves are smaller, often hypostomatous and dorsiventral, and can be a distinctly different shape to mature leaves. The reasons for this are unknown but may reflect differences in the light environment of saplings and that of adult trees. The needles of conifers can be round, elliptical or triangular in cross-section and have a much smaller total surface area per leaf than most angiosperms. Needles generally have very thick waxy cuticles and sunken stomata to reduce water loss. Resistance to freezing temperatures is a common adaptation in needles and this is discussed in Chapter 14, in the context of Boreal forests.

2.1.1 Leaf Attributes, with Particular Reference to Remote Sensing

The dominant class of pigments in terms of concentration, function, and distribution in plant canopies, are chlorophylls. While ten variants are known, the major types in higher plants are chlorophyll *a* and *b*. Chlorophyll is found in chloroplasts, which are principally found in mesophyll cells of leaves, although chloroplasts have a major role in guard cells and trunks of many eucalypt species.

Chlorophyll *a* has a long 20-carbon tail that attaches the chlorophyll-protein complex to chloroplast lamellar membranes. This tail has a minimal effect on the optical properties of chlorophyll. The optical activity of chlorophyll lies in its ability to absorb solar radiation, better known as light, and it is especially effective at absorbing within the visible region of light, at a waveband of 400–700 nm (Fig. 2.3). Absorption of the energy contained in photons of light is the first step in the photosynthetic fixation of CO_2. Such absorption is possible because of the flat porphyrin "antennae" or "head" or "molecular solar panel" attached to the 20-carbon tail. In the centre of this solar panel is a magnesium atom. Around the magnesium atom are four pyrrole rings and each pyrrole ring has its N atom bound to the magnesium atom. The key to light absorption is the presence of nine double bonds in the four pyrrole rings. These double bonds contain many delocalised π electrons that can absorb the energy in photons within the photosynthetically active radiation (PAR) waveband. Chlorophyll *a, b,* and *c* differ slightly in chemical structure and differ slightly in their absorption spectra (see Fig. 2.3). However, chlorophylls are all very good at absorbing in the red/orange and blue parts of the spectrum and have low absorbance in the green and yellow part of the spectrum, which explains why healthy leaves are green.

Absorption of red light (680 nm) results in a π electron in the ground state of the chlorophyll molecule (that is, the lowest energy state of all the electrons in the

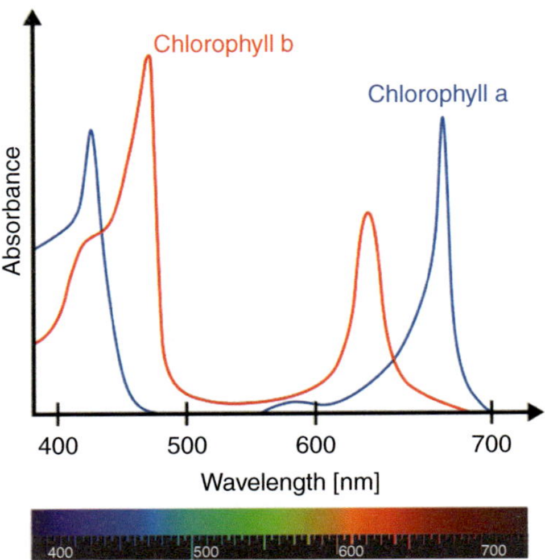

Figure 2.3. Absorption spectra for chlorophylls *a* and *b*. From: http://en.wikipedia
.org/wiki/File:Chlorophyll_ab_spectra-en.svg

molecule; all electrons are in pairs in the lowest possible energy orbitals) being raised to a higher, or more energetic, orbital, the π^*. Since electrons can only move between a limited number of orbitals and energy states, molecules such as chlorophyll can only absorb a limited number of wavelengths of light – hence the absorption spectrum of chlorophyll has only a few peaks at specific wavelengths. The photon interacting with the electrons in the double bonds of the pyyrole rings must have one of a limited number of wavelengths to resonate with the electron and transfer its energy to that electron and lift it to the excited π^* state. This process is extremely rapid, taking about 10^{-15} s to occur. Once the energy of a photon has been absorbed by a chlorophyll molecule, it has three possible fates. The energy can be released as heat and, thus, the temperature of the leaf increases. Second, the energy can be released as fluorescence. Third, the energy can be used in the subsequent photobiology of photosynthesis.

Leaves contain several other pigments in addition to chlorophylls. Carotenoids are tetraterpenoid compounds (having 40 C atoms found in four rings of 10), and are found in chloroplasts of higher plants, algae, and some bacteria. There are two main classes of carotenoids: the xanthophylls, which contain oxygen, and the carotenes, which do not contain oxygen. Zeaxanthin is a xanthophyll with an important role in chloroplasts; α-carotene and β-carotene are also commonly found in plants. These pigments are effective at absorbing blue light and have two roles: they absorb light in photosynthesis and pass on absorbed energy to drive electron transport; alternatively they act to protect chlorophyll from photo-damage arising from excessive light levels.

Anthocyanins are water-soluble pigments found in the vacuoles of leaves and many other parts of plants. They are likely to have a photo-protective role, absorbing UV wavelengths. They are typically red but can change colour according to the pH of the solution. The young leaves of many species of tree are red in colour due to the presence of anthocyanins (Choinski *et al.* 2003) and their presence in the canopy of newly expanding leaves presents a challenge to remote sensing. For example, as eucalypt leaves mature, chlorophyll contents increase whilst anthocyanin contents decrease. Rates of photosynthesis and transpiration increase substantially as leaf area increases whilst leaf absorbance increases a little. Quantum yield of photosynthesis similarly increases substantially with increasing leaf area. These patterns do not mirror the delayed greening pattern of leaf development typically seen in rainforest trees or high-altitude tropical trees; rather, they mirror the transient red colour phase seen in open deciduous woodland species (e.g. *Quercus* spp.). Application of remote sensing to leaf canopies and the influence of leaf pigments on remotely sensed canopy reflectance and absorption is discussed more extensively in Chapter 5.

2.2 Vascular Tissues

Export of fixed carbon from a leaf and transpirational water flux to leaves occurs via vascular tissues. There are two main pathways for long-distance transport in plants. The first occurs in xylem and transports water, plant hormones and inorganic ions from roots to the canopy. Although traditionally considered unidirectional (up the plant), recent work shows that transport in the xylem can occur in the reverse direction (see discussion of hydraulic redistribution and hydraulic lift in Chapter 3). The second pathway for long-distance transport occurs in the phloem and is responsible for moving carbon, hormones, amino acids, some inorganic ions, and water, from sources to sinks. A source is defined as an organ (or part of an organ) that is exporting carbon, usually in the form of sucrose (a disaccharide composed of glucose and fructose) although raffinose (a trisaccharide composed of galactose, sucrose, and fructose) and stachyose (a tetrasaccharide composed of two galactose units and glucose and fructose) are transported in some species, while a sink is defined as an organ (or part of an organ) that is importing carbon. The most common source is a mature leaf, which exports most of its daily fixed carbon to support the growth of new leaves, roots, and respiration. The process of carbon fixation (photosynthesis) is discussed in detail later in this chapter. Immature small leaves, fruit in their growth phase, and roots, are all sinks. A developing storage organ (e.g. a potato tuber) is a sink for most of its life and then becomes a source when it supports the early growth of the newly sprouting plant.

2.2.1 Xylem

The xylem of higher plants is essentially a collection of interconnected pipe-like dead cells where the conductance to water flow along the pipe is much enhanced by the absence of plant cell membranes, organelles, cytoplasm and vacuole. Because the cells are dead they have no need of any of these. Thus the lumen of xylem cells is a highly conductive pathway for water flow from roots to canopies.

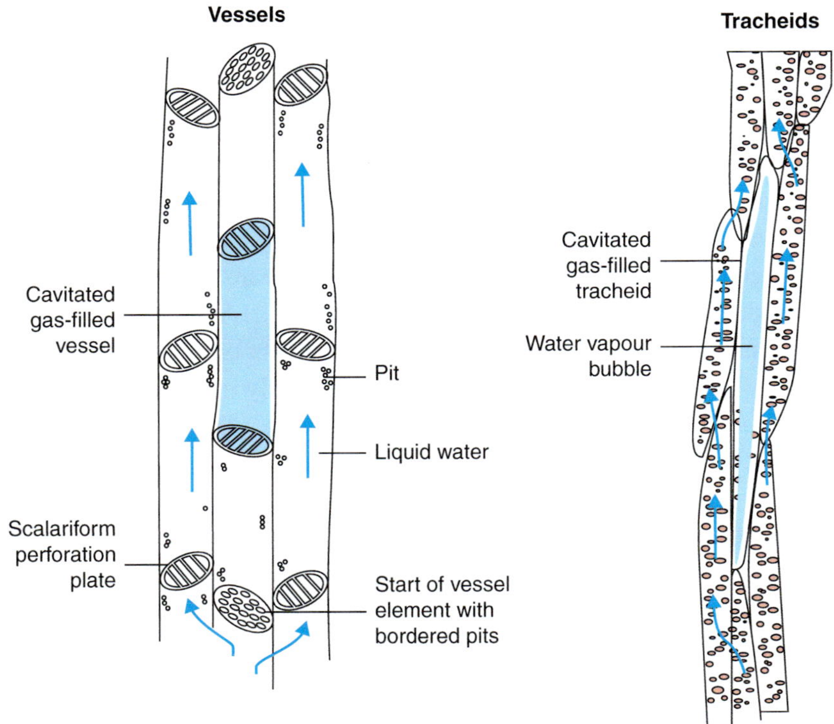

Figure 2.4. A diagrammatic representation of tracheids and vessel elements of plants. In both figures an embolised gas filled cell is indicated by blue shading.

Four types of cells are seen in xylem tissue. These are tracheids and vessel elements, which conduct water, and fibres, and xylem parenchyma. Tracheids are long and thin (diameter 10–25 μm, length 100–200 μm) and they are found in gymnosperms and angiosperms (Fig. 2.4). They are more resistant to flow than vessels. Xylem vessels are shorter and fatter (diameter to 100 μm, occasionally 500 μm) and they are found only in angiosperms. The larger vessel diameter means that they exert a smaller resistance to flow than tracheids (see Hagen-Poiseulle equation later in this chapter).

The end-walls of xylem vessels are partially broken down to produce large pores enabling connection to superior (above) and inferior (below) adjacent cells. These perforated end-walls are called perforation plates and constitute the largest resistance to flow of the entire xylem vessel. Lateral flow between adjacent vessels is facilitated by the extensive presence of lateral pores, called pits (Fig. 2.5). Pits form simultaneously at the same place in two adjacent vessels. A long (2–3 μm), thin channel from the lumen of each vessel passes through the secondary cell wall of each vessel and then this channel widens considerably to form a chamber. This chamber is crossed by the so-called pit membrane which is made up of the middle lamella plus the primary cell wall. Both the middle lamella (a layer of pectin glue holding adjacent cells together) and primary cell wall are very permeable to water. These pit membrane pores have a very important role in plant water relations: they prevent air bubbles

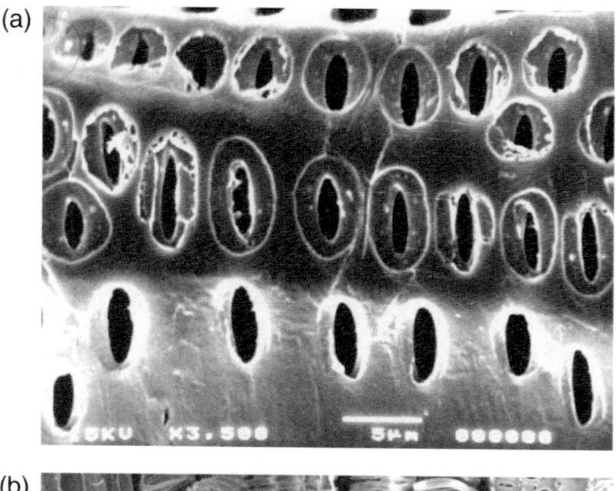

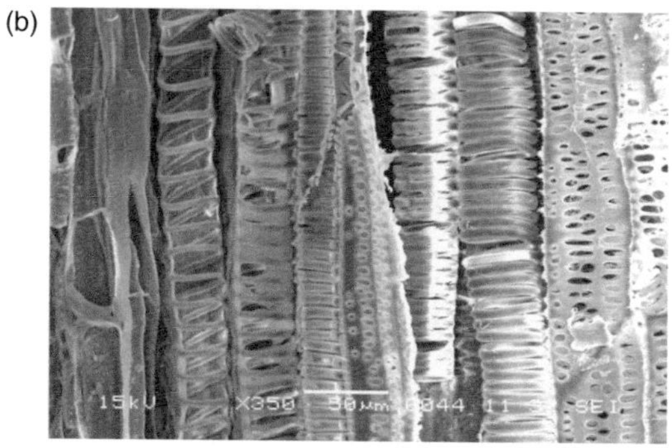

Figure 2.5. A scanning electron micrograph of (a) bordered pits with non-vestured pits on the vessel wall of *T. ferdinandiana*, and (b) vessel elements of plants illustrating the spiral pattern of cell wall thickening and the large number of pits on the side walls of the vessels. Images by Ms Sankam.

or emboli (singular = embolism; see Chapter 3) within the xylem of water-stressed plants, from moving from vessel to vessel. Plants have evolved several adaptations to prevent the development of an embolism as the loss of hydraulic conductance arising from embolism results in a reduction in stomatal conductance, hence photosynthesis and hence growth. Prevention of the propagation of emboli occurs at the pit membrane through the presence of a meniscus at the water-air interface which, given its small radius of curvature, is able to withstand pressure differences of about 1.5 MPa. This is sufficient for most short-term periods of water stress. Pit membranes may also assist in the repair of embolised xylem vessels during periods of low transpiration, for example, at night.

That the conductance of the lumen of xylem vessels is much larger than that of a plant cell membrane can be demonstrated by determining the pressure gradient required to cause a given rate of flow (J_v) across a membrane compared to that required for the same flow along the lumen of the xylem. For a xylem cell that is 1 mm long with a 20 μm radius, using the Hagen-Poiseuille equation (Eq. 2.1):

$$J_v = (\pi r^4 / 8\eta)(\Delta P / \Delta x) \tag{2.1}$$

where J_v = volume flow of sap (m³ m⁻² s⁻¹), r = lumen radius (m), η = sap viscosity (Pa s) and $\Delta P/\Delta x$ is the pressure gradient along the lumen. When we solve Equation 2.1 for $\Delta P/\Delta x$ and set J_v equal to 1 mm s⁻¹, then the pressure gradient required to maintain this flow is 0.02 MPa m⁻¹ (when viscosity = 0.001 Pa s). In contrast, the pressure gradient required to maintain the same J_v across a plasma membrane with a hydraulic conductivity coefficient of 7×10^{-7} m s⁻¹ MPa⁻¹ is 3000 MPa m⁻¹. Clearly the plasma membrane, which is living and very hydrophobic (repels water; but see discussion of aquaporins in Chapter 3) has a much lower ability to conduct water than the open lumen pathway of the xylem cell. Even when the Hagen-Poiseuille equation is applied to the pit pore diameter rather than the lumen diameter, on the assumption that this limits flow in the xylem rather than the lumen diameter itself, the pressure gradient is in the order of 2 MPa m⁻¹, still considerably smaller than that required to move water at the same rate across a cell membrane.

The mechanism by which water moves from soil, through the plant and out of the canopy during transpiration is described in Chapter 3.

2.2.2 Phloem: Structure and Function

Phloem is the tissue that conducts sugars, amino acids, some inorganic ions and water from sources (e.g. mature leaves) to sinks (e.g. fruits, developing leaves, roots, and stems). Phloem is usually located on the outer edge of vascular bundles of leaves but can be on the inner edges of some species. Phloem tissue has three main cell types. The first component is sieve element cells, which is the collective term for sieve tube cells (Fig. 2.6; STC; highly differentiated cells of angiosperms) and sieve cells (less specialised cells of gymnosperms). These cells are the specialised cells engaged in long-distance (e.g. leaf to root) translocation (= transport) in plants. The second type of cell is a companion cell (Fig. 2.6). These are very important for long-distance translocation although they themselves don't conduct solutes for long distances; their role is in phloem loading. The third group of cells is composed of fibres, sclereids, and latex cells, which are not involved in transport but have structural and defensive roles.

Mature sieve elements are alive but lack a nucleus, vacuole, Golgi apparatus, and ribosomes. These elements are missing to increase the ease of bulk flow, i.e. to reduce resistance to flow, through the cell. Large pores are seen in the end cell walls of adjacent STCs; this is the so-called "sieve plate". These pores facilitate flow between adjacent STCs and are typically 1–15 μm in diameter. In addition, pores are found across many areas of the lateral side walls, concentrated into lateral sieve areas. These facilitate lateral transfer of solutes and water between adjacent STCs.

P-protein is found in STCs and this acts to block the flow out of STCs if they become damaged, e.g. through herbivory. Thus the P-protein is responsible for the "blood clot" effect in plants. After damage occurs, P-protein blocks the sieve plate

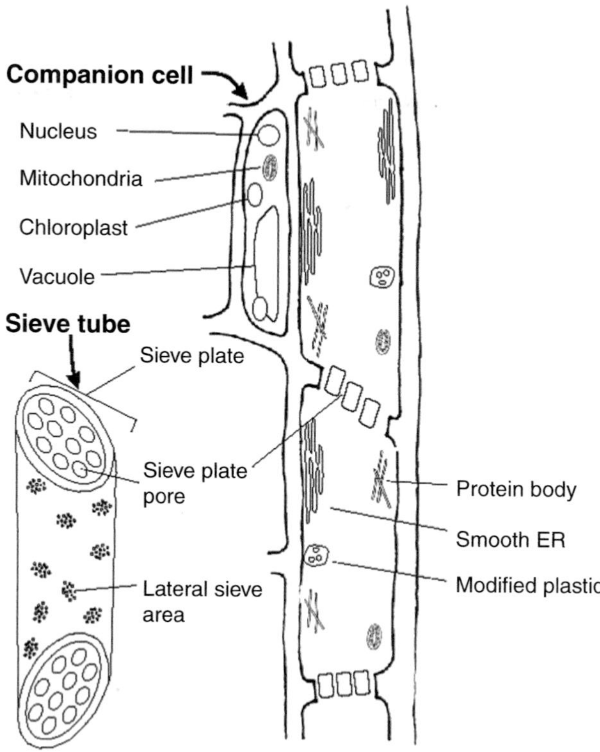

Figure 2.6. A diagrammatic representation of a sieve tube cell showing lateral sieve areas, the sieve plate and the typical array of organelles present in STCs and a companion cell.

and then callose is laid down for longer-term closure. Once STCs are repaired, the callose is dissolved.

2.2.3 Phloem Loading

Phloem sap contains water, sugars, amino acids, organic acids, protein, inorganic ions (mostly cations but some monovalent anions) and plant hormones. Table 2.1 summarises a typical phloem sap composition. Phloem sap moves as a bulk flow so that all constituents travel at the same speed. A gradient in turgor pressure (high turgor in the source end of the phloem, lower turgor in the sink end) is the driving force for long-distance translocation. The turgor gradient is generated by the formation of a gradient of water potential between source and sink through the accumulation of solutes (principally sugars) in the source phloem (phloem in mature leaves that export sugar) and the removal (unloading) of solutes in the sink. A full discussion of water potential and its components is given in Chapter 3. Sink regions unload the solutes and use them to support respiration and growth. Phloem loading is an active process and two mechanisms exist for loading. Apoplast loading is when sugars move

Table 2.1. Representative concentrations of the major components of phloem sap

Component	Concentration (mg L^{-1})
Sugars	80–110
Amino acids	4–6
Organic acids	2–3
Protein	1–2
Cations (e.g. K^+, Mg^{2+})	2.5–5
Anions (e.g. Cl^-, PO_4^{2-})	0.7–1.4

from the mesophyll cells via plasmodesmatal links towards the minor veins in mature leaves. Sugars leave the symplast and enter the apoplast adjacent to companion cells by crossing the plasmalemma of mesophyll cells close to the phloem. Active loading from the apoplast into the companion cells occurs by an energy-driven transport protein (a sucrose-H^+ symport protein; symport means moving two solutes in the same direction; antiport means the movement of two solutes in opposite directions) located in the plasma membrane. Sugars then move via plasmodesmata into the sieve-tube cells. The solute potential of a typical mesophyll cell might be −1.3 MPa but the solute potential of the sieve-tube cell and companion cell might be −3.0 MPa and therefore sugar transport is against a concentration gradient. This fact explains the need for energy input in the loading process as movement of sugar against a concentration gradient is energetically "uphill".

An alternative loading mechanism involves an entirely symplastic pathway whereby sugars are transported from mesophyll cells to the STC via the symplast. The absence of a membrane barrier precludes energetically uphill movement against a concentration gradient. Thus this mechanism relies on polymer trapping to ensure that loading can occur in the absence of a membrane barrier between mesophyll and STC. Polymer trapping means that sucrose (produced by mesophyll cells) is converted to raffinose or stachyose molecules (polymers of 3 or 4 hexose sugars) in intermediary cells and this maintains a concentration gradient of sucrose to drive the diffusion of sucrose from mesophyll to intermediary cells. These polymers diffuse into the STC, thereby reducing the water potential of the STC in the source leaf, thereby inducing the movement of water into the STC of the source leaf, thereby producing the required gradient in turgor potential to drive long-distance translocation.

Phloem unloading occurs in sink regions and involves three steps. The first is unloading *per se*, whereby diffusion of sugars out of the STC occurs symplastically down a concentration gradient, or by apoplastic unloading. The second is short distance transport and the third is storage and metabolism of phloem sugars. Unloading can be entirely symplastic, including the unloading out of the sieve element-companion cell followed by diffusion to a nearby sink cell. Plasmodesmata are needed along the entire pathway for symplastic unloading. Transport is passive during symplastic unloading, no membranes are crossed and a concentration gradient is needed to maintain the diffusion of sucrose to the sink.

Alternatively, unloading can be apoplastic. The transition from symplast to apoplast can occur either at the sieve tube cell/companion cell plasma membrane, or it can occur in cells further removed from the sieve tube cell. At some stage, sugars unloaded apoplastically must re-enter the symplast for storage or for metabolic use. Because sugars cross at least one plasma membrane, symport proteins are likely to be important in the movement of sucrose across the membranes.

2.2.4 Phloem Sap Velocity and Coupling Photosynthesis to Soil Respiration

Phloem transports sugars to roots. Some of this sugar supports respiration and growth while some is lost to the soil where it supports fungal and microbial respiration. It has long been accepted that soil respiration was primarily driven by decomposition of old soil organic matter and litter (i.e. heterotrophic respiration), but it is now known that autotrophic respiration (which includes respiration of plant roots, mycorrhizal fungi and rhizospheric bacteria that rely on root exudates) can be the major component of total soil respiration. Changes in rates of canopy carbon fixation can be reflected in changes of rates of soil respiration, typically within 1–4 days. However, the velocity of bulk flow in the phloem is normally in the order of 1 m h^{-1}. Therefore for a tree that has a transport distance of 102 m (100 m tall plus 2 m root depth) it would take 102 hours or almost 5 days for a sucrose molecule to travel from the uppermost leaf to the roots. However, the coupling of soil respiration and photosynthesis is much faster than that. How can this be explained?

Mencuccini and Holtta (2010) proposed a model in which hydrostatic pressure pulses and waves of sugar concentration travel much faster than individual sugar molecules within phloem sap. These authors make an analogy with the movement of electrons in copper wires. In copper wires, electrons move with a velocity of about 1 mm s^{-1}. When an electrical switch is closed, a light bulb that is 20 m away from the switch lights up immediately. If electrons are moving at 1 mm s^{-1} it would take 5.56 h for the light to come on. However, it is the electromagnetic pulse associated with the voltage applied when the switch is closed which travels at a significant fraction of the speed of light, not the electrons *per se*. This pulse pushes the first electron through the light filament almost instantaneously, thus making the light come on "immediately". Mencuccini and Holtta (2010) show how increased rates of sugar loading into the phloem of leaves (arising from increased canopy photosynthesis, which might arise from increased light supply after a storm front has moved away, for example) results in rapid declines of sieve tube cell solute potential, which induces water flow into the sieve tube cell of the source leaves and hence increases turgor potential of the source phloem. Because water is essentially inelastic (although phloem cells are not) a pressure pulse is very quickly transmitted down the length of the phloem in even the tallest trees. The increase in turgor in the phloem of roots (the sink) causes a rapid increase in sugar unloading out of the sieve tube cell into root cells and consequently root exudation is increased. Taken together, root, mycorrhizal, and microbial respiration respond to the increased supply of carbon substrates by increasing respiration (Kuzyakov and Gavrichkova 2010). It is the speed of the pressure pulse that allows rapid links between canopy photosynthesis and soil respiration, rather than the slow speed of transport of individual sugar molecules in the phloem.

2.3 Root Anatomy

Roots have four functions. First, they act as anchors for the aerial parts of the plant, holding the plant firm against lateral forces generated by wind and flooding water. Second, they serve as stores of water, nutrients and fixed carbon in, for example, perennial plants that die-back entirely at the start of winter. Third and fourth, they are sites of water and nutrient uptake from soil. A fifth function is observed in some aquatic or marine species where the diffusion of oxygen to deep roots located in anaerobic conditions is facilitated by the presence of aerenchyma–tissue that has a large number of large air-filled channels to ease the diffusion of air down into the anaerobic region of the root distribution.

To maximise water and nutrient uptake, roots are highly branched. Roots of crop plants can have several kilometres of total root length. Only fine roots (< 2 mm diameter) are functional in the uptake of water and nutrients. Larger, older roots primarily function as anchors and transport paths for movement of water and solutes to the aerial parts of the plant.

The epidermis of the terminal portions of fine roots contains roots hairs, single celled structures that increase the surface area for uptake of water and nutrients. The exodermis is a layer of cells found just beneath the epidermis and these usually contain a suberized lamella on the inside of the primary cell wall, similar to that of the Casparian strip which is located in the endodermis. The cortex is the layer of cells that are typically 2–5 cells thick and is located between the epidermis and the endodermis. Cortical cells provide pathways for movement of water and ions from the soil solution to the xylem in the roots. Water can move symplastically, especially via the plasmodesmata that link adjacent cells; or water can move apoplastically through the hydrated cell walls of the cortex. However, movement of water and solutes all the way to the xylem is impeded by the presence of the Casparian strip in the endodermis. The Casparian strip has a hydrophobic layer of suberin which stops apoplastic water flow into the stele. The stele is the tissue found within the bounds of the endodermis (that is, the endodermis encircles a tissue called the stele; Fig. 2.7). Within the stele lie the xylem and phloem which are the long distance transport pathways for water, inorganic ions, solutes such as sugars and amino acids, and hormones. An apical meristem is located at the tip of roots which allows extension growth. The apical meristem is protected by a root cap which lubricates the passage of the root tip through the soil and is involved in sensing the water content of the soil.

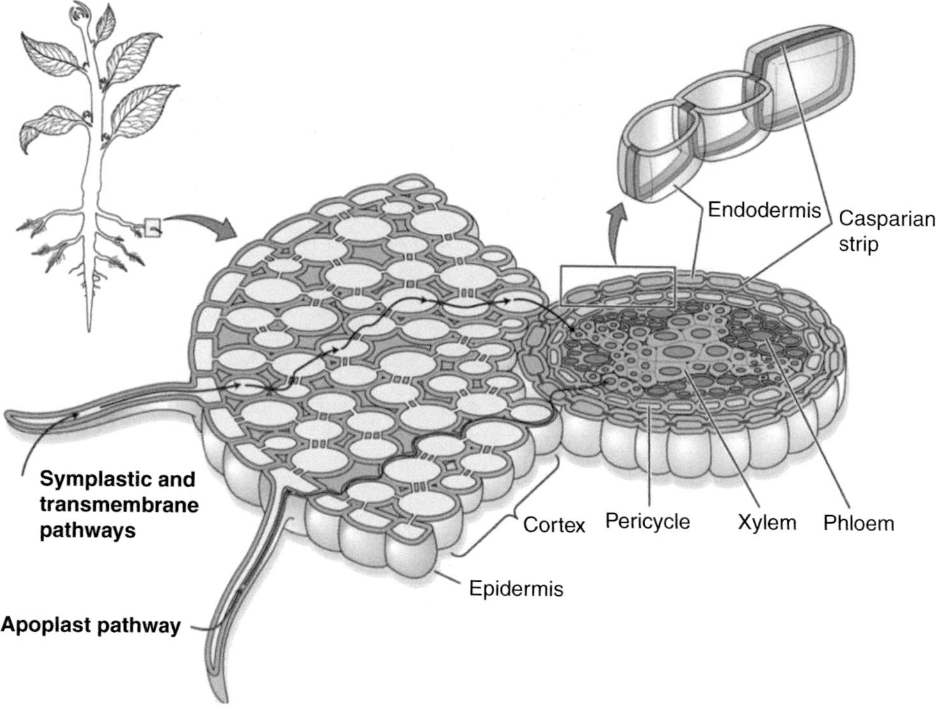

Figure 2.7. Stylised representation of the anatomy of a root. Reproduced with permission from Sinauer Publishing.

2.4 Stomatal, Mesophyll and Canopy Conductances

2.4.1 Stomatal Structure, Physiology and Conductance

Stomata are the pores in the surface of leaves (and fruit) through which CO_2 and water diffuse, along with O_2 and gaseous pollutants such as O_3. They provide the fast response (able to adjust their aperture within minutes) point of control of diffusion into and out of the leaf. Guard cells (GC), differ from all other epidermal cells in that they contain chloroplasts, lack plasmodesmatal links to adjacent cells and change their size in response to a range of biotic (e.g. the hormone abscisic acid, and leaf water status) and abiotic (e.g. light flux density and vapour pressure deficit) factors. In grasses their shape resembles that of a dumb-bell (narrow in the middle, wide at the two ends) but in dicotyledons (dicots) their shape is reminiscent of the shape of a kidney (Fig. 2.8). In dicots, when GC volume increases, the thicker part of the cell wall lining the pore plus the radial micellation of the cell wall results in bending of the GCs such that a pore is formed between the two adjacent GCs. In grasses, the middle portion of each GC remains fixed in volume but the end regions swell and push against each other, thereby creating an opening between adjacent GCs. Surrounding each pair of GCs is a variable number, typically between 2 and 8, of subsidiary cells. These cells store large amounts of potassium and chloride ions when the pore is closed. When the pore opens, this

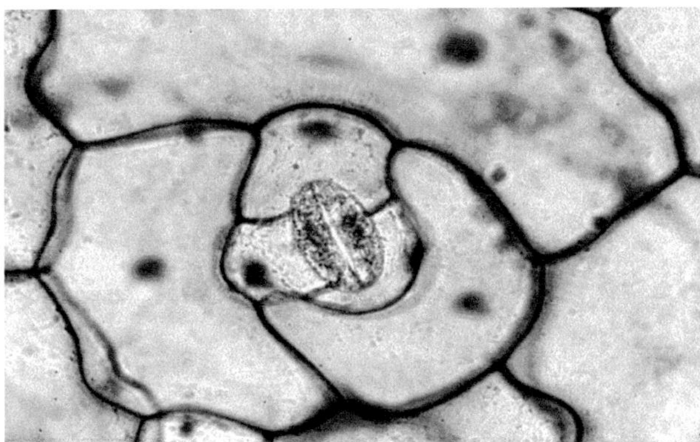

Figure 2.8. Surface view of a pair of guard cells surrounded by four specialised subsidiary cells.

potassium moves into the GCs. Stomata can be randomly distributed across a leaf surface, or they may occur in rows across the leaf, ranging in density from 50–300 mm^{-2}. Immediately beneath a stoma is the sub-stomatal cavity–an air space that allows rapid diffusion of water vapour and CO_2 out of and into the leaf.

2.4.2 *Stomatal Opening and Closing*

Stomata of C3 and C4 plants are generally closed at night (although partial opening at night does occur, giving rise to nocturnal transpiration; Chapter 3). Stomatal opening is triggered by sunlight in the morning and also by a reduction in the concentration of CO_2 inside the leaf (C_i = concentration of CO_2 in the sub-stomatal cavity). A reduction in C_i occurs in the morning when photosynthesis in the mesophyll consumes CO_2. We can experimentally induce stomatal opening in the dark by reducing the CO_2 concentration in the air surrounding the leaf. Blue light is especially effective in causing stomatal opening, although red light will also induce opening but to a smaller final aperture. The blue light response of GCs relies upon phototropin 1 and phototropin 2; these are blue light receptors located on the plasma membrane (plasmalemma) of the GC. The red light response is determined by the absorption of light by GC chloroplasts and the reduction in C_i that occurs as a result of photosynthesis.

Stomatal opening is initiated by the activity of a proton pumping ATPase (an example of an electrogenic pump; a pump that contributes to the generation of a voltage across a membrane). This pump is located in the GC plasma membrane and uses ATP to move protons from the inside of the GC to the outside. Protons (positively charged) accumulate in the apoplast (the cell wall of a plant cell is part of the apoplast); hence the GC membrane is hyperpolarised (becomes more negative on the inside). This activates the opening of voltage-gated influx channels that allow K^+ ions to diffuse into the GC down their respective electrochemical gradients (this is energetically downhill). Voltage gated influx channels are closed at night when

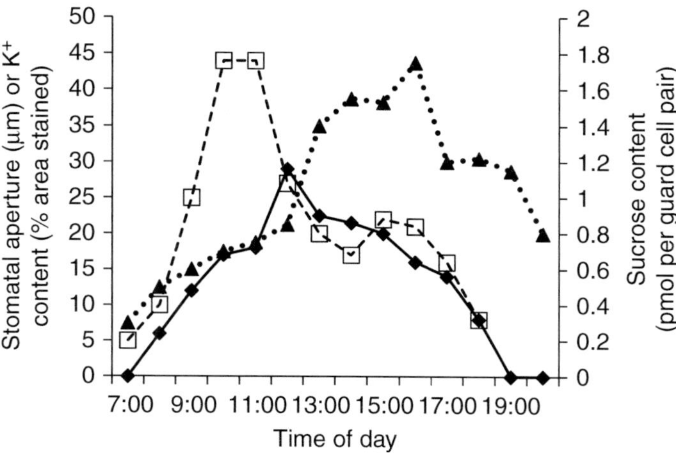

Figure 2.9. Stomatal aperture (solid line, diamonds) varies throughout the day and the guard cell concentration of potassium ions (dashed line, open squares) peaks in the morning while the concentration of sucrose (dotted line, triangles) peaks in the afternoon.

the membrane voltage is too low to open them. As the membrane voltage increases (hyperpolarises), the voltage crosses a threshold that induces the channels to open. Chloride ions also accumulate in GCs due to the activation of an H^+–Cl^- symporter. The movement of H^+ from the outside of the GC to the inside is energetically down-hill because the concentration of H^+ is larger on the outside than the inside and also because the membrane voltage is negative on the inside (and protons have a positive charge). Simultaneously phospho-enol pyruvate is carboxylated (a CO_2 molecule is attached enzymatically) and malic acid accumulates, while starch is broken down to yield glucose and fructose which are then used to synthesise sucrose. The accumulation of sucrose is especially strong in the afternoon when the chloroplasts of GC have accumulated starch through the process of photosynthesis (Fig. 2.9). Sucrose is a potent osmotically active solute, as are K^+, Cl^-, and malic acid. As the number of solute molecules in GCs increase the solute potential and hence the water potential of the GCs decline (Chapter 3). A new gradient of water potential is established between the GC and the surrounding apoplast and subsidiary cells, and water diffuses into the GC. Consequently GC volume increases and turgor potential increases. This causes the pore to widen and stomatal conductance increases. There is a strong positive correlation between GC turgor and stomatal aperture (Fig. 2.10) and there is a linear relationship between GC aperture and stomatal conductance.

A summary of the changes in K^+, Cl^-, pH and membrane potentials for open and closed stomata, for epidermal cells, subsidiary cells and GCs, is shown in Table 2.2. It is clear that proton pumping out of the GC, hyperpolarisation and influx of K^+ and Cl^- are central to the process of opening of the pore. Closure at night occurs as a reversal of these processes. Proton pumping ceases, the membrane voltage increases, K^+ influx channels close and starch breakdown ceases. Water is lost from GCs and

Table 2.2. Changes in potassium, chloride, pH and membrane potential (E_m) of epidermal, subsidiary and guard cells when stomata are open or closed

	[K] mEq L^{-1}	[K] mEq L^{-1}	[Cl] mEq L^{-1}	[Cl] mEq L^{-1}	pH	pH	E_m (mV)	E_m (mV)
	Open	Closed	Open	Closed	Open	Closed	Open	Closed
Epidermal cell	73	450	85	120	5.1	5.7	−31	−34
Subsidiary cell	300	160	60	40	5.6	5.6	−30	−30
Guard Cell	450	95	120	30	5.6	5.2	−45	−30

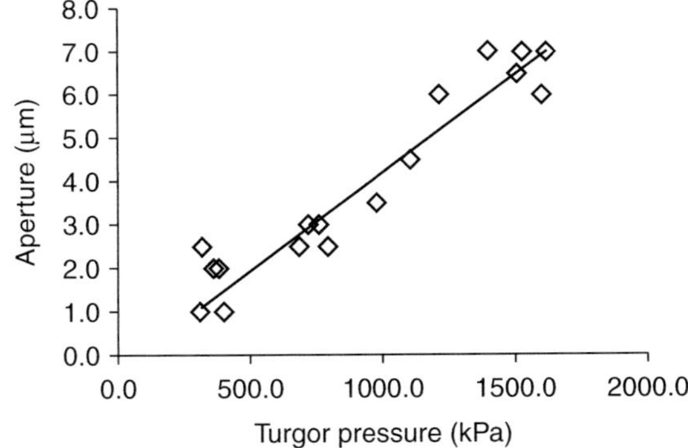

Figure 2.10. There is a strong positive correlation between stomatal aperture and guard cell turgor pressure.
Redrawn from Eamus *et al.* (2008).

consequently turgor is lost and deflation occurs. The pore closes. Simultaneously photosynthesis within the mesophyll stops while respiration continues and hence C_i increases. This also induces stomatal closure, possibly through acidification of the apoplast around the GCs.

2.4.3 *Stomatal Behaviour Under Controlled Conditions and in the Field*

Stomata respond to several biotic and abiotic factors, including light flux density, atmospheric vapour pressure deficit, soil moisture content, leaf water potential and the concentration of CO_2 within the leaf (C_i) (Figs. 2.11–2.14). An understanding of the generalised patterns of behaviour of stomatal conductance (g_s) in the field is central to many ecophysiological studies of plant function and is central to all case studies presented in the last section of this book.

2.4.4 *Light, Vapour Pressure Deficit and Soil Moisture Content Regulate Stomatal Conductance*

Stomatal conductance in C3 and C4 plants is generally assumed to be zero at night (see discussion of nocturnal transpiration below for examples to the contrary). When

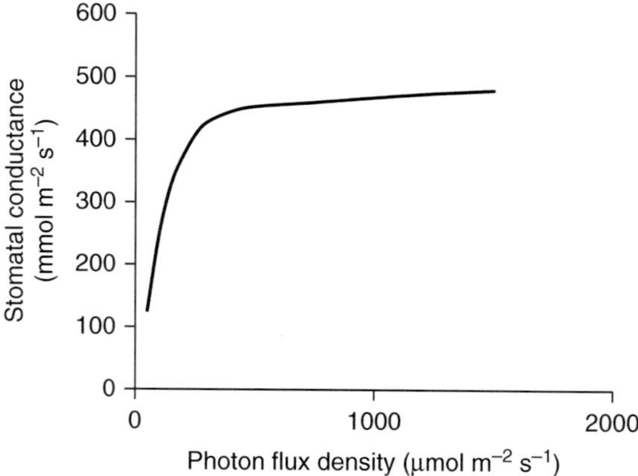

Figure 2.11. Stomatal conductance increases almost linearly with increasing light levels when light levels are very low, but approaches a maximum value at approximately one-quarter full sunlight.

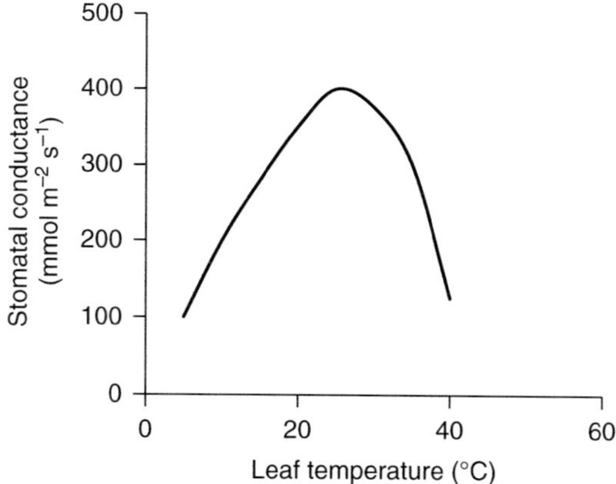

Figure 2.12. Stomatal conductance is maximal at some optimal temperature but decreases when temperature increases or decreases above this optimum temperature.

the sun rises, g_s increases rapidly and reaches a maximal value. This maximal value differs among species and plant functional types. Crops tend to have the largest g_s under optimal conditions (400–800 mmol m^{-2} s^{-1}), while the g_s of conifers tends to be much smaller (100–250 mmol m^{-2} s^{-1}) and deciduous trees have intermediary maximal values of g_s. As air temperature and solar radiation supply increase in the early part of the morning, photosynthetic rates increase and C_i declines, further stimulating stomatal opening. As temperature increases further, atmospheric vapour pressure deficit (D) continues to increase even if photon flux density remains relatively constant. This is because air temperature continues to increase throughout the morning and much of the afternoon. Increasing D has a negative influence on g_s once D is larger than 1.0–2.0 kPa (approximately; in semi-arid or arid zones, this may not occur until $D > 3+$ kPa). The decline of g_s to increasing D (with constant light flux

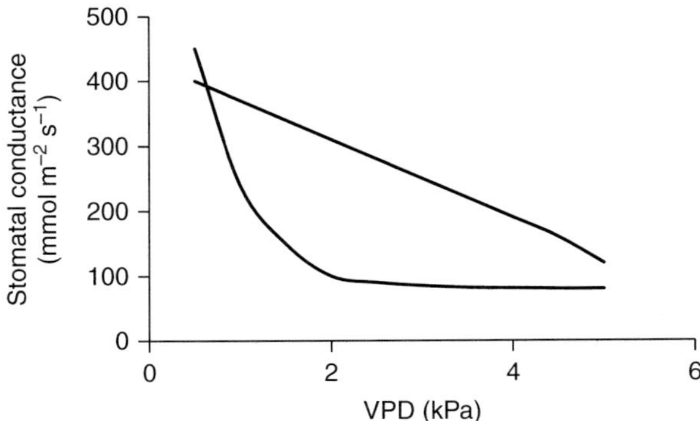

Figure 2.13. Stomatal conductance declines with increasing vapour pressure deficit (VPD). Some species show an approximately linear decline whilst others show an approximately exponential decline, with increasing VPD.

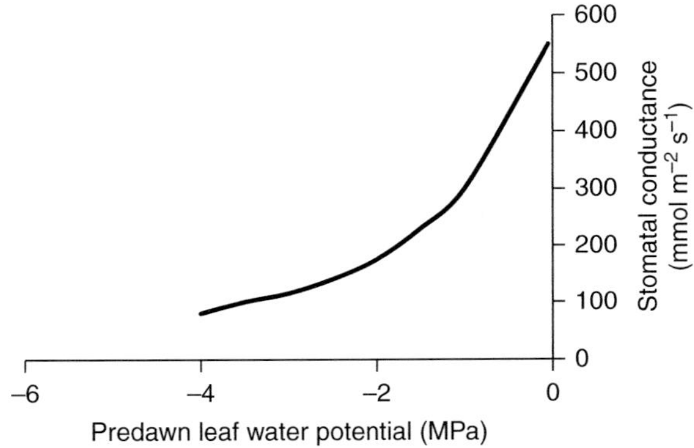

Figure 2.14. Stomatal conductance decreases approximately exponentially with declining pre-dawn leaf water potential. The decline in pre-dawn leaf water potential is indicative of a drying soil.

density and temperature) is generally either linear or exponential (Fig. 2.13; Thomas and Eamus 1999).

More detail on the abiotic (non-living) controls of g_s is given in Section 2.4.9 on the development and application of a modified Jarvis-Stewart model of tree water-use. Biome-specific examples of the behaviour of stomatal and canopy conductance are given in the case-study chapters.

2.4.5 *Three-Phase Response of Stomata and the Interaction of Soil Moisture Content and VPD in the Control of g_s*

The behaviour of stomatal conductance (g_s) is central to the maintenance of plant water status (Roelfsema and Hedrich 2005) since it is the diffusion of water vapour at the leaf epidermis, predominantly through stomata but also across the cuticle, which determines the rate of canopy water-use. At a constant leaf-to-air vapour pressure difference (D), increased stomatal or cuticular conductance increases the rate of transpiration. However, increasing D causes a decline in g_s and several decades of research have been dedicated to determining the mechanism of this response. Mott and Parkhurst (1991) were the first to show that it is the rate of transpiration that influences g_s rather than D *per se*. They did this by making artificial air by replacing N_2 with Helium (He). At any given D, g_s was always smaller in air containing He than air containing N_2. This was because water vapour diffuses more quickly through He than N_2 and therefore transpiration was faster (and consequently g_s smaller) in artificial air than normal air.

When g_s is plotted as a function of transpiration (E) through manipulation of D, a 3-phase response is observed (Fig. 2.15). The interpretation of this is challenging, but a recent model and experimental test of the model suggest that feedback processes based upon cellular hydraulics may explain this pattern (Eamus *et al.* 2008).

At low values of D (for example 0.1 to 2 kPa), transpiration increases as D increases (phase C) and the limitation on E exerted by stomata is minimal. The reason why E increases is because D is increasing and g_s shows very little change over this region. At intermediate values of D (for example 2–3.5 kPa) stomata begin to increasingly limit E, but E remains almost constant as D increases (phase A). The increase in D is offset almost perfectly by decreasing g_s. At larger values of D (for example 3.5–5.5 kPa), increases in D cause large declines in g_s, leading to reduced E (phase B). Monteith (1995) provides a review of some of the data that display this pattern. The three-phase response is not immutable and can be influenced by previous exposure to water stress (Thomas and Eamus 1999). As water stress develops the three phases collapse into two, and then one, phase (phase B; Fig. 2.15).

How is this complex behaviour explained? Guard cells are strongly linked hydraulically to epidermal cells since the water for stomatal opening is derived from the epidermis. The epidermis is covered in a waxy cuticle to minimise water loss, but no cuticle is 100% impermeable to water. Furthermore, the underside of guard cells in many species is also covered with a cuticle. As D increases, the rate of cuticular transpiration (water loss across the leaf's cuticle) increases linearly. In a detailed mechanistic model, Eamus and Shanahan (2002) showed that the three-phase behaviour of g_s in response to increasing D could be explained on the basis of feedback processes such that as cuticular transpiration increased, the supply of water to guard cells decreased and guard cell turgor declined and hence stomatal aperture declined. More importantly, four predictions based upon this model were examined experimentally (Eamus *et al.* 2008). The feedback model predicts that:

(a) increasing cuticular conductance to water vapour will decrease supply of water to guard cells, thus g_s will decline and the response curve of g_s to D will tend to become linear;

(b) increasing cuticular conductance to water vapour will decrease the supply of water to guard cells and the three-phase response of g_s to E will be lost;

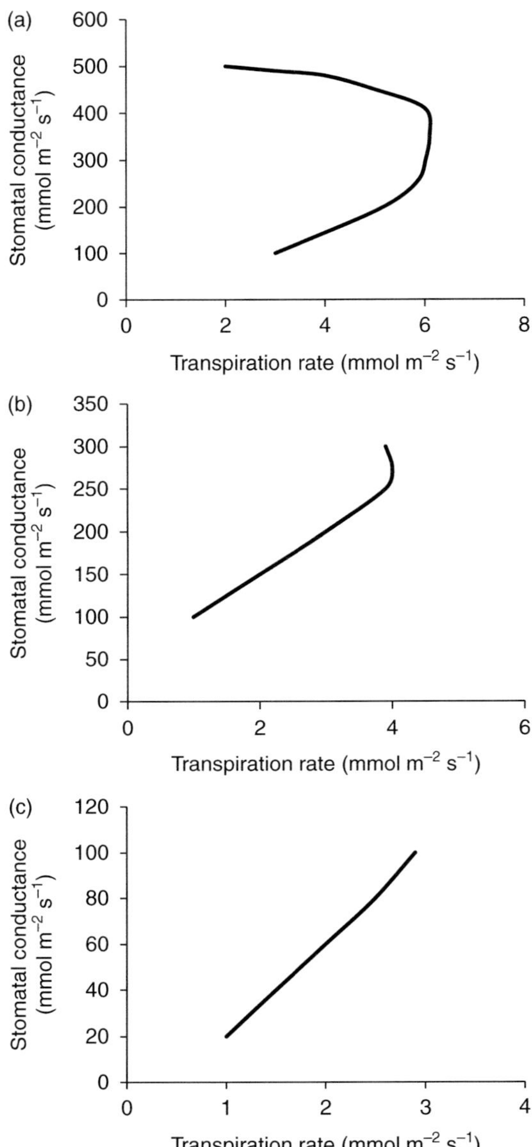

Figure 2.15. An example of a three-phase response of stomatal conductance to transpiration rate. In the upper panel, the plant is well-watered and leaf water potential –0.1 MPa. In the middle panel a mild water stress has developed. In the third panel a severe water stress has developed.
Redrawn from Thomas and Eamus (1999).

(c) increasing leaf temperature will cause cuticular conductance to increase exponentially but rates of transpiration through stomatal pores will not increase;

(d) decreasing the local cuticular conductance in close proximity to guard cells will increase guard cell turgor.

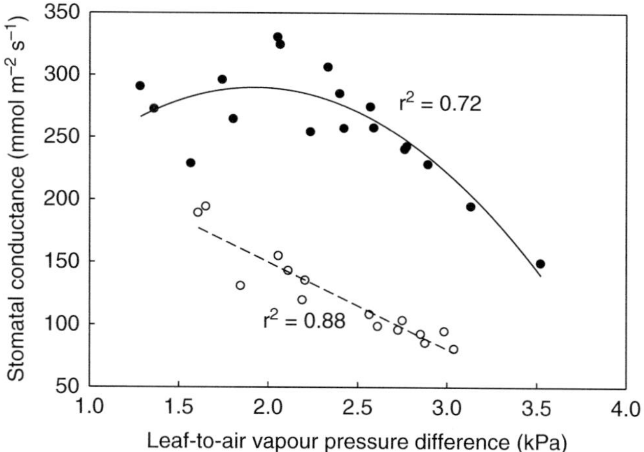

Figure 2.16. The stomatal conductance (g_s) of leaves with intact cuticles (solid line) show a convex response to increasing D, but when the cuticle is removed by organic solvent, the absolute magnitude of g_s is reduced and the convexity is lost. Redrawn from Eamus *et al.* (2008).

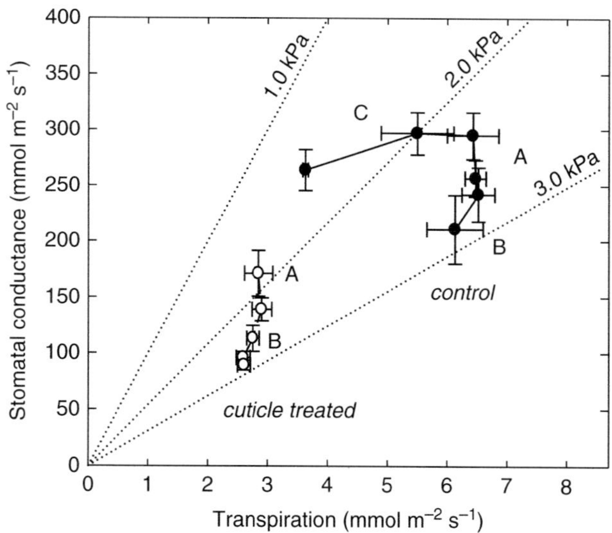

Figure 2.17. The three-phase response of stomatal conductance (g_s) of leaves with intact cuticles (solid line) is lost when the cuticle is removed by organic solvent and only two-phases are then observed. Redrawn from Eamus *et al.* (2008).

Prediction (a) is supported by Fig. 2.16, where it can be seen that the absolute value of g_s was decreased in response to dissolving much of the cuticle using a solvent and the convexity of the response of g_s to increasing D was lost. The second prediction is supported by Fig. 2.17, where the three-phase response of intact leaves is lost and the cuticle-treated leaves display only a two-phase response.

Over the temperature range 19–40°C, transpiration (and hence stomatal conductance as D was constant) showed an approximately parabolic response (Fig. 2.18).

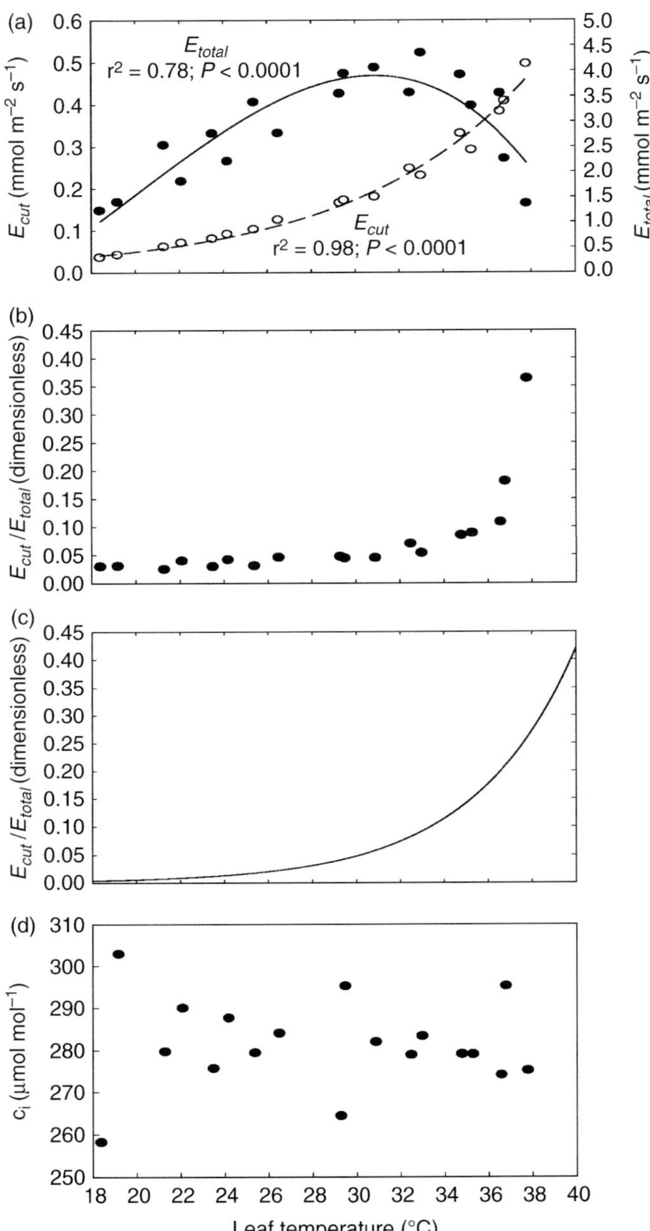

Figure 2.18. (a) Transpiration through stomata of leaves of *Eucalyptus haemastoma* increases to a maximum and then declines as temperature increases at constant D (solid line). In contrast cuticular transpiration increases approximately exponentially with increasing temperature (dashed line). However, (b) the ratio of E_{cut} to E_{total} increases with increasing temperature at constant D. Similarly the model shows (c) an increase in this ratio with increasing temperature at constant D, whilst simultaneously (d) C_i remains relatively constant across the full range of temperatures used.

Table 2.3. Stomatal aperture increases after the application of a light oil to the cuticle of epidermal cells in close proximity to guard cells

Species	Light level	Stomatal aperture + SE (µm) before application of oil	Stomatal aperture + SE (µm) after application of oil	% Change in aperture
Commelina cyanea	High	9.72 + 0.37	11.40 + 0.36	17.28
	Moderate	8.84 + 0.34	11.36 + 0.32	28.51
Vicia faba	High	11.76 + 0.48	14.12 + 0.38	20.06
	Moderate	11.31 + 0.41	13.12 + 0.58	16.01

Note: This is because the decrease in cuticular conductance increases the supply of water to the guard cells, which causes guard cell turgor to increase.
Source: From Eamus *et al.* (2008).

Thus transpiration was maximal at a temperature of approximately 26–32°C, and declined at lower or higher temperatures. Over the full temperature range of the experiment, cuticular transpiration rate increased approximately exponentially (prediction (c) above). The ratio of cuticular transpiration to total transpiration increased gradually over the range 18–30°C, but above *ca* 32°C the ratio increased rapidly and substantially. This pattern agreed well with the model prediction (Fig. 2.18). The concentration of CO_2 within the leaf (C_i) remained constant across the entire temperature range, thereby indicating that the response was not a function of any changes in internal CO_2 concentration (Eamus *et al.* 2008). Finally, when cuticular conductance was decreased by applying a very small quantity of oil to the cuticle in the vicinity of guard cells, stomatal aperture increased because local cuticular conductance declined and consequently there was an increased supply of water to the guard cells and hence an increase in stomatal turgor (prediction (d) above; Table 2.3).

Although the model discussed in 2.4.5 provides a mechanistic explanation of the three-phase behaviour of stomata to changes in *D*, debate about alternate mechanisms continues. Recent developments include two new mechanisms, summarised in the following paragraphs.

Mott and Peak (2013) tested the conceptual model of Peak and Mott (2011). This model combines the influence of both atmospheric humidity and temperature on g_s and has three novel features:

1. It predicts that feedforward and feedback behaviours may occur within the same plant, at different times.
2. Water transport to guard cells is predominantly in the vapour phase, not liquid phase because guard cells are assumed to be hydraulically isolated from the epidermis. Under steady-state conditions, guard cells are assumed to be in near-equilibrium with the water vapour in the bottom of the stomatal pore.
3. It includes a temperature dependence and a humidity dependence of g_s within a single model.

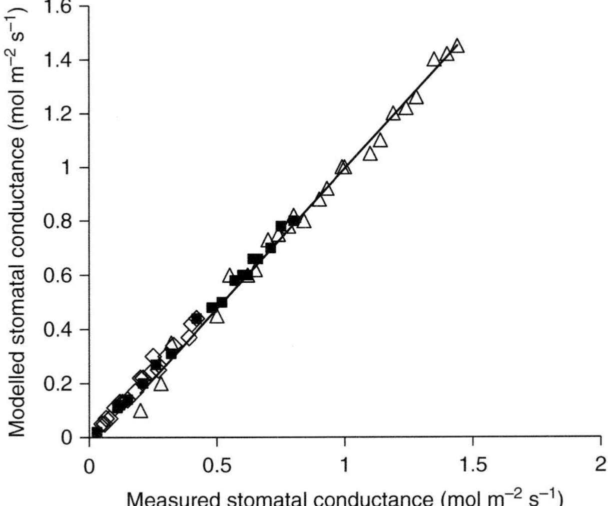

Figure 2.19. For three species of plants differing in Z/θ (Z/θ = 75 for *Pastinaca sativum*, squares; Z/θ = 17 for *Xanthium strumarium*, triangles; Z/θ = 864 for *Nerium oleander*, diamonds), the model was able to simulate g_s with a 1:1 fit to observed data. Redrawn from Mott and Peak (2013).

The model is written as:

$$g_s = \frac{g_s^0 - \theta \dfrac{\Delta w}{w_{es}}}{1 + Z\Delta w} \qquad (2.1a)$$

and has three independent variables: g_s^0 is the value of g_s when Δw is zero, Z is a model parameter that is determined mostly by the thermal resistance (resistance to heat transfer) between the epidermis and the site of evaporation and θ is a model parameter determined mostly by the resistance to the diffusion of water from the evaporating site to the guard cells. Δw is the difference in water vapour mole fraction between the air space adjacent to the evaporating surface and the bulk atmospheric air and Δw_{es} is the saturated mole fraction of water vapour in ambient air. In a test of this model using three species that differ substantially in their ratio Z/θ they found that the ratio of observed to modelled g_s all three species collapsed to a single regression (Fig. 2.19) with a slope of 1. Differences in the response curves of transpiration *versus* air and Δw were accounted for by differences in g_s^0 and the ratio Z/θ was constant within a single species.

More recently Duursma *et al.* (2014) focus on the fact that changes in rates of photosynthesis can affect g_s via changes in C_i (Wong *et al.* 1979). Strong linear correlations between photosynthesis and g_s are observed across a range of environmental stresses. Consequently, in the field, where changes in D are generally correlated with

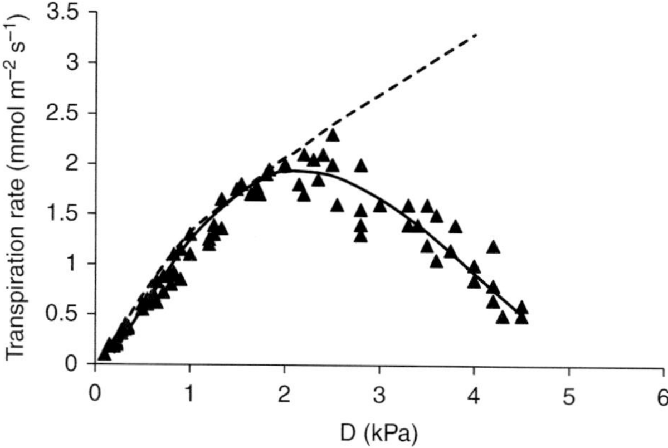

Figure 2.20. Variation in measured rates of transpiration (triangles) as a function of *D* is simulated well with a coupled g_s-assimilation model when both temperature and *D* are varied in the model (solid line) but when the effect of temperature on photosynthesis is omitted (dashed line) the model is unable to simulate observed changes in transpiration. Based upon Duursma *et al.* (2014).

changes in temperature and changes in temperature directly influence photosynthesis and hence C_i they postulate that the apparent feed forward response of E to *D* in the field can be explained by the response of photosynthesis to temperature. Using the coupled leaf gas exchange model of Medlyn *et al.* (2011; see Chapter 11) they were able to show that when *D* was varied alone, the response of transpiration to increasing *D* did not simulate observations but when the temperature response of assimilation (and hence C_i) was included, the model was able to capture the apparent feed forward response of transpiration to increasing *D* (Fig. 2.20; Duursma *et al.* 2014).

2.4.6 Mesophyll Conductance

The rate of diffusion of CO_2 from the atmosphere to the sites of carboxylation in chloroplasts is determined by the total resistance to diffusion. There are three resistances in series between the atmosphere and chloroplasts. The first is produced by the presence of a boundary layer (a layer of stationary air) near the leaf surface. The second resistance is encountered at the stomatal pore. The third resistance occurs during diffusion from the sub-stomatal cavity to sites of carboxylation in chloroplasts and includes both gaseous and liquid phases. The partial pressure of CO_2 in the chloroplast is approximately 50% of that of ambient air and about 60% of this difference (which arises because of the resistances to the diffusion of CO_2) can be attributed to the boundary layer and stomatal aperture. The remainder is because of the resistance encountered between the sub-stomatal cavity and the site of CO_2 fixation in the chloroplast. Gaseous diffusion occurs through the air spaces of leaves but liquid diffusion occurs when CO_2 dissolves in the water of cell walls, the cytoplasm of mesophyll cells

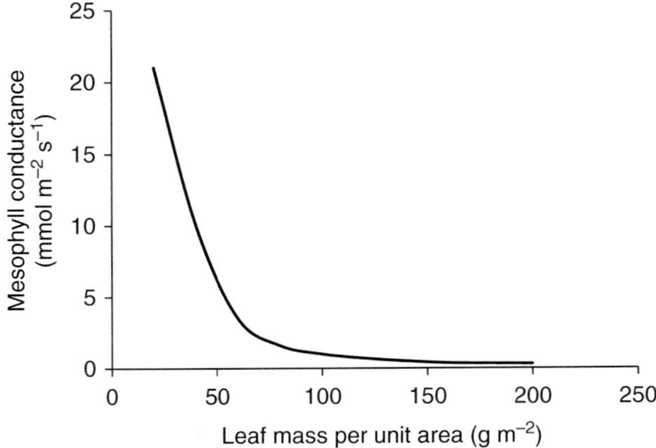

Figure 2.21. There is a highly non-linear negative correlation between leaf mass per unit area and mesophyll conductance (from Niinemets *et al.* 2009).

and the intra-chloroplast stroma. The sum of all of the resistances downstream of the sub-stomatal cavity is called the mesophyll resistance, although resistance is replaced by conductance in modern terminology, hence mesophyll conductance (g_m). This is sometimes called the "internal conductance" because mesophyll conductance was a term used historically to refer to the slope of the initial linear portion of an A/C$_i$ curve.

Mesophyll conductance is now recognised as an important factor limiting the diffusion of CO_2 into leaves. Estimating g_m requires measurements of gas exchange of CO_2 and water vapour and calculation of C_i and C_c (the concentration of CO_2 in the chloroplast). Using an Ohm's law analogy, g_m can be calculated from the ratio of CO_2 flux into the leaf (due to photosynthetic fixation in the chloroplast) to the gradient in CO_2 concentration as per Equation 2.2:

$$g_m = A / (C_i - C_c) \qquad (2.2)$$

where g_m is mesophyll conductance, A is rate of photosynthesis and C_i and C_c are the concentrations of CO_2 in the sub-stomatal cavity and chloroplast respectively. Typical values of g_m range from 20 to 150 mmol m^{-2} s^{-1} which is in the lower range of typical values of g_s observed in broad leaf species (100–500 mmol m^{-2} s^{-1}). Changes in leaf structure and physiology (e.g. specific leaf area; changes in the fraction of lignified support structures) occur in response to environmental stresses and such changes alter g_m. For example, there is a strong negative correlation between mass based mesophyll conductance and leaf mass per unit area (Fig. 2.21). From equation two, it can be seen that changes in g_m have the potential to alter the maximum rate of photosynthesis of leaves (Niinemets *et al.* 2009).

Plant water stress can be simulated by experimental additions of abscisic acid to leaves or induced by the addition of a non-permeable solute (e.g. polyethylene glycol) around the roots. When such treatments are applied a rapid decline of g_m

is observed. Similarly, slow cycles of soil drought can decrease g_m. Following the removal of water stress, rates of photosynthesis frequently remain slightly lower than pre-stress levels despite g_s reaching pre-stress levels. This may reflect a long-term reduction in g_m arising from long-term changes in cell wall thickness and increased bulk leaf elastic modulus (which means the walls are less elastic; this is a mechanism for drought tolerance). Salt stress can reduce g_m, possibly because salt stress is also associated with reduced water availability for the plant. Reductions in leaf temperature during periods of leaf expansion are generally associated with reductions in average leaf area and increased leaf thickness and density. Consequently g_m tends to be smaller and it's limiting effect on photosynthesis larger when growth temperatures are smaller (Niinemets *et al.* 2009). Finally, as leaves age in evergreen species, or as a tree ages and grows taller, g_m tends to decline by up to 70%. Interestingly, photosynthesis is least sensitive to changes in g_m when g_s is reduced, for example when plants are water stressed. Leaves with low g_m experience a smaller reduction in assimilation rate as a result of stress induced reductions in g_s than leaves with a large g_m. This of course is not true when the stress is removed and g_s subsequently increases and this reflects the "cost" of having robust leaves that can tolerate and survive stressful environments.

Mechanisms to explain rapid changes in g_m remain elusive. While a role for aquaporins and carbonic anhydrase have been proposed previously, recent analyses by Tholen *et al.* (2012) suggest that at least some of the variation in g_m arises because of changes in the rate of photorespiration, or more specifically, changes in the ratio of photosynthesis to photorespiration. Photorespiration produces CO_2 which can be refixed by the chloroplasts and this re-fixation has an impact on the observed rate of photosynthesis as a function of apparent concentration of CO_2 within the chloroplast.

2.4.7 Canopy Conductance

At the leaf-scale, the flux of water and CO_2 from/to a leaf is regulated by changes in g_s. In contrast, fluxes in reasonably open plant canopies (e.g. the canopy of a stand of savanna trees) are regulated by changes in canopy conductance (G_c). When the canopy is closed and dense (e.g. a short, dense sward of grass), changes in G_c are less important because the canopy is decoupled from the atmosphere. The meaning of decoupling in this context is discussed in 2.4.10.

G_c is a measure of the ease with which gases can diffuse into and out of a canopy. There are many ways to calculate G_c. However, generally all of the equations can be understood in terms of the classic Ohm's Law, which states that the resistance of an electrical circuit is equal to the voltage across it divided by the current flowing along it. Thus Equation 2.3:

$$V = IR \tag{2.3}$$

where V = voltage across the circuit, I = current flowing in the circuit, and R is the resistance of the circuit to flow of electricity. Re-arrangement of Equation 2.3 provides us with Equation 2.4:

$$R = V/I \qquad (2.4)$$

Since conductance is the inverse of resistance, conductance = 1/R; hence conductance is equal to I/V, or flux divided by driving force.

This Ohm's Law analogy can be applied to water-use by canopies. Canopy conductance is equal to canopy water flux divided by the driving force causing water to move. In the liquid phase the gradient driving water flow is the gradient of water potential between the roots and the canopy (Chapter 3). In the gas phase the gradient is usually expressed in terms of the vapour pressure difference between leaf and air, although this is often simplified to be the equivalent to the vapour pressure deficit of the air assuming that leaf and air temperatures are identical and the leaf internal air spaces are saturated with water vapour. Both assumptions are reasonably correct under many but not all circumstances.

Classically, calculation of G_c requires inversion of the Penman-Montheith (PM) equation to give Equation 2.5:

$$G_c = [\lambda\gamma E \, G_a]/[(\Delta R_n) + (k\rho C_p D G_a) - (\lambda([\Delta + \gamma) E] \qquad (2.5)$$

where λ is the latent heat of vaporisation (2.45 MJ kg^{-1}), Δ is the slope of the relationship between saturation vapour pressure and temperature (kPa °C^{-1}), R_n is net radiation intercepted by the forest canopy (MJ m^{-2} hr^{-1}), ρ is density of dry air (kg m^{-3}), C_p is the specific heat of moist air (1.013 MJ kg^{-1} ° C^{-1}), D is the vapour pressure deficit (kPa), G_a is aerodynamic conductance (m s^{-1}), γ is the psychometric constant (0.066 kPa °C^{-1}), E is transpiration rate, and k is a factor to convert values from seconds to hours.

Hourly or mean daily values of R_n and D and tree water-use (derived from sapflow measurements or eddy covariance estimates) are inputs that allow G_c to be calculated at an hourly or daily time-step. However, estimates of G_c *per se* are not usually the desired research goal; rather it is an estimation of canopy water-use that is desired for hydrological and water resource management purposes. Consequently the non-inverted version of the PM equation is used and user-defined functional relationships between G_c and *Rn, D* and soil moisture (θ) are required so that the PM equation can be used to calculate canopy water-use. In order to determine the relationships among D, R_n, θ, and G_c, the Jarvis-Stewart (JS) method is applied (Jarvis 1976, Stewart 1988). In the JS method, three normalised functions are used to describe the individual response of g_s to changes in D, R_n or θ, assuming that for each response there is only one limiting factor (i.e. D, R_n, or θ). This is written as Equation 2.6:

$$g_s = g_{s,max} f(R_n) \, f(D) \, f(\theta) \qquad (2.6)$$

where $g_{s,\max}$ is the maximum stomatal conductance under non-limiting conditions and f denotes the function describing the upper boundary response of conductance (i.e. the response of the maximum values of conductance measured as a single environmental variable changes) to a change in one or other of the variables. This methodology has been widely applied (Harris *et al.* 2004, Komatsu *et al.* 2006) but requires extensive spatial and temporal replication of measurements of g_s and subsequent use of the PM equation to derive estimates of canopy water use. Alternatively G_c can be derived from g_s using the relationship $G_c = \Sigma(g_s\mathrm{LAI})$ where canopy conductance is the parallel sum of values of g_s sampled throughout the layers of the canopy (Whitehead 1998).

2.4.8 Alternate Methods to Calculate Canopy Conductance

O'Grady *et al.* (2008) calculated G_c on a leaf area basis using rates of sapflow (i.e. water flux through the canopy) and measurements of vapour pressure deficit, following Ewers *et al.* (2005) (Eq. 2.7):

$$G_c = \frac{K_g(T_a)E_1}{D} \tag{2.7}$$

where K_g is a conductance coefficient that varies with temperature (K_g = 115.8 $\pm$ 0.4236 (T_a) kPa m^3 kg^{-1}). This coefficient accounts for temperature dependence of the psychrometric constant, latent heat of vapourization, specific heat of air at constant pressure and the density of air. T_a is air temperature (°C), E_l is transpiration rate (expressed per unit leaf area) (kg of water s^{-1} m^{-2}), and D is vapour pressure deficit (kPa). Equation 2.7 calculates G_c in m s^{-1} and assumes boundary layer conductance to be high (i.e. a canopy coupled to the atmosphere) so that there are only small gradients in D through the boundary layer and hence strong coupling between the vapour pressure deficit of the leaf and atmosphere. It also assumes that there are no vertical gradients in D throughout the canopy and negligible water stored above the point at which sapflux is measured.

Yunusa *et al.* (2010) and Nicolas *et al.* (2008) used the expanded version of Equation 2.7 as per Equation 2.8:

$$G_c = \lambda E_c \gamma / \rho_a C_p D \tag{2.8}$$

in which E_c is equivalent to E_l of O'Grady *et al.* (2008) and actual values of γ, the psychrometric constant (Pa °C^{-1}) and λ is the latent heat of evaporation of water (J kg^{-1}); ρ_a is the density of moist air (kg m^{-3}) and C_p is the specific heat of moist air at constant pressure (J kg^{-1} °C^{-1}). λ is often taken as constant (2.45 MJ kg^{-1}).

Pataki *et al.* (1998) calculated G_c and expressed it on a leaf area basis as opposed to a ground area basis (which is the usual method of expression). They termed this "canopy stomatal conductance", G_t (mmol m^{-2} s^{-1}), and expressed its derivation using Equation 2.9:

$$G_t = G_c/A_l = \lambda E_c \gamma /[(\rho C_p D)(A_l/A_b)] \tag{2.9}$$

where A_l is leaf area (m^2), γ is the psychrometric constant, λ is the latent heat of vaporization, ρ is the density of moist air, C_p is the volumetric heat capacity of moist air at constant pressure, D is vapour pressure deficit (kPa), and A_l/A_b is the ratio of leaf area to cross-sectional stem area. When D is measured above the canopy, boundary layer and aerodynamic conductance are both included in this estimation of G_t (and indeed in G_c in all other methods too). However, where boundary layer conductance is large relative to stomatal conductance, leaf temperature does not deviate significantly from air temperature and D can be used as a proxy for the leaf-to-air vapour pressure difference. Under such conditions G_t closely approximates the mean stomatal conductance of the canopy.

Perhaps the simplest method was used by Zeppel and Eamus, (2008). In order to convert leaf-scale stomatal conductance (g_s) to canopy conductance (G_c), the following equation (Eq. 2.10) was used (Lu *et al.* 2003):

$$G_c = g_s \times \text{LAI} \tag{2.10}$$

where LAI is the leaf area index of the stand of trees. The major problem with this method is obtaining a reasonable estimation of the average g_s of a large canopy of trees that includes sunlit and shaded leaves.

2.4.9 Behaviour of G_c and Transpiration in the Field: A Modified Jarvis-Stewart Model

In the past, large numbers of measurements of g_s in the field were needed to use the PM equation to estimate canopy transpiration. However, recently an alternative methodology has been developed (Whitley *et al.* 2009) which is conceptually identical to a Jarvis-Stewart (JS) approach but which removes the need to measure g_s in the field and the need to use the PM equation. The PM equation, although it performs well in wet soils, does not incorporate the influence of changes in soil moisture on canopy water-use and consequently performs poorly under dry or drying conditions.

In the modified version of the JS approach, canopy water use is calculated as a function of the three principle drivers of water use, namely R_n, D, and θ. An estimate of the maximum rate of canopy water use is required, derived either from sapflow sensors or eddy covariance data. Thus Equation 2.11:

$$E_c = E_{s,max} f(R_n) f(D) f(\theta) \tag{2.11}$$

This approach is much simpler to fit, requires far fewer field measurements and avoids the circularity of inverting the PM to calculate G_c and then using the PM with this value of G_c to calculate water use (Whitley *et al.* 2009). The functional forms of the three equations used by Whitley *et al.* (2009) are shown below:

$$f_1(R_S) = \left(\frac{R_s}{1000} \right) \left(\frac{1000 + k_1}{R_s + k_1} \right)$$

(2.12a)

$$f_3(\theta) = \begin{cases} 0 & ,\theta < \theta_w \\ \dfrac{\theta - \theta_w}{\theta_c - \theta_w} & ,\theta_w < \theta < \theta_c \\ 1 & ,\theta > \theta_c \end{cases}$$

(2.12b)

A modification (Whitley *et al.* 2009) of the original response function for D is:

$$f_2(D_v) = \exp\left\{ -\frac{k_{D1}(D_v - D_{peak})^2}{D_v + k_{D2}} \right\}$$

(2.12c)

The graphs in Figure 2.22 represent field measurements of transpiration (symbols) and simultaneous measurements of solar radiation, D and soil moisture content (Whitley *et al.* 2009). Similar relationships can be found in Ewers *et al.* (2005), Forrester *et al.* (2010), Kutsch *et al.* (2008), and Quentin *et al.* (2011). The lines on each graph are the boundary regressions defined by the three functions for transpiration or G_c shown in Equations 2.12 a-c above. The data points generally fall below the boundary line because a factor other than the one defined by the respective x axis was limiting G_c and hence transpiration. It is clear that G_c responds to solar radiation, D and soil moisture content in a way that closely resembles the responses discussed for g_s previously. In many temperate and boreal zone studies, the range of D experienced is frequently too small (< 2 kPa) to show the decline in transpiration that occurs at moderate to large values of D. In contrast small changes in soil moisture content, which can be approximated by measurements of pre-dawn leaf water potential (Chapter 3), can exert significant effects on G_c. An example of changes in maximum canopy conductance for a plantation of *E. globulis* is presented in Figure 2.23, where the impact of declining soil moisture content (measured as declining pre-dawn leaf water potential; Chapter 3) is evident as a decline in canopy conductance (From O'Grady *et al.* 2008).

2.4.10 The Decoupling Factor

The decoupling factor or decoupling coefficient (Ω) is a measure of the dependence of canopy transpiration on physical (boundary layer) or physiological (stomatal) factors. This coefficient expresses the relative sensitivity of canopy transpiration to

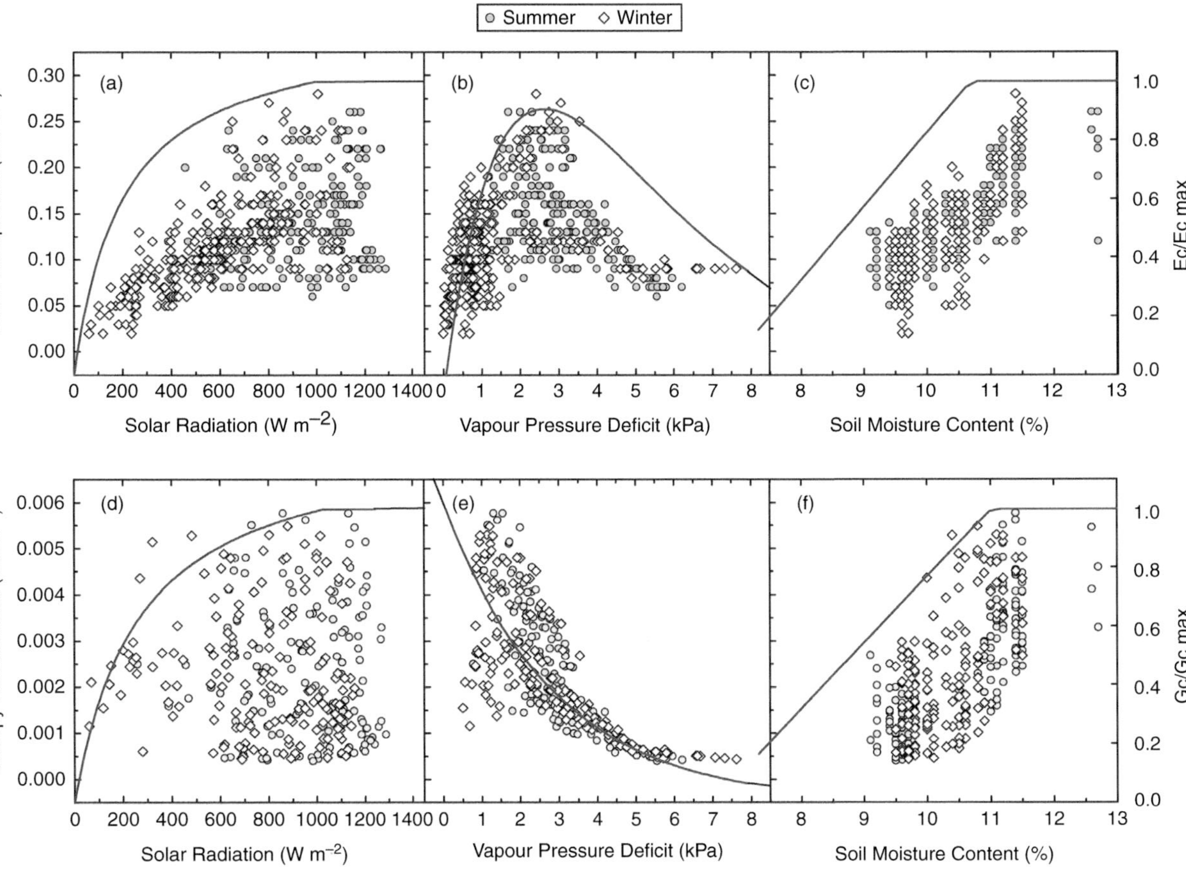

Figure 2.22. The response of stand water-use (transpiration) and canopy conductance (G_c) to changes in solar radiation, vapour pressure deficit (*D*) and soil moisture content. The symbols represent measured data for stand transpiration rate or G_c calculated from the PM equation whilst the lines represent the upper boundary value derived from the modified Jarvis-Stewart equations of Whitley *et al.* (2009). Used with permission.

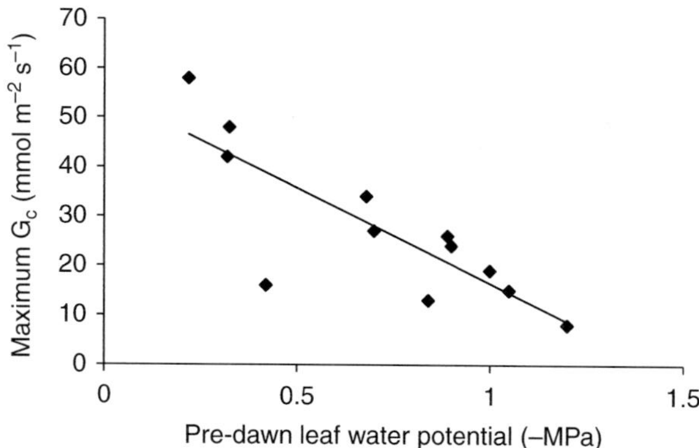

Figure 2.23. Changes in maximum canopy conductance with declining soil moisture content, as measured through declining pre-dawn xylem water potential, in a plantation of *Eucalyptus globulis*.
Redrawn from O'Grady *et al.* 2008).

a marginal change in stomatal conductance (Jarvis and McNaughton 1986) and is calculated thus (Eq. 2.13):

$$\Omega = 1/\{1 + [\gamma/(\Delta + \gamma)](g_a/g_c)\} \tag{2.13}$$

where Δ is the rate of change of saturation water vapour pressure with temperature and g_a and g_c are the aerodynamic and canopy conductances respectively. Essentially Ω is a measure of the ratio of g_a to g_c. If g_a is much larger than g_c then stomata exert a strong regulation on canopy water-use, and *vice versa*.

Canopies that are well-coupled to the atmosphere have Ω values near zero and changes in g_s have a large impact on rates of canopy water-use. For example, an ecosystem with tall, widely-spaced trees (e.g. an open savanna woodland) has a small decoupling coefficient (e.g. 0.1 – 0.4; Hutley *et al.* 2001). In contrast a short dense grass sward has a large decoupling coefficient (approaching a value of 1.0) which means that even large changes in D of the bulk air have very little impact on the rate of transpiration of the grass because the canopy is decoupled from the atmosphere. It is boundary layer conductance and not canopy conductance that most determines rates of canopy water-use in this instance. Similarly a dense, closed rainforest canopy has a large Ω, typically 0.5–0.9. A physical expression of decoupling can be seen in the wind profile through a canopy. For closed tall rainforests (large Ω), the air under the tree canopy is stationary and wind does not readily penetrate the upper canopy. In contrast, in an open savanna, turbulence (i.e. wind) can readily be felt when standing beneath the tree canopy. Values of Ω are not static at a site but can vary seasonally. Stomatal control of transpiration increases in the dry season in savannas, which is evident by a consistent decline in Ω. Thus, the ratio of aerodynamic to canopy surface conductance is significantly smaller in wet seasons compared to dry seasons, further highlighting the increasing importance of stomatal control of transpiration in dry seasons. These issues are further discussed in the savanna (Chapter 16) and arid-zone grassland (Chapter 15) case studies.

2.5 Photosynthetic Processes: Leaf-Scale

Leaves are the principle organs of C fixation (photosynthesis; C assimilation). Stems of some species (including the stems of some eucalypts, for example, and stems of many herbaceous species) also fix C, while the epidermis of some fruits (for example, Kiwi fruit and green apples) also photosynthesise. In a previous section the absorption of photons of light by chlorophyll was discussed. In this section we discuss how this absorbed energy is used, how electrons are transported within the chloroplast to generate adenosine tri-phosphate (ATP; the "energy currency" of all living cells) and nicotinamide adenine dinucleotide phosphate (NADPH; a reduced coenzyme that has a central role in photosynthetic C fixation), and how C is fixed.

Globally, photosynthesis exceeds all other chemical synthetic processes. Approximately 100 billion tonnes (8.3 billion moles) of C are fixed through photosynthesis each year and in the process, 8.3 billion moles of O_2 are released. Fixed C in plants supports biomass growth in forests, plantations and crops, which in turn supports the bacteria, fungi, insects, herbivores and all animal and human populations of the world.

Photosynthesis can be considered as occurring in three linked and simultaneously occurring processes:

(1) Photochemistry–the absorption of light by pigments in chloroplasts
(2) Electron transport–the transport of electrons to produce ATP and NADPH
(3) Biochemistry–the fixation of CO_2 and production of the first stable products of photosynthesis.

Processes 1 and 2 are often termed "light reactions" and process 3 is often termed "dark reactions". This is a slight misnomer as both require light to proceed, but C fixation can be made to occur experimentally in the absence of light by providing the products of the light reactions.

2.5.1 Photochemistry

Photochemistry is the process of trapping photons of light in leaf pigments. Specifically we are concerned with electron absorption by chlorophylls in chloroplasts. A photon of light is absorbed by a chlorophyll molecule located in the thylakoid membrane of chloroplasts. Chlorophyll a is the dominant form of the two major types of chlorophylls in higher plants (chlorophyll a and chlorophyll b). The structural difference between chlorophyll a and chlorophyll b is relatively minor ($C_{55}H_{72}O_5N_4Mg$ and $C_{55}H_{70}O_6N_4Mg$ respectively) but chlorophyll a has absorption peaks of 430 nm and 664 nm while the peaks for chlorophyll b are 460 nm and 647 nm (in a 90% acetone-water mix).

Thylakoid membranes contain a lot of chlorophyll a but also contain accessory pigments, including chlorophyll b and carotenoids. Carotenoids increase the ability of leaves to absorb yellow wavelengths, thereby increasing the ability of leaves to capture solar radiation. Accessory pigments (chlorophyll b, carotenoids) absorb solar radiation (photons) but this energy is transferred to chlorophyll a within the thylakoid membrane. The reverse (transfer from chlorophyll *a* to the accessory pigments) doesn't occur because it is energetically "uphill". These accessory pigments and chlorophyll *a* molecules are arranged in an array that has evolved such as to optimise the absorption of solar radiation and the resonant transfer of energy towards the two reaction centres of photosynthesis. Resonant transfer of energy means that an excited electron (i.e. one that has moved from the ground state to the excited state by absorbing a photon) in chlorophyll *b* or another accessory pigment is transferred very efficiently (i.e. with very little loss of energy during the process) to chlorophyll *a*. There are about twice as many chlorophyll *a* molecules as there are chlorophyll *b* and carotenoid molecules to ensure this process is not limited by the availability of a recipient chlorophyll *a* molecule. The close spacing of pigment molecules (about 2–3 nm) also facilitates resonant transfer of energy between adjacent molecules; resonant transfer also occurs between two adjacent chlorophyll *a* molecules, not only between accessory and chlorophyll *a* molecules. Although resonant transfer between adjacent chlorophyll *a* molecules is very efficient (approaching 100%) the transfer between accessory molecules and chlorophyll *a* is less efficient and energy is lost as heat.

Once energy has reached chlorophyll *a* molecules, it is transferred (directly or via an intermediate chlorophyll *a* molecule) to one of two special chlorophyll *a* molecules, called P680 or P700. P680 has a peak absorbance at 680 nm and P700 has a peak absorbance at 700 nm. P680 and P700 are also called trap chlorophylls because once they absorb energy (by resonant transfer) from chlorophyll *a*, they do not pass any energy back–the energy is trapped in the sense that it can only go in one direction, towards the electron transport chain. P680 and P700 form the core of two reaction centres, called photosystem I and photosystem II (PSI and PSII). P700 is found exclusively in PSI and P680 is the dominant chlorophyll *a* in the reaction centres of PSII.

Each photosystem contains a large number of protein subunits that are associated with particular pigments. PSI has 14 sub-units which bind 100 chlorophyll *a* molecules, within a reaction centre that has the P700 molecule. PSII has 20 + sub-units and about 35 chlorophyll *a* molecules, plus the P680 molecule. Around each reaction centre core lie a set of light-harvesting antenna molecules that channel energy to P680 and P700 of PSII and PSI respectively.

2.5.2 Electron Transport and the Synthesis of ATP and NADPH

Once P680 has absorbed energy and has an excited electron, it is able to pass this on to the primary electron acceptor of PSII. The loss of the electron (P680 is thus oxidised) requires P680 to quickly gain an electron. This electron comes from a water molecule. Thus:

$$2H_2O \rightarrow O_2 + 4H$$

and

$$4H = 4H^+ + 4e^-$$

where e^- is an electron and H^+ is a proton. Oxygen is released and protons are produced in this process. The electrons are used to reduce P680 after it has donated an electron to the primary electron acceptor of PSII. Four photons are required to release one molecule of O_2. The primary electron acceptor of PSII is plastoquinone A. Once this has accepted the electron it passes it down the electron transport chain, a series of linked proteins embedded in the chloroplast lamellar membranes. Cytochrome b_6 and cytochrome f are iron-containing tetrapyrolles called hemes which are attached to a protein. Cytochromes b_6 and f occur as a cytochrome $b6f$ supramolecule, a complex protein embedded in the chloroplast lamellar membranes. Because PSII tends to be located in the granal stacks of chloroplasts but PSI is located in the stromal lamellae, components of the electron transport chain diffuse laterally to link the two.

Electron movement down the electron transport chain in chloroplasts is mostly energetically downhill and occurs spontaneously once energy input (through absorption of light energy at PSII and PSI) has occurred at two points in the Z-scheme (the name given to the arrangement of P680, the electron transport chain, P700 and the final electron acceptor, NADP$^+$; Fig. 2.24). Linear electron flow consists of electron transport from water to NAPD$^+$. Cyclic electron flow occurs when electrons move from ferrodoxin to the cytochrome b_6f complex (Fig. 2.24).

There are two key stable products of the light reactions of photosynthesis. These are NADPH and ATP. Both are used in the "dark reactions" of photosynthesis, that is, the process of fixation of CO_2 by Ribulose bisphosphate carboxylase-oxygenase (Rubisco). The production of ATP (in the process called photophosphorylation) requires the formation of a H^+ concentration gradient across the thylakoid membrane. This gradient is produced by the accumulation of H^+ in the lumen because of: (a) the splitting of water occurs in the lumen and produces H^+; and, (b) H^+ are pumped across the thylakoid membrane by the action of the plastoquinone shuttle. The plastoquinone shuttle takes protons from the stromal side of the thylakoid membrane, plus an electron from PSII. The H atom is moved across the membrane, the H^+ is released into the lumen and the electron transferred to the cytochrome b_6f complex. Both the membrane potential difference across the thylakoid membrane and the H^+ concentration gradient (typically about 3–3.5 pH units) contribute to the free energy available to synthesise ATP. This mechanism (photophosphorylation requiring the existence of a voltage gradient across the membrane and a H$^+$ concentration gradient) is called the chemiosmotic mechanism and is perhaps one of the most central unifying processes of life. It was proposed by Peter Mitchell in the 1960s and is known as the Mitchell chemiosmotic theory.

The key to the chemiosmotic mechanism is the generation of a proton motive force (PMF), a force acting on H^+ ions to cause their movement "energetically downhill". The release of energy is coupled to ATP synthesis. The force arises from the action of both the membrane voltage gradient (directed from the lumen towards the stroma as the lumen has a more negative voltage relative to the stroma) and the concentration gradient (lower pH in the lumen than the stroma). However, the contribution of the membrane voltage is relatively small and most of the PMF arises because of the large gradient in pH across the thylakoid membrane. Electrical neutrality to counter the H^+ translocated by the plastoquinone shuttle is maintained by the movement of Mg^{2+} ions out of the lumen and Cl^- into the lumen across the thylakoid membranes, and this is why the membrane voltage is relatively small. The PMF can be calculated from Equation 2.14:

$$PMF = \Delta E - 59(\Delta pH) \tag{2.14}$$

where PMF is the proton motive force, ΔE is the voltage gradient across the thylakoid membrane, and ΔpH is the pH gradient across the thylakoid membrane. Every pH unit difference across the membrane is equivalent to a membrane voltage gradient of 59 mV.

ATP synthesis requires H^+ to pass through a channel within the enzyme ATP synthase (also called ATPase or $CF_o - CF_1$). ATPase consists of two sub-units, CF_o and CF_1. CF_o is hydrophobic, contains four different polypeptides and lies across (is embedded within) the thylakoid membrane and contains the pore through which H^+ diffuse. CF_1 contains five different polypeptides, with one particular polypeptide present as 14 copies and each copy can transport a H^+ ion independently of the others. CF_1 is the "head" that protrudes into the stroma (and only the stromal side) and is the enzymic portion of ATPase. It is the rotation of CF_1 with the passage of H^+ that converts ADP to ATP (with the input of inorganic phosphate, P_i, from the stroma). For every 14 H^+ transported, 3 ATP molecules are produced.

2.5.3 Linear and Cyclic Electron Flow

Linear electron flow is deemed to occur when electrons move from water to $NADP^+$ to produce NADPH via PSII and PSI in the Z-scheme (Fig. 2.24). Such electron transport is coupled to ATP production at two sites: the splitting of water (which releases $4H^+$) and the action of the platoquinone shuttle (which releases $4H^+$ into the lumen). In contrast cyclic electron transport involves only PSI electrons which are recycled from either reduced ferrodoxin or NADPH to plastoquinone and then on to cytochrome b_6f complex. Such cyclic electron transport is coupled to ATP synthesis only at the plastoquinone shuttle and does not produce NADPH as a product of photochemistry. For many years the apparent inefficiencies of cyclic electron flow have been difficult to explain: why has natural selection not resulted in the loss of this pathway and its associated reduction in the rate of synthesis of NADPH? Munekage *et al.* (2004) have shown that cyclic electron transport might be required to adjust the ratio of ATP:NADPH production in chloroplasts to optimise the overall rate of photosynthetic C fixation in chloroplasts by preventing the over-production of NADPH in the stroma.

2.5.4 Photosynthetic C Fixation by Leaves

The light reactions of photosynthesis produce ATP, the "energy currency of cells" and NADPH, the "reducing power" of cells. These are required for CO_2 to be fixed in photosynthesis. Photosynthetic C fixation occurs in the Calvin cycle, a three phase process involving carboxylation (the addition of C to another molecule), reduction

                           Plant Structure and Ecophysiology

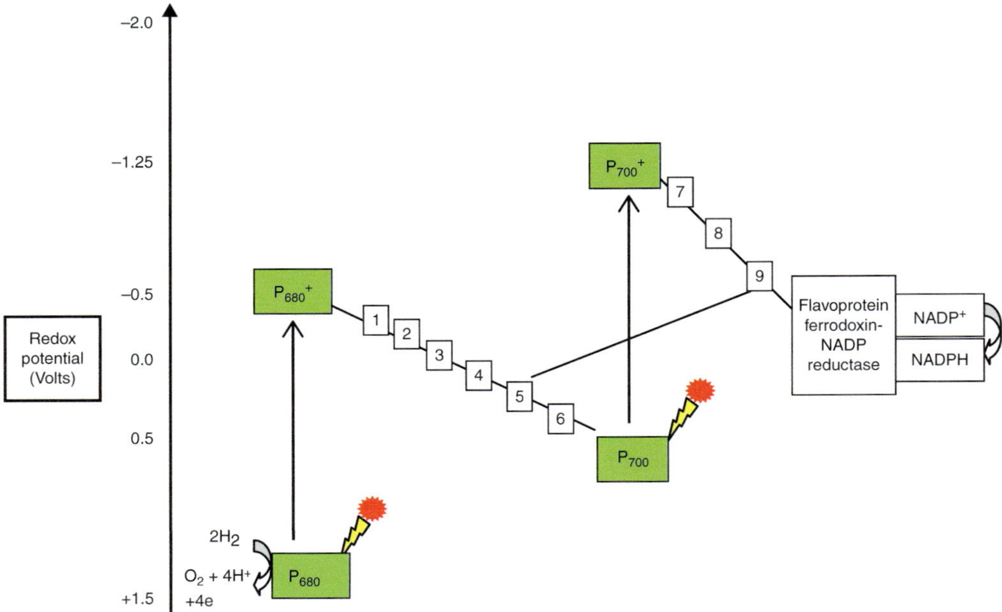

Figure 2.24. A representation of the Z-scheme in chloroplast thylakoid membranes and the stroma and lumen. Thick black vertical lines represent what happens to P680 and P700 of PSI and PSII respectively after photon absorption (represented by the red stars and yellow lightning bolts). Small numbered boxes represent the electron transport proteins. 1 = Electron acceptor pheophytin; 2 = electron acceptor Q_A; 3 = electron acceptor Q_B; 4 = electron acceptor Q; 5 = cytochrome b_6f complex; 6 = plastocyanin; 7 = electron acceptors A_o and A_1; 8 = a series of three electron acceptor iron-sulphur proteins; 9 = electron acceptor ferrodoxin (located in the stroma). Ferrodoxin-NADP reductase is located on the stromal side of the thylakoid membrane where $NADP^+$ is reduced to NADPH. The ultimate source of electrons is water, which is split by the oxygen evolving complex located on the lumen side of the thylakoid membrane.

(the addition of hydrogen atoms to a molecule), and regeneration (whereby the initial molecule to which C is added in carboxylation is produced to ensure the cycle can continue). Figure 2.25 summarises the Calvin cycle.

The enzyme Ribulose bisphosphate carboxylase-oxygenase (Rubisco) is the most ubiquitous and most important enzyme in the world and constitutes about 50% of the protein content of a leaf. It is responsible for the primary carboxylation step in C3 photosynthesis. C3 photosynthesis is the photosynthetic pathway used by all trees, most shrubs and herbs and many grasses, including most cereal crops (e.g. wheat, oats, barley). An alternate photosynthetic pathway, C4, occurs in several commercial crops, including *Zea mays* (sweet corn), sugar cane (*Saccharum* spp.), millet and sorghum (*Sorghum* spp.). C4 photosynthesis differs from C3 photosynthesis in several important ways, including:

(1) Maximum rates of photosynthesis are much larger in C4 than C3 plants; quantum efficiencies of C4 photosynthesis is larger than that of C3 plants;

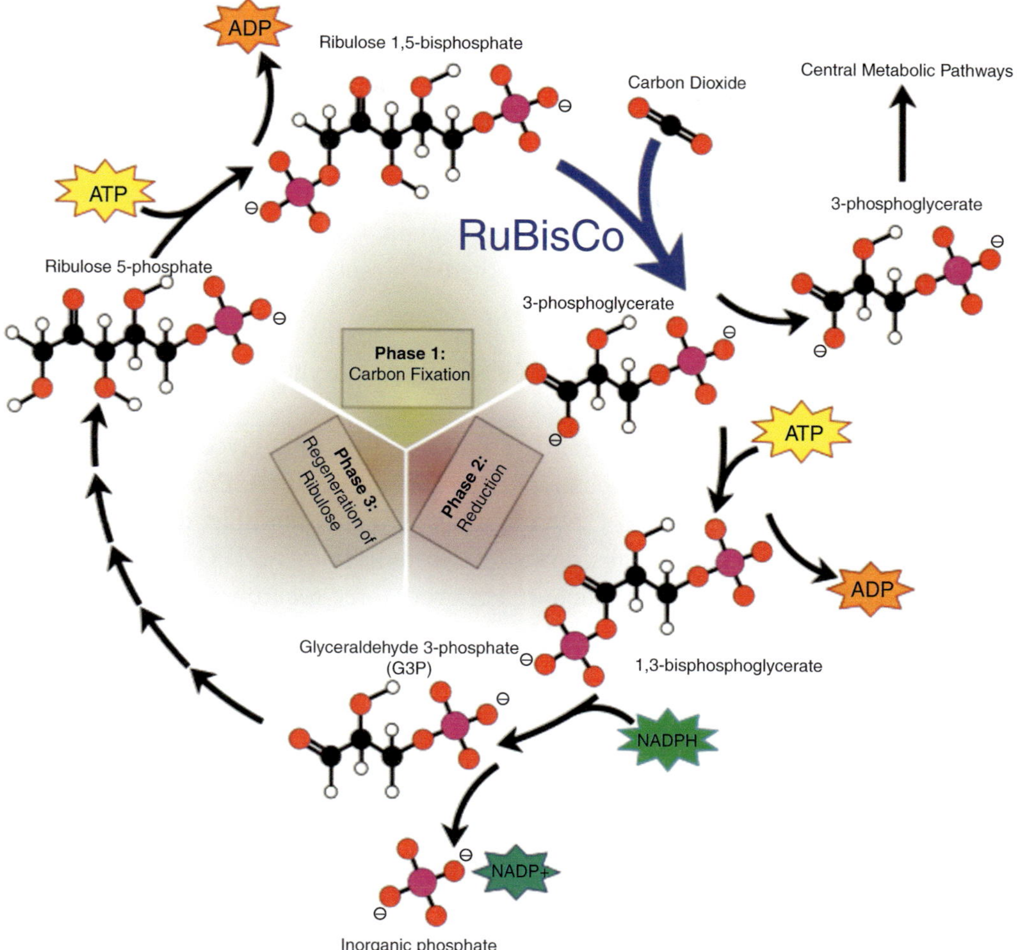

Figure 2.25. The Cavlin cycle summarised to its key features: C fixation, reduction and regeneration. From Wikipedia.

(2) The temperature optimum of C4 photosynthesis is higher than that for C3 photosynthesis;

(3) C3 plants fix C once using Rubisco in the Calvin cycle. In contrast C4 plants fix C twice. First CO_2 is fixed using phosphoenol-pyruvate carboxylase in the chloroplasts of mesophyll cells, but the resultant C4 products move into bundle sheath cells (cells surrounding the vascular bundle of leaves of C4 plants; this is called the Kranz anatomy) and CO_2 is released and subsequently re-fixed via Rubisco and the Calvin cycle;

(4) The concentration of CO_2 in bundle sheath cells is increased substantially and this suppresses photorespiration (the wasteful process whereby Rubisco uses O_2 instead of CO_2 and therefore doesn't produce useful photosynthetic product). This suppression of photorespiration increases the photosynthetic efficiency of C4 plants compared to C3 plants;

(5) There are three sub-types of C4 photosynthesis which differ in the enzyme responsible for decarboxylation in bundle sheath cells. In *Zea mays* the enzyme is NADP-malic enzyme; in millet it is NAD-malic enzyme; in the tropical forage plant *Panicum maximum* it is PEP carboxykinase. In contrast, there is no decarboxylation step in C3 photosynthesis;

(6) Some C4 species transport malate and some transport aspartate from mesophyll to bundle sheath cells. Both molecules contain 4 C atoms: hence the name "C4 photosynthetic pathway". In contrast, there is no such movement of C4 molecules in C3 plants.

For the remainder of this chapter, all discussion of photosynthesis relates to C3 photosynthetic processes.

In C3 photosynthesis the first *stable* product is 3-phosphoglyceric acid. This can also be called glycerate 3-phosphate or 3-phosphoglycerate, (3PGA). The first real product of carboxylation is a 6 C atom but this is highly unstable, doesn't leave the enzyme and immediately breaks down to two molecules of 3PGA, which diffuse away from the enzyme after production. For every three rotations of the Calvin cycle (that is, 3 CO_2 molecules are fixed; Fig. 2.23), one molecule of 3PGA continues to take part in the Calvin cycle and one molecule is the fixed product of photosynthesis and moves away to take part in the rest of the cell's metabolic processes (e.g. amino acid synthesis, lipid synthesis, etc.).

Adenosine tri-phosphate is used to add an inorganic phosphate group to 3PGA to produce 1,3 bisphosphoglycerate (bisphospho means "two phosphate groups" on a molecule). Nicotinamide adenine dinucleotide phosphate (NADPH) is used as a source of electrons to reduce 1,3 bisphosphoglycerate to glyceraldehyde 3-phosphate (G3P; also called phosphoglyceraldehyde, PGAL). Five out of six molecules of G3P are used to regenerate three molecules of the 5C sugar ribulose 1,5 bisphosphate (requiring the input of additional ATP molecules) so that the Calvin cycle can continue. Note that five lots of a 3C molecule are used to produce three molecules of a 5C molecule.

To summarise photosynthesis: the light reactions produce the ATP and NADPH required for the Calvin cycle to function and the Calvin cycle has two functions. First, to produce a stable 3C product that can be used metabolically elsewhere and second, to regenerate the 5C molecule that is the primary acceptor molecule that receives the C from CO_2.

2.5.5 *A/C$_i$ Curves and Their Interpretation*

When the rate of photosynthesis of leaves is measured as a function of the concentration of CO_2 inside the leaf and the data plotted, an A/C$_i$ curve is obtained (Fig. 2.26). These curves provide a significant amount of useful information about the biochemistry and functioning of leaves. In particular these curves can be used to calculate

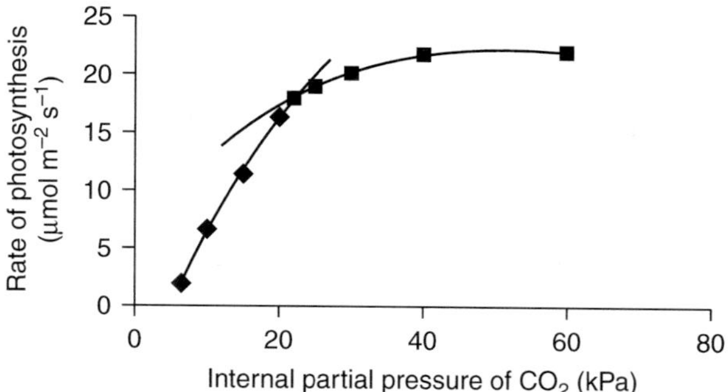

Figure 2.26. Changes in rates of photosynthesis as a function of the CO_2 concentration inside the leaf. Two curves are represented here: the curve describing the response for low levels of CO_2 concentration where Rubisco activity limits photosynthesis and the curve describing the response at high levels of CO_2 where electron transport limits photosynthesis. See text for a full explanation.

V_{cmax}, J_{max} and partition limitations on rates of photosynthesis into stomatal and non-stomatal limitations. V_{cmax} and J_{max} are the maximal rates of carboxylation (i.e. enzymic fixation of CO_2 by Rubisco) and electron transport (through the electron transport chain of chloroplasts), respectively. These are very important parameter values used in leaf-, canopy-, regional–and global-scale models of landscape function and vegetation productivity and this is discussed extensively in the chapter on modelling leaf–and canopy-scale photosynthesis (Chapter 10).

The rate of photosynthesis of a leaf is jointly limited by: (a) the activity of the primary photosynthetic enzyme, Rubisco; (b) the rate of electron transport, which is influenced by the absorption of light by chloroplasts; and, (c) stomatal limitation to CO_2 diffusion into the leaf. The original Farquhar-von-Caemmerer model has been modified to allow application of the model to water, nutrient and ozone stressed leaves (Dewar 2002, Karnosky *et al.* 2003, Hikosaka *et al.* 2006). See Chapter 10 for a more complete treatment of the Farquhar-von Caemmerer model of photosynthesis.

When CO_2 concentration is limiting to photosynthesis, that is, at the left side of the A/C_i curve (and light and water supply are not limiting), the rate of photosynthesis is limited by the activity of Rubisco. This can be determined using Equation 2.15:

$$A = \{[V_{cmax}(C_i - \Gamma^*)] / (C_i - K_m)\} - R_d \qquad (2.15)$$

where A is the rate of CO_2 assimilation (μmol CO_2 m^{-2} s^{-1}), V_{cmax} is the maximum rate of Rubisco carboxylase activity (μmol CO_2 m^{-2} s^{-1}), C_i is the intercellular CO_2 partial pressure, Γ^* is the CO_2 compensation point (the point where rates of respiration and photosynthesis are equal; on an A/C_i curve this corresponds to where the line crosses the x-axis), K_m is the Michaelis-Menten constant for Rubisco, and R_d is the rate of dark respiration.

When CO_2 supply is not limiting (the right side of the A/C_i curve), the rate of regeneration of RuBP is limiting and this is controlled by the supply of ATP and

NADPH which in turn are governed by the rate of electron transport through the electron transport chain of chloroplasts. In this part of the A/C_i curve, the following applies (Eq. 2.16):

$$A = \{J(C_i - \Gamma^*)/(4.5C_i + 10.5\ \Gamma^*)\} - R_d \tag{2.16}$$

where J is the rate of electron transport, and all other symbols are as per equation 2.15. V_{cmax} is calculated from Equation 2.17:

$$V_{cmax} = \{(C_i + K_m)(A + R_d)\}/(C_i - \Gamma^* K_m) \tag{2.17}$$

This is an alternate method to calculation of the slope in the initial part of the A/C_i curve for determining V_{cmax}. The rate at which Rubisco is carboxylated (i.e. V_c) depends upon: (i) the amount, activity, and kinetic properties of Rubisco; (ii) the rate of ribulose-1,5-bisphosphate (RuBP) regeneration via electron transport; and, (iii) the rate of triose phosphate utilisation.

J is calculated from:

$$J = \{(4.5C_i + 10.5\ \Gamma^*)(A + R_d)\}/(C_i - \Gamma^*) \tag{2.18}$$

where symbols are as per Equation 2.15.

Thus, an A/C_i curve can be used to estimate the important biochemical parameters of V_{cmax} and *J* which are fundamental aspects of models of canopy gas exchange and landscape function (Chapter 10).

2.5.6 Light Response Curves of Photosynthesis

The rate of photosynthesis increases curvilinearly as the amount of light absorbed increases. At low levels of light (light flux densities in the region 30–100 μmol m^{-2} s^{-1}) the rate of photosynthesis increases linearly with increasing absorbed light levels (Fig. 2.27). The quantum yield or photon yield is calculated as the slope of this linear region. Data at very low light levels, where respiration exceeds photosynthesis, are not used in the regression because of the Kok effect. This is manifest as an increase in the slope for light levels lower than the light compensation point. The light compensation point is the light level at which respiration and photosynthesis occurs at the same rate and therefore the rate of CO_2 production through respiration is equal to the rate of CO_2 fixation in photosynthesis. The light compensation point is typically between 5–25 μmol m^{-2} s^{-1} and is lower for shade-adapted plants than plants adapted to high light conditions. At higher light levels (typically 400–1000 μmol m^{-2} s^{-1}) photosynthesis approaches light saturation, and further increases in light supply either have zero or minimal effects on rates of photosynthesis of isolated leaves held in leaf cuvettes.

For canopies of leaves, self-shading is observed through much of the canopy and significant increases in rates of photosynthesis can be recorded over almost the full

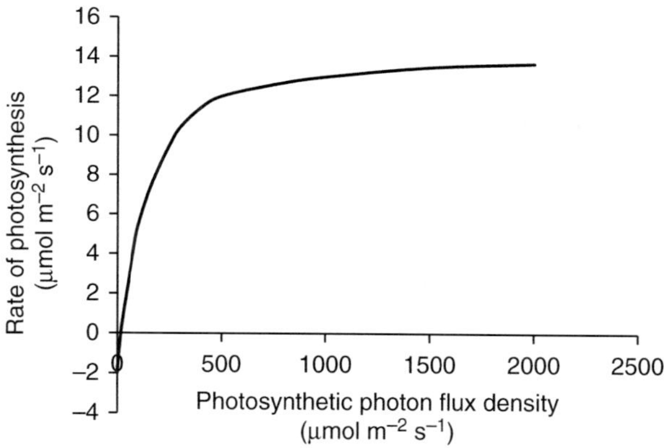

Figure 2.27. A light response curve for an isolated light in a cuvette shows a typical non-rectangular or saturating response curve.

range of solar radiation. Diffuse beam radiation (radiation that has been scattered by clouds and particles in the atmosphere) can increase the rate of canopy photosynthesis because it is more effective at illuminating leaves lower in the canopy than direct beam radiation (radiation that has not been scattered). As the supply of light increases from zero to about 500 μmol m^{-2} s^{-1} the rate of electron transport in chloroplasts and the rate of RuBP regeneration increases. Consideration of sunlit and shade leaves and their behaviours and representation in models of canopy photosynthesis are given in Chapter 10.

Many photosynthetic enzymes are regulated by light. A short period of induction is required before a steady state rate of photosynthesis is attained.

2.5.7 Canopy Uptake of CO_2

The concentration of CO_2 in well-mixed air above a canopy is currently about 400 μmol mol^{-1}, also written as 400 ppm. The concentration of CO_2 inside the leaf might be of the order 250–325 μmol mol^{-1}. Thus there is a gradient of concentration from the bulk turbulent air to the interior of the leaf. However, diffusion down this gradient is only important at very small scales, namely, at the scale of the boundary layer and within the sub-stomatal cavity, where diffusion is the dominant process driving movement of CO_2. However, if this is true, what replenishes the CO_2 that is taken from the boundary layer and fixed by a canopy each minute?

Wind speed above a canopy is maximal (at a local scale) several tens of metres above the canopy and minimal adjacent to the leaves where the boundary layer occurs. The boundary layer is the layer of relatively still air that exists adjacent to all solid objects in air (and water). Solid objects (leaves, the ground) exert a frictional drag on wind as it passes over it thereby slowing its movement and wind velocity tends to zero at the base (closest to the leaf) of the boundary layer. Transport of CO_2 (and water vapour) through the boundary layer occurs by diffusion rather than bulk transport.

Bulk transport of CO_2 and water vapour occurs outside of the boundary layer because of swirling and mixing (eddies) that occurs in the lower atmosphere. This can transport large amounts of CO_2 (and water vapour) to and from a leaf.

The average thickness of the boundary layer for an isolated leaf is approximated by the following Equation (Eq. 2.19):

$$D_{bl} = 4\sqrt{l/v} \tag{2.19}$$

where D_{bl} is the thickness of the boundary layer (in mm), l is length of the leaf downwind (in m), and v is wind velocity (in m s^{-1}). Clearly, as wind speed increases, boundary layer thickness decreases and as leaf length increases, boundary layer thickness increases. Thus for a wind speed of 1 m s^{-1} and a leaf length of 10 cm, boundary layer thickness is about 1.2 mm.

For a canopy, (as opposed to an isolated leaf), the velocity profile above the canopy is approximated by Equation 2.20:

$$V = [v_*/k] \, [\ln(z - d)/z_o] \tag{2.20}$$

where V is velocity at height z, v_* is the friction velocity, k is the von Karman constant (about 0.41), z is the height above the ground, d is the zero plane displacement, and z_o is the roughness length. The zero plane displacement is the height in the canopy where wind speed tends to zero and apparent $z = 0$, while the roughness length is a measure of the surface roughness of the canopy itself. Zero plane displacement is about 70% of canopy height whilst roughness length is about 10–15% of canopy height for dense canopies (or less for open canopies).

During a warm summer morning with moist soil, photosynthesis is occurring rapidly and the flux of CO_2 to the canopy (the flux density) can be calculated from Equation 2.21:

$$F_c = -K_c \, [\Delta C/\Delta z] \tag{2.21}$$

where F_c is the flux density of CO_2, K_c is the eddy diffusion coefficient of CO_2 (or the transfer coefficient), and, $[\Delta C/\Delta z]$ is the difference in concentration of CO_2 with respect to distance (height, z) above the canopy.

The eddy diffusion coefficient is not a constant; it varies with wind speed and also the rate of change in wind speed and temperature as a function of height above the canopy. Increases in wind speed or temperature result in increased K_c. While K_c is analogous to the diffusion coefficient of Fick's first law, it does not relate to the random diffusion of molecules through a gas. Rather it relates to the movement arising from the motion of eddies in turbulent air above a canopy. Because of the large scales of eddies or air compared to the molecular scale of diffusion of molecules of CO_2, K_c is 10,000 to 100,000 times larger than molecular diffusion coefficients. Thus the answer to the question 'What replaces the CO_2 fixed by a canopy of leaves?' is that the turbulent transfer of CO_2 from air tens of metres above a canopy is a very effective process for replacing the CO_2 lost from the leaf boundary layer as a result of photosynthesis.

2.6 GPP, NPP and NEE

In the absence of respiration, all the CO_2 entering a canopy would be photosynthetic flux, that is a flux arising only from photosynthetic fixation of CO_2. Such a flux is termed gross photosynthesis or gross primary productivity (GPP). However, in reality all vegetation and soils respire and respiration produces (releases) CO_2. Such respiration is a combination of autotrophic plus heterotrophic respiration. The total amount of C respired in one day by vegetation can account for about 50% of the total C fixed by the vegetation in one day. The difference between GPP and autotrophic (i.e. plant) respiration is the net accumulation of C and is called net primary productivity (NPP). NPP is therefore typically about 50% of GPP. NPP is the net accumulation of biomass by vegetation, usually calculated over an annual timeframe, and is large for a young, rapidly growing plantation but is small or zero for an old mature forest. Total global GPP is about 120 Gt C y^{-1}, while a hectare of forest may have a GPP of 20 t C ha^{-1} y^{-1}. Seasonal patterns of GPP and NEE and the environmental controls of GPP and NEE are discussed extensively in the boreal, savanna and arid-zone grassland case studies later (Chapters 14, 15 and 16).

Some of the C accumulated by vegetation on one day is lost through the activity of herbivores or through the death and decay of plant parts (roots, branches, leaves) on another day. Respiration by herbivores, plus the respiration of fungi and bacteria that cause the decay of dead plant material releases CO_2 into the atmosphere and this is termed heterotrophic respiration. The loss of C from vegetation due to the death of plants represents a loss of C from the NPP of the vegetation. Thus, NPP minus heterotrophic respiration equals net ecosystem production, or net ecosystem exchange (NEE). Eddy covariance methods for measuring CO_2 fluxes above landscapes measure NEE. From the above discussion the following relationships hold:

$$\text{GPP} = \text{total C fixed by a canopy, in the absence of respiration}$$

$$\text{GPP} - R_a = \text{NPP}$$

where R_a is autotrophic respiration.

Furthermore, $\text{NPP} - R_h = \text{NEE} = \text{NEP}$ where R_h is heterotrophic respiration. NPP is strongly correlated with rates of vegetation water-use (Fig. 2.26) because of the commonality of the pathway for diffusion of both CO_2 and water (via stomatal pores). Similarly, increased annual rainfall tends to be correlated with increased NPP, for the same reason (Fig. 2.28). These issues are discussed extensively in the savanna and arid-zone grassland case studies (Chapters 15 and 16).

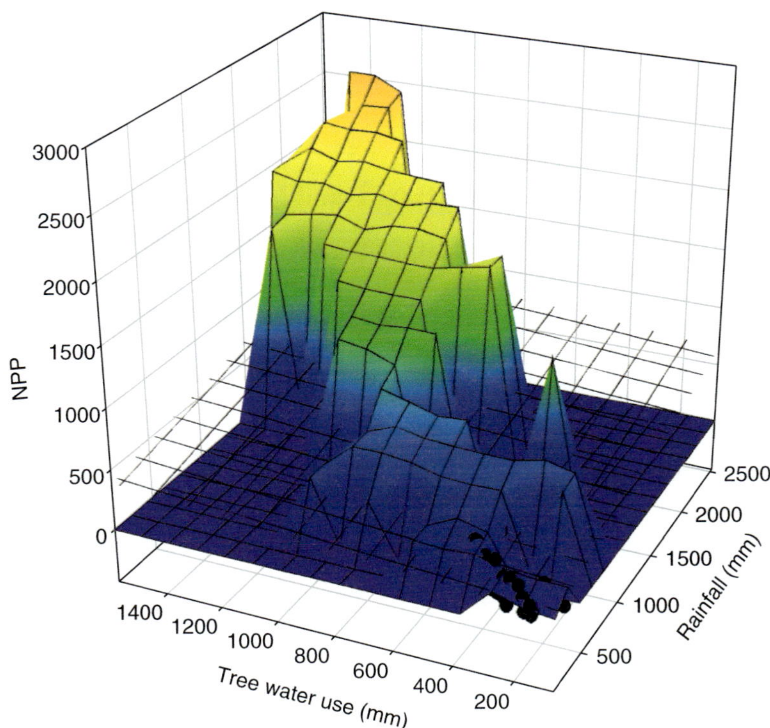

Figure 2.28. Rates of tree water-use and net primary productivity (NPP) are largest when rainfall is largest and both decline with declining annual rainfall.

2.6.1 Patterns of CO$_2$ Flux in the Field

The flux of CO$_2$ above vegetation shows distinct daily and seasonal patterns. Canopy-scale uptake of CO$_2$ is under the same controls as leaf-scale photosynthesis. Thus, soil moisture content (θ), vapour pressure deficit (D), solar radiation (R$_s$), and temperature, all influence rates of canopy photosynthesis. A full description of the patterns and drivers of changes in CO$_2$ fluxes is given in several of the case study chapters in the last section of this book.

2.6.2 Simple Modelling of Carbon Flux and LAI

Baldocchi and Meyers (1998) developed a simple globally applicable empirical model using data from boreal, temperate and tropical evergreen biomes. For these biomes, LAI could be predicted using annual equilibrium evaporation (E$_{eq}$), annual precipitation (P) and foliar nitrogen concentration ([N]), since they showed that (Eq. 2.22):

$$LAI = [N]\ P/E_{eq}$$

(2.22)

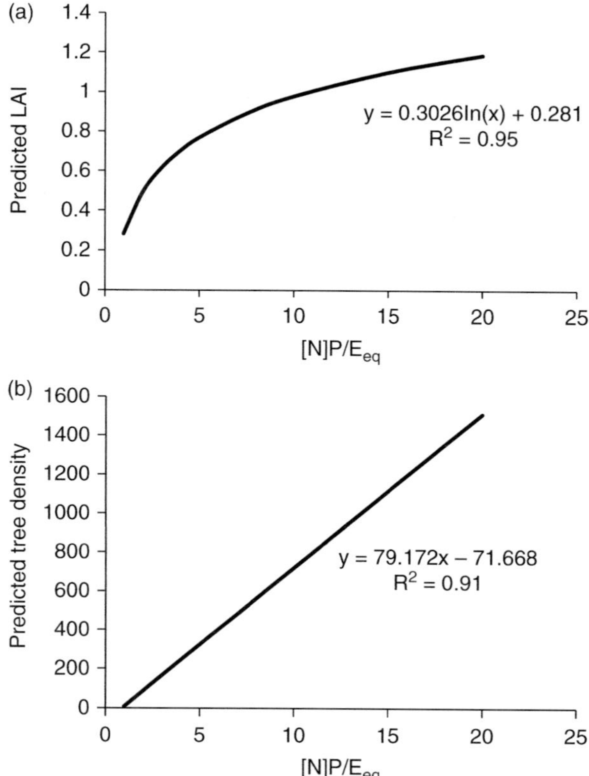

Figure 2.29. The simple model of Baldocchi and Meyers (1998) yields valuable predictive correlations. Here it has been used to predict the LAI and tree density of several diverse ecosystems. The equations and R^2 values are taken from Eamus *et al.* (2001). Redrawn from Eamus *et al.* (2001).

Leaf area index is a major determinant of maximum rates of C fluxes to canopies. Leaf area index is also highly correlated with a range of ecologically significant variables, including NPP, site water balance, and annual temperature (Waring and Schlesinger 1985, Neilson 1995).

Baldocchi and Meyers' equation (Eq. 2.22) can also be used to estimate C fluxes and it was successfully applied to savannas (Eamus *et al.* 2001). Such seasonally dry forests were absent from Baldocchi and Meyers' analyses. Using previously published data, Eamus *et al.* (2001) found that across a range of seasonally dry forests there is a significant relationship between [N], P/E_{eq} and LAI or tree density (Fig. 2.29).

Thus for savannas of north Australia covering a rainfall gradient of almost 1500 mm, and for sites in Africa, tree density (which is highly correlated with leaf area and stem volume) or LAI can be predicted ($r^2 = 0.98$) from Equation 2.23:

$$\text{Tree density} = [N]P/E_{eq} \qquad (2.23)$$

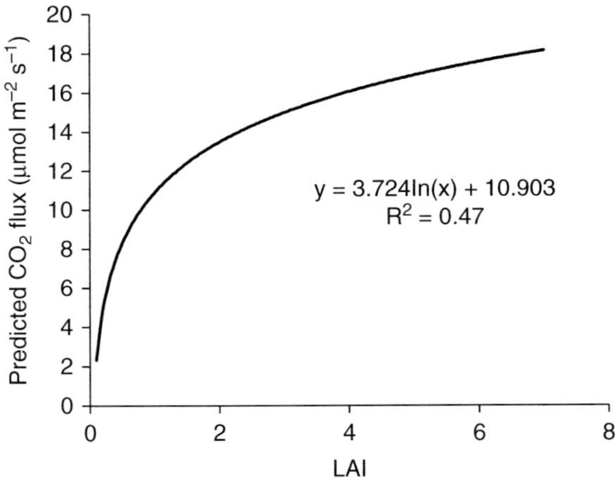

Figure 2.30. The simple model of Baldocchi and Meyers (1998) has been used to predict the CO_2 flux to the canopy as a function of leaf area index (LAI), for a range of diverse ecosystems. The equations and R^2 values are taken from Eamus *et al.* (2001).

where [N] is foliar N concentration in mg g^{-1}, *P* is annual precipitation, and E_{eq} is annual equilibrium evaporation. Foliar N concentration provides a simple, albeit crude, estimate of nutrient availability and is highly correlated with assimilation rate whereas the term P/E_{eq} is a surrogate measure of water availability. The [N] P/E_{eq} relationship for tree density and LAI may be easier to use than recently developed models based on plant available water and plant available nutrients, because the required data are far more available (Walker and Langridge 1997). These researchers also found a significant relationship between LAI and canopy CO_2 exchange rate for several disparate ecosystems (Fig. 2.30). It is noteworthy that the few savanna sites to have been studied appear to fit the same line as sites with a far larger LAI. Given the simple relationship between [N] P/E_{eq} and LAI (2.29), it appears, from a range of published data, that canopy CO_2 exchange rate can be predicted from knowledge of foliar N concentration and the ratio of precipitation to annual equilibrium evaporation.

Thus, because:

$$F_c = 3.724 \ Ln(LAI) + 10.903 \tag{2.24}$$

and:

$$LAI = 0.3026 \ Ln([N]P/E_{eq}) + 0.2814 \tag{2.25}$$

Then:

$$F_c = 3.724 \ Ln(0.3026 \ Ln([N]P/P_{eq}) + 0.2814) + 10.903 \tag{2.26}$$

Equation 2.26 (from Eamus et al. 2001) yields a prediction of peak growing season CO_2 flux for canopies over a large range of values of $[N]P/E_{eq}$ that include values exhibited by seasonally dry woodlands and forests.

2.7 Optimisation Theory as Applied to Leaf-Scale CO$_2$ and Water Fluxes

Stomatal pores regulate CO$_2$ influx and transpirational water efflux. It is not possible to simultaneously *maximise* CO$_2$ influx (photosynthetic rates) and *minimise* the rate of transpiration because the former requires a large stomatal conductance while the latter requires a small (zero) conductance. It has been proposed that stomatal behaviour is optimised such that photosynthesis is maximised per unit transpiration (Cowan and Farquhar 1977). This means that the ratio of the gain in assimilation (A) to the loss of water as transpiration (E) is optimised. This occurs when the marginal unit water cost of photosynthesis ($\partial E/\partial A$) is a constant, which means that the ratio of the sensitivities of E and A to changes in stomatal conductance (g_s) remain constant. An increasing $\partial E/\partial A$ means that stomata are not behaving optimally. Hence more water is being transpired per unit C gain.

There is evidence in laboratory studies that leaves do maintain a constant (or near constant) $\partial E/\partial A$ when D is varied experimentally (see Table 2.4 for a summary of several studies). However, the range of D applied is usually smaller than that experienced by many plants growing in tropical and sub-tropical climates or arid climates and therefore it was unclear for many years whether a constant $\partial E/\partial A$ is observed in all biomes or climates. Similarly the influence of drought on $\partial E/\partial A$ remains contradictory, with both increased and decreased values observed.

Lloyd and Farquhar (1994) found that in biomes with regular or periodic water supplies the relationship between $\partial E/\partial A$ and D was strongly linear but in biomes with episodic water supply (such as in savannas), the relationship between $\partial E/\partial A$ and D was much reduced or absent. Lloyd *et al.* (1995) used this approach to model the productivity of an Amazonian rainforest and concluded that a constant value of $\partial E/\partial A$ could not be successfully applied to explain observed changes in gas exchange.

In an attempt to expand our understanding of the behaviour of $\partial E/\partial A$ under realistic field conditions, Thomas *et al.* (1999) compared seven methods of calculating $\partial E/\partial A$ across five different species and compared the influence of changes in soil moisture content, a large range in D, temperature, and light levels, on $\partial E/\partial A$. Few differences in the absolute value of $\partial E/\partial A$ were observed across species or treatments but the different methods of calculation altered the absolute values of $\partial E/\partial A$. More importantly, $\partial E/\partial A$ was not constant across the range of D applied; stomata became increasingly less efficient (that is $\partial E/\partial A$ increased) as D increased. An example of the change in marginal unit water cost of photosynthesis with increasing D is given in Fig. 2.31.

In contrast to the results observed when D was manipulated, as drought developed, stomata became more efficient and $\partial E/\partial A$ decreased exponentially with declining soil moisture content. Thomas *et al.* (1999) concluded that stomata did not behave in a manner which optimised the marginal unit water cost of C gain when D increased or soil moisture content declined. However, $\partial E/\partial A$ was constant across a large range of leaf temperatures.

Table 2.4. Reported values of the carbon gain ratio ($\partial E/\partial A$) from the literature in various species as functions of leaf-to-air vapour pressure difference (D), leaf temperature (T_l), light levels (photosynthetic photon flux density; PPFD) and soil water parameters

Species	Treatment	Value of $\partial E/\partial A$ (mol mol^{-1})	Reference
Nicotiana glauca	D of 0.8 to 3.0 kPa	800 to 1250	Farquhar *et al.* 1980
Corylus avellana	D of 0.8 to 3.3 kPa	550 to 600	Farquhar *et al.* 1980
Pseudostuga menziesii	D from 0.6 to 1.8 kPa	500 to 300 in one individual and 250 to 150 in a second individual as D increased	Meinzer 1982
Ambrosia chamissonis	D from 1.1 to 3.6 kPa	350 to 400	Mooney *et al.* 1983
Eriogonum latifolium	D from 1.2 to 3.7 kPa	400 to 450	Mooney *et al.* 1983
Fragaria chiloensis	D from 1.2 to 3.9 kPa	700 to 1000	Mooney *et al.* 1983
Vigna unguiculata	D from 1.0 to 4.0 kPa over the T_l range of 30 to 40 °C. D constant over smaller ranges at particular T_l	800 to 1000 over these ranges	Hall and Schulze 1980
Macadamia integrifolia	D from 1.0 to 5.0 over temperature range of 15 to 35 °C. D constant over smaller ranges at particular T_l	Incline with T_l at a constant D; from 400 to 650. Stable with D changes at any T_l. Alternative method (3) showed decline from 800 to 650 as T_l increased or as D increased	Lloyd 1991
Many species from various biomes	D from 0.8 to 3.7 kPa and T_l from 10 to 31 °C	250 to 1500 with lower values associated with biomes having lower D	Lloyd and Farquhar 1994
Corylus avellana	Soil water potential from −1.5 to −1.8 MPa and D from 0.8 to 3.3 kPa	100 to 400 over D range and an incline in maximum attained from 300 to 700 as soil dried	Farquhar *et al.* 1980
Vigna unguiculata	Soil drought from 20 to 70% available water depletion as D ranged from 1.0 to 3.0 kPa	Constant with D at 500 and 300 respectively at 20% and 60% water depletion. Incline from 100 to 250 as D increased at 70% water depletion	Hall and Schulze 1980

Species	Treatment	Value of $\partial E/\partial A$ (mol mol^{-1})	Reference
Rhamnus californica	Diurnal time course covering T_l from 15 to 35 and PPFD from 0 to 1600 μmol m^{-2} s^{-1} in irrigated and droughted plants	Range from 100 to >2000. Most variable at low PPFD and D. Stable values of 600 in irrigated and 450 in droughted with variability between replicate leaves of 200 during the midmorning to mid afternoon period	Williams 1983
Various tropical savanna tree species	D from 2 to 5 kPa at 33 °C and from 2 to 6 kPa at 38 °C. T_l from 25 to 42 °C while D remained at 2 kPa. Soil drought from pre-dawn leaf water potential of −0.1 MPa to −3.2 MPa while D ranged from 2 to 4.5 kPa. Leaves expanded in wet and dry seasons	Mainly between 500 and 1500. Increased with D. Stable with T_l. Declined with soil drought. No substantial differences between season of leaf expansion. Usually similar values with all calculation methods but one method (from Lloyd *et al.* 1995) showed least response to changing environmental conditions	Thomas *et al.* 1999

Source: From Thomas *et al.* (1999).

Despite some conflict among answers to the question: "do leaves maintain a constant marginal water cost of photosynthesis as the abiotic environment varies?" application of optimisation theory *per se* to canopies in other contexts has proven useful and insightful. Schymanski *et al.* (2007) tested an optimality approach for modelling canopy properties and CO$_2$ uptake in native woodlands and compared modelled and observed fluxes of CO$_2$ at a north Australian savanna. These researchers assumed that natural vegetation has evolved over time to produce a species composition that is optimised with respect to carbon gain and transpiration. By making assumptions about costs incurred in root function, they were able to use canopy attributes and processes alone, to predict canopy carbon uptake. They also assumed that the rate of electron transport of leaves was a linear function of irradiance and that electron capacity was exponentially distributed over the canopy. Rates of electron transport and hence C uptake was calculated for both sunlit and shade fractions within each modelled layer of the canopy. Net C gain was calculated as the difference between absolute rates of

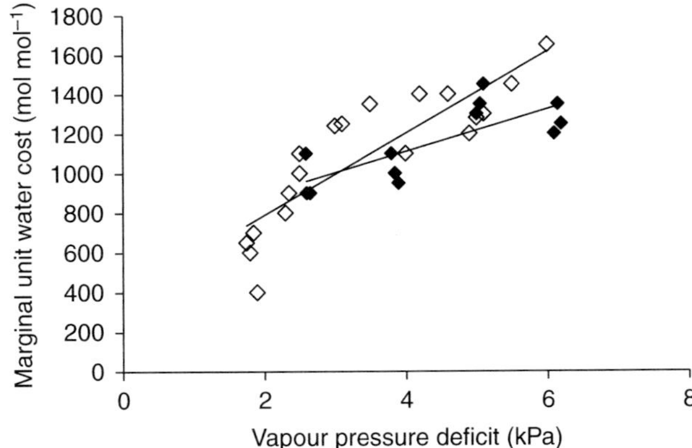

Figure 2.31. Changes in the marginal unit cost of water for two tropical tree species (*Maranthes corymbosa* (black symbols) and *Myristica insipida* (hollow symbols)) shows that with increasing vapour pressure deficit (*D*), stomata become less efficient (they spend more water per unit carbon gained at increasing values of *D*). Both linear regressions are statistically significant.
Redrawn from Thomas *et al.* (1999).

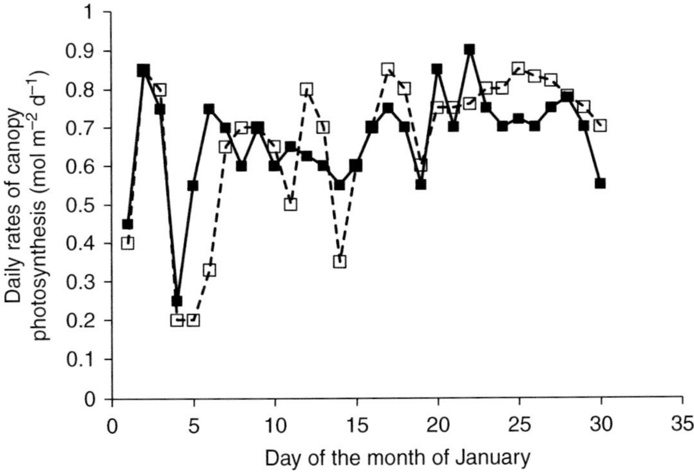

Figure 2.32. Application of an optimisation approach to modelling canopy gas exchange resulted in reasonable agreement between observed (solid line, solid squares) and modelled estimates (dashed line, hollow squares).
Redrawn from Schymanski *et al.* 2007.

C gain minus the cost of growing, maintaining and replacing leaves, computed on a monthly time-step. An example of their output is given in Fig. 2.32 where inter-daily variation in observations is captured by the model. From this theoretical insight it was possible to analyse daily patterns of stomatal conductance in the field as light levels and *D* change.

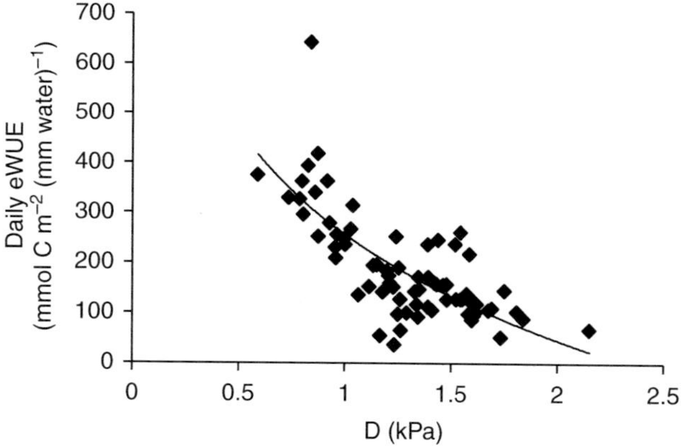

Figure 2.33. As vapour pressure deficit (D) increases, ecosystem water-use-efficiency declines. Data from Eamus *et al.* (2013) for an arid-zone Mulga woodland in Central Australia.

2.8 Water-Use-Efficiencies of Leaves and Canopies

Leaf-scale water-use-efficiency (WUE) is the ratio of C gain to transpired water (A/E; Barton *et al.* 2011). However, the diffusion of water (as transpiration; E) and CO_2 (as photosynthetic uptake; A) are the determinants of both WUE and stomatal conductance (g_s) *per se* because stomata respond to both C_i (the concentration of CO_2 inside the leaf air spaces) and E (Eamus *et al.* 2008). Consequently *intrinsic* WUE (the ratio of A to g_s; WUE_i) is a better measure of WUE at the leaf-scale. At larger scales, using eddy covariance data, calculation of *inherent* WUE (IWUE) as the ratio of NEE $*$ D to evapotranspiration (ET) (or GPP $*$ D to ET; Beer *et al.* 2009) can be used to normalise the effect of D upon ET out of the calculation (Jongen *et al.* 2011). Ecosystem WUE (eWUE) and IWUE are responsive to changes in soil moisture content and show seasonal and daily variation (Kuglitsch *et al.* 2008).

Leaf-scale WUE and eWUE decline curvilinearly, typically exponentially, with increasing D (Fig. 2.32). Such decreasing eWUE is the result of the differential effects of increasing D on transpiration and photosynthesis. As D increases from low to moderate values, E increases while the rate of photosynthesis is minimally affected or unaffected. Transpiration increases because the increased atmospheric evaporative demand is not matched by a sufficiently large decrease in stomatal conductance (g_s) to prevent E increasing with increasing D (Eamus *et al.* 2008). Only when D increases from moderate to larger values does g_s (at leaf-scales) or canopy conductance (G_c) decline sufficiently to reduce gas exchange. At this stage A begins to decline with increasing D but the magnitude of the decline in E is larger than that of A (Wharton *et al.* 2009). Thus eWUE declines with increasing vapour pressure deficit (Fig. 2.33).

In some studies an interaction of soil moisture content with D occurs to complicate the response of eWUE to D. For example, during the wet season in an arid zone

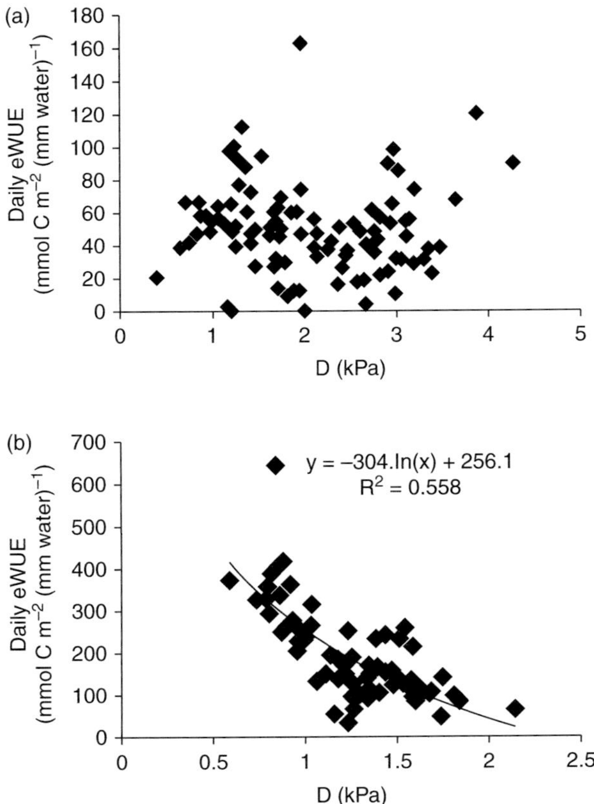

Figure 2.34. Ecosystem water-use-efficiency is either independent of increases in vapour pressure deficit (*D*) (upper panel; wet season) or declines with increasing *D* (lower panel; dry season). Data from Eamus *et al.* (2013) for an arid-zone Mulga woodland in Central Australia.

Mulga woodland, eWUE is independent of *D* (Fig. 2.34) but during the dry season eWUE declines with increasing vapour pressure deficit (Fig. 2.34). The reason for this disparate behaviour between seasons is the influence of changes in soil moisture content on eWUE (Fig. 2.34).

In arid zones large fluctuations in soil moisture content occur with high frequency due to the sporadic nature of rainfall in the wet season. In contrast during the dry season, soil moisture content varies little, with a trend of declining water content across the season. At leaf–and canopy-scales, declining soil moisture content causes WUE or eWUE to increase (Bruemmer *et al.* 2012) because the proportional decline in *E* is larger than that of photosynthetic CO_2 flux.

In both the wet and dry seasons in the Mulga study, eWUE increases significantly as θ declines (Fig. 2.35). Since θ and vapour pressure deficit are inversely correlated, a larger θ is generally associated with a smaller vapour pressure deficit and *vice versa*. This explains the difference in the response in eWUE observed between seasons. During the dry season changes in θ were small and slow thereby

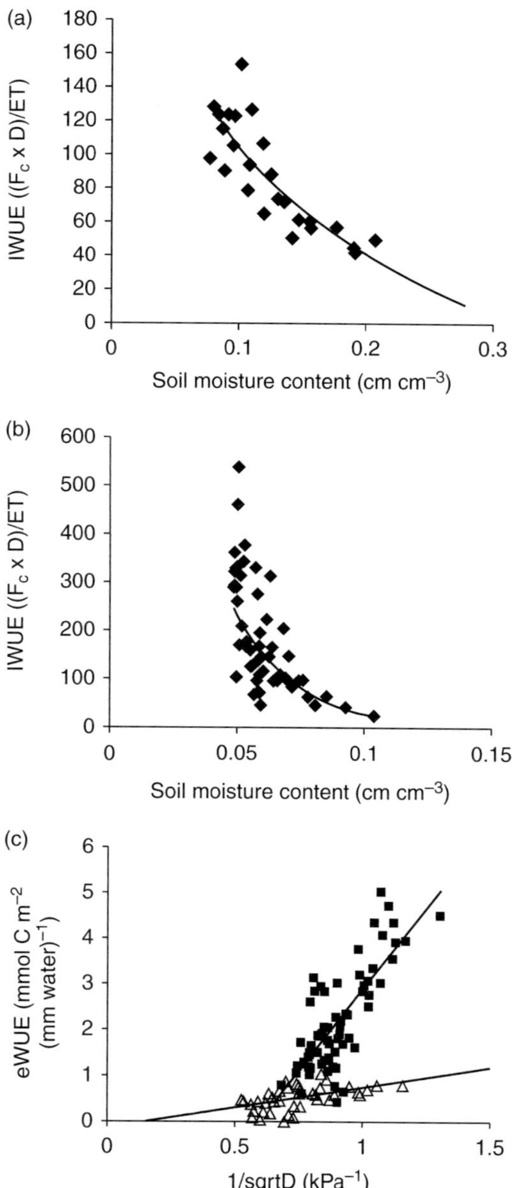

Figure 2.35. The slope of the relationship between inherent WUE (IWUE) and soil moisture content is smaller in the wet season (a) than the dry season (b). (c) eWUE as an inverse function of $\sqrt{D}$ in the dry (black square) and wet (open triangle) seasons also shows a larger slope in the dry season than the wet season so the water cost of C gain was larger in the dry season than the wet season. From Eamus *et al.* (2013).

minimising the interaction of θ with D. Consequently, a declining eWUE with increasing D was consistent with previous observations (Boulain *et al.* 2009; Linderson *et al.* 2012). In contrast during the wet season, large and rapid declines in moisture content following each rainfall event interacted with the response of eWUE to declining vapour pressure deficit such that the overall response of eWUE to vapour pressure deficit was zero.

Inherent WUE (an ecosystem-scale attribute) increases significantly as soil moisture content decreases (Fig. 2.35) with a steeper slope in the dry season than the wet season. This has been observed across multiple biomes. In parallel with increasing IWUE, the water cost of C gain declines with θ, while the C cost of water increases (Medlyn *et al.* 2011). This conclusion is further supported in the analyses of the slopes of eWUE (calculated as F_c/ET) *versus* $1/\sqrt{D}$ (Fig. 2.35). The slope, which is proportional to the marginal C cost of water, increases in the dry season. IWUE increases with increasing D, indicating that unit water cost of C gain increases with increasing D.

2.8.1 Using Stable Isotopes to Assess WUE and WUE$_i$

CO_2 in the atmosphere occurs predominantly as $^{12}CO_2$ but also present are the stable isotope $^{13}CO_2$ and the unstable isotope $^{14}CO_2$. The presence of the unstable isotope is ignored in this discussion because it is present in such minute quantities. The lighter $^{12}CO_2$ isotope diffuses more rapidly than the heavier $^{13}CO_2$ and thus diffusion of $^{12}CO_2$ through stomata is favoured above that of $^{13}CO_2$. From this fact alone, we would expect plant tissues to contain slightly less $^{13}CO_2$ than is found in the atmosphere, and this is indeed the case. However, there is an additional discrimination step: Rubisco discriminates against $^{13}CO_2$. In C3 plants, this discrimination (Δ) against ^{13}C is linked to photosynthesis via C_i/C_a, the ratio of intercellular (C_i) to atmospheric (C_a) CO_2 concentrations, as expressed in Equation 2.27:

$$\Delta = a + (b - a)\frac{C_i}{C_a} \tag{2.27}$$

where a and b are fractionation factors (the change of ratio in molar concentrations) that occurs during diffusion of CO_2 through stomata (4.4‰) and enzymatic fixation by Rubisco (27‰), respectively; the symbol ‰ means per thousand. Environmental factors such as water availability cause variations of C_i/C_a, mainly through their effects on stomatal conductance and photosynthetic activity. Thus, when drought occurs and g_s declines, there is a measurable increase in $\partial^{13}C_{leaf}$ or a decrease in Δ. Note that $\delta^{13}C_{leaf} = \delta^{13}C_{atm} - $ (diffusion effect + enzyme effect).

In addition, intrinsic WUE (WUE$_i$) for C3 plants is given by Equation 2.28:

$$WUE_i = \frac{C_a(b - \Delta)}{1.6(b - a)} \tag{2.28}$$

WUE and WUE$_i$ decrease as mean annual rainfall increases and as $\delta^{13}C_{leaf}$ increases (gets less negative) stomatal limitation increases and WUE increases (Fig. 2.36).

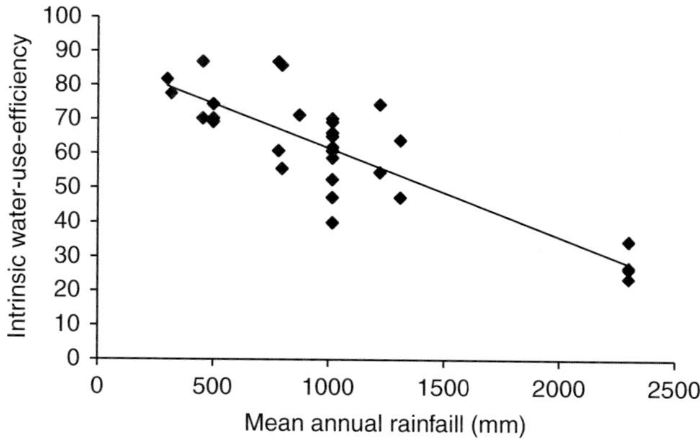

Figure 2.36. Intrinsic water-use-efficiency declines with increasing rainfall across continental Australia. Eamus unpublished.

2.9 Trade-off of N *versus* Water Allows Maintenance of High Rates of Photosynthesis in Arid Sites

The pathway of water movement from soil to atmosphere is made up of a series of liquid and gas phase conductances (Chapter 8). The largest resistance to water loss must be stomatal (Parkhurst 1994), because if the main limit to water supply to the canopy occurred in roots or stems, leaves would be forever desiccated. Positive correlations between hydraulic conductance (the inverse of resistance) of different parts of the water conducting pathway and g_s are frequently observed (Rust and Roloff 2002), as is expected if there is coordination between the conductance of the various parts of the hydraulic pathway between root and leaf. Reductions in g_s with decreasing hydraulic conductance of the upstream (petioles, stem xylem, roots) components of the hydraulic pathway prevent excessive water loss and limit the development of xylem embolism (Chapter 3) and leaf damage caused by very low leaf water potentials. However, reductions in g_s also restrict carbon dioxide uptake and limit plant and stand productivity (Santiago *et al.* 2004). Thus in a global meta-analyses increasing leaf-scale g_s is correlated with increasing leaf-scale rates of photosynthesis (Fig. 2.37).

Vegetation growing in arid zones, which might be expected to have a smaller g_s than vegetation growing in mesic environments, would therefore also be expected to have a lower rate of leaf-scale photosynthesis than vegetation growing in mesic environments. However, this is frequently not observed. Resource substitution can explain how leaf-scale assimilation rates can be maintained in arid zones, despite a low g_s, and this is now discussed.

Despite the relatively consistent observation that reduced g_s results in reduced rates of photosynthesis, there are several examples in comparisons across aridity gradients where photosynthesis appears to be minimally reduced despite reduced g_s. Thus Macinnis-Ng *et al.* (2004) reported a negative relationship between mass-based photosynthetic rate and sapwood specific hydraulic conductivity, noting

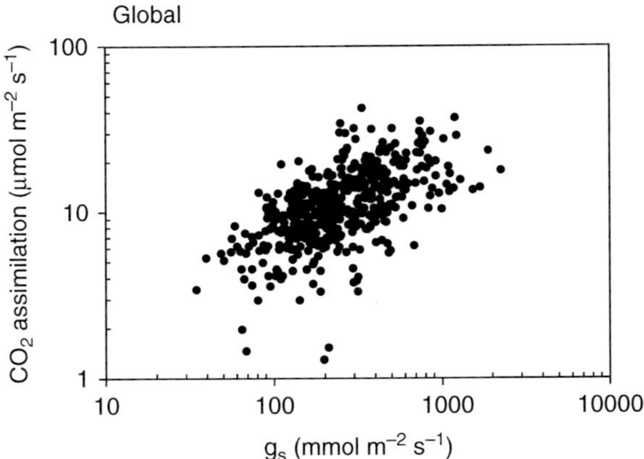

Figure 2.37. As leaf-scale stomatal conductance (g_s) increases, leaf-scale rates of photosynthesis also increase.

that hydraulic conductivity and g_s are expected to be positively correlated. These researchers proposed that the negative slope arose because of the need for leaves on branches with low hydraulic conductivity and large Huber value (high leaf area to sapwood area ratio; Chapter 3) to fix sufficient C to maintain the large Huber values. Thus they inferred low g_s but a large rate of C fixation. More direct evidence of the lack of decline in photosynthesis as a result of a small g_s was obtained in a detailed study across an aridity gradient in Australia. Taylor and Eamus (2008) examined relationships among sapwood hydraulic conductance (Chapter 3), foliar N content, rate of photosynthesis and stomatal conductance, to examine possible mechanisms that may underlie this intuitively unexpected result. In particular they asked the questions:

(1) Is it possible that photosynthesis–hydraulic architecture relationships can be mediated by coordinating water transport capacity with leaf traits that control or limit photosynthesis (e.g., foliar *N* or SLA)?
(2) Is there a trade-off between g_s (and hence water-use) and foliar *N* such that reductions in g_s can be offset by increases in foliar *N*, such that rates of photosynthesis remain relatively unaffected by the decline in g_s? (This is the crux of the resource substitution theory). Put another way: Can resource substitution allow C assimilation rates to be maintained across a gradient of water availability?

In their study, Taylor and Eamus (2008) examined relationships amongst branch hydraulic conductivity, g_s, foliar *N* and rates of photosynthesis at four temperate woodland sites across a rainfall gradient (464–1350 mm). They found the following key results:

(1) The rate of light saturated photosynthesis (expressed per unit leaf area) was positively correlated with both foliar *N* and g_s, as is generally observed globally (Fig. 2.38).

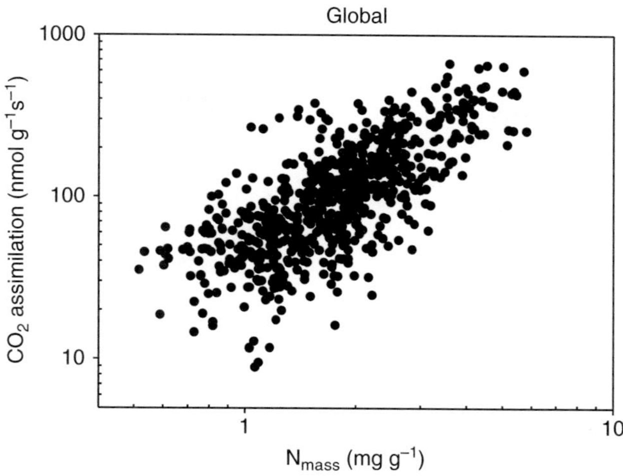

Figure 2.38. As foliar N content increases, rates of leaf-scale photosynthesis increase.

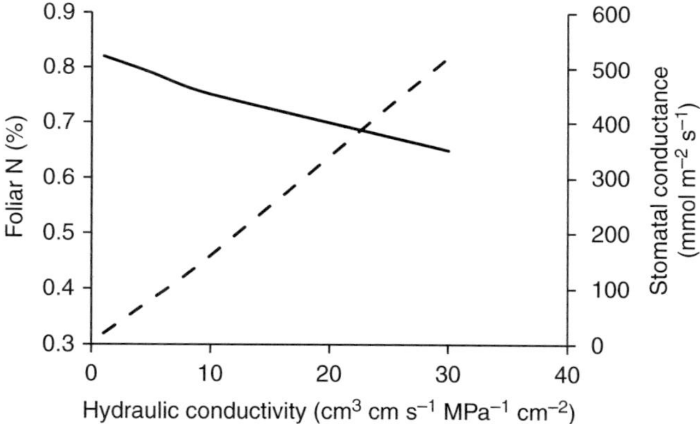

Figure 2.39. There is a positive correlation between hydraulic conductivity (k_s) and stomatal conductance (g_s) (dashed line) but a negative correlation between k_s and foliar *N* content (solid line).

(2) When variation in SLA was accounted for, multiple regression analyses showed that there was a negative correlation between g_s and foliar *N* such that at any given rate of photosynthesis leaves with a high g_s contained less foliar *N* than leaves with a low g_s.

(3) Foliar *N* was negatively correlated with branch hydraulic conductivity (k_s) but g_s was positively correlated with k_s (Fig. 2.39).

(4) Trees growing in arid sites had a low water transport capacity (low k_s) and also had small g_s but maintained a high rate of photosynthesis by investing more *N* in photosynthetic machinery (Rubisco, light harvesting, and electron transport proteins). In contrast, trees growing at mesic sites invested less *N* but spent more water through having a larger k_s and g_s than trees at arid sites (Fig. 2.40). Thus trees growing at mesic sites had less foliar *N* but a larger g_s.

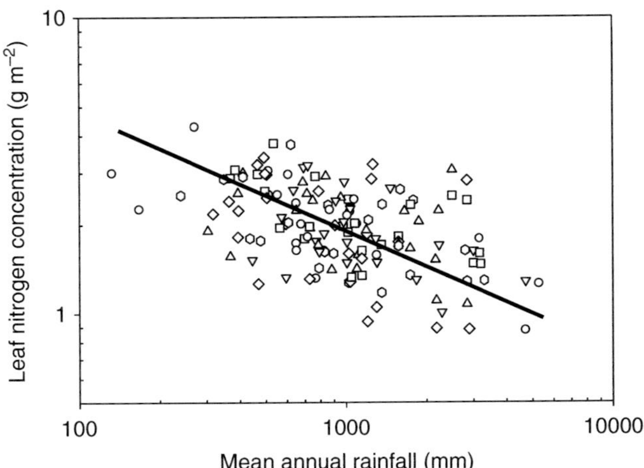

Figure 2.40. Foliar N content declines with increasing rainfall. The higher *N* investment in arid sites allows leaf-scale rates of photosynthesis to be larger than would otherwise be expected from a consideration of their low stomatal conductance (g_s) alone.

(5) Resource substitution occurs when there is a trade-off between g_s and N in relation to k_s. The mechanism by which a high foliar *N* but low g_s and k_s can result in a large rate of photosynthesis is through the C_i/C_a ratio. When a large C_i/C_a ratio occurs there is only a small diffusional gradient driving CO_2 flux into the leaf and therefore rates of photosynthesis are low. In contrast when a small C_i/C_a ratio occurs the diffusional gradient is large and high rates of photosynthesis can be sustained, despite low rates of transpiration (arising from the low g_s).

(6) Foliar N and g_s exert independent effects on rates of photosynthesis.

(7) Generally, water-*N* substitution is adaptive for evergreen species, but not deciduous species. This is because deciduous species generally maintain leaves when temperature and water supplies are not limiting to photosynthesis. During periods of sub-optimal temperatures or limiting water supplies, deciduous species tend to drop their leaves. In addition, deciduous species must maximise photosynthetic carbon gain during a short leaf life-span (see Chapter 1 for a discussion of cost-benefit analyses of evergreen *versus* deciduous strategies).

2.10 Nitrogen, Phosphorous and Drought

Woody, nitrogen fixing species (for example *Acacia, Prosopis, Casuarina, Alnus*) are important components of canopy trees in lowland mesic and wet tropical forests globally. Such biomes are frequently nutrient limited, especially in the availability of phosphorous (*P*). There is an explicit link between nitrogen (*N*) fixation and *P* availability in soil. This is because the activity of phosphatase enzymes is much larger in soils in locations that contain *N*-fixing species compared to locations without *N*-fixing species (Fig. 2.41; Houlton *et al.* 2008). Whether the phosphatases are

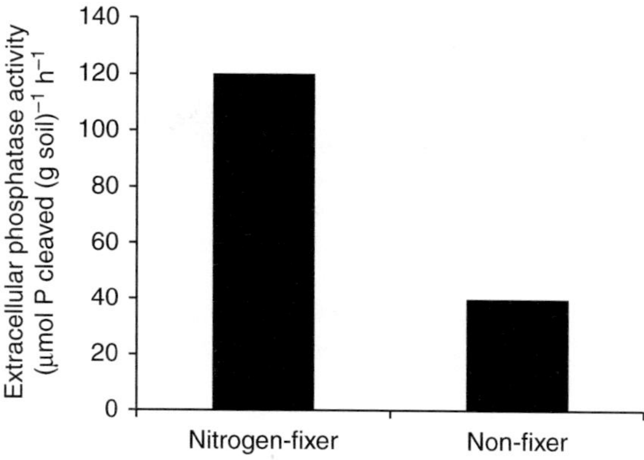

Figure 2.41. Phosphatase enzyme activity is much larger in soils with N-fixing plant species present than soils without *N*-fixing species. Data taken from the literature. Redrawn from Houlton *et al.* (2008).

derived from the fungus or the plant root remains debated (Houlton *et al.* 2008). Adams *et al.* (2010) have also noted that across vast areas of the continents of the southern hemisphere, *N*-fixing trees (especially *Acacia*) are dominant in non-tropical climates where soils are conspicuously *P*-depleted. Increased availability of *P* in soils is captured through the activity of mycorrhizal fungi associated with the roots of *N*-fixing species. Hyphae of mycorrhizal fungi extend for large distances from roots, absorbing *P* and transporting it back to roots where it is exchanged for sugars released by plant roots, in a highly mutualistic relationship for the two species. It is expected that some of the *N* fixed by *N*-fixing species is used in a *N*-water trade-off to increase rates of photosynthesis in water-limited environments. In Australia, Africa and North and South America, such environments are often dominated by *N*-fixing trees, including *Acacia*, *Brachystegia*, and *Prosopis*.

The maintenance of a positive turgor pressure (Chapter 3) is required for sustained growth as it provides the driving force for cell expansion. Osmotic adjustment (the accumulation of osmotically active solutes to reduce water potential and maintain uptake of water into cells) is a commonly observed response to drought, especially in *Acacia* and *Eucalyptus* species. Osmotically active solutes can include fructose, glucose and sucrose, but also often include quaternary ammonium compounds such as glycine and betaine. In *Acacia*, accumulation of pipecolic acid, trans-4 hydroxypipecolic acid and proline occurs, all of which contain *N*. Thus it is likely that *N*-fixing trees have a two-fold strategy for coping with drought: osmotic adjustment and *N*-water trade-off.

This chapter has provided an introduction to some of the key topics of plant ecophysiology, from a predominantly carbon-focussed perspective (because C fluxes drive ecology). However, understanding the water relations of plants is of equal importance to any understanding of the ecology and productivity of vegetation. Consequently the next chapter provides a detailed description of plant water relations at cellular, whole plant and canopy perspectives.

2.11 References

Adams MA, J Simon and S Pfautsch, (2010). Woody legumes: a view from the south. *Tree Physiology* 30, 1072–1082.

Baldocchi DD and T Meyers, (1998). On using eco-physiological, micrometeorological and biogeochemical theory to evaluate carbon dioxide, water vapor and trace gas fluxes over vegetation: a perspective. *Agricultural and Forest Meteorology* 90, 1–25.

Barton CVM, RA Duursma, BE Medlyn, DS Ellsworth, D Eamus, DT Tissue, MA Adams, J Conroy, KY Crous, M Liberloo, M Low, S Linder and RE McMurtrie, (2011). Effects of elevated atmospheric [CO_2] on instantaneous transpiration efficiency at leaf and canopy scales in *Eucalyptus saligna*. *Global Change Biology* 18, 585–595.

Beer C, P Ciais, M Reichstein, D Baldocchi, BE Law, D Papale, JF Soussana, C Ammann, N Buchmann, D Frank, D Gianelle, IA Janssens, A Knohl, B Kostner, E Moors, O Roupsard, H Verbeeck, T Vesala, CA Williams and G Wohlfahrt, (2009). Temporal and among-site variability of inherent water use efficiency at the ecosystem level. *Global Biogeochemical Cycles* 23, GB2018 DOI: 10.1029/2008GB003233.

Boulain N, B Cappelaere, D Ramier, HBA Issoufou, O Halilou, J Seghieri, F Guillemin, M Oi, J Gignoux and F Timouk, (2009). Towards an understanding of coupled physical and biological processes in the cultivated Sahel-2. Vegetation and carbon dynamics. *Journal of Hydrology* 375, 190–203.

Bruemmer C, TA Black, RS Jassal, NJ Grant, DL Spittlehouse, B Chen, Z Nesic, BD Amiro, MA Arain, AG Barr, CPA Bourque, C Coursolle, AL Dunn, LB Flanagan, ER Humphreys, PM Lafleur, HA Margolis, JH McCaughey and SC Wofsy, (2012). How climate and vegetation type influence evapotranspiration and water use efficiency in Canadian forest, peatland and grassland ecosystems. *Agricultural and Forest Meteorology* 153, 14–30.

Choinski JS, P Ralph and D Eamus, (2003). Changes in photosynthesis during leaf expansion in *Corymbia gummifera. Australian Journal of Botany* 51, 111–118.

Cowan IR and GD Farquhar, (1977). Stomatal function in relation to leaf metabolism and environment. In *Integration of activity in the higher plant*, pp. 471–505, Cambridge University Press.

Dewar RC, (2002). The Ball-Berry-Leuning and Tardieu-Davies stomatal models: synthesis and extension within a spatially aggregated picture of guard cell function. *Plant, Cell and Environment* 25, 1383–1398.

Duursma RA, CVM Barton, Y-S Lin, BE Medlyn,D Eamus, DT Tissue, DS Ellsworth, RE McMurtrie, (2014). The peaked response of transpiration rate to vapour pressure deficit in field conditions can be explained by the temperature optimum of photosynthesis. *Agricultural and Forest Meteorology* 189–190, 2–10.

Eamus D, J Cleverly, N Boulain, N Grant, R Raux and R Villalobos-Vega, (2013). Carbon and water fluxes in an arid zone *Acacia* savanna woodland: an analyses of seasonal patterns and responses to rainfall. *Agricultural and Forest Meteorology* 182, 225–238.

Eamus D, LB Hutley and AP O'Grady, (2001). Daily and seasonal patterns of carbon and water fluxes above a north Australian savanna. *Tree Physiology* 21, 977–988.

Eamus D and S Shanahan, (2002). A rate equation model of stomatal responses to vapour pressure deficit and drought. *BMC Ecology* 2(8), doi 10.1186/1472–6785.

Eamus D, DT Taylor, CMO Macinnis-Ng, S Shanahan and L De Silva, (2008). Comparing model predictions and experimental data for the response of stomatal conductance and guard cell turgor to manipulations of cuticular conductance, leaf-to-air vapour pressure difference and temperature: feedback mechanisms are able to account for all observations. *Plant, Cell and Environment* 31, 269–277.

Ewers BE, ST Gower T, B Bond-Lamberty and CK Wang, (2005). Effects of stand age and tree species on canopy transpiration and average stomatal conductance of boreal forests. *Plant, Cell and Environment* 28, 660–678.

Farquhar GD, ED Schulze and M Kuppers, (1980). Responses to humidity by stomata of *Nicotiana glauca* L. and *Corylus avellana* L. are consistent with the optimization of carbon dioxide uptake with respect to water loss. *Australian Journal of Plant Physiology* 7, 315–327.

Forrester DI, JJ Collopy and JD Morris, (2010). Transpiration along an age series of *Eucalyptus globulus* plantations in southeastern Australia. *Forest Ecology and Management* 259, 175–760.

Hall AE and ED Schulze, (1980). Stomatal responses to environment and a possible inter-relation between stomatal effects on transpiration and CO_2 assimilation. *Plant, Cell and Environment* 3, 467–474.

Harris PP, C Huntingford, PM Cox, JHC Gash, Y Malhi, (2004). Effect of soil moisture on canopy conductance of Amazonian rainforest. *Agricultural and Forest Meteorology* 122, 215–227.

Hikosaka K, K Ishikawa, A Borjigidai, O Muller and Y Onoda, (2006). Temperature acclimation of photosynthesis: mechanisms involved in the changes in temperature dependence of photosynthetic rates. *Journal of Experimental Botany* 57, 291–302.

Holmes MG and DR Keiller, (2002). Effects of pubescence and waxes on the reflectance of leaves in the ultraviolet and photosynthetic wavebands: a comparison of a range of species *Plant, Cell and Environment* 25, 85–93.

Houlton BZ, Y-P Wang, PM Vitousek and CB Field, (2008). A unifying framework for dinitrogen fixation in the terrestrial biosphere. *Nature* 454, 327–331.

Hutley LB, AP O'Grady and D Eamus, (2001). Monsoonal influences on evapotranspiration of savanna vegetation of northern Australia. *Oecologia* 126, 434–443.

Jarvis PG, (1976). Interpretation of variations in leaf water potential and stomatal conductance found in canopies in the field. *Philosophical Transactions of the Royal Society of London* 273, 593–610.

Jarvis PG, and KG McNaughton, (1986). Stomatal control of transpiration: scaling up from leaf to region. *Advances in Ecological Research* 15, 1–49.

Jongen M, JS Pereira, LMI Aires and CA Pio, (2011). The effects of drought and timing of precipitation on the inter-annual variation in ecosystem-atmosphere exchange in a Mediterranean grassland. *Agricultural and Forest Meteorology* 151, 595–606.

Karnosky DF, DR Zak, KS Pregitzer, CS Awmack, JG Bockheim, RE Dickson, GR Hendrey, GE Host, JS King, BJ Kopper, EL Kruger, ME Kubiske, RL Lindroth, WJ Mattson, EP McDonald, A Noormets, E Oksanen, WFJ Parsons, KE Percy, GK Podila, DE Riemenschneider, P Sharma, R Thakur, A Sober, J Sober, WS Jones, S Anttonen, E Vapaavuori, B Mankovska, W Heilman, and JG Isebrands, (2003). Tropospheric O_3 moderates responses of temperate hardwood forests to elevated CO_2: a synthesis of molecular to ecosystem results from the Aspen FACE project. *Functional Ecology* 17, 289–304.

Komatsu H, KH Kang, T Kume, N Yoshifuji and N Hotta, (2006). Transpiration from a *Cryptomeria japonica* plantation, part 1: aerodynamic control of transpiration. *Hydrological Processes* 20, 1309–1320.

Kuglitsch FG, M Reichstein, C Beer, A Carrara, R Ceulemans, A Granier, IA Janssens, B Koestner, A Lindroth, D Loustau, G Matteucci, L Montagnani, EJ Moors, D Papale, K Pilegaard, S Ramba, C Rebmann, ED Schulze, G Seufert, H Verbeeck, T Vesala, M Aubinet, C Bernhofer, T Foken, T Grünwald, B Heinesch, W Kutsch, T Laurila, B Longdoz, F Miglietta, MJ Sanz and R Valentini, (2008). Characterisation of ecosystem water-use efficiency of European forests from eddy covariance measurement. *Biogeosciences Discussion* 5, 4481–4519.

Kutsch WL, N Hanan, B Scholes, W Kubheka, H Eckhardt and C Williams, (2008). Response of carbon fluxes to water relations in a savanna ecosystem in South Africa. *Biogeosciences* 5, 1797–1808.

Kuzyakov Y and O Gavrichkova, (2010). Time lag between photosynthesis and CO_2 flux from soil: a review. *Global Change Biology* 16, 3386–3406.

Linderson ML, TN Mikkelsen, A Ibrom, A Lindroth, H Ro-Poulsen and K Pilegaard, (2012). Up-scaling of water use efficiency from leaf to canopy as based on leaf gas exchange relationships and the modelled in-canopy light distribution. *Agricultural and Forest Meteorology* 152, 201–211.

Lloyd J, (1991). Modelling stomatal responses to environment in *Macadamia integrifolia*. *Australian Journal of Plant Physiology* 18, 649–660.

Lloyd J and GD Farquhar, (1994). ^{13}C discrimination during carbon assimilation by the terrestrial biosphere. *Oecologia* 99, 201–203.

Lloyd J, J Grace, AC Miranda, P Meir, SC Wong, BS Miranda, IR Wright, JHC Gash, J Mcintyre, (1995). A simple calibrated model of Amazon rain-forest productivity based on leaf biochemical properties. *Plant, Cell and Environment* 18, 1129–1145.

Lu P, IAM Yunusa, RR Walker and WJ Muller, (2003). Regulation of canopy conductance and transpiration and their modelling in irrigated grapevines. *Functional Plant Biology* 30, 68–98.

MacInnis-Ng C, K McClenahan and D Eamus, (2004). Convergence in hydraulic architecture, water relations and primary productivity amongst habitats and across seasons in Sydney. *Functional Plant Biology* 31, 429–439

Martin G, DA Myers and TC Vogelmann, (1991). Characterisation of plant epidermal lens effects by a surface replicate technique. *Journal of Experimental Botany* 42, 581–587.

Medlyn BE, RA Duursma, D Eamus, DS Ellsworth, IC Prentice, CVM Barton, KY Crous, P Angelis, M Freeman and L Wingate, (2011). Reconciling the optimal and empirical approaches to modelling stomatal conductance. *Global Change Biology* 17, 2134–2144.

Meinzer FC, (1982). The effect of vapour pressure on stomatal control of gas exchange in douglas fir (*Pseudotsuge menziesii*) saplings. *Oecologia* 54, 236–242.

Mencuccini M and T Holtta, (2010). The significance of phloem transport for the speed with which canopy photosynthesis and belowground respiration are linked. *Global Change Biology* 185, 189–203.

Monteith JL, (1995). A reinterpretation of stomatal responses to humidity. *Plant, Cell and Environment* 18, 357–364.

Mooney HA, C Field, WE Williams, JA Berry, O Blorkman, (1983). Photosynthetic characteristics of plants of a Californian cool coastal environment. *Oecologia* 57, 38–42.

Mott KA and DF Parkhurst, (1991). Stomatal responses to humidity in air and helox. *Plant, Cell and Environment* 14, 509–515.

Mott KA and D Peak, (2013). Testing a vapour phase mechanism for stomatal responses to humidity. *Plant, Cell and Environment* 36, 936–944.

Munekage Y, M Hashimoto, C Miyake, K-I Tomizawa, T Endo, M Tasaka and T Shikanai, (2004). Coupled electron flow around photosystem I is essential for photosynthesis. *Nature* 429, 579–582.

Neilson RP, (1995). A model for predicting continental-scale vegetation distribution and water balance. *Ecological Applications* 5, 362–385.

Nicolas E, VL Barradas, MF Ortuno, A Navarro, A Torrecillas and JJ Alarcon, (2008). Environmental and stomatal control of transpiration, canopy conductance and decoupling coefficient in young lemon trees under shading net. *Environmental and Experimental Botany* 63, 200–206.

Niinemets U, A Diaz-Espejo, J Flexas, J Galmes and CR Wanner, (2009). Role of mesophyll diffusion conductance in constraining potential photosynthetic productivity in the field. *Journal of Experimental Botany* 60, 2249–2270.

O'Grady AP, D Worledge, M Battaglia, (2008). Constraints on transpiration of *Eucalyptus globulus* in southern Tasmania, Australia. *Agricultural and Forest Meteorology* 148, 453–465.

Parkhurst DF, (1994). Diffusion of CO_2 and other gases inside leaves. *New Phytologist* 126, 449–479.

Pataki DE, R Oren, G Katul and J Sigmon, (1998). Canopy conductance of *Pinus taeda*, *Liquidambar styraciflua* and *Quercus phellos* under varying atmospheric and soil water conditions. *Tree Physiology* 18, 307–315.

Peak D and DK Mott, (2011). A new vapour-phase mechanism for stomatal responses to humidity and temperature. *Plant, Cell and Environment* 34, 162–178.

Quentin AG, AP Grady, CL Beadle, D Worledge and EA Pinkard, (2011). Responses of transpiration and canopy conductance to partial defoliation of *Eucalyptus globulus* trees. *Agricultural and Forest Meteorology* 15, 1356–364.

Roelfsema MR and R Hedrich, (2005). In the light of stomatal opening: new insights into 'the Watergate'. *New Phytologist* 167, 66–91.

Royer DL, (2001). Stomatal density and stomatal index as indicators of paleoatmospheric CO_2 concentration. *Review of Palaeobotany and Palynology* 114, 1–28.

Rust S and A Roloff, (2002). Reduced photosynthesis in old oak *(Quercus robur)*: the impact of crown and hydraulic architecture. *Tree Physiology* 22, 597–601.

Santiago SL, G Goldstein, FC Meinzer, JB Fisher, K Machado, D Woodruff and T Jones, (2004). Leaf photosynthetic traits scale with hydraulic conductivity and wood density in Panamanian forest canopy trees. *Oecologia* 40, 543–550.

Schymanski SJ, ML Roderick, M Sivapalan, LB Hutley and JA Beringer, (2007). A test of the optimality approach to modelling canopy properties and CO_2 uptake by natural vegetation. *Plant, Cell and Environment* 30, 1586–1589.

Stewart JB, (1988). Modelling surface conductance of pine forest. *Agricultural and Forest Meteorology* 43, 19–35.

Taylor D and D Eamus, (2008). Coordinating leaf functional traits with branch hydraulic conductivity: resource substitution and implications for carbon gain. *Tree Physiology* 28, 1169–1177.

Tholen D, G Ethier, B Genty, S Pepin and X-G Zhu, (2012). Variable mesophyll conductance revisited: theoretical background and experimental implications. *Plant, Cell and Environment* 35, 2087–2103.

Thomas DS and D Eamus, (1999). The influence of predawn leaf water potential on stomatal responses to atmospheric water content at constant C_i and on stem hydraulic conductance and foliar ABA concentrations. *Journal of Experimental Botany* 50, 243–251.

Thomas DS, D Eamus and D Bell, (1999). Optimization theory of stomatal behaviour I. A critical evaluation of five methods of calculation. *Journal of Experimental Botany* 50, 38–92.

Vogelmann TC, JF Bornman and DJ Yates, (1996). Focusing of light by leaf epidermal cells. *Physiologia Plantarum* 98, 43–56.

Walker BH and JL Langridge, (1997). Predicting savanna vegetation structure on the basis of plant available moisture (PAM) and plant available nutrients (PAN): a case study from Australia. *Australian Journal of Biogeography* 24, 81–25.

Waring RH and WH Schlesinger, (1955). *Forest Ecosystems: Concepts and Management.* Academic Press, San Diego California USA.

Wharton S, M Schroeder, K Bible, M Falk and UKT Paw, (2009). Stand-level gas-exchange responses to seasonal drought in very young *versus* old Douglas-fir forests of the Pacific Northwest, USA. *Tree Physiology* 8, 959–974.

Whitehead D, (1998). Regulation of stomatal conductance and transpiration in forest canopies. *Tree Physiology* 18, 63–44.

Whitley R, B Medlyn, M Zeppel, C Macinnis-Ng and D Eamus, (2009). Comparing the Penman Monteith equation and a modified Jarvis-Stewart model with an artificial neural network to estimate stand-scale transpiration and canopy conductance. *Journal of Hydrology* 373, 256–266.

Williams WE, (1983). Optimal water-use efficiency in a Californian shrub. *Plant, Cell and Environment* 6, 145–151.

Wong SC, IR Cowan and GD Farquhar, (1979). Stomatal conductance correlates with photosynthetic capacity. *Nature* 282, 424–426.

Yunusa IAM, CD Aumann, MA Rab, N Merrick, PD Fisher, PL Eberbach and D Eamus, (2010). Topographical and seasonal trends in transpiration by two co-occurring *Eucalyptus* species during two contrasting years in a low rainfall environment. *Agricultural and Forest Meteorology* 150, 1234–1244.

Zeppel MJ and D Eamus, (2008). Coordination of leaf area, sapwood area and canopy conductance leads to species convergence of tree water use in a remnant evergreen woodland. *Australian Journal of Botany* 56, 97–108.

3

Water Relations, Hydraulic Architecture and Transpiration by Plants

Plate 3.1. A tall dense evergreen eucalypt woodland in temperate New South Wales, Australia.

Water has many fundamental roles in the life of all organisms. Typically 70–90 percent of the fresh weight of plants is water. Exceptions include the stems and roots of trees which might contain 50 percent water and for seeds it might be as low as 3–10 percent. Water is essential for plant survival and function and consequently there are many regulatory mechanisms used by plants to stabilise water content to within the limits that ensure function and survival when water supply is reduced. Such regulatory processes include regulation of stomatal conductance, leaf abscission, increased root growth to deeper soils to access deeper stores of water, changes in hydraulic conductivity of sapwood and changes in the Huber value (the ratio of leaf area to sapwood area). In this chapter the basics of plant water relations and the processes of water uptake and transpiration are reviewed. The chapter includes examination of the hydraulic architecture of plants as a major determinant of the functioning of plant canopies.

3.1 Functions and Properties of Water

Water fulfils several functions, including that of:

- A solvent: all biochemical processes within plants occur with water as the solvent for the reactants.
- Transport: short-distance (within cells and between adjacent cells) transport of solutes occurs within water; gaseous diffusion in leaves and roots is the only major transport process that does not occur in water. Similarly, long-distance transport within the xylem and phloem (Chapter 2) occurs as bulk flow in water. When water supply becomes limiting, long-distance transport is reduced.
- Structural support and the driver of growth: leaves remain erect and turgid because of the hydrostatic pressure of water within the phloem and leaf cells. Wilting is the result of a loss of hydrostatic pressure (also called turgor pressure) within leaves and serves to reduce the leaf area exposed directly to incoming direct beam solar radiation (this is not true for the pendulous leaves of eucalypts). Cell expansion (i.e., growth) is dependent on cells exerting a significant turgor pressure directed outwards against cell walls. A drop in turgor of cells that are not fully expanded results in a decrease in growth rate. A simple analogy is that of putting air into a partially deflated car tyre. As the pressure in the tyre increases, the volume of the tyre increases, just as increasing the turgor pressure in young cells causes them to expand and grow.
- Cooling: significant cooling occurs of leaves as a result of transpirational loss of water. Converting liquid water to water vapour inside the leaf requires energy (heat) and this heat is lost when water vapour diffuses through the stomatal pore, thus cooling the leaf. A leaf with closed stomata on a sunny day will be several degrees warmer than an equivalent leaf with open stomata.
- Physiological and biochemical processes: the opening and closing of stomata is the result of changes in turgor of guard cells. Changes in turgor arise because of the movement of water into (increased guard cell turgor) and out of (decreased turgor) guard cells (Chapter 2). Long-distance translocation (movement) of sugars and other constituents of phloem sap occurs because of gradients of turgor in sieve-tube cells, which arise because of the influx of water into the phloem in source cells and the removal of water from the phloem in sink-cells (Chapter 2). Finally, water is an active participant (a reactant) in some biochemical processes, including photosynthesis (it is the source of the O_2 released during photosynthesis (Chapter 2) and the source of the hydrogen atoms used in reduction reactions of photosynthesis) and hydrolysis reactions.

Water has several important physico-chemical properties which explain much of its behaviour. The oxygen atom (O) of a water molecule is strongly electronegative and therefore tends to attract electrons from the hydrogen (H) atom. Consequently the O atom has a small but significant partial negative charge (δ^-). The H atom, which has "lost" some negative charge, has a partial but significant positive charge (δ^+). Because of this, water molecules are "sticky", which means that adjacent water molecules are attracted to each other through electrostatic interactions. Very large numbers of *H*-bonds exist between water molecules as the δ^- of an *O* atom of one molecule of water attracts the δ^+ of an *H* atom of another water molecule. These bonds have large implications for the properties of water (Table 3.1). Such properties include:

i. Large heat of vapourisation: It takes a lot of energy (2.26 MJ kg^{-1} at 100°C) to convert liquid water to water vapour. This means that much of the energy absorbed by leaves (which greatly exceeds the capacity of photosynthetic machinery to use it) can be dissipated through evaporative cooling.

ii. Large heat capacity: The heat capacity of a substance is defined as the energy required to raise the temperature of a unit mass (a gram or a kg or a mole) of a substance by 1°C (or 1 K, which is the same rise in temperature). Heat capacity can have units of J kg^{-1} K^{-1}. The heat capacity of water is very large (1.9 kJ kg^{-1} K^{-1}), which means that it is a good thermal buffer. This also means that the temperature of plants changes less for a given input of heat than objects with a lower heat capacity. This increases the thermal stability of plant tissues, thereby smoothing out the diurnal changes in air temperature that occur every day.

iii. Large surface tension: This allows pond-skaters to glide across the water surface of ponds. It also means that water is very good at filling the empty voids present in dry soil and so it disperses very effectively through a soil volume. Water also does this in plant tissues including sapwood and cell walls thereby keeping the entire internal volume of plant tissues hydrated.

iv. Capillary rise: water molecules are sticky and the movement of one water molecule attracts adjacent water molecules. This water-water attraction is called cohesion. However, water molecules are also attracted to the surface of many solids, including glass, xylem cell walls, and soil particles. Water-solid surface attraction is called adhesion and explains the capillary rise of water up narrow glass tubes or empty xylem vessels. When a water molecule adheres to the wall, it is pulled up against gravity and *H*-bonding pulls adjacent water molecules up. As long as the adhesive force exceeds the cohesive force of water, water can move up the tube/xylem vessel until the weight of the water column (being pulled down by gravity) is sufficient to cancel out the upward adhesive force arising from the water-tube-wall interaction. For a vessel of 40 μm diameter, water can rise to a height of 0.75 m by capillary action. In reality, the xylem walls of plants are composed of a very fine network of cellulose microfibrils with pore diameters in the order of 10 nm. Capillary rise could move water to a height of several kilometres. Thus cell walls are very good wicks, capable of moving water (albeit slowly) through capillary action across long distances. Capillary action does not, however, explain transpirational water fluxes in xylem (Section 3.6) but it does assist in overcoming the gravitational pull on water in the water column and it assists in maintaining hydration of cell walls in leaves and roots.

Table 3.1. Typical values of several properties of water

Property	Value
Cohesive strength	> 25 MPa
Surface tension at 20 °C	0.073 N m^{-1}
Specific heat at 20 °C	4.167 kJ kg^{-1} K^{-1}
Latent heat of vaporisation at 20 °C	2.5 MJ kg^{-1}
Latent heat of melting	0.34 MJ kg^{-1}
Density	1 g cm^{-3}

v. Low viscosity: this assists in maintaining high flow rates of sap in both the phloem and the xylem. High viscosity liquids require more energy to move them through narrow pores than low viscosity liquids. The viscosity of water increases significantly as temperature decreases and it is likely that increases in viscosity associated with declining air and soil temperature reduces the hydraulic conductivity of surface roots, stems and leaves (Cochard *et al.* 2000). This may influence stomatal conductance directly.

vi. Water is polar and so is an effective solvent for biochemical reactions: water is polar because it has two δ^+ *H* atoms and a δ^- *O* atom. Consequently it acts as a very effective solvent for salts, sugars and many of the solutes essential to the biochemistry of plant cells.

vii. Ice floats: it is unusual for the density of a solid to be less than the density of the same material as a liquid. Ice is less dense than liquid water and therefore floats. This is important as ice that forms when air temperatures are sufficiently cold acts as an effective insulator to the water beneath. If ice didn't float, it is likely that the total volume of liquid water on the earth would be much smaller than it is today as much more would be frozen than is currently the case.

3.2 The Water Relations of Plant Cells

The water potential in plants is the sum of two principal components: solute potential and turgor potential. In addition there are two minor components: matric and gravity potentials.

In this section these terms are explained with particular reference to our understanding of the movement of water in the soil-plant-atmosphere continuum.

3.2.1 Free Energy and Work

In order to understand water potential it is necessary to first understand the nature of free energy and work. Consider a beaker of pure water at standard temperature and pressure (STP; 0.1 MPa of air pressure and 20 °C temperature). By definition, the free energy of the water is set at zero. Assigning a zero value to the free energy is arbitrary in the same way that measuring the height of mountains is made by reference to "sea-level". A 3000 m high mountain is only 3000 m higher than the arbitrary choice of "sea-level" as a reference point which happened historically. There is no theoretical reason why mountains could not be measured with reference to some other point, for example, the deepest trench in the world's oceans. However, "sea-level" was chosen instead. Similarly, assigning a value of zero to the chemical potential of pure water at STP was chosen in the past to be the reference point.

Free-energy means the energy that is available to do work. Work is done when a force acts on a mass and the mass moves. If a force of 10 newtons acts on an object and the object moves 4 m then the work done equals $10 \text{ N} \times 4 \text{ m} = 40 \text{ N m} = 40$ joules (since $1 \text{ J} = 1 \text{ N m}$). That pure water has free energy (energy available to do work) can be illustrated by showing that pure water can make a solution (which has mass) move vertically against gravity. Thus, consider the following experiment:

a) Have a thistle funnel inverted with the lower end in a beaker of pure water (Fig. 3.1). The lower end of the funnel is covered with a semi-permeable membrane that allows the passage (diffusion) of water molecules but not the diffusion of large solute molecules (for example polyethylene glycol, PEG; molecular weight = 2000). At the start of the experiment, pure water is in the thistle funnel and the height of the water in the funnel and beaker are identical (Fig. 3.1).

b) The free energies of water on both sides of the semi-permeable membrane are both zero and hence are equal.

c) The rate of diffusion of water into the funnel is the same as the rate of diffusion of water out of the funnel and so the water level in the funnel doesn't change. No work is being done at this stage.

d) Now add the solute polyethylene glycol into the funnel.

e) The free energy of the water in the funnel is decreased relative to that of the pure water in the beaker. This is because the water molecules interact (through *H*-bonds and van der Waals forces) with the PEG molecules and their ability to freely diffuse is reduced. At a molecular level, the random motion of some water molecules is reduced so the free energy of the water overall is reduced. There is now a difference in free energy between the water in the beaker and the water in the funnel. This difference in free energy can be made to do work. The rate of diffusion of water into the funnel now exceeds the rate of diffusion of water out of the funnel and the water level in the funnel rises.

f) The water will keep rising until the weight of the water downwards is equal to the difference in free energy between the water in the beaker and the water in the funnel. If the funnel radius is 1 cm and the height of the water column at equilibrium is 10 cm, the volume of water that has moved into the funnel is $3.142 \times 1 \text{ cm} \times$

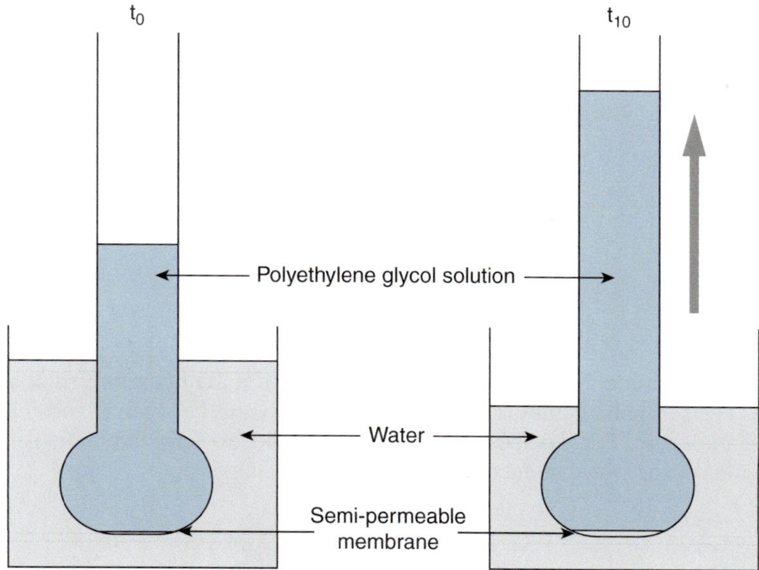

Figure 3.1. Water enters the thistle funnel on the left because of the concentration gradient between the water in the beaker and the water in the funnel. This pushes water upwards within the funnel, against the force of gravity, thereby doing work.

$1\ cm \times 10\ cm = 31.42\ cm^3$ or 31.42 g. Thus, work has been done in moving 31.42 g of water up an average height of 5 cm (half the total height as this is the average height) (Fig. 3.1) and this work is possible because of the free energy available in the water in the beaker.

g) If more solute is added to the funnel, the difference in free energy between the water in the beaker and the water in the funnel is increased and so more energy is available to do work and the column of water rises in the funnel, so water is being moved uphill again, further demonstrating that work is being done.

The presence of the solute in the funnel reduces the free energy of the water in the funnel relative to that of the water in the beaker. A simple example is illustrated in Figure 3.2. If a small amount of *KCl* is dissolved in a beaker of water, the *K* and *Cl* disassociate and the *K* becomes K^+ and the *Cl* becomes Cl^-. The K^+ attracts a number of water molecules such that the δ^- of two *O* atoms of each of several water molecules are close to K^+ and have their free mobility (vibrational movement known as Brownian motion) significantly reduced. Similarly the Cl^- attracts the δ^+ of the *H* atoms in several molecules of water. A cage-like structure (called the hydration shell) of water molecules surrounds each K^+ and Cl^- ion and the free mobility of these water molecules is reduced. Consequently the free energy of the water in each hydration shell in the beaker is reduced. Consequently the overall free energy of the water is reduced. The following is a crude representation of the hydration shell around a K^+ ion.

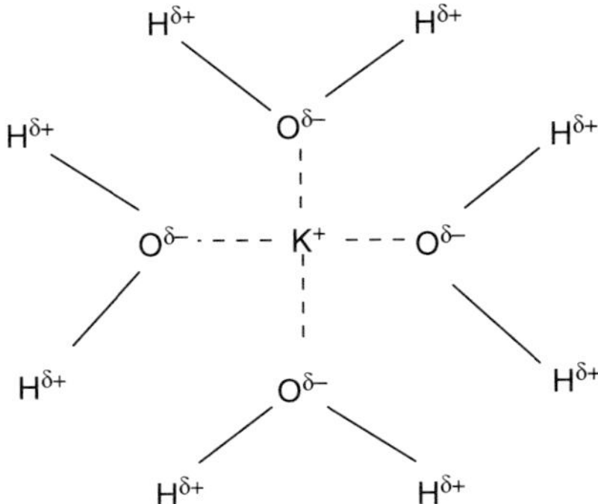

Figure 3.2. A crude 2-dimensional representation of a 3D hydration shell around a K^+ ion. The solid black lines represent the covalent bonds between O and H atoms in a water molecule. The dashed lines represent the electrostatic interactions between the dipoles of the water molecules and the positive charge of the K^+ ion.

3.2.2 Solute Potential

The relationship between the amount of solutes present in the solution (in the funnel) and the effect on the potential energy of the solution is given by van't Hoff's equation:

$$\Pi_s = cRT \tag{3.1}$$

where Π_s is the solute pressure or osmotic pressure (both terms are used in different texts), c is the concentration of all the solutes present, R is the gas constant (= 8.3134 J mol^{-1} K^{-1}), and T is the temperature (in Kelvin).

The van't Hoff equation only applies to dilute solutions that behave as ideal solutions. Ideal solutions are those where the activity coefficient of water is equal to 1; that is, the effective concentration of water has not been affected by the presence of the solute. For the purpose of this text, cells are assumed to behave as ideal solutions and the van't Hoff equation applies.

The solute pressure (or osmotic pressure) is the pressure that would need to be exerted on the water in the funnel (Fig 3.1) to bring the water level in the funnel back to its original level, that is, the same level as the water in the beaker. If the solute pressure is calculated to be 0.1 MPa the air pressure above the funnel (but not the beaker) would need to be increased by 0.1 MPa (or 1 bar) to push the water in the funnel back down to be the same level as the water in the beaker.

In plant physiology, the term solute potential (or osmotic potential) is generally used rather than solute pressure or osmotic pressure. Solute potential is equivalent to the negative value of the solute pressure. The symbol for solute potential is Ψs (pronounced *Psi s*). The key point to gain from all of this is that the presence of solutes in a cell (or soil solution) reduces the solute potential of the solution from its maximum value of zero (no solutes present). Thus, solute potentials are equal to zero (pure water with no solutes, at STP) or negative (e.g. −1.2 MPa).

3.2.3 Turgor Potential

Consider the beaker of pure water sitting on a bench at standard temperature and pressure. By definition the free energy is set to zero for this system. Now increase the atmospheric pressure above the water. This forces the water molecules to be compressed (by an extremely small amount). This results in an increase in the free energy of the water. This means that more work can be done by the system, that is, the water in the beaker. We know that water under pressure can do work because when we turn on a water tap at home, water gushes out of the tap as we have released the pressure in the pipe. Attaching a hose to the tap allows us to direct the water upwards. The water will spray upwards to a height determined by the pressure in the pipe. The higher the pressure in the pipe, the higher the water sprays into the air. Water has mass, so work is being done to lift that mass up into the air. The pressure in the pipe provides the energy for work to be done. Similarly, work is done when guard cells open as the guard cells have mass and move when they open. The driving force for opening of guard cells is the turgor generated in the cell (Chapter 2). Compressing water (subjecting it to an external pressure) forces water molecules to come closer together. This increases the repulsive forces between nuclei, just as compressing a spring increases the energy stored in the spring. Release the compressed spring and the energy stored in the coils can be used to do work (for example, firing an arrow through the air using energy stored in a spring-operated toy gun).

Fully hydrated living plant cells all have positive hydrostatic pressures (i.e., turgor pressures) in them. Typical values of turgor pressures range from 1 to 5 MPa (10 to 50 bar). By comparison, the air pressure in a typical car tyre is about 0.25 to 0.3 MPa (2.5 to 3 bar). This turgor pressure (generally called turgor potential in plant physiology and ecophysiology texts) arises because of the presence of: (a) the relatively rigid cell wall against which the pressure is exerted; and, (b) the presence of solutes in the cell which cause water to move into the cell from the adjacent cell walls. When cells lose water, their turgor potential declines and at wilting point turgor potential is zero. Plasmolysis of cells occurs when more water is lost from a cell that already has a zero turgor potential. The symbol for turgor potential is Ψp (pronounced *Psi p*).

3.2.4 Water Potential

The solute potential of a solution is always zero (pure water) or negative (solutes are present). Since all plant cells have solutes in them, all cells have a negative solute potential. No cells in a plant have a zero solute potential. The turgor potential of a cell is either zero (a water stressed cell) or positive.

Water potential is the algebraic sum of solute potential and turgor potential. Thus:

$$\Psi w = \Psi p + \Psi s \tag{3.2}$$

where Ψw is water potential. Note that since Ψs is negative the equation becomes turgor potential minus solute potential. Thus, if $\Psi p = 2.0$ MPa and $\Psi s = -3.2$ MPa then Ψw is $2 - 3.2 = -1.2$ MPa. Water potentials are always zero (if $\Psi s = \Psi p$) or negative (because Ψs is always lower than or equal to Ψw). If a cell's solute and turgor potentials are of equal value (but one is negative and one is positive) then water potential of the cell is zero.

3.2.5 Matric and Gravity Potentials

Gravity potentials exist as a result of the earth's gravitational pull on water. It is only of significance for tall trees, where a static hydrostatic pressure gradient equal to gravity (0.01 MPa m^{-1}) exists. Thus, for a stationary column of water in the xylem of a 10 m tall tree, a pressure potential of 0.1 MPa exists at the base of the tree, a result of the weight of the column of water 10 m tall. For leaf-scale and branch-scale considerations of water relations, the gravity potential component is generally ignored.

Matric potentials are considered when dealing with soils and also in theoretical considerations of water at the molecular level. Because water is bipolar and the surfaces of clay particles and proteins contain fixed charges, water can be associated with these surfaces through electrostatic interactions. This reduces the free energy of the water and therefore reduces the water potential of the system under consideration. For soils, this tends only to be significant in dry soils. For leaf and root tissues, this tends only to be significant at very low relative water contents–so low that the tissue is no-longer viable and functional. One exception is the water held in seeds, where matric potentials are important. However, for all other instances of plant water relations, matric potentials are generally ignored as their contribution to the water potential of a cell/tissue is very small.

3.2.6 Hofler Diagrams

As a fully hydrated cell (or leaf) loses water, its relative water content declines. Relative water content (RWC) is simply calculated as:

$$RWC = [(FW-DW)/(TW-DW)] \times 100 \qquad (3.3)$$

Where FW is the fresh weight of a cell (or more usually, a leaf), DW is the dry weight of a cell (or leaf), and TW is the turgid weight of a cell (or leaf). The turgid weight is the weight when fully hydrated, which means the cell/leaf cannot absorb any more water.

The relationships between water, solute and turgor potential and RWC can be easily represented in a Hofler diagram. Hofler diagrams are useful plots for summarising the water relations of leaves and have been used extensively in plant ecophysiological studies.

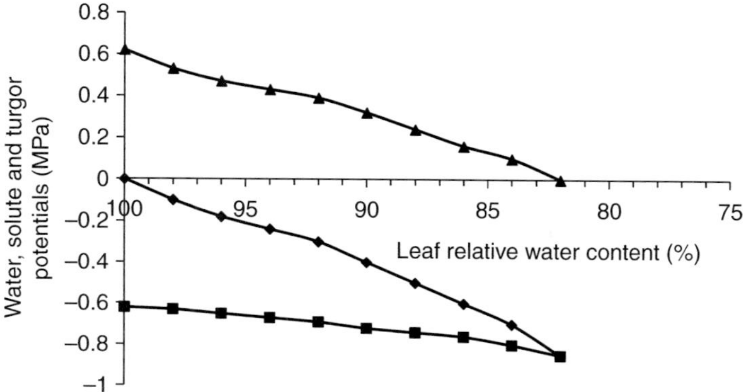

Figure 3.3. An example of a Hofler diagram showing the relationship between RWC and water (diamonds), solute (squares) and turgor (triangles) potentials.

This diagram (Fig. 3.3) clearly shows that as the RWC of a leaf declines (because of water stress over several days, for example), the turgor potential (hydrostatic pressure) of cells in the leaf decline with a steep curve. The slope of this decline can differ among species or between plants that have never experienced stress and plants that have experienced water stress (the stress hardening effect). The slope of the turgor *versus* RWC can be used to determine the average volumetric elastic modulus (ε) of the leaf. The volumetric elastic modulus of a cell or tissue is a measure of its elasticity. A large value for ε means that the cell walls are relatively inelastic and that large changes in turgor occur for only small declines in RWC. Changes in ε occur in response to environmental stress, including chilling stress and water stress, and examples can be found in Eamus and Narayan (1990) where the impacts of water stress on these parameters were examined. Osmoregulation (the accumulation of solutes to reduce cell and leaf water potentials as drought develops) was observed in leaves of *Solanum melongena* when whole plants were rehydrated following drought. An increase in the maximum turgor (from 0.56 to 0.65 MPa) and volumetric elastic modulus of leaves (from 13.37 to 15.62 MPa) was also observed and these too are common adaptations to drought.

In contrast to the steep decline in Ψp with declining RWC, only a shallow decline in Ψs occurs with reductions in RWC. The reduction in Ψs occurs because as water is lost the concentration of cellular solutes increases and thus Ψs declines (becomes more negative) according to the van't Hoff equation. The combination of the steep decline in Ψp and the shallow decline in Ψs results in a steeper decline in Ψw. Where Ψs equals Ψw, Ψp is zero (the point of incipient plasmolysis). At the point where Ψp is zero, $\Psi s = \Psi w$.

3.3 Water in the Atmosphere

The atmosphere is the recipient of transpired water from leaves and canopies and evaporation from soil. The rate at which water evaporates from a wet surface (such as

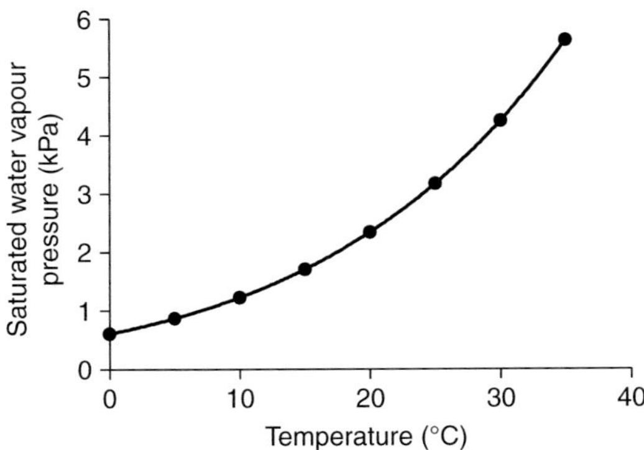

Figure 3.4. The saturated water vapour pressure of air increases curvilinearly with increasing temperature.

a lake surface or a wet canopy) is strongly influenced by the evaporative demand of the air. Air that is saturated with water vapour is not able to contain more water vapour and so the rate of evaporation into a saturated atmosphere is zero. Conversely evaporation into air increases as the concentration of atmospheric water vapour declines. The amount of water that air can hold increases significantly with air temperature. At 10 °C the saturated water vapour concentration is 9.41 g m^{-3}. At 20 °C the saturated water vapour concentration is 17.31 g m^{-3}.

An alternative way to express atmospheric water content is to use water vapour pressure. Air that is totally devoid of water vapour, but at atmospheric pressure, has a water vapour pressure of zero Pascals (the unit of pressure is the Pascal; Pa). As water vapour is introduced into the air, and if atmospheric pressure is constant, the partial pressure of the water vapour increases. At some point the air becomes saturated with water vapour and this corresponds to the saturated water vapour pressure. As with water vapour concentration, the saturated water vapour pressure increases with temperature (Fig. 3.4).

Humidity is another way of expressing water content of air. When humidity is 50 percent, the water vapour concentration is 50 percent of saturated value, at the temperature as defined when humidity was measured. Similarly, a humidity of 50 percent means that the water vapour pressure (WVP) is 50 percent of the saturated value at the temperature defined when humidity was measured. Knowledge of the temperature at which humidity was measured is important since temperature strongly influences the saturated water vapour pressure (and concentration). Thus, 50 percent humidity at 10 °C corresponds to a WVP of 0.614 kPa (or a concentration of 4.7 g m^{-3}) but at 30 °C a 50 percent humidity corresponds to a WVP 2.123 kPa (or a concentration of 15.2 g m^{-3}).

The water potential of air can be calculated from the WVP thus:

$$\psi = \frac{RT}{\overline{V}} \ln \frac{e}{e^*} \tag{3.4}$$

Table 3.2. The relationships between humidity, the ratio of observed to saturated water vapour pressure (*e/e**), and atmospheric water potential

Humidity (%)	Ratio of *e/e**	Atmospheric water potential (MPa)
100	$e/e^* = 1$	0.0
99	$e/e^* = 0.99$	−1.357
94	$e/e^* = 0.94$	−8.35
90	$e/e^* = 0.9$	−14.2
50	$e/e^* = 0.5$	−93.6

Note: This table assumes temperature remains constant at 20°C.

Table 3.3. Indicative values of water potential within the soil-plant-atmosphere continuum

Location	Water potential (MPa)	Humidity (%)
Bulk soil in a wet profile	−0.25	
Soil adjacent to a root in a wet profile	−0.4	
Root xylem	−0.6	
Leaf xylem	−1.0	
Leaf mesophyll cell	−1.1	
Air inside a leaf	−1.35	99.9
Bulk air on a "typical" warm day	−93.6	50

where e is the water vapour pressure, e^* is saturated water vapour pressure, and $\bar{V}$ is the partial molar volume of water.

At 20°C this can be simplified to:

$$\psi = 135 \ln \frac{e}{e^*} \tag{3.5}$$

Water potential of air is very sensitive to changes in water vapour pressure. Atmospheric water potential is zero when the air is saturated with water vapour and declines substantially with only small changes in humidity (or WVP) (Table 3.2).

Since plant water potentials are generally in the range of −0.1 to −5 MPa for most species in most environments (exceptions do occur, for example arid zone trees such as *Acacia aneura* in central Australia can experience water potentials as low as −12 MPa), it is clear that for most of the time the atmosphere exerts a large evaporative demand (has a much lower water potential) on leaves.

Table 3.3 provides representative values of water potentials for the soil-pl ant-atmosphere-continuum. The decline in water potential evident between soil, root and leaf is the driving force for the flow of water from soil to the leaf via the roots and xylem, as water moves energetically downhill via this gradient. The resistance to flow of water exists because water has to pass through plant membranes, xylem walls and the lumen of xylem elements, all of which impose a frictional drag on the movement of water molecules. This is the reason for differences in water potential along the soil-plant-atmosphere pathway.

3.4 Daily and Seasonal Patterns of Leaf Water Potential

Typical patterns of pre-dawn leaf water potential and diurnal and seasonal patterns in leaf water potential are presented below. Further examples are discussed in the biome case study chapters.

Diurnal patterns of leaf water potential (Ψ_w) show a reasonably consistent pattern. Leaf water potential is generally high (close to zero) during the night (usually measured immediately before dawn) and declines when transpiration starts at sun-up. Minima occur typically around solar noon or soon thereafter and then increase in the late afternoon and early night. In wet-dry seasonal climates, the absolute values of Ψ_w in the dry season are usually lower than those in the wet season (Fig. 3.5). This is because both the soil and the atmosphere are drier in the dry season than the wet season. Consequently roots, and hence leaves, must maintain a lower water potential in the dry season to maintain a gradient of water potential between soil and root and between root and leaf in order for water uptake and subsequent transport to occur.

In a study of walnut (*Juglans regia*) trees in northern China (Li *et al.* 2008), pre-dawn leaf water potential declined from maximal values (closest to zero) in April, when soil moisture contents were maximal and air temperatures were *ca* 15 °C, to minimum values in August when soil moisture contents were lowest and air temperatures maximal (*ca* 30 °C). Recovery of pre-dawn leaf water potentials occurred in September after the start of the wet season (Fig. 3.5).

In addition to seasonality of rainfall *per se* influencing pre-dawn leaf water potentials, the amount of rainfall can also influence the magnitude of the seasonal decline

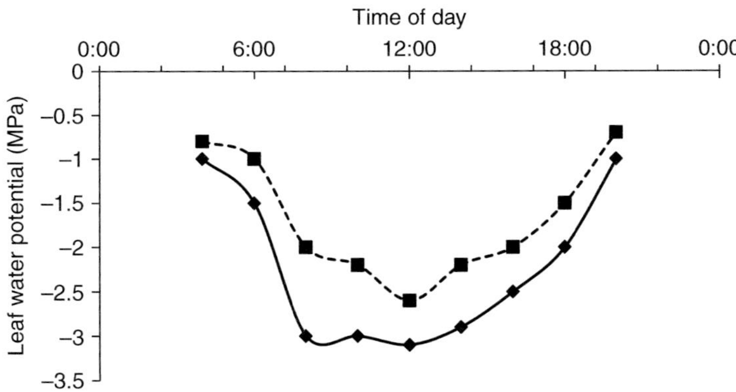

Figure 3.5. Leaf water potential show diurnal patterns, with maxima occurring at night and minima occurring around mid-day. The dashed line (squares) represents wetter season values and the solid line (diamonds) is drier season values. Redrawn from Li *et al.* (2008).

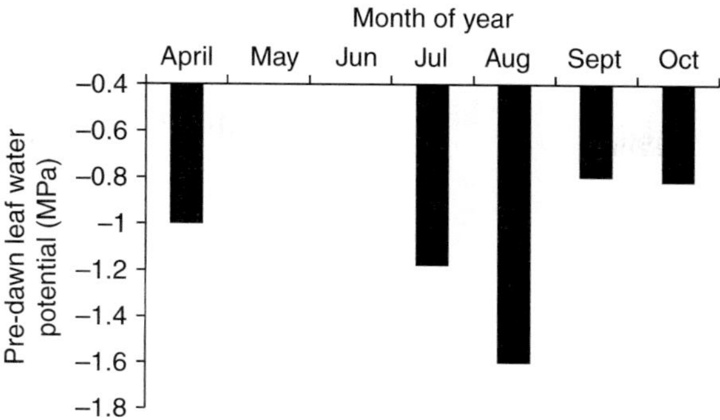

Figure 3.6. Pre-dawn leaf water potentials decline between the wet season (November–March) and the dry season May–September.
Redrawn from Sun *et al.* (2011).

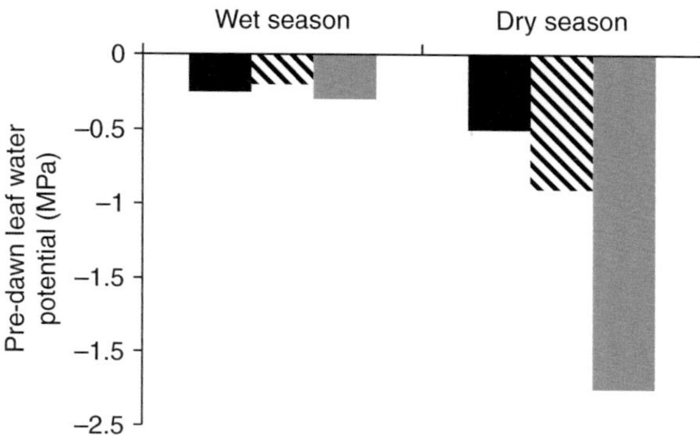

Figure 3.7. Pre-dawn leaf water potential declines between the wet and dry seasons of seasonally wet-dry tropical sites. However, as annual rainfall declines along a rainfall gradient, pre-dawn water potentials also decline. Annual rainfall is 1700 mm (black bar), 1000 mm (striped bar) and 650 mm (dark grey bar) along an aridity gradient in north Australia.
Redrawn from Eamus *et al.* (2000).

in pre-dawn water potential (Sun *et al.* 2011; Fig. 3.6). Similarly Eamus *et al.* (2000) observed in a 3-site comparison across north Australia that in the wet season sufficient rain occurs to allow pre-dawn water potentials to remain higher (closer to zero) than −0.25 MPa. In contrast at the end of the dry season pre-dawn water potentials are much lower at the driest site because: (a) the volume of water stored in the soil profile is much smaller at the driest site; and, (b) the evaporative demand of the air is larger at the driest site (Eamus *et al.* 2000; Fig. 3.7).

Such patterns are commonly observed across aridity gradients or across seasonal gradients of rainfall (see Chapters 15 and 18). For example, in a 12 species comparison of tree water relations across wet and dry forests in Costa Rica, Gotsch *et al.* (2010) observed significant declines in both pre-dawn leaf water potential and mid-day water potential in the dry season compared to the wet season and trees growing in dry

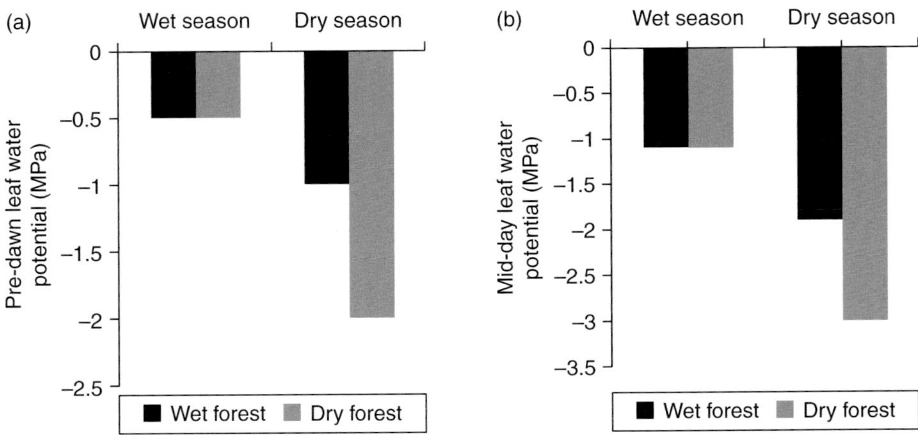

Figure 3.8. A comparison of wet and dry season average (a) pre-dawn and (b) mid-day leaf water potentials for several tree species growing in Costa Rica wet and dry forests.
Redrawn from Gotsch *et al.* (2010).

forests maintained lower leaf water potentials in the dry season compared to wet forest sites (Fig. 3.8). There were no such differences apparent in the wet season because soil moisture content and atmospheric water content were both large and non-limiting to canopy hydration. In contrast, in the dry season, differences in soil moisture content arising from the very different annual rainfall (3269 mm *versus* 1528 mm for the wet and dry forests respectively) caused leaf water potential to decline differentially.

3.5 Anisohydric *versus* Isohydric Leaves

Throughout the day, the key abiotic factors of temperature, VPD and solar radiation levels fluctuate. Similarly, soil moisture content can vary substantially over a period of several days and weeks. Consequently, transpiration rates can vary substantially over a period of hours-to-days-to-weeks and stomata act to regulate transpiration in the short-term to prevent leaf water potential from declining too low. Excessive declines in leaf water potential result in reduced growth rates and the development of xylem emboli (Section 3.7.4). If this is maintained for too long leaves are lost and competitive outcomes among species influenced.

Historically two patterns of changes in leaf water potential in response to changes in abiotic factors have been identified. Some species are identified as being anisohydric, others as isohydric. However, these two terms represent the two ends of a continuum of responses. Isohydric species are those that exert a strong control on transpiration and maintain a relatively constant diurnal minimum Ψ_w despite large fluctuations in soil moisture content and VPD, even during drought. In contrast, anisohydric species exert less control of transpiration and much larger fluctuations in Ψ_w occur in response to, for example, declining soil moisture content. An

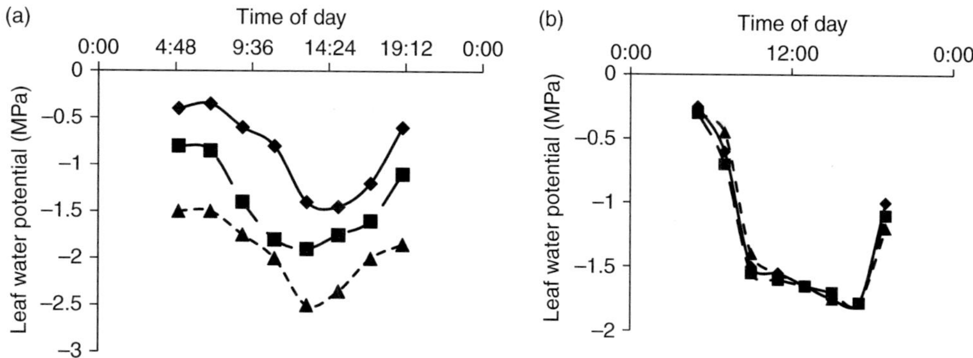

Figure 3.9. Comparisons of isohydric and anisohydric behaviour in leaf water potentials of: (a) sunflower, an anisohydric species; and, (b) maize, an isohydric species. Solid lines with diamond symbols are control, well-irrigated plants; long dashed lines with square symbols are mildly water stressed; short dashed lines with triangular symbols are severely water stressed plants. Redrawn from Tardieu and Simonneau (1998).

example of isohydric and anisohydric behaviour is shown in Figure 3.9. For the anisohydric sunflower plants, leaf water potential declines with increasing water stress from control (well irrigated) plants through to severely water stressed plants. By reducing leaf water potential during drought a gradient of water potential from soil to root is maintained and water uptake thereby maintained for longer. For the isohydric species maize, the diurnal range of leaf water potentials do not differ between control, mildly, and severely water stressed, plants because stomatal regulation is so strong as to increasingly limit transpiration as the soil dries out. The role of isohydric and anisohydric behaviour and the influence of stomatal behaviour in contributing to tree mortality during extended drought is discussed in Chapter 20. It is possible that isohydric behaviour is the result of an interaction of hydraulic (evaporative demand and leaf water potential) and chemical (abscisic acid) signals while anisohydric behaviour is the result of the absence of the interaction of these signals in the regulation of stomatal conductance (Tardieu and Simonneau 1998).

Similar contrasts in isohydric and anisohydric behaviour of stomata and leaf water potentials during water stress for two contrasting grape varieties are discussed by Schultz (2003). In isohydric species there is often a more clearly defined threshold Ψ_w that is correlated with rapid stomatal closure; in anisohydric species, such a threshold is often absent (see Maseda and Fernandez 2006).

The benefits of isohydry are the avoidance of water potentials that are too low, thus avoiding the inhibition of growth and C uptake associated with this condition. Isohydry might be relatively common in neo-tropical savanna trees (Bucci *et al.* 2004) and Amazonian forest species (Fisher *et al.* 2006). Anisohydry appears more commonly in annual crops and therefore might be a means of fixing as much C as possible in

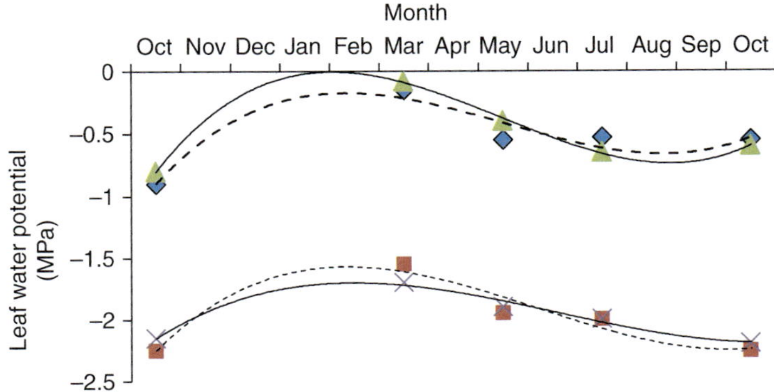

Figure 3.10. The difference between pre-dawn leaf water potential of *Eucalyptus miniata* (blue diamonds; dashed lines) or *E. tetrodonta* (green diamonds; solid line) and their respective midday leaf water potentials (red square, dashed line and purple cross, solid line) is relatively constant across the wet (Dec–Mar) and dry seasons (May–Oct) despite an almost 1 MPa difference across the year.

the short growing season in which annual crops have evolved, although anisohydry has been observed in seasonally dry shrubland and desert shrubs too. The high risk of embolism is presumably associated with an efficient repair mechanism for such emboli (Brodribb and Holbrook 2004). Repair mechanisms for emboli on a daily basis have been well documented (Clearwater and Clark 2003) and are discussed in 3.7.6.

In its most extreme and passive formulation, anisohydric species reduce leaf water potentials continually as soil moisture content (and hence soil water potential) declines, until leaf death occurs. This is a rare event in anything except annual broad leaved crop species. More commonly observed is that the difference between pre-dawn and mid-day leaf water potentials are maintained within narrow constraints; that is, as pre-dawn leaf water potential declines during drought the difference between pre-dawn and mid-day and Ψw remains approximately constant (Fig. 3.10). The reasons for such behaviour probably reflects the need to maintain the gradient between root and leaf Ψw to maintain the flow of water to the canopy despite a drying soil.

It is likely that changes in leaf-specific hydraulic conductance account for the relatively tight homeostatic control of Ψw observed in isohydric species (Maseda and Fernandez 2006).

3.6 Transpiration at the Leaf and Plant Scale

3.6.1 Water Uptake and Drivers of Transpiration

Two mechanisms drive water uptake in plants. The first occurs predominantly during the day and accounts for almost all transpirational water flow. This is the cohesion-tension model of uptake and explains the development of the large gradients in water potential observed between leaf and soil during the day. The second is root-pressure driven flow. Both mechanisms are now described.

When the sun rises, stomata open and water vapour diffuses out from the sub-stomatal cavity. This occurs because of the large gradient in water vapour pressure between the saturated air in the sub-stomatal cavity and the lower water vapour pressure of the air surrounding the leaf. Water vapour that is lost from the sub-stomatal cavity is replaced by water that evaporates from the cell walls of mesophyll cells lining the sub-stomatal cavity. The energy driving this evaporation is the solar radiation absorbed by the leaf. Even at night, however, water can evaporate from these cell walls because of the thermal energy in warm leaves (i.e., leaves that are not at absolute zero temperature). When water evaporates from cell walls, the Ψ_w of the cell wall decreases (because of the presence of solutes and the matric forces in the cell wall). The cell wall develops a Ψ_w lower than the Ψ_w of the cytoplasm of an adjacent cell and consequently liquid water moves from inside the cell, across the cell membrane and into the cell wall. This reduces the Ψ_w of the cells lining the sub-stomatal cavity to a value lower than that of the cells adjacent to these cells lining the sub-stomatal cavity. Water moves down this gradient in Ψ_w and this process is transmitted all the way to the minor veins of the leaf. Water is now drawn from the xylem cells and the water potential of the xylem in the leaf declines. As water exits the minor veins of the leaf, the turgor potential of the xylem declines and becomes negative. The cohesion between water molecules in the column of water in the xylem is sufficient to pull water molecules up the xylem against the force of gravity. A tension develops in the xylem, but as long as the gradient in water potential between leaf and root xylem is not too large the tension can be tolerated without rupture (xylem caviation). In the root, the gradient in Ψ_w is propagated radially across the root until the Ψ_w of the root cells outside of the endodermis (the cortical cells) is lower than that of the soil, at which point water from the soil enters the root cells down the gradient of Ψ_w.

Several features are apparent in this explanation of water flow up plants. These can be listed thus:

1. The gradient of Ψ_w required to cause water uptake originates in leaves.
2. Energy is required to vapourise water in the leaf but the plant itself does not expend energy in the process of transpiration nor water uptake. Water uptake is therefore a passive process (i.e., does not require expenditure of *plant-derived* energy).
3. It takes time for the gradient that originates in leaves to reach the soil and therefore transpiration precedes water uptake by a period of up to several hours for tall trees (Fig. 3.11). Root water uptake also occurs for many hours after transpiration has ceased in order to re-fill stores of water in the stem and rehydrate the canopy to full hydration (RWC = 100%).
4. Water flow from root to atmosphere is driven by serial reductions in Ψ_w along the pathway. For example: $\Psi_{w,soil} = -0.05$ to -0.1 MPa; $\Psi_{w,root} = -0.15$ to -0.25 MPa; $\Psi_{w,stem} = -0.5$ to -1.0 MPa; $\Psi_{wl,eaf} = -1.5$ to -2.0 MPa; $\Psi_{w,air} = -25$ MPa, as shown in Figure 3.12.

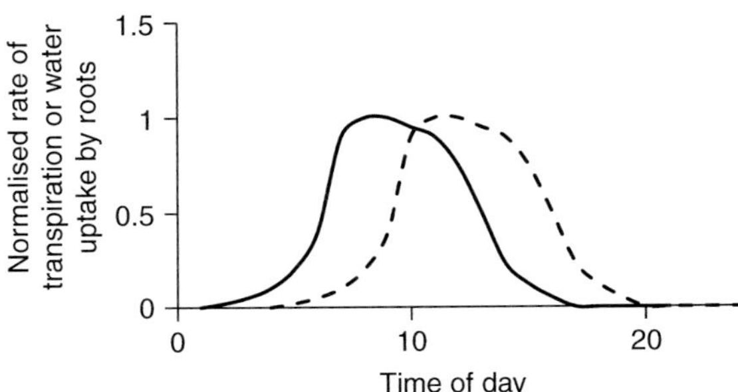

Figure 3.11. Stylised graph of rates of transpiration (solid line) and root water uptake (dashed line), normalised so that the peak rate of both equals one. Note the time-lag in uptake between transpiration starting early in the morning but uptake by roots is delayed by several hours (in trees) because of the time taken for the gradient in water potential to be transmitted from the leaves to the root surface.

5. Tension in the xylem causes the diameter of plant stems to decrease during the day and for water to be drawn from stores in the stem (termed capacitance). These changes in diameter (Fig. 3.14) are measured using sensitive stem dendrometers (see Fig. 3.13).

6. Stem capacitance can be large in large trees and most of the water transpired early in the morning is derived from storage sites in the stem. Water capacitance (C_p) can be calculated from the following:

C_p = (change in water content of the stem)/(change in water potential of the stem).

If the water potential of the storage site changes by 0.5 MPa diurnally and the change in water volume is 0.008 m^3 per day, then the capacitance is $-0.008/-0.5 = 0.16$ m^3 MPa^{-1}. Water lost from storage in the day is replenished during the night and the stem increases in diameter and this increase is the result of both refilling of storage sites and a small amount of growth of the stem. Stem capacitance and its importance in hydraulic safety margins is discussed in Section 3.7.6.

7. Because water flows "energetically downhill", that is, water moves down gradients of water potential from soil to leaf, and because of the resistance to flow of water at all points along the transport path (across roots, up xylem, across leaves, out through stomata), there are gradients of water potential generated all along the flow path.

8. Just before dawn and in the absence of nocturnal transpiration (but see Section 3.6.2), transpiration is zero and leaf and xylem water potentials equilibrate to that of the soil, especially if the soil is moist. Measuring pre-dawn leaf water potential is a relatively simple way of estimating the average soil water potential across the root profile. However, if significant rates of nocturnal transpiration occur, equilibration of the xylem and leaf water potential to that of the soil does not occur overnight.

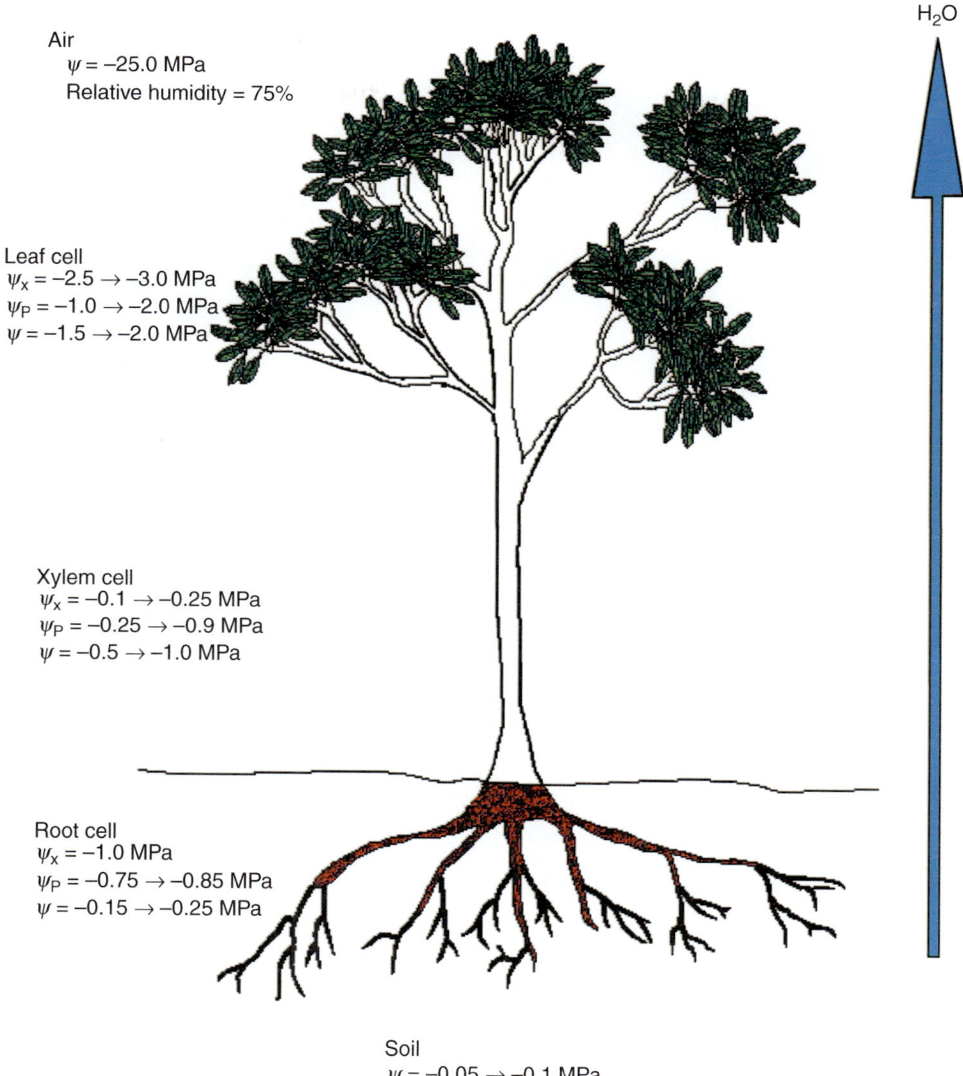

Figure 3.12. A gradient in water potential exists between the soil, root, canopy and atmosphere.

The second mechanism of water uptake by plants occurs at night and can only raise water a short distance vertically. During the night, ion uptake by roots occurs and solutes are loaded into the xylem in roots. This reduces the solute potential and hence the water potential of the xylem. Water enters the xylem from root cells because of the presence of a gradient of water potential. This generates a positive turgor potential in root xylem which pushes water up the xylem. However the magnitude of the turgor pressure generated is generally small, less than 0.1 MPa, and therefore the height to which water can move vertically is less than 10 m. Guttation is the result

(a)   (b)

Figure 3.13. A high precision point dendrometer installed on a pine tree. The cabling supplies excitation voltage and carries the signal from the transducer to the data logger.
Photographs courtesy of Stephen Fujiwara.

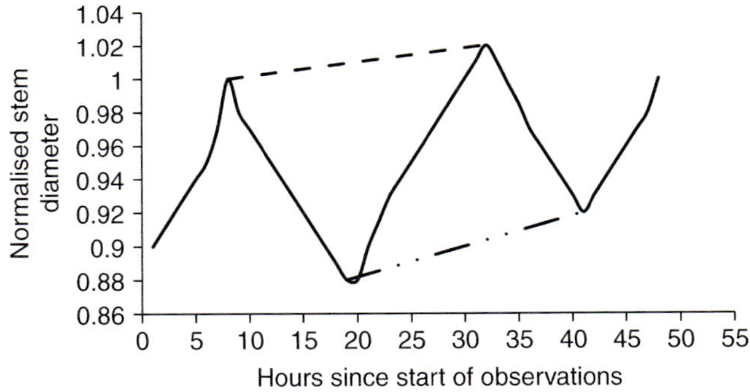

Figure 3.14. Idealised changes in stem diameter of a tree over a 48 hour cycle. Night time corresponds to the period 0–6 h and 18–30 h. Stem shrinkage starts several hours after sunrise and stem refilling starts several hours after sunset. The two dashed lines show that both the maximum and minimum stem diameters are larger on the second day than the first day, indicating that stem growth has occurred.

Figure 3.15. Strawberries show guttation in the morning because of the presence of a strong root pressure during the night. Picture kindly supplied by Professor Roni Aloni, Tel Aviv University.

of a positive hydrostatic pressure in the xylem and causes water to exude from specialised structures called hydrathodes. Strawberries often exhibit guttation at the margins of their leaves early in the morning (Fig. 3.15).

3.6.2 Nocturnal Transpiration

Nocturnal sap flow and nocturnal transpiration occurs across a wide range of species and ecosystems (Dawson *et al.* 2007; Phillips *et al.* 2010). Nocturnal fluxes can be a high proportion of a site's annual water budget, and the contribution that nocturnal transpiration makes to the 24-h sap flow can range from 0 to 60 percent (Bucci *et al.* 2004, Dawson *et al.* 2007, Novick *et al.* 2009). The proportion of daily sapflow that occurs at night tends to be largest in arid ecosystems (30–60%), although in most ecosystems it is commonly around 7 to 10 percent. It wasn't until very sensitive sapflow sensors were readily available that the extent and magnitude of nocturnal sapflow was established. Nocturnal transpiration is defined as water loss from the canopy at night and is distinct from the movement of sapflow that is associated with the refilling of water storage sites in the stem which occurs at night. This recharge of stem water stores is especially large in trees where the contribution of stored stem water to the early stages of daytime (i.e., sunlit) transpiration can be large. Differentiation of recharge flow and nocturnal transpiration can be achieved by comparing the timing and rates of sapflow measured at the base of a stem with the timing and rate of flow measured at the base of the canopy (i.e., at the top of the stem). Sapflow at the base of the stem includes water that is destined for recharge of the stem and water that is destined for transpiration, while sapflow at the top of the stem reflects

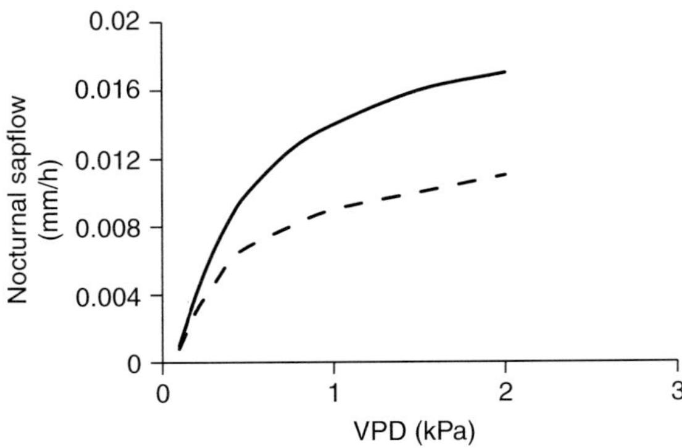

Figure 3.16. Rates of nocturnal sapflow increase with increasing vapour pressure deficit (VPD) in *Angophera bakeri* (solid line) and *Eucalyptus sclerophylla* (dashed line).
Redrawn from Zeppel *et al.* (2010).

transpirational water flux, assuming that recharge of the branches in the canopy is relatively small.

Rates of nocturnal transpiration can differ between co-occurring species. In a study of co-occurring tree species in a North American deciduous forest, one species exhibited moderate (13%) and two species exhibited low (2–7%) nocturnal fluxes despite the same micrometeorological conditions (Daley and Phillips 2006). The function of nocturnal transpiration remains unknown. It may facilitate nutrient uptake at night or hydraulic redistribution from deep (i.e., wet) soil water stores to the upper (i.e., dry) soil profile. Alternatively it may assist in refilling of embolised xylem at night, a process that has been frequently documented. Vapour pressure deficit is a key determinant of rates of nocturnal transpiration (Fig. 3.16), although wind speed also plays a role. The contribution of cuticular transpiration appears to be small and nocturnal transpiration is associated with incomplete stomatal closure at night (Zeppel *et al.* 2010).

3.7 Hydraulic Architecture

Hydraulic architecture refers to the set of attributes of a plant that specifically relate to the transport of water. Hydraulic architecture is usually taken to include discussion of the following:

1. Hydraulic conductance and hydraulic conductivity of sapwood of trees, or more generally the water conducting pathway of plants (xylem), including root, stem, petiole and leaf xylem
2. Leaf hydraulics and relationships with photosynthesis and leaf lifespan
3. Xylem vulnerability to embolism
4. Huber value (the ratio of sapwood area to leaf area)

3.7.1 Hydraulic Conductance and Conductivity

Hydraulic conductance (L_p) is defined as the rate of flow of water per unit pressure difference causing the flow. The driving force inducing flow of water from roots to leaves is the difference in water potential between root and leaf. Thus:

$$L_p = J_v/(\Psi_r - \Psi_l) \tag{3.6}$$

where L_p is hydraulic conductance, J_v is flow rate of water, and Ψ_r and Ψ_l are root and leaf water potential respectively. Measures of conductance do not take account of the distance between the root and leaf, and conductance values are usually normalised (or scaled) with sapwood area or leaf area by dividing L_p by sapwood or leaf area. Volume flow (J_v) is often estimated using sapflow sensors and the gradient of water potential is estimated by measuring leaf water potential at dawn and then throughout the day. It is assumed that pre-dawn Ψ_l is a reasonably accurate measure of the root water potential (it is assumed that the entire sapwood pathway has equilibrated to the water potential of the roots). Furthermore it is assumed that this value does not change during the following day. Hydraulic conductance values range from about 10 g m^{-2} s^{-1} MPa^{-1} to about 125 g m^{-2} s^{-1} MPa^{-1} expressed on a sapwood area basis, or from about 20 to about 200 g m^{-2} s^{-1} MPa^{-1} expressed on a leaf area basis. Reductions in soil moisture content during drought result in reduced hydraulic conductance, partly because of the increase in soil-to-root resistance for water flow and partly because of a loss of sapwood conductance arising from xylem embolism (see (3.7.4).

Hydraulic conductivity (k_h) is defined as the rate of flow or water per unit pressure gradient, where pressure gradient is defined as the difference in water potential per unit length of pathway (i.e., $\Delta\Psi/l$) where $\Delta\Psi$ is the difference in water potential between root and leaf and l is the distance (in metres) between the root and leaf. Thus:

$$k_h = J_v / (\Delta\Psi/l) \tag{3.7}$$

When k_h is scaled by leaf area it is called leaf specific conductivity (k_l). Scaling by sapwood area results in either sapwood specific conductivity or just specific conductivity (k_s). The word "specific" should only be used when a value is expressed per unit mass and therefore the correct terms should be "conductivity on a leaf area basis" or "conductivity on a sapwood area basis". Typical values of k_l range from 10 to 12000 g m^{-2} s^{-1} MPa^{-1} while k_s ranges from about 0.5 to 200 g m^{-2} s^{-1} MPa^{-1}. Deciduous trees tend to have larger values of k_s than evergreen trees. This is because they tend to have wider diameter vessels and tracheids in order to support their larger stomatal apertures compared to that of evergreen species. This arises because they need to fix as much C in their short leaf lifespan (see Section 1.7; Eamus and Prichard 1998). Similarly, k_l and k_s tend to be smaller in seasonally dry environments (e.g. savannas) than mesic environments. This is because trees in seasonally dry environments tend to experience lower leaf water potentials, and,

(a)   (b)

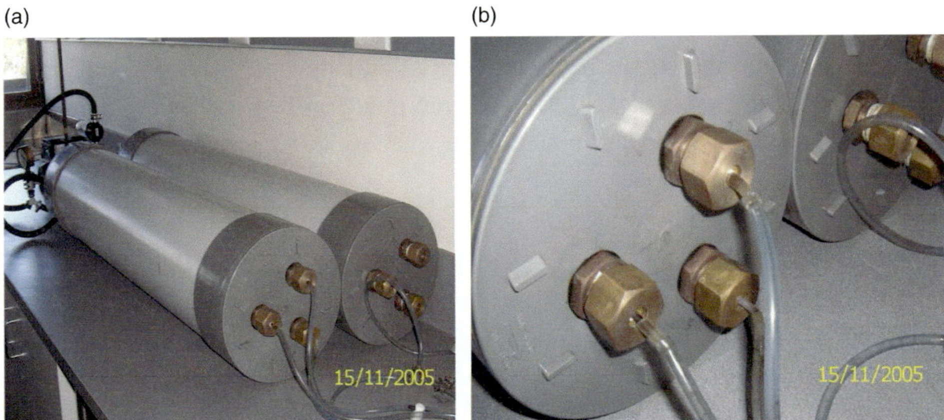

Figure 3.17. An example of equipment used to measure branch hydraulic conductance by applying a vacuum inside the chamber and inducing flow into the chamber via three excised branches.

in order to prevent emboli developing in the xylem, vessel and tracheids diameters tend to be narrower (thus the trend to larger sapwood densities in trees from arid and semi-arid sites compared to mesic sites).

Hydraulic conductivity is frequently measured in isolated (i.e., excised) branches (Eamus and Prior 2001). Water flow through excised branches can be induced by applying a vacuum to one end of a branch whilst the other is attached to a reservoir of high-purity degassed acidified water (Fig. 3.17). Alternatively, a pressure is applied to the reservoir to induce flow through the branch. The rate of flow per unit time per unit pressure is measured with a balance or mass flow meters. There has been increasing attention recently given to measuring the hydraulic architecture of leaves and how this is related to photosynthetic C fixation. This is now discussed.

3.7.2 *Leaf Hydraulics, Stomatal Conductance and Photosynthesis*

Although photosynthesis and transpiration can be viewed as separate processes they are in fact tightly coupled (linked) through the fact that when stomata open to allow CO_2 diffusion into the leaf for photosynthesis hundreds of molecules of water for each molecule of CO_2 are lost at the same time (as transpiration).

Water movement in leaves involves transport through major veins and then minor veins and then loss from the veins into the walls of the bundle sheath cells and then into the walls of palisade and mesophyll cells. Because minor veins account for most of the total vein length (> 70%), more water is lost from minor veins than major veins. The effect of having a large minor vein length is to minimise the distance between water in the xylem and mesophyll cells. The presence of aquaporins (protein lined pores in plant cell membranes that allow relatively easy transport of water; aquaporins can increase and decrease their conductance and are under endogenous control

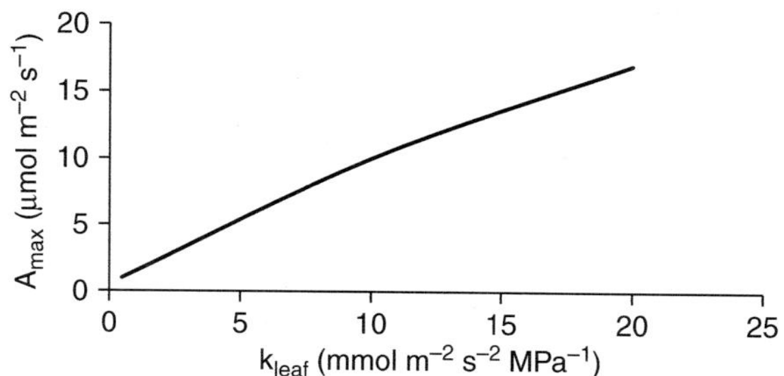

Figure 3.18. Maximum rates of photosynthesis (A_{max}) increase with increasing hydraulic conductance (k_{leaf}) of a leaf.
Redrawn from Brodribb (2009).

by the plant) reduces the resistance to flow across cell membranes, allowing flow to occur transcellularly (cell-to-cell across membranes, as opposed to symplastically which is cell-to-cell via plasmodesmata). There is some evidence that water flow in the leaf is principally apoplastic during the daytime when transpiration is occurring and symplastical and transcellular at night or when transpiration rates are very low. It is likely that the resistance (R) to flow in leaf xylem is about half of the total leaf resistance to water flow, the other half is found in the extra-xylem flow paths, although there can be large differences between species in this ratio. Furthermore, the ratio $R_{xylem}/R_{extraxylem}$ is not fixed. Xylem embolism significantly increases xylem resistance, for example, and $R_{extraxylem}$ increases when light levels fall or when temperature declines (Sack *et al.* 2004, Tyree *et al.* 2005).

Transpirational water loss is always associated with a reduction in leaf water potential and the propagation of gradients of water potential from the leaf to the root. This creates tension in the water column in the xylem and excessive tension results in cavitation of the column (xylem embolism; the formation of air bubbles within the xylem). Such hydraulic dysfunction is to be avoided as it reduces the rate of C fixation (photosynthesis) by the canopy. Coordination between stomatal conductance (g_s) and hydraulic conductance reveals the importance of leaf hydraulic traits to leaf function. Thus, in Figure 3.18, there is a strong positive correlation between hydraulic conductance of a range of species (gymnosperms and angiosperms; Brodribb 2009) and maximum rate of photosynthesis. In order to fix CO_2 quickly, a large g_s is required, which results in a large transpiration rate (E). To support this large E, a large hydraulic conductance is required.

Similarly as leaf water potential declines, leaf hydraulic conductance declines because xylem becomes increasingly embolised as water potential declines (Fig. 3.19; Brodribb and Holbrook 2003) and this decline in conductance is associated with a decline in maximum E (Fig. 3.20), arising from a decline in g_s.

Although xylem embolism is a significant factor causing reduced hydraulic conductance, it is not the only factor responsible. Physical collapse of xylem conduits can precede cavitation (Brodribb and Holbrook 2005).

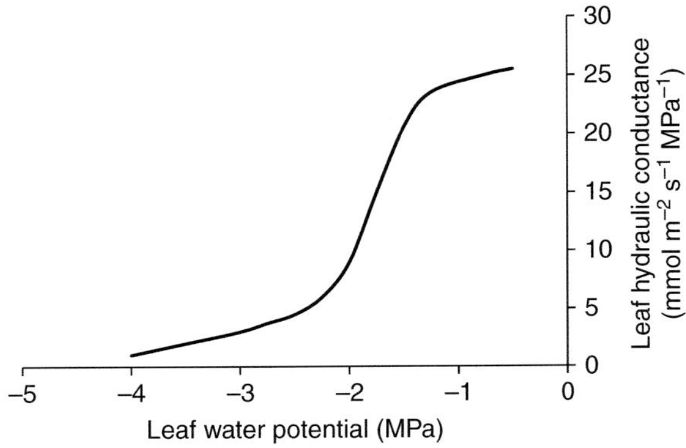

Figure 3.19. As leaf water potential declines, leaf hydraulic conductance also declines. Redrawn from Brodribb and Holbrook (2003).

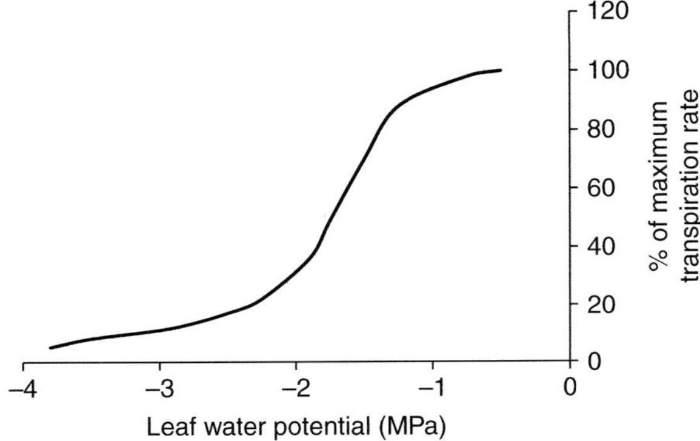

Figure 3.20. Transpiration rate declines as leaf water potential declines because of the closure of stomata. Redrawn from Brodribb and Cochard (2009).

3.7.3 Leaf Hydraulic Conductance Correlates with Leaf Lifespan

Leaves engage in both transpirational water loss and photosynthetic C gain. There is a strong positive correlation between stomatal conductance (g_s) and the hydraulic architecture of leaves because any increase in average g_s without increased capacity to supply water to leaf cells will result in the desiccation of the leaves. Thus g_s and leaf hydraulic conductance are positively correlated and because of the correlation of g_s with rates of photosynthesis, there is a correlation of photosynthesis and hydraulic conductance. However, the hydraulic architecture of leaves includes more than the xylem vessels involved in long-distance transport. Water movement outside of the xylem vessels relies on apoplastic (cell wall), symplastic (cell-to-cell transport via plasmodesmata), and transcellular (cell-to-cell across the plasma membrane) movement, and such pathways may account for about 50 percent of the total resistance to water flow through a leaf.

The chemical composition of cell walls and the thickness of xylem and mesophyll cell walls contribute to the longevity of leaves by increasing or decreasing the mechanical

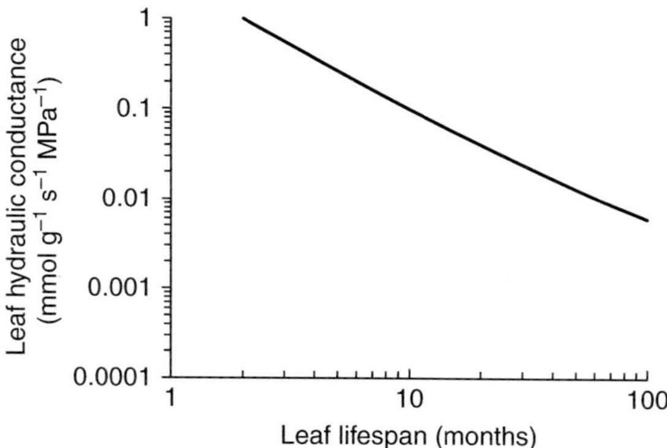

Figure 3.21. Leaf hydraulic conductance declines with increasing leaf lifespan because of the increased investment in leaf resources (especially C) required for increasing leaf lifespan. This decreases the C resources available for increased investment in xylem. The slope is −1, indicating that a change in leaf lifespan results in an equal proportional change in conductance. Note the log-log scale. Data redrawn from Simonin *et al.* (2012).

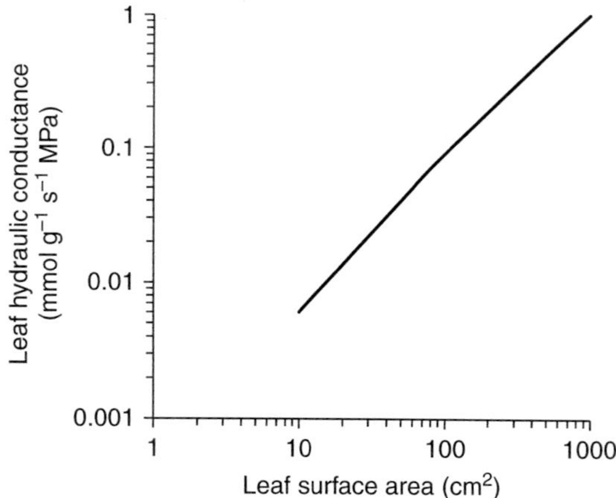

Figure 3.22. Leaf hydraulic conductance increases with increasing leaf surface area. The slope is not different from one, indicating that a change in leaf lifespan results in an equal proportional change in conductance. Redrawn from Simonin *et al.* (2012).

strength and palatability of leaves. Thicker cell walls are more costly to synthesise and maintain and therefore the leaf must increase the total C gained during its lifetime to repay this investment (see Section 1.7). Similarly, increased concentrations of toxins in leaves to decrease palatability and increase leaf lifespan reduce the C available for the construction and maintenance of conducting tissue and concomitant reductions on hydraulic conductance of leaves (Fig. 3.21).

Leaf-lifetime water-use (the total water used over the life of the leaf) is strongly optimised to maximise leaf C gain. Thus across a large range of leaf areas, positive isometric (1:1) scaling between hydraulic conductance and leaf area (Fig. 3.22) means that any increase in leaf area is accompanied by a proportionally equal increase in conductance. Similarly the isometric and negative relationship between conductance and leaf-lifespan means that increased leaf lifespan requires a reduced conductance, but the increased lifespan means that the payback interval for the C invested in conducting tissue and toxins is increased.

There is a negative scaling of leaf conductance (expressed on a dry weight basis) with leaf lifespan but there is a positive scaling of leaf conductance expressed on a dry weight basis with leaf area (Simonin *et al.* 2012). Consequently when $\Delta\Psi_{stem\text{-}leaf}$ (the difference in water potential between stem and leaf and hence the driving force for water flow between the stem and the leaf) is constant, then lifetime C gain per unit dry weight and lifetime water use per unit dry weight are constant, and both are independent of leaf lifespan.

3.7.4 Xylem Vulnerability to Embolism

Assessing sensitivity to embolism is now routinely undertaken in many labs and a common measure of sensitivity is P_{50}, the leaf water potential that causes a decline of 50 percent in hydraulic conductance. Some researchers now use P_{80}, the leaf water potential that causes a decline of 80 percent in hydraulic conductance (Markesteijn *et al.* 2011) An alternate measure of sensitivity is discussed in Section 3.7.5. Air injection equipment (Fig. 3.23) is used to measure xylem vulnerability curves of excised branches. In leaves, P_{50} has been correlated with plant survival across different habitats (Brodribb and Cochard 2009) and the P_{50} of stems is an adaptive trait for survival across sites that differ in water availability. In dry sites, P_{50} in stems is correlated with the minimum water potential experienced at that site.

In a recent comparison of 13 tropical tree species Markesteijn *et al.* (2011) found that pioneer species (those that come into a newly created open space within a forest and are sun adapted; P_{50} and P_{80} water potentials of −1.83 and −3.13 MPa respectively) were more vulnerable to xylem cavitation than shade-tolerant species (P_{50} and P_{80} water potentials of −4.76 and −6.22 MPa respectively). Pioneer species are fast growing, exhibiting large rates of photosynthesis and stomatal conductances and hence large transpiration (E) rates. To support these large rates of E, large hydraulic conductances of xylem are required. Larger stem and leaf hydraulic conductivities were observed in pioneer species compared to shade-tolerant species and, as would be expected, smaller sapwood densities (reflecting the larger vessel sizes required to support larger E). Markesteijn *et al.* (2011) also found close coordination of hydraulic traits across multiple spatial scales. P_{50} was negatively correlated with Huber value (H_v; the ratio of sapwood area to leaf area) and leaf dry mass and negatively correlated with leaf size. A small H_v is commonly observed in species with large hydraulic conductivities. The smaller H_v in pioneer species reflects a smaller sapwood area composed of larger diameter vessels that were more vulnerable to cavitation (hence the larger P_{50} and P_{80} values). The positive correlation of vulnerability to embolism and leaf dry mass occurs because leaf dry mass is inversely correlated with the

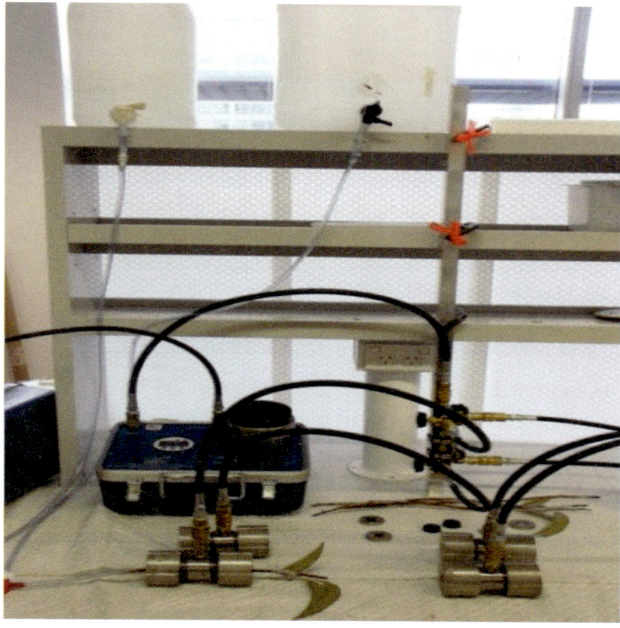

Figure 3.23. Vulnerability to xylem embolism can be measured with the air injection apparatus shown here. The four steel chambers enclose a portion of four replicate branches. The rate of water flow through each branch is recorded; flow is induced by the small hydraulic head imposed by the fact that the reservoir of water is higher than the branch. Once a steady flow rate is achieved the pressure inside the steel chambers is increased slightly for a period of time, which induces a small increase in embolism. Flow rate is then measured again. This procedure is repeated about 10 times to produce a vulnerability curve.

elasticity of leaf cell walls. Pioneer species experience higher leaf temperatures and larger vapour pressure differences, than shade-tolerant species and consequently need thicker cell walls (less elastic) as an adaptation to the more arid environment experienced as a pioneer species.

Much of the resistance to water flow between root and leaf occurs within the leaf (about 50%) and this resistance imposes significant constraints on the maximum value of g_s. Unlike xylem in stems, vessels and tracheids of leaves lack significant thickening of the cell wall and have a larger vulnerability to hydraulic failure than stems. There is a strong correlation between conduit dimensions (the ratio of lumen breadth, b and cell wall thickness, t) such that as the ratio $(t/b)^3$ increases (indicating stronger conduits and hence a larger resistance to embolism), P_{50} increases linearly (Fig. 3.24).

3.7.4 Hydraulic Safety

The relationship (coupling) between hydraulic conductance and leaf gas exchange is mediated via leaf water potential (Ψ_w). The Ψ_w at which stomatal closure occurs is

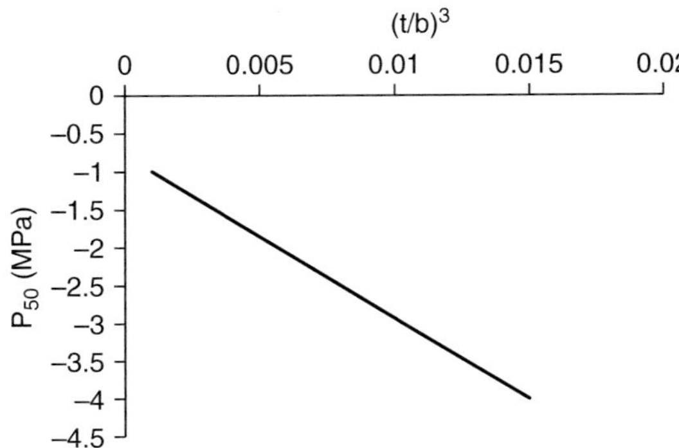

Figure 3.24. As conduit strength increases (as indicated by the increasing ratio of wall thickness to lumen diameter; $(t/b)^3$), the pressure required to induce 50 percent loss of conductance increases. Redrawn from Blackman *et al.* (2010).

higher (closer to zero) than the Ψ_w at which irreversible xylem embolism occurs and the difference between these two values is one definition of a "safety margin". Some species have a large safety margin and stomatal closure occurs at values of Ψ_w much higher than those required to induce irreversible embolism. In contrast, some species live "close-to-the-edge" with narrow margins of safety. There is a trade-off between the two strategies. A wide margin means that damage is avoided but gas exchange is curtailed at relatively high values of Ψ_w. In contrast, small margins allow photosynthesis to occur for longer periods of time and increase carbon gain, with the inherent risk of damage occurring. Irreversible embolism can cause loss of parts of leaves (through loss of the functioning of minor veins) or the loss of whole leaves, minor (terminal) branches or major branches, depending on the magnitude and duration of the drought that induces the embolism. Further discussion of hydraulic safety and alternative measures of safety are given in Section 3.7.5. An alternate definition of hydraulic safety margin is also discussed more extensively in the Forest Mortality case study (Chapter 20).

Is there a relationship between maximum hydraulic conductance and resistance to drought induced embolism? Whilst it might be thought that a large hydraulic conductance results in a large sensitivity (or low resistance) to drought induced embolism (because of the large xylem vessel diameters associated with large conductance values), this is not observed in leaves (Sack and Holbrook 2006).

3.7.5 Xylem Hydraulic Safety Margins, Sapwood Capacitance and Mid-Day Stomatal Closure

Reductions in midday stomatal conductance ($g_{s,m}$) and resulting reductions in rates of photosynthesis are commonly observed in herbaceous and woody species. The causes of this reduction in $g_{s,m}$ are poorly understood, but the magnitude of the reduction tends to increase with declining soil moisture content. While reductions in leaf water potential during drought are associated with declining g_s and leaf water potential during non-drought periods is usually lowest (furthest from zero) around the same time as reductions in $g_{s,m}$ occur, the causal relationship between leaf water potential and g_s is unclear (Zhang *et al.* 2013). Partly this is because not all species show reductions in $g_{s,m}$, and even within a species reductions are not observed on all days, but all species show diurnal declines in leaf water potential between pre-dawn and midday. The differential behaviour of isohydric and non-isohydric further complicates any putative relationship between patterns of change in diurnal leaf water potential and g_s for non-drought conditions.

Stomata do, however, regulate transpiration rates in a manner to minimise the frequency and extent of xylem embolism and declines in $g_{s,m}$ can reduce transpiration and serve to protect xylem against embolism. Leaf hydraulic conductance shows diurnal depression in many species (Johnson *et al.* 2009) but recovery is rapid and it is likely that reductions in $g_{s,m}$ are more important in minimising excessive reductions in stem xylem water potential rather than protecting leaf hydraulics. Reductions in leaf hydraulic conductance may serve to amplify the signal arising from changes in stem hydraulics in order to accelerate the stomatal response (Brodribb and Holbrook 2004). The hypothesis that declines in $g_{s,m}$ are correlated better with stem rather than leaf water status was recently tested (Zhang *et al.* 2013) in a multi-species comparison on sub-tropical trees. Zhang *et al.* (2013) showed that:

1. Species with higher stem K_s (sapwood specific hydraulic conductivity) and larger daily reliance on sapwood hydraulic capacitance (the volume of water stored to buffer the sapwood during periods of high transpiration) maintained a higher stem water potential and therefore a larger g_s (and hence a smaller decline in g_s) at midday
2. As sapwood density increased (and so sapwood hydraulic conductivity decreased) sapwood capacitance decreased and the percentage mid-day decline in g_s increased
3. Strong correlations between midday stem water potential and sapwood capacitance occurred across all species. Thus species with a large sapwood capacitance maintained a high mid-day stem water potential and hence exhibited a smaller decline in mid-day g_s
4. Evergreen species had smaller stem capacitances and smaller sapwood hydraulic conductivities and lower midday stem water potential and higher midday leaf water potential and smaller midday g_s and smaller maximum daily values of g_s than deciduous species.

From these correlations Zhang *et al.* (2013) concluded that a decline in $g_{s,m}$ is an important mechanism for preventing stem embolism and that a reliance on sapwood capacitance to buffer against changes in stem sapwood water potentials is largest in deciduous species, which have lower wood densities and larger hydraulic conductivities. The maintenance of high stem water potentials is dependent on both a large hydraulic conductivity of the sapwood and a large sapwood hydraulic capacitance.

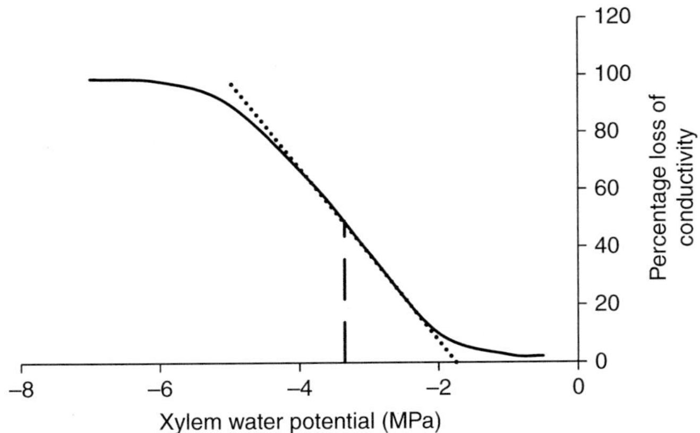

Figure 3.25. As xylem water potential declines within a single species, an increasing proportion of conduits experience embolism and hence the percentage loss of conductivity increases and a unique vulnerability curve is produced. The thick vertical dashed line corresponds to the water potential at which 50 percent of conductivity has been lost (P_{50}). The thin dotted line is the tangent to the P_{50} point on the vulnerability curve and the intercept on the x-axis corresponds to P_e, the "air entry threshold". The difference between these two water potentials is one measure of hydraulic safety. Redrawn from Meinzer *et al.* (2009).

Hydraulic safety margins can also be defined as the difference between daily minimum leaf water potential and some measure of the vulnerability of xylem to embolism. The most commonly used measure of xylem sensitivity to embolism is P_{50}, the xylem water potential that causes 50 percent loss of hydraulic conductivity (Maherali *et al.* 2006). However, as Meinzer *et al.* (2009) demonstrate, P_{50} is probably not the most appropriate measure for assessing the role of stomatal regulation of transpiration to minimise hydraulic failure of xylem because by the time the water potential corresponding to P_{50} is reached, the canopy is well progressed toward catastrophic loss of xylem function. Furthermore, the role for the buffering capacity of sapwood is probably exhausted by the time P_{50} is attained. Instead of P_{50}, Meinzer *et al.* (2009) used the first inflection point on a xylem vulnerability curve, which corresponds to the water potential at which there is a rapid increase in the rate of embolism as water potential declines (P_e; the so-called "air entry threshold"; Fig. 3.23) as a measure of the vulnerability of xylem to cavitation. This is determined by taking the tangent to the vulnerability curve at the water potential for P_{50}. Figure 3.25 shows these relationships graphically on a vulnerability curve. Meinzer *et al.* (2009) also measured the minimum xylem water potential of the branches used for determination of vulnerability curves. From this they showed how hydraulic safety margins are better estimated as the difference between stem water potential and P_e. Note the use of stem (xylem) water potential and not leaf water potential.

Several important features are apparent in the results of Meinzer *et al.* (2009):

1. The size of the hydraulic safety margin, defined as the difference between P_e and P_{50}, showed significant relationships with stem hydraulic capacitance and the

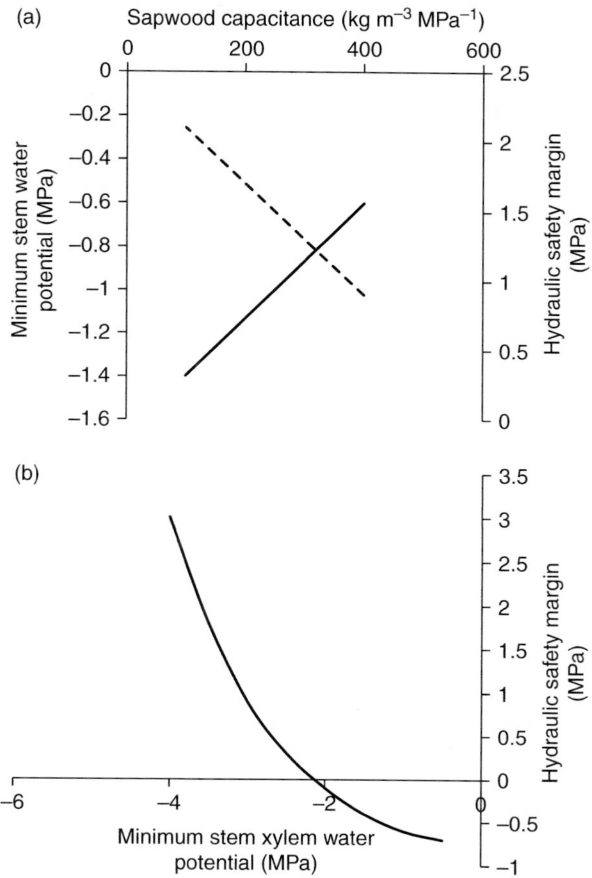

Figure 3.26. (a) Across a range of tree species, as sapwood capacitance increases, the minimum stem water potential increases (tends to zero; solid line) but the hydraulic safety margin (P_e–P_{50}) declines (dashed line); (b) As the minimum stem xylem water potential experienced across a range of species declines, the size of the hydraulic safety margin (minimum stem xylem water potential minus P_e) increased. Note the use of two different ways of calculating hydraulic safety margins. Redrawn from Meinzer *et al.* (2009).

minimum stem water potential (Fig. 3.26a). Thus as stem (sapwood) capacitance increases the minimum stem water potential experienced increases (tends to zero) but the hydraulic safety margin (P_e–P_{50}) declines

2. The hydraulic safety margin increased with decreasing minimum stem xylem water potential (Fig. 3.26b). This suggests a declining role for stem hydraulic capacitance in slowing or reducing the fluctuations in xylem water potential as the minimum stem xylem water potential decreased

3. Species with a low hydraulic capacitance and more dense sapwood experience larger maximum xylem tensions (i.e., more negative water potentials) and consequently rely on xylem structural attributes (thicker walls, narrower diameters) to increase resistance to embolism.

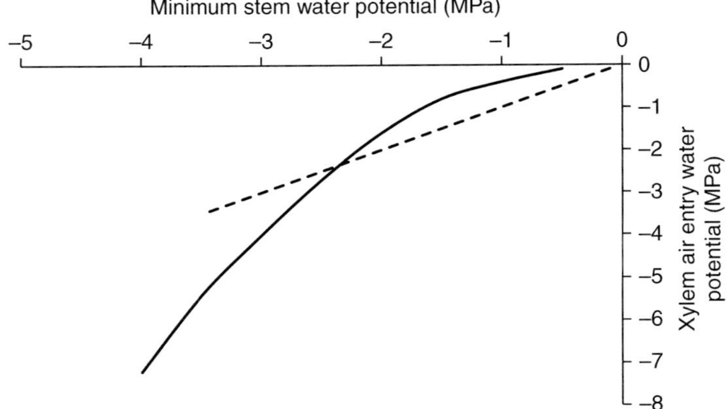

Figure 3.27. As the minimum stem water potential observed in the field declines across a range of tree species, the xylem "air entry" water potential declines. "Air entry" water potential is the water potential at which xylem embolism begins to rapidly accelerate with declining stem water potential. The dashed line is the 1:1 line. Note that the field data (solid line) are above and then below this 1:1 line in different ranges. See text for explanation. Redrawn from Meinzer *et al.* (2009).

4. Species with a large hydraulic capacitance and less dense sapwood rely on hydraulic capacitance to avoid embolism through the rapid and short-term release of water from storage, thereby buffering the changes in xylem water potential arising from transpirational water flow.

5. As the minimum stem xylem water potential experienced in the field by a species declines, P_e declines non-linearly (Fig. 3.26b). Thus as the degree of water stress experienced increased, the P_e value declined, representing an adaptive shift in hydraulic architecture. For very low values of water stress (stem xylem water potentials > -1.5 MPa), the value of P_e-P_{50} was positive but for large values of water stress (stem xylem water potentials < -3 MPa) the value of P_e-P_{50} was negative (Fig. 3.27) indicating that the safety margin is larger for species experiencing more arid environments.

6. For many species, when soil moisture content is high, stomata are able to constrain stem xylem water potentials to values that are at, or slightly less negative (closer to zero) than, P_e. This implies a large risk of run-away embolism as stem xylem water potentials approach values where the vulnerability curve is steepest and small declines in stem xylem water potential cause large increases in rates of embolism. For other species however, stem minimum xylem water potentials are much lower and stomatal regulation of stem xylem water potential was more conservative: the safety margin increased significantly from close to zero to nearly 4 MPa (Fig. 3.27).

7. These relationships showed convergence among a range of angiosperm and coniferous species, with single regressions describing the relationships for both sets of species.

3.7.6 Xylem Repair

Recovery from embolism occurs in the field. One mechanism relies on positive root pressure to force water into embolised xylem vessels at night. However, this is effective for short distances, and certainly cannot repair xylem that is 10's of meters up a tree because the root pressure is insufficient to push water that far up. Whilst there is debate about whether recovery from xylem embolism can occur while the xylem is under tension, there is extensive evidence that this does happen (Cochard and Delzone 2013) and can happen rapidly (that is, overnight; Brodersen and McElrone 2013).

There are three hypotheses to explain repair of emboli and all three require the generation of a positive turgor potential gradient to force water into embolised xylem. The osmotic hypothesis requires accumulation of solutes (inorganic ions, sugars) in embolised xylem vessels. This induces a gradient of water potential directing water flow into the embolised vessel thereby refilling the vessel. In contrast the reverse osmotic hypothesis requires solute accumulation in living cells surrounding vessels to generate large turgor in these cells, which drives water flow into embolised vessels. Experimental and theoretical considerations suggest this mechanism to be unlikely. The third theory relies on accumulation of solutes in phloem cells to generate a turgor gradient towards the xylem, which refills embolised vessels. Extensive experimental data support this. A fundamental role for non-structural carbohydrates (for example starch and sugars) is central to all mechanisms.

3.7.7 Huber Value

The Huber value of a branch, tree or stand of trees is the cross sectional sapwood area (of the branch, tree or entire stand) divided by the leaf area distal to where sapwood area was measured. Typical values for H_v are in the range $1{-}15 \times 10^{-4}$ m^2 m^{-2}. Plants in arid zones tend to have larger H_v values than plants from mesic climates because the leaf area that can be maintained (that is the transpirational demand on the sapwood) tends to be larger at sites having a larger or more consistent supply of water than at sites where water supply is more limited. Huber values are under environmental control and can change across seasons or across sites systematically. This is discussed in the following section.

3.8 Field Studies of Hydraulic Architecture of Stands of Trees

A detailed study of hydraulic architecture was conducted in arid central Australia across three contrasting habitats: a riparian habitat, a floodplain habitat, and an open woodland (O'Grady *et al.* 2009). Annual rainfall is highly variable (typically 100 mm to 700 mm pa) with a long-term average of about 300 mm, and depth to groundwater varies between 7 m and 50 m at the sites examined. Mid-day temperatures and vapour pressure deficits vary between about 15–35°C and 2–6 kPa. It is therefore a hot, dry site.

Huber values (H_v) across four species varied between about 1×10^{-4} m^2 m^{-2} and 14.5×10^{-4} m^2 m^{-2} but was consistently larger (about 100% larger) in November (start of the summer) than April (end of autumn) across all four species examined across the 3 habitats (O'Grady *et al.* 2009). This reflects a decline in leaf area between April and November. In April, soil moisture content was relatively high because there had been 235 mm of rain in the preceding 3 months (typical annual rainfall is 300 mm pa). However, in November soil moisture content was much smaller because there had been only 70 mm of rain in the period April to November and much of this was in events of less than 10 mm, ineffective in wetting-up the soil profile. Changes in H_v reflecting changes in leaf area are commonly observed over monthly and seasonal time-frames in response to changes in soil moisture content and atmospheric water content (Bucci *et al.* 2008). For example, Mencuccini and Grace (1995) found that Scots pine trees growing at a warm dry site produced less leaf area per unit sapwood, that is, had a larger H_v, than Scots pine trees growing at a cool wet site. Similarly Macinnis *et al.* (2004) observed a consistently smaller H_v in summer compared to winter across four contrasting habitats in temperate Australia. Increasing the evaporative demand of the atmosphere, with or without reduced soil moisture content results in increased H_v as an adaptive response to ensure the water potential of the canopy does not decline to low and thus reduces the risk of severe embolism.

Strong correlations between attributes of hydraulic architecture were observed by O'Grady *et al.* (2009). Thus whole-plant hydraulic conductance (calculated from the slope of tree water use *versus* soil-to-leaf water potential gradients) was negatively correlated with sapwood density and H_v but positively correlated with specific leaf area (O'Grady *et al.* 2009; Fig. 3.28). Similarly Stratton *et al.* (2000) and Bucci *et al.* (2004) showed negative relationships between leaf-specific conductance and wood density in Hawaiian forest and Brazilian Cerrado (savanna) forests. This relationship makes intuitive sense. A high wood density is associated with thick xylem cell walls and narrow lumen diameters, attributes that make sapwood more resistant to collapse with decreasing water potential but also reduce the theoretical maximum conductance of the lumen (based on the Hagen-Poiseuille equation (Chapter 2) where conductance is proportional to the fourth power of the radius of the pipe).

3.8.1 Hydraulic Limits to Growth of Tall Trees

As trees grow taller, the total resistance to flow from roots to the top of the canopy increase. This is partly because the total length of the conducting pathway increases and therefore the frictional resistance to water flow increases, independently of the fact that the trees are vertically oriented. It is also because water is moving against gravity, which is pulling it downwards and the weight of water pulled upward increases as tree height increases. It is noteworthy that the gradient in water potential observed

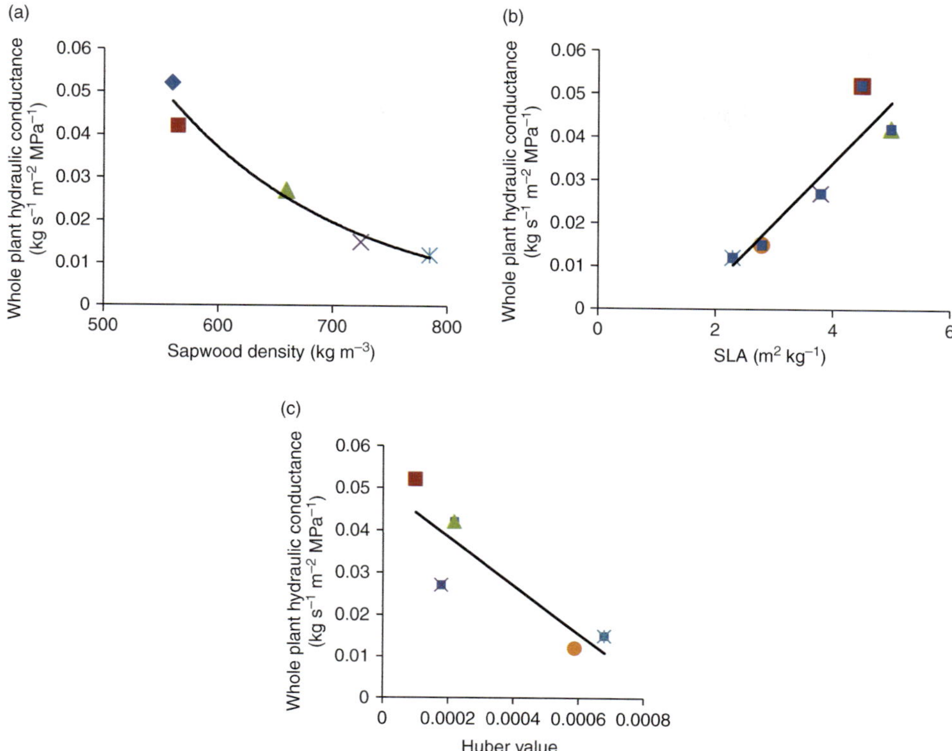

Figure 3.28. Whole plant hydraulic conductance shows a negative correlation with sapwood density and Huber value but a positive correlation with specific leaf area across five arid zone tree species of central Australia. Each symbol is the average for one species. Redrawn from O'Grady *et al.* (2009).

in giant redwoods just before dawn was 0.0096 MPa m^{-1}, almost identical to the gradient predicted to occur under the influence of gravity (0.0098 MPa m^{-1}; Koch *et al.* 2004). To overcome this increasing resistance to flow, leaves of the upper canopy must maintain a low water potential to ensure the gradient of water potential is large enough to pull water up from the soil. However, since there is a limit to how low leaf water potentials can fall and still be functional, there is a limit to how tall trees can grow. This "hydraulic limit to growth" model is under considerable debate (Bond and Ryan 2000, Mencuccini and Magnani 2000, Meinzer *et al.* 2005) but may explain why the productivity of forests (tonnes of carbon fixed per hectare per year) declines with tree age (Ryan and Yoder 1997, Koch *et al.* 2004). As trees grow taller, stomatal conductance of the upper canopy declines (to limit the minimum water potential experienced) and this causes a decline in rates of carbon fixation and hence productivity. A linear decline in leaf turgor with increased tree height was observed by Koch *et al.* (2004) in accordance with the hydraulic limits to growth hypothesis, and the increased $\delta^{13}C$ observed with height is compelling evidence of increased stomatal limitation to photosynthesis (Fig. 3.29; Koch *et al.* 2004). Conceptually, the foliage

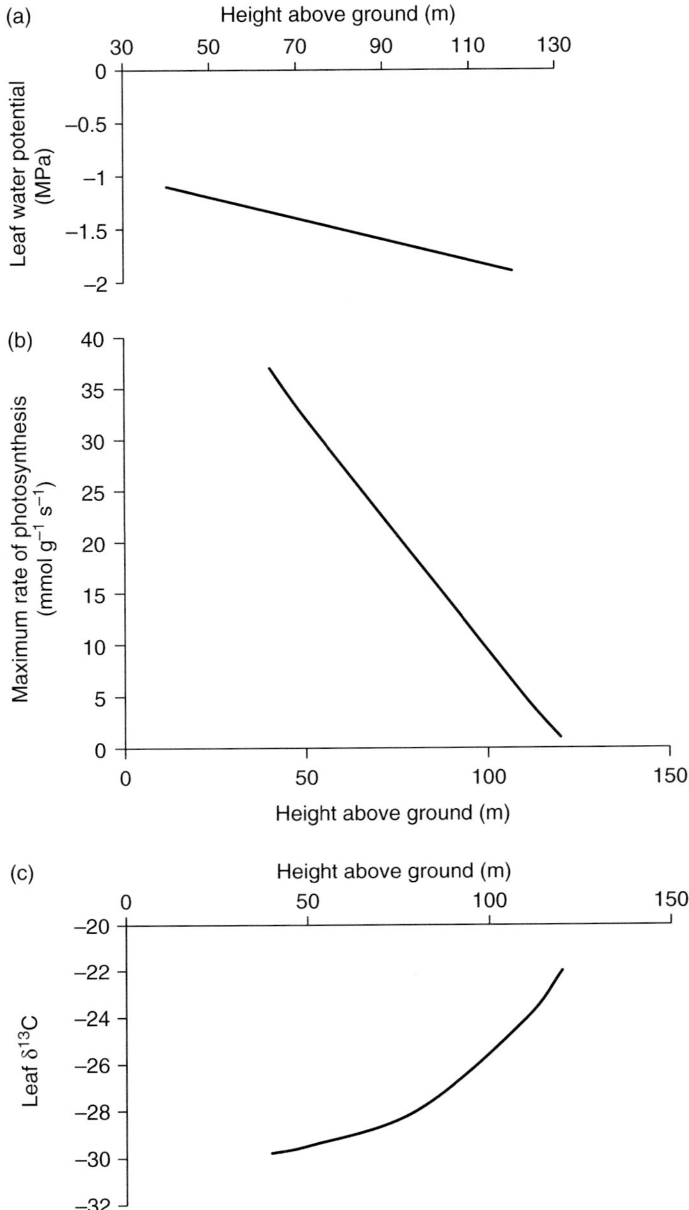

Figure 3.29. The leaf water potential (a) and maximum rate of photosynthesis (b) of leaves decline with increasing height through the canopy of giant redwood trees in California. Isotope discrimination (c) increases with height, indicating an increasing stomatal limitation to photosynthesis. Redrawn from Koch *et al.* (2004).

of tall trees behaves at increasing heights up the tree in a manner that is similar to that observed for plants growing along an aridity gradient.

In this chapter the water relations of cells, leaves and whole plants was discussed. Mechanisms of water uptake in plants and the influence of soil moisture content and vapour pressure deficit on plant water relations in the field were presented. The influence of hydraulic architecture and xylem embolism on plant water relations and correlations of hydraulic attributes and carbon fixation were also discussed. These issues will be recurrent themes within many of the case study chapters presented later.

3.9 References

Blackman CJ, TJ Brodribb and GJ Jordan, (2010). Leaf hydraulic vulnerability is related to conduit dimensions and drought resistance across a diverse range of woody angiosperms. *New Phytologist* 188, 1113–1123.

Bond DJ and MG Ryan, (2000). Comment on "Hydraulic limitation of tree height: a critique. *Functional Ecology* 14, 137–140.

Brodersen CR and AJ Mcelrone (2013). Maintenance of xylem network transport capacity: a review of embolism repair in vascular plants. *Frontiers in Plant Science* 24th April, doi: 10.3389/fpls.2013.00108.

Brodribb TJ, (2009). Xylem hydraulic physiology: The functional backbone of terrestrial plant productivity. *Plant Science* 177, 245–251.

Brodribb TJ and H Cochard, (2009). Hydraulic failure defines the recovery and point of death in water-stressed conifers. *Plant Physiology* 149, 575–584.

Brodribb TJ and NM Holbrook, (2003). Stomatal closure during leaf dehydration, correlation with other leaf physiological traits. *Plant Physiology* 132, 2166–2173.

Brodribb TJ and NM Holbrook, (2004). Stomatal protection against hydraulic failure: a comparison of coexisting ferns and angiosperms. *New Phytologist* 162, 663–670.

Brodribb TJ and NM Holbrook, (2005). Water stress deforms tracheids peripheral to the leaf vein of a tropical conifer. *Plant Physiology* 137, 1139–1146.

Bucci SJ, G Goldstein, FC Meinzer, FG Scholz, AC Franco and M Bustamante, (2004). Functional convergence in hydraulic architecture and water relations of tropical savanna trees: from leaf to whole plant. *Tree Physiology* 24, 891–899.

Bucci SJ, FG Scholz, G Goldstein, WA Hoffmann, FC Meinzer, AC Franco, TW Giambelluca and F Miralles-Wilhelm, (2008). Controls on stand transpiration and soil water utilization along a tree density gradient in a Neotropical savanna. *Agricultural and Forest Meteorology* 148, 839–849.

Clearwater MJ and CJ Clark, (2003). *In vivo* magnetic imaging of xylem vessel contents in woody lianas. *Plant, Cell and Environment* 26, 1205–1214.

Cochard H, R Martin, P Gross and MB Bogeat-Triboulot, (2000). Temperature effects on hydraulic conductance and water relations of *Quercus robur. Journal of Experimental Botany* 51, 1255–1259.

Cochard H and S Delzon, (2013). Hydraulic failure and repair are not routine in trees. *Annals of Forest Science* 70, 659–661.

Daley MJ and NG Phillips, (2006). Interspecific variation in nighttime transpiration and stomatal conductance in a mixed New England deciduous forest. *Tree Physiology* 26, 411–419.

Dawson TE, SSO Burgess, KP Tu, RS Oliveira, LS Santiago, JB Fisher, KA Sominin and AR Ambrose, (2007). Night-time transpiration in woody plants from contrasting ecosystems. *Tree Physiology* 27, 561–575.

Eamus D and AD Narayan, (1990). A pressure-volume analysis of aubergine leaves. *Journal of Experimental Botany* 41, 661–668.

Eamus D, AP O'Grady and L Hutley, (2000). Dry season conditions determine wet season water-use in wet-dry tropical savannas of north Australia. *Tree Physiology* 20, 1219–1226.

Eamus D and H Prichard, (1998). A cost-benefit analyses of 4 evergreen or deciduous species in a north Australian savanna. *Tree Physiology* 18, 537–546.

Eamus D and L Prior, (2001). The ecophysiology of tropical savannas, with special reference to phenology, water relations and photosynthesis. *Advances in Ecological Research* 32, 114–197.

Fisher RA, M Williams, RL Do Vale, AL Da Costa and P Meir, (2006). Evidence from Amazonian forests is consistent with isohydric control of leaf water potential. *Plant, Cell and Environment* 29, 151–165.

Gotsch SG, JS Powers and MT Lerdau, (2010). Leaf traits and water relations of 12 evergreen species in Costa Rican wet and dry forests: patterns and intra-specific variation across forests and seasons. *Plant Ecology* 211, 133–146.

Johnson DM, DR Woodruff, KA McCulloh and FC Meinzer, (2009). Leaf hydraulic conductance measured *in situ* declines and recovers daily: leaf hydraulics, water potential and stomatal conductance in four temperate and three tropical tree species. *Tree Physiology* 29, 879–887.

Koch GW, SC Sillett, GM Jennings and SD Davis, (2004). The limits to tree height. *Nature* 428, 851–854.

Li J, Y-B Gao, Z-R Zheng and ZL Gao, (2008). Water relations, hydraulic conductance and vessel features of three Caragana species in the Inner Mongolia Plateau of China. *Botanical Studies* 49, 127–137.

Macinnis-Ng C, K McClenahan and D Eamus, (2004). Convergence in hydraulic architecture, water relations and primary productivity amongst habitats and across seasons in Sydney. *Functional Plant Biology* 31, 429–439.

Maseda PH and RJ Fernandez, (2006.) Stay wet or else: three ways in which plants can adjust hydraulically to their environment. *Journal of Experimental Botany* 57, 3963–3977.

Maherali H, CF Moura, MC Caldeira, CJ Wilson and RB Jackson, (2006). Functional co-ordination between leaf gas exchange and vulnerability to xylem cavitation in temperate forest trees. *Plant, Cell and Environment* 29, 571–583.

Markesteijn L, L Poorter, H Paz, L Sack and F Bongers, (2011). Ecological differentiation in xylem cavitation resistance is associated with stem and leaf structural traits. *Plant, Cell and Environment* 34, 137–148.

Meinzer FC, BJ Bond, JM Warren and DR Woodruff, (2005). Does water transport scale universally with tree size? *Functional Ecology* 19, 558–565.

Meinzer FC, DM Johnson, B Lachenbruch, KA McCulloh and DR Woodruff, (2009). Xylem hydraulic safety margins in woody plants: coordination of stomatal control of xylem tension with hydraulic capacitance. *Functional Ecology* 23, 922–930.

Mencuccini M and J Grace, (1995). Climate influences the leaf area/sapwood area ratio in Scots pine. *Tree Physiology* 15, 1–10.

Mencuccini M and F Magnani, (2000). Comment on "Hydraulic limitations of tree growth". *Functional Ecology* 14, 135–137.

Novick KA, R Oren, PC Stoy, MBS Siqueira and GC Katul, (2009). Nocturnal evapotranspiration in eddy covariance records from three co-located ecosystems in the Southeastern USA: implication for annual fluxes. *Agricultural and Forest Meteorology* 149, 1491–1504.

O'Grady A, P Cook, D Eamus, A Duguid, J Wischusen, T Fass and D Worldege, (2009). Convergence of tree water use of an arid-zone woodland. *Oecologia* 160, 643–655.

Phillips NG, JD Lewis, BA Logan and D Tissue D, (2010). Inter- and intra-specific variation in nocturnal transport in *Eucalyptus*. *Tree Physiology* 30, 586–596.

Ryan MJ and BJ Yoder, (1997). Hydraulic limits to tree height and tree growth. *Bioscience* 47, 235–242.

Sack L and NM Holbrook, (2006). Leaf hydraulics. *Annual Review of Plant Biology* 57, 361–381.

Sack L, CM Streeter and NM Holbrook, (2004). Hydraulic analysis of water flow through leaves of sugar maple and red oak. *Plant Physiology* 134, 1824–1833.

Schultz HR, (2003). Differences in hydraulic architecture account for near-isohydric and anisohydric behaviour of two field-grown *Vitis* cultivars during drought. *Plant, Cell and Environment* 26, 1393–1405.

Simonin KA, EB Limm and TE Dawson, (2012). Hydraulic conductance of leaves correlates with leaf lifespan: implications for lifetime C gain. *New Phytologist* 193, 939–947.

Stratton LC, G Goldstein and FC Meinzer, (2000). Temporal and spatial partitioning of water resources among eight woody species in a Hawaiian dry forest. *Oecologia* 124, 309–317.

Sun S-J, P Meng, J-S Zhang and X Wan, (2011). Variations in soil water uptake and its effect on plant water status in *Juglands regia* during dry and wet seasons. *Tree Physiology* 31, 1378–1389.

Tardieu F and T Simmonneau, (1998). Variability among species of stomatal control under fluctuating soil water status and evaporative demand: modelling isohydric and anisohydric behaviour. *Journal of Experimental Botany* 49, 419–432.

Tyree MT, A Nardini, S Salleo, L Sack, B El Omari, (2005). The dependence of leaf hydraulic conductance on irradiance during HPFM measurements: any role for stomatal response? *Journal of Experimental Botany* 56, 737–744.

Zeppel M, D Tissue, D Taylor, C Macinnis-Ng and D Eamus, (2010). Rates of nocturnal transpiration in two evergreen temperate woodland species with differing water-use strategies. *Tree Physiology* 30, 988–1000.

Zhang YJ, FC Meinzer, JH Qi, G Goldstein and KF Cao, (2013). Midday stomatal conductance is more related to stem rather than leaf water status in sub-tropical deciduous and evergreen broadleaf trees. *Plant, Cell and Environment* 36, 149–158.

Section Two
Remote Sensing

4

An Overview of Remote Sensing

4.1 Introduction

The surveillance of the Earth and its terrestrial and aquatic ecosystems, land use activities and snow and ice cover constitute the broad framework of 'remote sensing', defined as discerning information about the Earth's surface from afar without direct physical contact. Remote sensing employs non-destructive measurement techniques for monitoring the Earth's structurally and functionally diverse landscapes in a consistent and robust manner.

From early days, aerial photography from balloons and airplanes was widely adopted for its synoptic viewing of vast and often inaccessible areas. This greatly facilitated spatial surveys and monitoring of our ecosystems, soils, hydrologic features and land cover conditions. Today we routinely observe the entire Earth as a planet utilising spacecraft sensors that circle the Earth every 90 minutes and remote sensing technologies have become widely integrated within ecosystem sciences and agriculture and natural resource management communities (Kerr and Ostrevsky 2003).

Satellite data offer unprecedented capabilities to capture spatial and temporal details of dynamic ecosystems, functional processes and characterisation of ecosystem structure and biological properties (Pettorelli *et al.* 2005; Glenn *et al.* 2008). When properly integrated with ecological principles and models, remote sensing is a powerful tool for landscape process studies, ecosystem assessments of health and productivity and vegetation -climate interactions at local-to-global scales. This has enabled new discoveries about our planet and has improved our understanding of its state, health and functioning (e.g. Lefsky *et al.* 2002, Tucker *et al.* 2005; Fig. 4.1).

In this book the principles of remote sensing and their application to ecophysiological and environmental studies are coupled to modelling of landscape-atmosphere interactions within the Earth science community. We address the recent challenges in biosphere science, climate change, diminishing biologic diversity, large-scale droughts and land sustainability. These issues have exerted increasingly complex demands for more quantitative, longer-term and accurate satellite-based remote

Figure 4.1. The Earth as a single integrative ecosystem from space (*courtesy of NASA*).

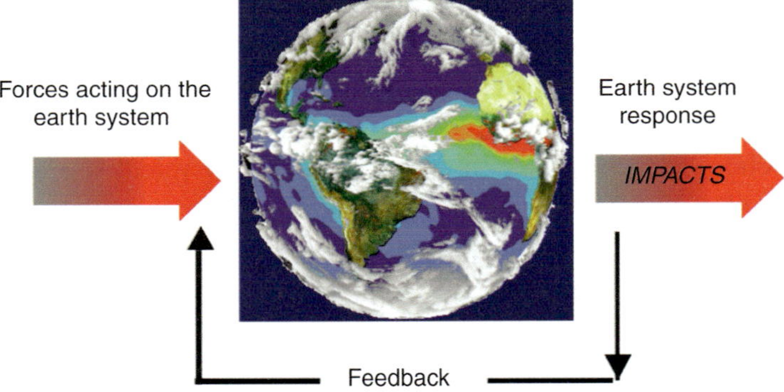

Figure 4.2. Diagram illustrating the concept of forcings, impacts and feedbacks in Earth system science (*Courtesy of NASA*).

sensing data to better diagnose ecosystem health and status and assess how the Earth is changing over time (Fig. 4.2).

4.2 A Framework of Remote Sensing

Remote sensing encompasses a multitude of activities that include the development, operations and calibration of satellite sensor systems; data acquisition and storage; image interpretation and analyses; and visualisation of spatial data. The process of

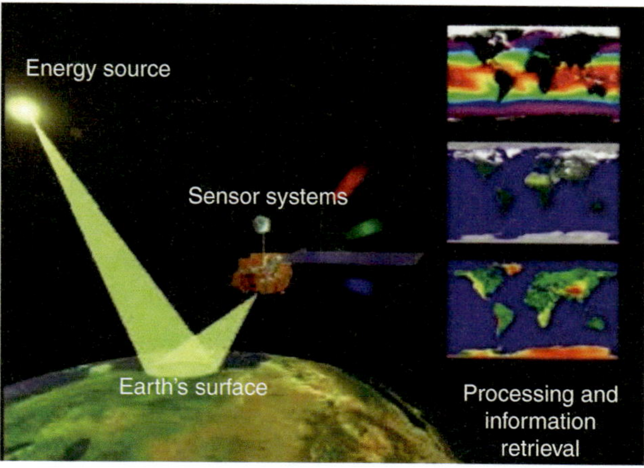

Figure 4.3. Fundamentals of the remote sensing process involving energy source, surface detection, sensor systems and derivation of thematic imagery (*Image courtesy of the NASA Terra Project*).

remote sensing generally includes a source of radiant energy that illuminates and interacts with the spatially and temporally varying Earth's surface; orbiting sensor systems that detect and record the radiant energy signals resulting from surface–light interactions; and the processing, translation, interpretation and visualisation of remotely-sensed data (Fig. 4.3).

Components of the remote sensing process can be further described as:

(i) An energy source, or source of electromagnetic radiation. The most common energy source is the Sun as it illuminates the Earth. This encompasses the realm of passive remote sensing since the (solar) energy is generated naturally. The Earth is also a passive energy source as it naturally emits energy in the thermal and microwave portions of the electromagnetic spectrum. This is in contrast to *active* remote sensing, in which the energy source is artificially generated, as in Radar (*RAdio Detection And Ranging*) or Lidar, (*Light Detection And Ranging*).

(ii) The Earth's surface, which consists of biochemical and structural materials (vegetation, soils, water, rocks, snow) that interact with the radiant energy source and subsequently reflect, absorb and emit portions of this energy back toward the top of the Earth's atmosphere. The relationship between the solar energy incident at the surface and the wavelength dependent spectral composition of the reflected and emitted energy provides a wealth of information about the biogeochemical nature (leaf pigments, canopy chemistry, soil mineralogy), moisture status and the physical characteristics of the surface (leaf area, canopy structure, soil roughness, etc.) (Ustin *et al.* 2004; Baumgardner *et al.* 1985).

(iii) The sensor instruments that measure and record the spectral signals emanating from the Earth's surface. The detected signals may be solar energy that is reflected by the Earth's surface; self-emitted energy from the Earth's surface (for example long wave thermal energy); or the signal measured may have been artificially generated and backscattered following its interactions with the Earth, as with Radar. There are an increasing number of different satellite sensor systems in the optical, thermal and microwave portions of the spectrum. These are utilized to observe and monitor the Earth over a range of spatial, temporal and spectral resolutions.

(iv) The raw digital signals measured and recorded by the sensor are transmitted to various receiving stations, located in various countries, for processing and calibration prior to the distribution of the imagery to the user communities. In some cases, satellite imagery is processed and made available in near real-time for fire detection, floods and other disturbance events.

(v) The user community, including ecologists, earth scientists and land resource managers, who utilise and interpret the remotely-sensed data and images in support of science, management applications and decision support systems.

As remote sensing imagery exists in digital formats, the data are readily processed into thematic–and biophysical satellite image products of interest to the end-user. These products may depict, for example, land cover, vegetation greenness, leaf area index, land surface temperature, albedo, snow cover and facilitate the monitoring and assessment of ecosystems and enable model integration of spatial information (Fig. 4.3). Large quantities of remote sensing data are now generated from numerous sensors orbiting the Earth, resulting in a variety of new ways to study the complex dynamics of the Earth's surface. The effective utilization of these data is challenging and is aided with the use of complex algorithms, high speed computers, spatial analysis techniques and data visualisations.

4.3 Advantages of Remote Sensing

Remote sensing has several advantages that complement environmental monitoring and sampling methods. These strengths include synoptic viewing capabilities of entire landscapes; observations in non-visible portions of the electromagnetic spectrum; measurements at multiple scales using different sensor systems; and repetitive observation capabilities in a consistent manner. These facilitate the creation of long-term, time-series datasets, crucial for quantifying changes within landscape environments and for assessment of climate–and human-induced changes.

4.3.1 Global and Synoptic Coverage of the Earth's surface

Satellite images are able to observe and depict phenomena that would be nearly impossible using ground and aircraft observations. Assembling global databases can be extremely difficult and is often plagued by disparate data sources that must be merged and compiled with different criteria and formats, particularly across political

Figure 4.4. Satellite image of a glacier-capped mountain chain in Tibet about 110 km from Lhasa. This false-colour image was acquired with the Advanced Spaceborne Thermal Emission and Reflection Radiometer (ASTER) on NASA's Terra satellite on January 5, 2007. The Tibetan Plateau contains the largest persistent ice mass outside the Arctic and Antartica (*Courtesy of National Geographic*).

boundaries (state or national). From space-based platforms, however, orbiting satellite sensors can provide complete coverage of the planet with a consistent source of uniformly collected data from the same instrument and platform. A single Landsat image may encompass 34,000 km^2 and a meteorological satellite image from NOAA covers up to 9 million km^2 in one image.

The global coverage afforded by satellites has been invaluable in mapping large-scale land surface transformations and in better understanding the impacts of anthropogenic factors on our environment. This has proven useful in monitoring and understanding the dynamic processes and stresses affecting our environment (Foody and Curran 1994). As examples, the tropical glaciers in the Andes of Peru, the Mt Kilimanjaro ice fields of East Africa and many Tibetan glaciers are melting and shrinking in extent due to global warming, with wide ranging impacts on runoff, flooding and the long term sustainability of seasonal fresh water supplies to millions of people in these regions (Fig. 4.4). The International Geosphere-Biosphere Programme (IGBP) has developed various global databases generated from satellite images and other international agencies, such as the UNEP-GRID program, are attempting to provide global datasets that are consistent and validated over large territories (http:// geodata .grid.unep.ch/).

4.3.2 Multi-Scale Observations

Monitoring and management of the Earth's natural resources require spatial data at various scales in order to incorporate land-use patterns, geomorphology, topography and seasonal hydrologic and vegetation parameters. Remote sensing may be the only feasible means of providing such spatially distributed data at multiple scales and on a consistent and timely basis for environmental studies. Satellite-based sensors have a wide range of spatial resolutions and coverage swaths. Fine resolution imagery with 1 meter or less pixels have very narrow coverage swaths, while coarser resolution sensors (>1 km pixels) generate imagery with much larger swaths that can cover entire continents (Fig. 4.5). Furthermore, the ability to

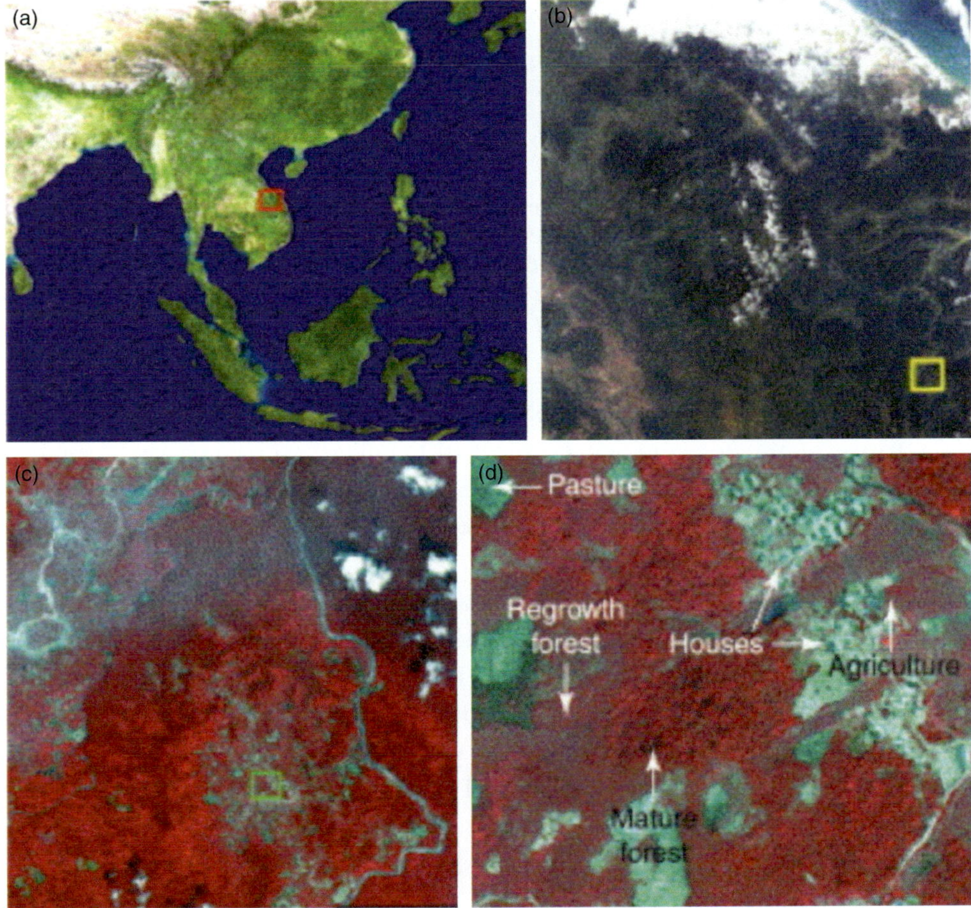

Figure 4.5. Satellite images from multiple sensors at different spatial resolutions. (a) coarse resolution MODIS of Southeast Asia. (b) medium resolution Landsat 7 ETM+ scene of central Vietnam. (c) fine resolution IKONOS false color scene for 109 km² region near Song Thon Dac Pring; (d) IKONOS closeup showing different land-use types. Figure courtesy of AMNH/Ned Gardiner; IKONOS imagery from Space Imaging (*Adapted from Turner et al. 2003*).

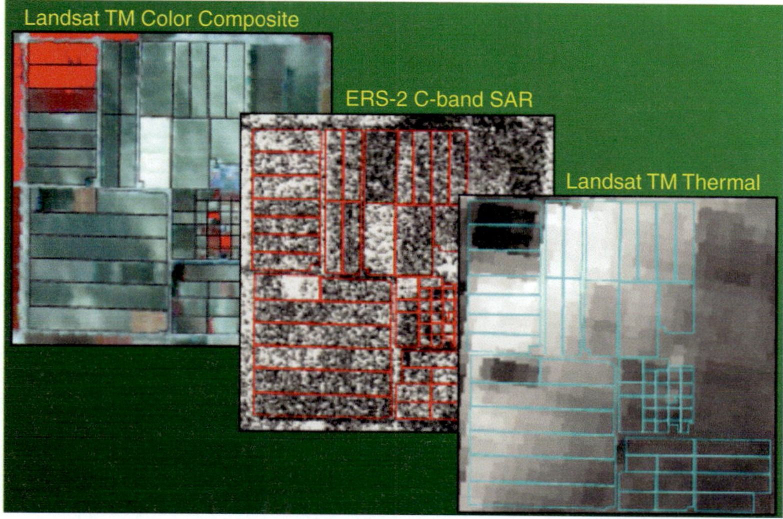

Figure 4.6. Examples and comparisons of remote sensing imagery acquired in various portions of the electromagentic spectrum. Landsat NIR-red-green imagery (left); ERS-2 C-band Radar (middle) and Landsat thermal (right) imagery acquired at the Maricopa Agricultural Center, Arizona in 1997 (*Courtesy of M.S. Moran, USDA*).

spatially extrapolate remotely-sensed observations is of great importance to the study of complex environmental phenomena in which relationships between variables often change with scale.

4.3.3 Observations Over the Non-Visible Regions of the Spectrum

Satellite sensors can sense and acquire image data over the non-visible portions of the electromagnetic spectrum. Sensor detectors and filters in the near-infrared (NIR), shortwave infrared (SWIR), thermal infrared (TIR) and microwave portions of the spectrum yield unique information about the Earth's surface, enabling one to discriminate features and better depict processes (Fig. 4.6). For example, thermal infrared imagery provide information about vegetation water stress symptoms well before it is detectable in the visible spectrum. In the microwave region, remote sensing is transparent to clouds and can sense vegetated canopies to different depths, as well as sense surface soil moisture status (Fig. 4.6).

Advancements in Laser altimetry from airborne sensors provide maps of the three-dimensional structural properties of vegetation canopies, including tree height and leaf cover (Lefsky *et al.* 2002). The fusion of these data with spectral data has been very successful in providing estimates of standing woody biomass, carbon stock assessments of tree stands, forest disturbance and selective logging studies, carbon accumulation through forest regrowth and the spread of invasive species (Asner *et al.* 2008; Tollefson 2009) (Fig. 4.7).

Figure 4.7. Example of airborne Lidar for mapping forest clearing and the 3-dimensional properties of tropical forest canopies in the Peruvian Amazon *(G. Asner, Carnegie Airborne Observatory, adapted from Tollefson 2009).*

4.3.4 Repeat Acquisitions

Remote sensing offer unique capabilities for monitoring the Earth's ecosystems on a timely and consistent basis. The orbital characteristics of most satellite sensors provide repeat observations of the same area of the Earth's surface at regular intervals, which range from as often as 30 minutes to as long as a month. This enables remote sensing methods for multi-temporal studies at daily, weekly, seasonal and inter-annual periods encompassing plant growing seasons, land cover changes and climate variability (Fig. 4.8). Such periodic observations are vital given the highly dynamic nature of the Earth's surface.

Multi-temporal applications of remote sensing include landscape phenology characterization, land-use and land cover change, deforestation assessments and drought and flooding patterns (Liang and Schwartz 2009; Zhang *et al.* 2003). The moderate resolution Landsat and coarser resolution NOAA-Advanced Very High Resolution Radiometer (AVHRR) programs have provided systematic satellite time series observations since the early and late 1970's, respectively, for ecosystem and agricultural productivity assessments, vegetation-climate studies and many other ecological applications.

4.3.5 Real-Time Observations

Real-time satellite transmission of digital imagery to ground receiving stations and end-users has now become feasible and operational. This is highly useful in providing

Figure 4.8. Remote sensing is highly useful in detection and mapping of fires along with burned areas and subsequent ecosystem recovery. The Landsat image at the top depict the extent of the Yellowstone National Park fires that burned approximately 50 percent of the park (>320,000 ha) in 1988, while a 2002 astronaut photograph (bottom) reveals the orange scars of plant-free sand caused by fire in the Simpson Desert of central Australia (*Courtesy of NASA Earth Observatory*).

timely information on the occurrence of fires and other disturbance events. The Moderate Resolution Imaging Spectroradiometer (MODIS) provides a Land Rapid Response system to provide near–real time alerts of active fire information via the internet on a global basis (Fig. 4.9). Applications of MODIS data are discussed extensively throughout the case study chapters in this book.

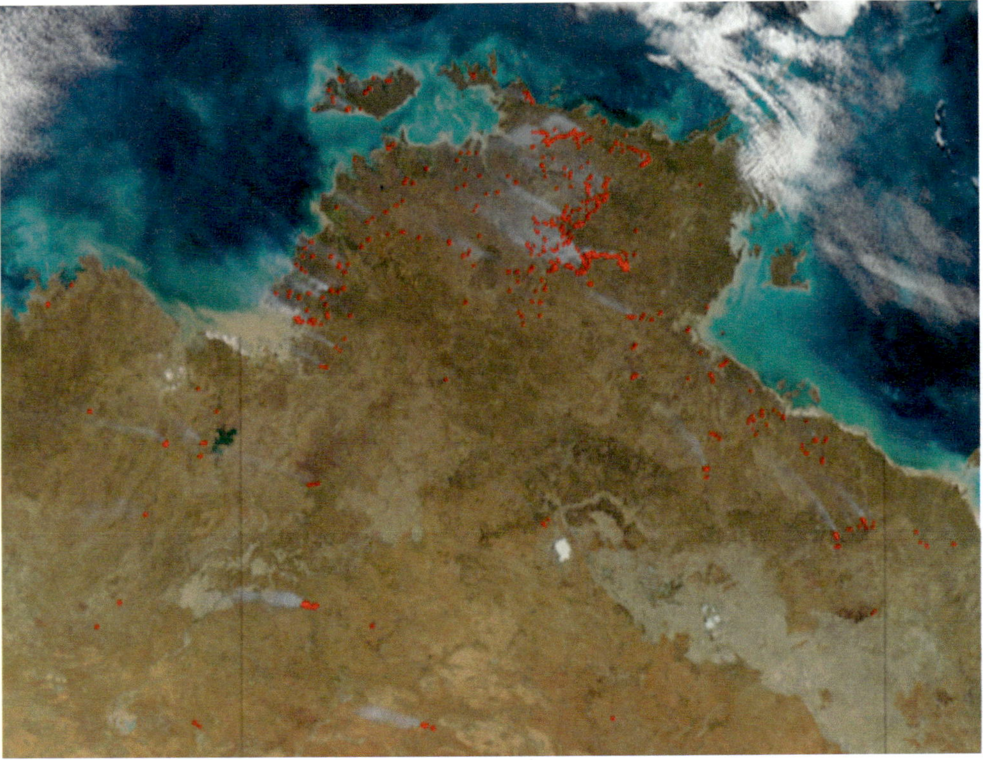

Figure 4.9. Fires in northern Australia for a single day (June 25, 2012). The 1 km resolution pixels that exceed a temperature threshold are labelled red for suspected presence of fires (*Courtesy of NASA*).

4.4 Conclusions

The international scientific community and numerous national space agencies have launched Earth orbiting satellites for environmental monitoring purposes, with remote sensing missions having increased exponentially since the 1980's. This has greatly increased the availability and diversity of remotely-sensed data worldwide. The private sector has also played an important role in promoting and advancing remote sensing capabilities, through the creation of consortiums for their design and development. The foreseeable trend in the growth and development of remote sensing is for imagery to be more widely available to the public and scientists of varying disciplines.

In summary, there are many advantages in using remote sensing data to study and monitor ecological landscapes. However, remote sensing also has significant limitations associated with limited spatial, spectral and temporal capabilities, that often require ground measurements to accurately interpret the optical image values. Remote sensing observations should be viewed as complementary tools in environmental monitoring and vegetation health assessments, to be combined with in situ measurements, e.g. flux tower data and meteorology.

Figure 4.10. Example of a field campaign, known as Bigfoot, in which ground–and tower based sensors were coupled with *in-situ* biophysical measurements and up-scaled to satellite imagery for the purposes of validation (*Courtesy of Bigfoot validation site*).

Remote sensing data also encompasses field, tower and airborne spectrometers, cameras, phenocams, and ground imaging systems. These sensor systems are often used, as 'ground truth measures' in research, development, validation and upscaling/downscaling satellite imagery with canopy biophysical variables and ecophysiological processes (Fig. 4.10). The utility of remote sensing data is demonstrably enhanced when combined with field measurements and other environmental data streams (Running *et al.* 1999; Cohen *et al.* 2003), as shown in the case study chapters of this book.

4.5 References

Asner GP, RF Hughes, PM Vitousek, DE Knapp, T Kennedy-Bowdoin, J Boardman, RE Martin, M Eastwood, RO Green, (2008). Invasive plants transform the three-dimensional structure of rain forests. *Proceedings of the National Academy Sciences* 105 (11), 4519–4523.

Baumgardner MF, LF Silva, LL Biehl and ER Stoner, (1985). Reflectance properties of soils. *Advances in Agronomy* 38, 1–44.

Cohen WB, TK Maiersperger, ZQ Yang, ST Gower, DP Turner, WD Ritts, M Berterretche and SW Running, (2003). Comparisons of land cover and LAI estimates derived from ETM+ and MODIS for four sites in North America: a quality assessment of 2000/2001 provisional MODIS products. *Remote Sensing of Environment* 88, 233–255.

Foody GM and PJ Curran, (1994). Estimation of tropical forest extent and regeneration stage using remotely sensed data. *Journal of Biogeography* 21, 223–244.

Glenn EP, AR Huete, PL Nagler and SG Nelson, (2008). Relationship between remotely-sensed vegetation indices, canopy attributes and plant physiological processes: What vegetation indices can and cannot tell us about the landscape. *Sensors* 8, 2136–2160.

Kerr JT and M Ostrovsky, (2003). From space to species: ecological applications for remote sensing. *Trends in Ecology and Evolution* 18, 299–305.

Lefsky MA, WB Cohen, GG Parker and DJ Harding, (2002). Lidar remote sensing for ecosystem studies. *BioScience* 52, 19–30.

Liang L and M Schwartz, (2009). Landscape phenology: an integrative approach to seasonal vegetation dynamics. *Landscape Ecology* 24, 465–472.

Pettorelli N, JO Vik, A Mysterud, J-M Gaillard CJ Tucker and NC Stenseth, (2005). Using the satellite-derived NDVI to assess ecological responses to environmental change. *Trends in Ecology and Evolution* 20, 503–510.

Running SW, DD Baldocchi, DP Turner, ST Gower, PS Bakwin and KA Hibbard, (1999). A global terrestrial monitoring network integrating tower fluxes, flask sampling, ecosystem modeling and EOS satellite data. *Remote Sensing of Environment* 70, 108–127.

Tollefson J, (2009). Climate: Counting carbon in the Amazon. *Nature* 461, 1048.

Tucker CJ, HE Dregne and WW Newcomb, (2005). Expansion and contraction of the Sahara desert from 1980 to 1990. *Science* 253, 299–300.

Turner W, S Spector, N Gardiner, M Fladeland, E Sterling, M Steininger, (2003). Remote sensing for biodiversity science and conservation. *Trends in Ecology and Evolution* 18(6), 306–314.

Ustin SL, DA Roberts, JA Gamon GP Asner and RO Green, (2004). Using imaging spectroscopy to study ecosystem processes and properties. *BioScience* 54, 523–534.

Zhang X, MA Friedl, CB Schaaf, AH Strahler, JFC Hodges, F Gao. BC Reed and A Huete, (2003). Monitoring vegetation phenology using MODIS. *Remote Sensing of Environment* 84, 471–475.

5

Fundamentals and Physical Principles
of Remote Sensing

An understanding of the physical principles of how electromagnetic energy interacts with the Earth's surface is essential for accurate interpretations and effective use of remote sensing images and datasets. In this chapter we highlight the main physical processes of radiant energy interactions with the Earth's surface as they relate to satellite measurements and the generation of remote sensing imagery for use in ecology and environmental sciences.

5.1 Fundamentals of the Remote Sensing Signal

Remote sensing is defined as the acquisition of information about the biophysical state and condition of the Earth's surface through non-contact, sensor-based observations. The information is transmitted from the surface to the sensor in the form of electromagnetic radiation, providing us the opportunity to detect or *'sense'* this signal and derive information about the health, structure, and condition of objects on the land surface from afar.

The electromagnetic signal may be solar energy that is reflected from the Earth's surface or it may be emitted energy coming from the Earth's surface itself, irrespective of the presence or absence (night time) of sunlight (Fig. 5.1). Thermal and microwave radiation are two examples of emitted energy commonly measured by remote sensing. In certain cases, the signal received at a sensor may be a combination of solar-reflected and self-emitted energy, which may be useful, for example, in the detection of fires when portions of the shortwave infrared are detected. Alternatively, an 'active' sensor system generates the energy that interacts with the Earth's surface, which is then detected by the same sensor system, as in the case of Radar (Radio Detection and Ranging) and Lidar (Light Detection and Ranging) remote sensing (Fig. 5.1).

Regardless of the type of electromagnetic signal, the energy received and detected by remote sensing provides valuable information that can be used to interpret and characterize the physical, chemical, and biologic state and condition of the surface, along with surface processes involving soils, vegetation, water, and land-use

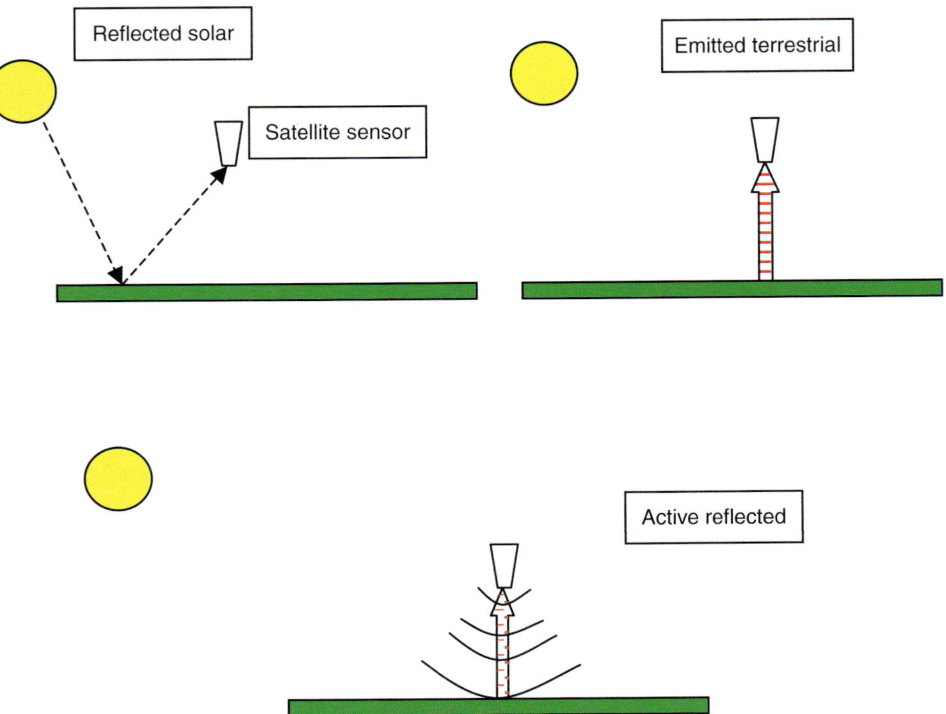

Figure 5.1. Illustration of observations involving solar reflected and emitted radiation and active *versus* passive sensors.

activities. Although, in remote sensing we focus primarily on the measurement and information content of the electromagnetic radiation, this signal may be coupled with other energy terms to yield estimates of evapotranspiration, photosynthesis, and soil moisture content.

An important goal of remote sensing is to understand how electromagnetic energy interacts with the Earth's surface in order to learn how to convert the electromagnetic signal into information for specific applications. In summary, for remote sensing to be useful, one needs (1) a source of electromagnetic energy, (2) sensors capable of detecting the electromagnetic signal emanating from the surface, and (3) trained users that are able to interpret the signal received by the sensors.

5.2 Properties of Electromagnetic Radiation

The physical properties of electromagnetic radiation can be explained by two seemingly contradictory theories, the wave theory of light (Huygens, Maxwell) and the quantum theory of light (Planck, Einstein). In the wave theory, electromagnetic radiation consists of oscillating magnetic and electrostatic fields that are mutually orthogonal to each other and to the direction of propagation (Fig. 5.2).

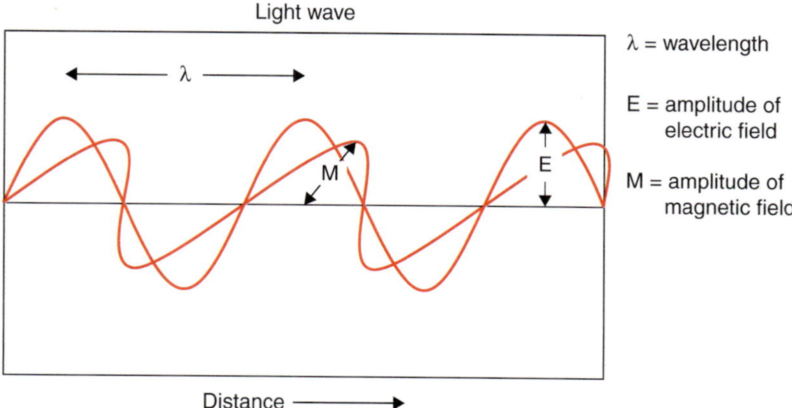

Figure 5.2. The oscillating electric and magnetic components of electromagnetic radiation propagation; the wave theory of light.
Source: http://en.wikipedia.org/wiki/Transverse_wave

Electromagnetic energy is transmitted from one place to another following a harmonic and continuous model with a velocity, $c = 3 \times 10^8$ m s^{-1} (speed of light). The properties of this wave energy can be described according to its wavelength (λ) and frequency (ν), which are related by:

$$c = \lambda \, \nu \qquad (5.1)$$

where c indicates the speed of light, λ expresses the wavelength or distance between two successive peaks (usually in micrometers, 1 μm = 10^{-6} m; or nanometers, 1 nm = 10^{-9} m), and ν is the frequency, or number of cycles that pass over a fixed point per unit of time (in hertz or cycles per second). In the quantum theory of light, radiation is described as a succession of particles or discrete packets of energy known as photons or quanta, with mass equal to zero. This theory allows us to calculate the amount of energy transported by a photon with knowledge of its frequency:

$$Q = h \, \nu \qquad (5.2)$$

where Q is the radiant energy of a photon (in joules, J), ν is the frequency, and h is Planck's constant (6.626×10^{-34} J s). Combining this with Eq. (1) results in,

$$Q = h \, (c \, / \, \lambda) \qquad (5.3)$$

which states that the larger the wavelength, or the shorter the frequency, the lower the energy content and *vice versa*. This demonstrates that it is more difficult to detect long wave radiation (longer wavelengths with lower energy) than shortwave energy, the former requiring more sensitive means of detection to achieve a reasonable signal to noise ratio. In summary, the wave/particle dual concept of light establishes that particles may sometimes behave like waves and waves may some-times behave like particles and that both wave and quantum theories of light are useful in describing light.

5.3 The Electromagnetic Spectrum

The last two equations provide a means to characterise the various types of radiant energy according to its wavelength or frequency. The entire set of wavelengths or frequencies makes up the electromagnetic spectrum (EMS) (Fig. 5.3). The EMS ranges from the shortest wavelengths (gamma rays, X-rays) up to the long wavelengths used in telecommunications (microwaves). Although most spectral regions are referred to in wavelength units (1 μm = 10^{-6} m), microwaves are more commonly expressed in frequency units (in Gigahertz, GHz = 10^9 Hz).

In remote sensing, various discrete spectral regions within the EMS are commonly identified for their specific interactions with components of the Earth's surface. These are as follows:

1. The Ultraviolet (UV) region (0.1 to 0.4 μm). Much of this energy is absorbed by the ozone content in the upper atmosphere, and is mostly unavailable for us to use for land observations. However this spectral region is used in atmosphere studies and include assessments of smokestack emissions.

2. The Visible (VIS) region (0.4 to 0.7 μm). This spectral region coincides with the EMS that our eyes are capable of sensing, i.e., we can *see* in this region. The visible region is only a very small part of the entire spectrum and is further divided into the three primary colours that our eyes can sense, namely the blue (0.4 to 0.5 μm), green (0.5 to 0.6 μm) and red (0.6 to 0.7 μm) spectral intervals. The mixing of these 3 primary colours results in the entire set of colours sensed by the human eye. This spectral region is of special interest in sensing of leaf pigments, surface water quality, and soil minerals.

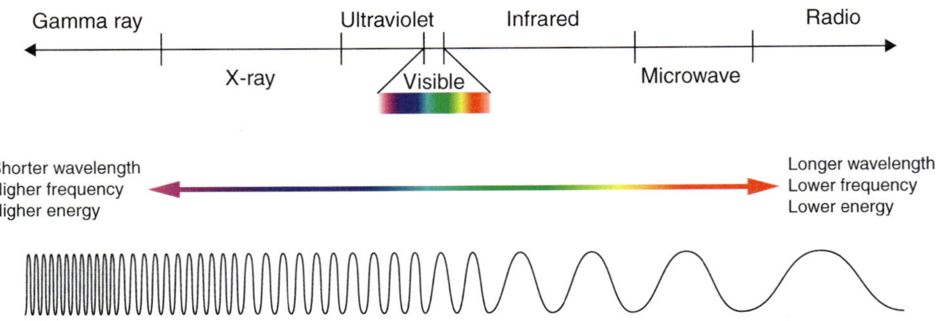

Figure 5.3. The electromagnetic spectrum from gamma rays to radio waves and their spectral energy classifications (ultraviolet, infrared, microwaves). Courtesy of NASA.

3. The Near-infrared region (NIR, 0.7 to 1.3 μm). This portion of the spectrum lies just beyond the human eye's sensing capability, and the NIR is of special interest due to its high sensitivity to leaf structure and morphology.

4. The Shortwave infrared region (SWIR, 1.3 to 8 μm). This spectral region extends beyond the NIR and is further divided into the reflected solar radiation sensed from the Earth's surface, from 1.3 to 3 μm, and surface *emitted* signal from 3 to 8 μm. The first region is very sensitive to the moisture content of vegetation and the upper soil surface. The second region includes both reflection and emission processes during daylight hours and emission signals during the night and is useful for detecting high temperature sources, such as fires.

5. The Thermal infrared region (TIR, 8 to 14 μm). This is the emitted energy from the Earth's surface that is often used to map surface temperatures. The thermal region has been shown useful in detecting vegetation stress, soil moisture, clouds, minerals, and in assessments of environmental contamination.

6. The Microwave region (>1 mm). This region of very long wavelengths is transparent to cloud cover and enables sensing through clouds. Microwaves can also penetrate forest canopies to various depths and are very useful in analyses of soil surface moisture and roughness, as well as plant canopy moisture and roughness.

The electromagnetic energy signals received by a sensor across these different spectral regions will vary uniquely with land cover conditions and with the biophysical and biochemical properties of the surface.

5.4 Basic Energy Concepts

There are various energy terms, formulas, and units commonly used in remote sensing to characterize the incoming and outgoing electromagnetic energy as they relate to the spectral composition, magnitude, and directional properties of the energy sensed (Slater 1980). These are summarised in Table 5.1 and described as follows:

- Albedo (A). This unitless energy measure is the ratio of Exitance (M) to Irradiance (E) over the entire solar reflective, or shortwave range at a given surface. Albedo = M / E, is a fundamental variable in energy balance studies, climate modelling and soil degradation studies (Gutman *et al.* 1989; Henderson-Sellers and Wilson 1982; Schaaf *et al.* 2002).

- Radiance (L). This is the energy exitance in a given direction, per unit area and solid angle of measurement. It is the most fundamental term in remote sensing since it describes exactly what the sensor measures, expressed in watts per square meter per steradian ($W\ m^{-2}\ sr^{-1}$) (Fig 5.5). Satellite and airborne sensors as well as field spectroradiometers, are unable to measure the outgoing exitance, or hemispherical reflected energy, but instead measure a much more restrictive directional radiance (L_λ) over a narrow angular field-of-view.

Table 5.1 Basic energy terms and units used in remote sensing to physically characterise radiant energy transfer among sensor, surface, and energy source

- Radiant energy (Q), measured in joules (J), is the most basic energy unit and refers to the total energy radiated in all directions away or toward a surface.
- Radiant flux (ϕ), measured in watts (W), represents the rate of energy transfer per unit of time ($J\ s^{-1} = W$).
- Radiant flux density is the rate of energy transfer per unit area measured in watts per square meter ($W\ m^{-2}$). This is commonly used to describe the energy arriving at the surface or leaving the surface, as follows (Fig. 5.4).
- Exitance (M). This is the radiant flux density leaving away from the surface in all directions per unit area and per unit time ($W\ m^{-2}$).
- Irradiance (E). This is the radiant flux density incident upon the surface per unit area and per unit time.
- Radiance (L). This is the radiant flux density leaving a surface in a specific direction, as for example toward the sensor overhead, and is measured in watts per square meter per steradian ($W\ m^{-2}\ sr^{-1}$).

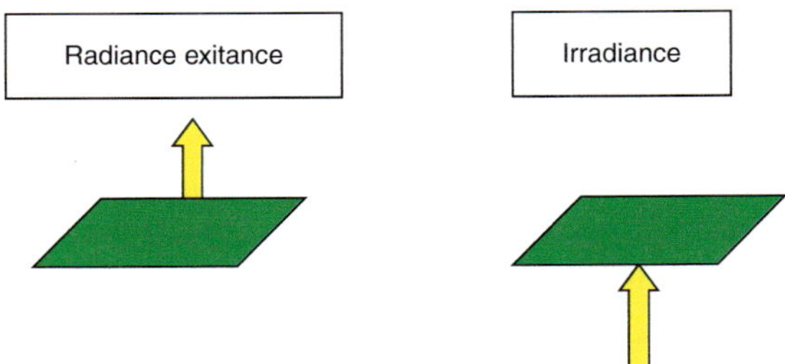

Figure 5.4. Concept of radiant exitance from the surface and irradiance energy onto a surface.

The spectral directional radiance (L_λ) is related to the hemispherical spectral exitance (M_λ) as follows (Slater 1980):

$$M_\lambda = \pi\ L_\lambda \tag{5.4}$$

Three processes may occur when the solar irradiance is incident upon and interacts with the surface (E or ϕ_i). This energy may be either reflected (ϕ_r), transmitted (ϕ_t), or absorbed (ϕ_a) according to:

$$\phi_i = \phi_r + \phi_a + \phi_t \tag{5.5}$$

with units of $W\ m^{-2}$. Often, these are expressed in relative terms by dividing each of the energy quantities by the incident energy, ϕ_i, yielding the terms,

- Reflectance (ρ). This is the relationship between the energy reflected by a surface to the energy incident upon that surface, $\rho = \phi_r\ /\ \phi_i$.
- Absorptance (α). This is the relationship between the energy absorbed by the surface to the energy incident upon that surface, $\alpha = \phi_a\ /\ \phi_i$.
- Transmittance (τ). This is the relationship between the energy transmitted by a surface to the energy incident upon that surface, $\tau = \phi_t\ /\ \phi_i$.

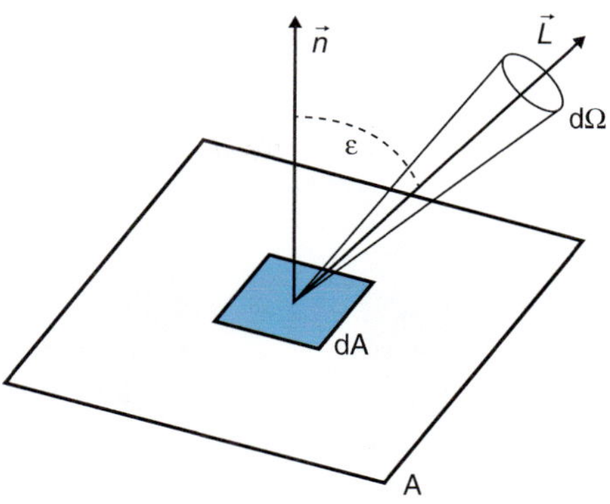

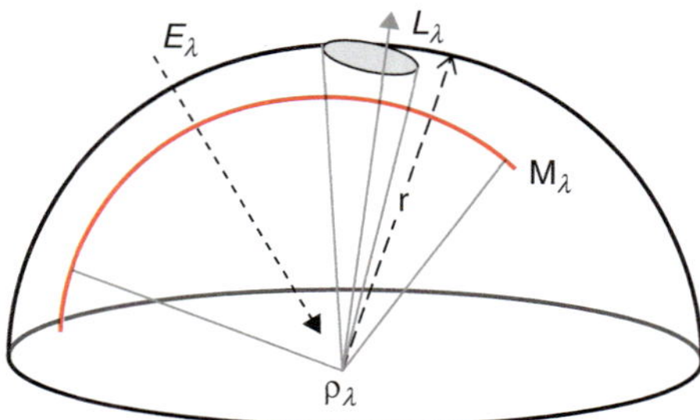

Figure 5.5. Concept of radiance leaving from the surface (top) and the relationship of radiance (L) with radiant exitance (M), reflectance (ρ), and irradiance (E) (bottom). Upper figure from: http://www.ptb.de/cms/en/fachabteilungen/abt4/fb-42/ag-42400/terms.html. Lower figure adapted from Slater 1980.

Equation 5.5 is now written as:

$$1 = \rho + \alpha + \tau \tag{5.6}$$

in which the sum of the reflectance ρ, absorptance α, and transmittance τ, is equal to one, in accordance with the law of conservation of energy.

5.5 Defining Spectral Units

The above energy terms are also expressed on a per unit wavelength basis with the λ prefix, such as spectral radiance, L_λ, which refers to the energy output from a unit area, unit solid angle, and unit wavelength, $L_\lambda = W\ m^{-2}\ sr^{-1}\ \lambda^{-1}$. Similarly, spectral irradiance, E_λ, refers to the energy incident upon a unit area of surface per unit wavelength, as $E_\lambda = W\ m^{-2}\ \lambda^{-1}$.

Spectral albedo refers to the spectral exitance divided by the irradiance for a specific wavelength or spectral interval as:

$$\text{Spectral Albedo} = M_\lambda / E_\lambda, \tag{5.7}$$

The proportions of incident energy that is reflected, absorbed, and transmitted are a function of the unique characteristics of the surface and vary with wavelength. For any given surface, the magnitudes of reflectance, absorptance, and transmittance are not constant, and will also vary with wavelength. It is therefore more appropriate to express Eq. 5.6 as:

$$1 = \rho_\lambda + \tau_\lambda + \tau_\lambda \tag{5.8}$$

These distinct variations in reflectance, transmittance, and absorptance with wavelength tell us much about the state and composition of the surface. For example, a leaf will appear green if its reflectance at green wavelengths is greater than its reflectance in the blue or red portions of the visible spectrum. A water body will appear blue if it reflects more energy in the blue region relative to the other wavelengths within the visible spectral region. Hence, the variation in reflectance behaviour of an object over the visible wavelengths results in what is called "colour".

A spectral reflectance signature refers to a plot of reflectances over a series of wavelengths, and one can also plot spectral transmittance and spectral absorptance spectral signatures (Fig. 5.6).

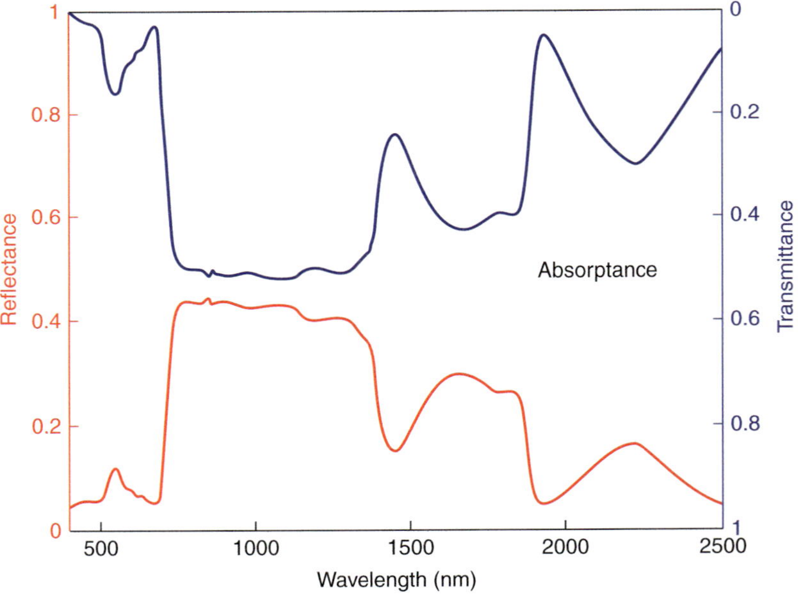

Figure 5.6. Example of a spectral reflectance, spectral absorptance, and spectral transmittance signature of a green leaf. The sum of the three terms equals one, or 100%. Reflectance (red) and transmittance (blue) spectra of a fresh Carolina poplar (*Populus canadensis*) leaf. Reproduced from http://www.photobiology.info/Jacq_Ustin.html

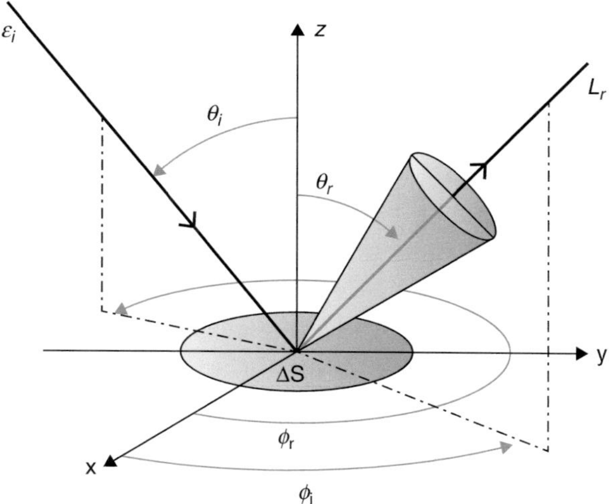

Figure 5.7. Definition of Solar Zenith and Azimuth (θ_i, ϕ_i) and View Zenith and Azimuth (θ_r, ϕ_r). Reproduced from: http://spie.org/Images/Graphics/Newsroom/Imported-2012/004047/004047_10_fig1.jpg

5.6 Defining Directional Quantities

It is important to note that satellite and airborne sensors as well as field spectroradiometers, are unable to measure the integrative, or hemispherical outgoing energy that is reflected and scattered in all directions. Rather, they measure a much more restrictive directional radiance (L_λ) leaving the surface towards the field-of-view of the sensor over a narrow angular field-of-view (Fig. 5.5). A sensor, hence will only sample a portion of the reflected energy and the surface reflectance values derived are more appropriately termed *directional* reflectances (or *bidirectional* reflectances) in that they pertain only to a specific measurement geometry of the satellite sensor and sun relative to the surface.

The amount and composition of reflected energy reaching a sensor is dependent on the geometric conditions of an observation, in particular, the angle of solar incidence, the angle of observation, and their relative azimuthal orientation (Fig. 5.7). Variations in observation geometry result in different radiance values at the sensor as well as computed surface reflectance values for a given surface condition. Seasonal and latitudinal variations in reflectance will be partly a function of the changing illumination conditions (sun angle and azimuth) with latitude that further vary seasonally over the year for a given location.

This relationship between the signals received by a sensor ($L_{sen,\lambda}$) and the surface spectral reflectance (ρ_λ) becomes:

$$\rho_\lambda = \pi\, L_\lambda\, /\, E_{i,\lambda} \tag{5.9}$$

where $E_{i,\lambda}$ is the solar irradiance arriving at the surface, composed of both direct sun irradiance and diffuse (sky) irradiance.

In computing surface reflectances using Eq. 5.9, the surface is often assumed to behave as a perfectly diffuse surface scatterer, known as a *Lambertian* surface, in which the reflected radiance, L_λ, is equal in all directions (isotropic) and largely independent of sensor viewing and illumination angles. However, this assumption is reasonable only as a first approximation in the derivation of reflectances, as most land surfaces scatter incident radiation anisotropically, a consequence of its three dimensional structure and the resulting scattering behaviour. This anisotropy is linked to, among many factors, the roughness of the surface and the way the surface is shadowed and illuminated by the sun.

A complete characterisation of the angular and directional dependencies of reflected energy is described by the Bidirectional Reflectance Distribution Function (BRDF), which provides the reflected radiance of a target as a function of illumination geometry and viewing geometry (Walthall *et al.* 1993; Schaaf *et al.* 2002):

$$\text{BRDF} = L\ (\theta_i, \phi_i, \theta_r, \phi_r, \lambda)\ /\ E\ (\theta_i, \phi_i, \theta_r, \phi_r, \lambda), \qquad (5.10)$$

with units of sr^{-1}; θ_i and ϕ_i refer to the solar zenith angle and solar azimuth, respectively, while θ_r and ϕ_r are the sensor view zenith and view azimuth angles, respectively (Fig. 5.5). This is an intrinsic property governing the reflectance behaviour of a surface and is a complex function of vegetation canopy structure and density, canopy shadow casting and mutual shading, soil surface roughness, and soil and vegetation optical properties (Kimes and Deering 1992; Deering *et al.* 1999; Holben and Fraser 1993).

The λ parameter is included to indicate that BRDF is also wavelength dependent, indicating that differences in the geometry of the observation will not have the same effect in all spectral wavelengths, and tends to be more pronounced at shorter wavelengths. When integrated across a hemisphere, the BRDF provides "albedo", the ratio of shortwave (0.4 μm to 4 μm) radiant energy scattered in all directions to the downwelling irradiance incident on the surface.

Albedo is a fundamental variable in energy balance studies, climate modelling and soil degradation studies (Dickinson *et al.* 1990; Ranson *et al.* 1991; Pinker *et al.* 2000). Often, the BRDF is defined simply as the bidirectional reflectance factor (BRF = πBRDF) at a multitude of view zenith and azimuthal angles for a given sun position (Walthall *et al.* 1985).

Figures 5.8 and 5.9 show examples of angular variations and anisotropies over bare soil and a black spruce forest from a constant view and sun zenith angles but from two different azimuthal angles, 180^0 apart. The conifer and soil surfaces are significantly brighter (more reflective) when viewed with the sun behind the sensor while the conifer/soil surface becomes much darker when viewed with the sun directly in front of the sensor. These two azimuthal directions are commonly labelled,

Bare soil along the Principal Plane

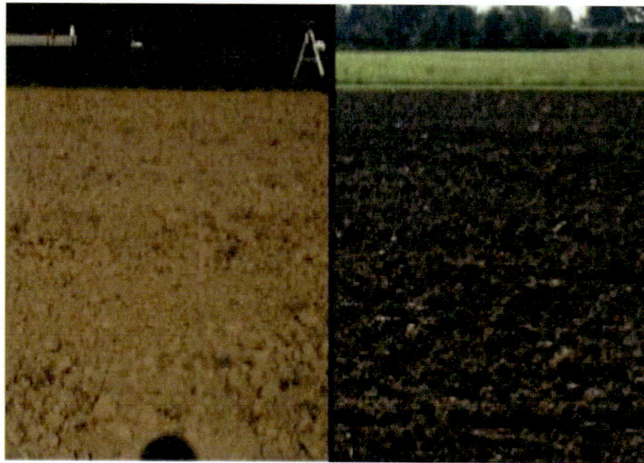

Figure 5.8. BRDF example of viewing a bare soil field at a constant view and sun zenith angles, but opposite azimuthal angles, that is, 180^0 apart, across the principal plane. Left: backscattering (sun behind observer) and most shadows are hidden. Right: forward scattering (sun opposite observer) with shadows dominant. Photograph by Don Deering.

Conifer Forest along the Principal Plane

Figure 5.9. Example of BRDF in a black spruce forest in Canada. Left: backscattering (sun behind observer), note the bright region (hotspot) where all shadows are hidden. Right: forward scattering (sun opposite observer), note the shadowed centres of trees and transmission of light through the edges of the canopies. Photograph by Don Deering.

the backscatter and forward scatter directions, respectively. When the view azimuths are along the same plane as the sun, we term this the "principal plane", that is, the relative difference in azimuth between sun and sensor, $\Delta\phi$ is 0 or 180^0 (Fig. 5.7). In the case of the orthogonal plane, the relative difference in azimuth between sun and sensor, $\Delta\phi$ is 90^0 or 270^0. One important feature of the principal plane is the case in which the sun and sensor are aligned such that their relative azimuth angle is zero degrees and the view angle is the same as the sun angle. This condition is known as the 'hot spot' and generally results in the brightest signal returned to the satellite sensor.

5.7 Introduction to Thermal Measurements

There are a set of physical laws that govern the behaviour and characteristics of electromagnetic radiation. From Eq. 5.3 we see that the energy content of electromagnetic radiation varies inversely with wavelength. The spectral distribution of electromagnetic radiation emitted by a blackbody (a perfect emitter) can be characterized through Planck's radiation Law:

$$M_{n,\lambda} = \frac{2\pi hc^2}{\lambda^5 \left\{ \exp\left(\frac{hc}{\lambda kT} \right) - 1 \right\}} \tag{5.11}$$

where $M_{n,\lambda}$ ($Wm^{-2}\mu m^{-1}$) indicates the spectral radiant exitance, at a certain wavelength (λ), emitted from a blackbody with absolute temperature, T (K, Kelvin degrees); h is Planck's constant (6.626×10^{-34} W s^2); k is Boltzmann's constant (1.38×10^{-23} W s K^{-1}); c is the speed of the light; and λ is the wavelength in meters. This equation can be simplified by replacing some of the terms with constants:

$$M_{n,\lambda} = \frac{c_1}{\lambda^5 \left\{ \exp\left(\frac{c_2}{\lambda T} \right) - 1 \right\}} \tag{5.12}$$

where λ is the wavelength, now in μm; c_1 becomes 3.741×10^8 W m^{-2} μm^4 and c_2 is 1.438×10^4 μm K.

This equation describes the spectral radiant exitance distribution of a blackbody at a certain temperature as a smooth curve with a single maximum. Planck's equation indicates that any object warmer than absolute zero ($-273°C$) emits radiant energy and that the energy increases in proportion to its temperature. With increasing temperature, an object will radiate with more intensity and the single maximum shifts toward shorter wavelengths. These curves are useful in approximating the spectral radiant exitance distribution of natural objects.

Integrating the spectral radiant exitance curve of a blackbody over all wavelengths results in the total radiant energy exitance per unit surface area. This is described by the Stefan-Boltzmann law:

$$M_n = \sigma T^4 \tag{5.13}$$

where σ is the Stefan-Boltzmann's constant (5.67×10^{-8} W m^{-2} K^{-4}) and T is the temperature in Kelvin. This law tells us that the overall radiant exitance of an object is a function of its temperature and that small changes in temperature result in large changes, to the 4th power, in radiant exitance.

Kirchoff's law enables us to extend the above relationships describing blackbody emission behaviour to naturally emitting surfaces through an emissivity correction:

$$M_\lambda = M_{n,\lambda}\, \varepsilon_\lambda \tag{5.14}$$

$$M_\lambda = \varepsilon_\lambda\, \sigma\, T^4 \tag{5.15}$$

where emissivity, ε, is defined as the ratio of the radiant exitance of an object to that of a blackbody at the same temperature. A blackbody is a perfect emitter (emissivity = 1) in that it absorbs and emits all the energy it receives. When an object does not absorb any of the incident energy, it is called a white body because it completely reflects all energy received (emissivity = 0). Grey bodies absorb and emit a certain amount of energy equally at all wavelengths. Most objects in nature have emissivity values that vary with wavelength and are referred to as selective radiators. This equation relates and provides a method to convert thermal radiation, M_λ, measured by sensors to kinetic or surface temperatures, T. The variation of emissivity values with wavelength also provides a mechanism for the discrimination of surface materials in the thermal infrared.

The single maximum or wavelength of peak spectral radiant exitance of a blackbody may be derived from the zero slope point of the derivative of Planck's radiation law, which is described by Wien's displacement law:

$$\lambda_{max} = (2898\ \mu m\ K)/T \tag{5.16}$$

with the temperature (T) in Kelvin. Wien's displacement law states that the higher the temperature of an emitting object, the shorter the wavelengths at which the object will radiate. An example of the use of Wien's law is in the determination of the best portions of the electromagnetic spectrum for the detection of forest fires. With combustion temperatures between 540 and 700 K, the maximum energy output from forest fires will occur at wavelengths between 5.28 and 4.30 μm (shortwave-infrared region).

In summary, the amount and spectral distribution of energy radiated by an object vary according to (1) the temperature of the object and (2) the nature of the material, as characterized by its emissivity. Thus, from the above equations, we can estimate the total radiant exitance and spectral distribution of exitance of an object by knowing its kinetic temperature and emissivity. As with blackbodies, the energy emitted from a natural object is primarily a function of its temperature, but with important modifications dependent on its emissivity. Lastly, knowing the temperature and emissivity of various surface features, allows us to determine the most sensitive portions of the spectrum for its discrimination and investigation.

5.8. The Role and Influence of the Atmosphere

The solar radiant exitance that illuminates the Earth's surface is primarily limited to the 0.3–2.5 μm portion of the spectrum. The solar portion of the spectrum refers to the wavelengths that are directly dependent on solar energy and in which reflection processes are dominant over emission ones. Not all solar radiation incident at the top of the Earth's atmosphere (top-of-atmosphere, TOA) arrives at the land surface. Between the sun and land surface as well as between the surface and the satellite sensor, there is an intervening atmosphere through which the solar irradiance and the surface reflected radiance must pass. The atmosphere consists of various gases and aerosols that selectively absorb and scatter solar radiation over different parts of the spectrum, thereby altering remotely-sensed measurements over such spectral regions. Thus, the atmosphere determines, in part, which regions of the spectrum are suitable for satellite observations of the Earth's surface.

The atmosphere influences and alters the radiant flux in different ways, but primarily through scattering and absorption processes. These processes alter the spectral irradiance at the Earth's surface and the resulting spectral reflectance signatures (Kaufman 1987; Vermote *et al.* 2002) and are further discussed in Sections 5.8.1 and 5.8.2. The scattering of solar radiation by the atmosphere illuminates the sky and a part of this scattered radiation will also illuminate the Earth's surface as diffuse or sky irradiance. Since atmospheric constituents and aerosols are variable in time and space, it is not necessarily straightforward to quantify the overall influence of the atmosphere on the irradiance reaching the Earth's surface as well as radiance reflected from the Earth's surface which has to pass through the atmosphere once again before reaching the space-borne sensor.

5.8.1 Atmospheric Absorptions

Atmospheric absorption processes in the optical region are related to 7 main gases present in the atmosphere, namely carbon dioxide (CO_2), oxygen (O_2), ozone (O_3), nitrous oxide (N_2O), carbon monoxide (CO), methane (CH_4), and water vapour (H_2O) (Fig. 5.10). These gases affect approximately half of the 0.4 to 2.5 μm, optical region.

Among these components, carbon dioxide, ozone, and water vapour are the main gases influencing the interaction of electromagnetic energy with the atmosphere. The main gases and absorption regions may be summarised as:

- Diatomic oxygen (O_2), which filters the ultraviolet radiation below 0.1 μm, as well as small portions in the thermal infrared.

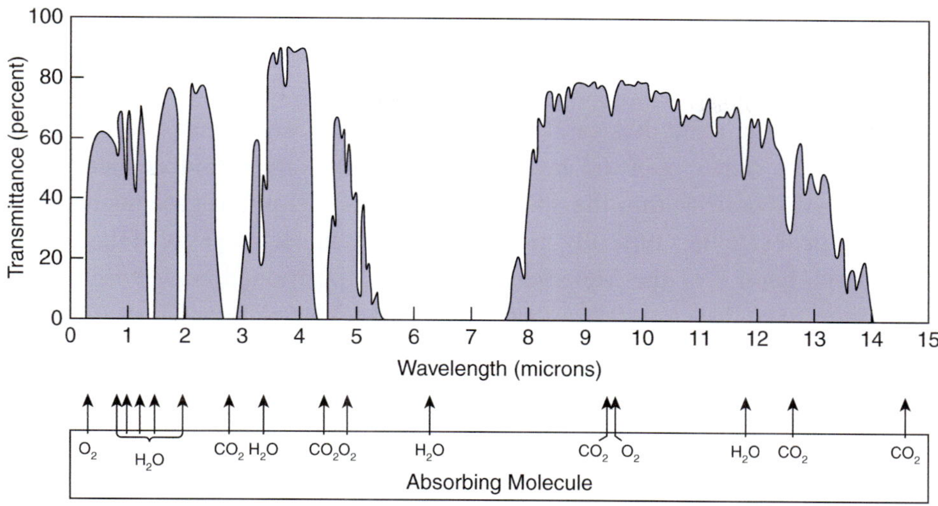

Figure 5.10. Atmosphere transmittance spectra and atmosphere "windows". From: http://en.wikipedia.org/wiki/Transmittance#/media/File:Atmospheric.transmittance .IR.jpg

- Ozone (O_3), which absorbs ultraviolet energy below 0.3 μm as well as in a portion of the microwave region at around 27 μm.
- Water vapour (H_2O) with strong absorptions at 1.45 μm, 1.95 μm, and 6 μm with minor absorptions in the near-infrared.
- Carbon dioxide (CO_2), which absorbs in the thermal infrared (15 μm) and with important absorption effects in the shortwave infrared, between 2.5 and 4.5 μm.

A consequence of these gaseous absorption processes is that we cannot remotely sense the Earth's surface in certain portions of the spectrum. The strong water absorption bands at 1.45 and 1.95 μm completely attenuate solar energy and render these wavelengths impractical for the study of the land surface from space. Ultraviolet sensing of the land surface also cannot occur due to strong absorption by ozone (O_3). Our observing capabilities become limited to certain spectral regions known as "atmospheric windows". Atmospheric windows are regions of the spectrum in which atmospheric transmission is sufficiently high that a significant amount of solar radiation is able to reach the terrestrial surface and be reflected towards the sensor (Fig. 5.10). The main atmospheric windows are the following:

- the visible spectrum and near infrared, located between 0.3 and 1.35 μm;
- several regions in the shortwave infrared from 1.5 to 1.8 μm, 2.0 to 2.4 μm, 2.9 to 4.2 μm, and 4.5 to 5.5 μm;
- the thermal infrared, between 8 and 14 μm, and
- the microwave region beyond 20 mm, where the atmosphere is practically transparent.

Satellite sensor systems designed for land surface applications will have spectral bandpasses that lie within these atmospheric windows in order to minimize sources of atmosphere contamination. Clouds are the exception in that they absorb throughout the optical spectrum and thus cannot be avoided. If we were interested in observing the atmosphere rather than the Earth's surface, then one would design the spectral bandpasses to lie within the spectral regions of maximum gaseous absorption. Meteorological satellites typically incorporate bands in such regions of the spectrum, for example, band 2 of the Meteosat satellite was positioned between 5.7 and 7.1 µm, in order to study the water vapour content in the atmosphere. The Global Ozone Monitoring Experiment (GOME) sensor on board the European satellite, ERS-2, and the Total Ozone Mapping Spectrometer (TOMS) have several bands in the ultraviolet for the purpose of monitoring the ozone layer.

5.8.2 Atmospheric Scattering

In general, atmospheric gas absorption effects are best minimized through placement of spectral bandpasses where atmospheric transmittance is high. Atmospheric scattering, on the other hand, is more complex in that it occurs in all optical imagery acquired with remote sensing, regardless of bandpasses. The scattering of electromagnetic radiation is caused by its interaction with gaseous molecules and atmospheric particles in suspension, namely aerosols and water vapour. An aerosol is a dispersion of solid particles (smoke, smog and dust) and/or liquid particles (haze and fog) suspended in air and include continental dust, oceanic salt spray, industrial activities, and biomass burning. Aerosol size distributions vary markedly with cloud droplets generally >6 µm in size; dust particles, 1–3 µm and; smoke and anthropogenic aerosol particles are submicron (0.2–0.4 µm).

Atmospheric particles scatter solar radiation in all directions creating an additional, diffuse irradiance component reaching the Earth's surface. As a result of scattering, the irradiance arriving at the surface has a direct (solar beam) and diffuse (sky) component. The diffuse component illuminates shadows and sunlit surfaces. On clear, dry days most of the irradiance is direct from the sun, with diffuse (sky) radiation making up as much as 10–15 percent of the sunlight. The more hazy (or turbid) the atmosphere, the stronger the diffuse, or sky, component and lower the direct, solar beam irradiance at the Earth's surface. Scattering within the atmosphere also creates an upward 'path radiance' that augments the signal received at the sensor.

Aerosol amounts and composition are highly variable in time and space, making it difficult to quantify this influence in satellite remote sensing images (Kaufman and Tanré 1996). The atmosphere simultaneously reduces a surface reflected signal through a wavelength dependent transmission (attenuation) function, and adds its own atmospheric signal, known as the 'upward path' (sky) radiance. Depending on the wavelength and the 'brightness' of the surface, as well as the turbidity of the atmosphere, the signal received at the sensor may be lower, higher, or unchanged relative to the ground signal. For a dark surface, atmospheric attenuation will be minimal but the path radiance contribution may far exceed the ground signal, particularly at shorter wavelengths. In addition there is an adjacency effect whereby the surface leaving radiance from an adjacent pixel may be viewed within the pixel being observed. Atmospheric corrections are particularly relevant in multi-temporal remote sensing studies, areas of biomass burning, and when remotely-sensed data are being used to derive important surface biophysical parameters.

5.9 References

Deering, DW, TF Eck and D Banerjee, (1999). Characterization of the reflectance anisotropy of three boreal forest canopies in spring-summer. *Remote Sensing of Environment* 67(2), 205–229.

Dickinson RE, B Pinty, MM Verstaete, (1990). Relating surface albedos in GCM to remotely sensed data. *Agricultural and Forest Meteorology* 52(1–2), 109–131.

Gutman G, G Ohring, D Tarpley and R Ambriozak, (1989). Albedo of the U.S. Great Plains as determined from NOAA-9 AVHRR data. *Journal of Climate* 2, 608–617.

Henderson-Sellers A, and MF Wilson, (1982). High resolution planetary albedos–values and variability. *Remote Sensing of Environment* 12, 479–484

Holben B and RS Fraser, (1993). Red and near-infrared sensor response to off-nadir viewing. *International Journal of Remote Sensing* 5, 145–160.

Kaufman YJ, (1987). Atmospheric effect on spectral signature measurements. *Advances in Space Research* 7(11), 203–206.

Kaufman YJ and D Tanré, (1996). Strategy for direct and indirect methods for correcting the aerosol effect on remote sensing: From AVHRR to EOS-MODIS. *Remote Sensing of Environment* 55(1), 65–79.

Kimes DS and DW Deering, (1992). Remote sensing of surface hemispherical reflectance (albedo) using pointable multispectral imaging spectroradiometers. *Remote Sensing of Environment* 39(2), 85–94.

Pinker RT, I Laszlo, D Goodrich and G Pandithurai, (2000). Satellite estimates of surface radiative fluxes for the extended San Pedro Basin: sensitivity to aerosols. *Agricultural and Forest Meteorology* 105(1–3), 43–54.

Ranson KJ, JR Irons and CST Daughtry, (1991). Surface albedo from bidirectional reflectance. *Remote Sensing of Environment* 35(2–3), 201–211.

Schaaf CB, F Gao, AH Strahler, W Lucht and X Li, (2002). First operational BRDF, albedo nadir reflectance products from MODIS. *Remote Sensing of Environment* 83(1–2), 135–148.

Slater PN, 1980. *Remote Sensing, Optics and Optical Systems*. Reading, MA: Addison-Wesley Pub. Co.

Vermote EF, NZ El Saleous CO Justice, YJ Kaufman, JL Privette. JL Remer and L Roger and D Tanré, (2002). Atmospheric correction of MODIS data in the visible to middle infrared: first results. *Remote Sensing of Environment* 83(1–2), 97–111.

Walthall CL, M Kim DL Williams, BW Meeson, PA Agbu, JA Newcomer and ER Levine, (1993). Data sets for modeling: A retrospective collection of bidirectional reflectance and forest ecosystems dynamics multisensor aircraft campaign data sets. *Remote Sensing of Environment* 46(3), 340–346.

Walthall CL, JM Norman, JM Welles, G Campbell and BL Blad, (1985). Simple equation to approximate the bi-directional reflectance from vegetative canopies and bare soil surfaces. *Applied Optics* 24, 383–387.

6

Satellite Sensors and Platforms

6.1 Introduction

There are numerous Earth orbiting satellite sensors that provide observations useful in assessing land cover conditions and landscape dynamics. These orbiting sensors measure spatial patterns of reflected and emitted energy from the land surface that can be used to generate geospatial image products of soil, vegetation, water and biogeochemical features. They measure changes over time through their repeat observations across a range of spatial and temporal scales. Satellite imagery extending back to the 1970s now provides a forty-plus year observation data record of dynamic ecosystem conditions and land surface changes. The synoptic coverage, higher-quality and consistency of satellite imagery have greatly improved the mapping of Earth resources compared with aerial photography.

A basic understanding of the variety of sensor designs and their characteristics is beneficial in correctly applying remote sensing tools to achieve various science and resource management objectives (Fig. 6.1). It is also important to know which sensors yield the appropriate type of remote sensing data to best answer specific ecological questions. Various sensor-dependent properties, such as pixel size, spectral bands, temporal repeat period, radiometric fidelity, polarization and viewing geometry are utilized to measure and characterize the Earth's surface. Each sensor system will have unique measurement strengths and limitations in its ability to characterise and retrieve land cover information.

There are many ways to classify the multitude of orbiting sensors and imagery available for ecosystem landscape studies. Important differentiating criteria may include the region of the electromagnetic spectrum (e.g., microwave, thermal, visible, near-infrared) being sensed, whether the energy source is active *versus* passive, sensor orbital characteristics (e.g., geostationary), frequency of image acquisition and sensor spatial resolution. In this chapter we introduce basic sensor principles, their design and properties, and discuss their respective capabilities and limitations in assessing land cover status, ecological variables, and landscape processes. Examples of the various types of sensor systems used in landscape studies are also highlighted.

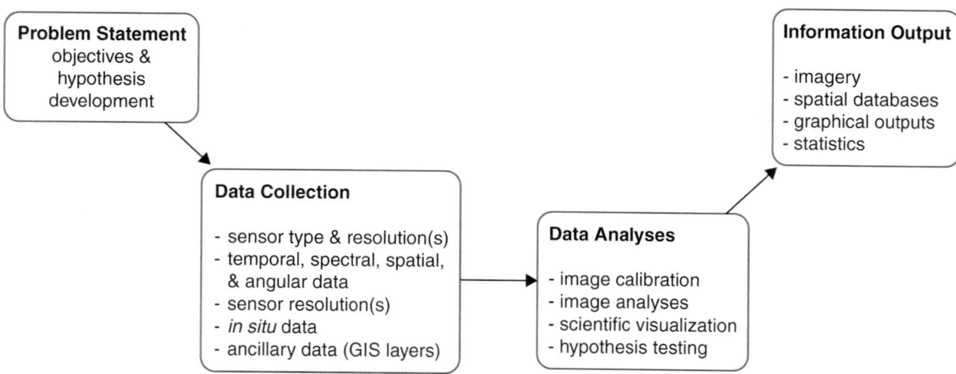

Figure 6.1. Concept of the remote sensing process involving problem statement to sensor data collection, calibration and spatial analyses and derivation of information.

6.2 Sensor Resolution

The resolution properties of a sensor define the magnitude and extent to which a sensor is able to discriminate variations and changes in landscape properties. In general, improved surface characterisations are achieved with finer resolution imaging capabilities. However, all sensors are limited by resolution constraints and signal noise limitations. Measurements are restricted to a finite surface area or pixel resolution, as well as by discrete samplings of the electromagnetic spectrum, and a restricted temporal capability for repeat observations of the same area.

6.2.1 Spatial Resolution

The spatial resolution of an image determines the level of surface detail that is provided by the sensor. Spatial resolution refers to the smallest area of the ground surface sampled by a sensor. This minimum sampling unit in an image is commonly referred to as a picture element or 'pixel', defined by the size of the sampling unit of the sensor at nadir (zero degree view angle). A pixel is also known as the "instantaneous field of view" (IFOV) of the imaging sensor, which are the sub-elements within the overall field-of-view (FOV), or size of an entire image (Fig. 6.2).

The more cells present in an image, the finer the resolution and more data available across an image with the same FOV. As all the energy within a pixel is integrated to produce a pixel value, there is a decrease in signal with smaller pixels and corresponding lower signal to noise ratio. Eventually the pixel size becomes limited when the detected signal is too weak relative to the noise present and no further meaningful information can be obtained.

Earth observing sensors in operation today cover a wide range of spatial resolutions, from < 1 meter pixel sizes to much coarser, km-scale pixel sizes. Fine resolution satellite sensors have pixel sizes in the 0.5 m to 5 m size range, which are particularly useful in urban studies, assessing hydrologic drainage patterns, discriminating tree

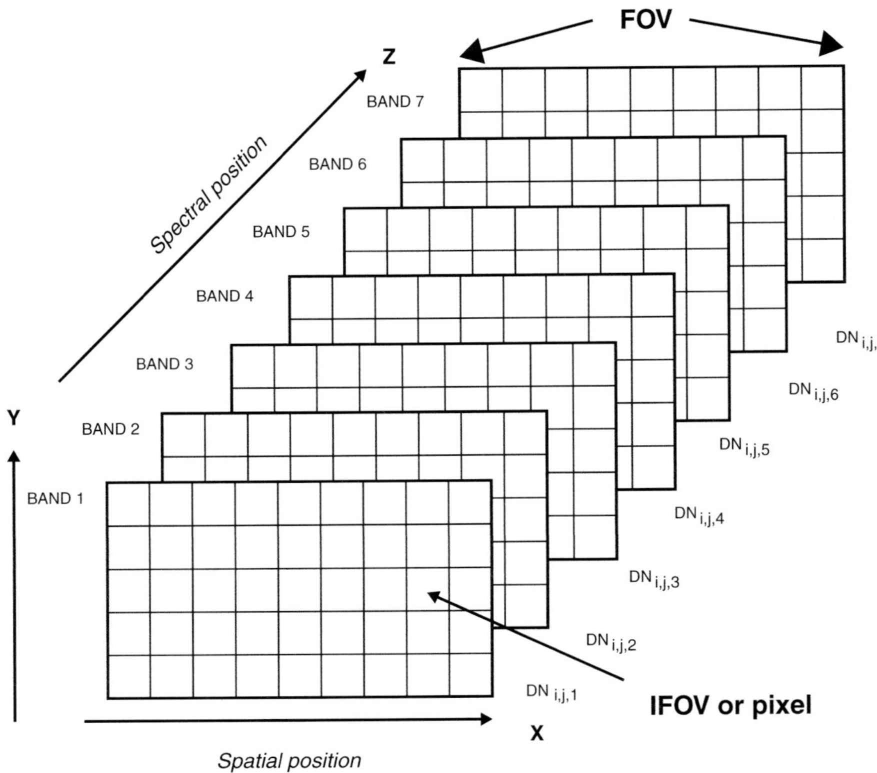

Figure 6.2. Basic image structure or data cube consisting of pixels within the spatial and spectral properties of an imaging sensor.

crowns, invasive species identification, and the scaling of *in situ* field data to larger scales (Townshend and Justice 1981). Moderate resolution sensors may range from 5 m to 100 m pixel size and are ideally suited to acquire land cover information at the community and landscape level. Coarse resolution sensors have pixel sizes between 0.10 km to 100 km and are utilized in regional and global applications.

Generally, moderate and coarse spatial resolution data consist of mixed pixels defined by an aggregate, weighted signal of several plant species or multiple land cover types within that pixel. In contrast, the finer the pixel size, the greater the pixel purity and lower the probability that the pixel will be a mixture of two or more cover types. Smaller pixels may consist of individual tree crowns or bare soil surfaces devoid of vegetation and hence enable a more accurate interpretation of an image (Hopkins *et al.* 1988).

6.2.2 Spectral Resolution

The spectral resolution of a sensor is determined by its spectral sampling characteristics across the electromagnetic spectrum. This is defined by the number of wavebands, their bandwidths, and overall spectral coverage (Fig. 6.2). Generally, fine spectral

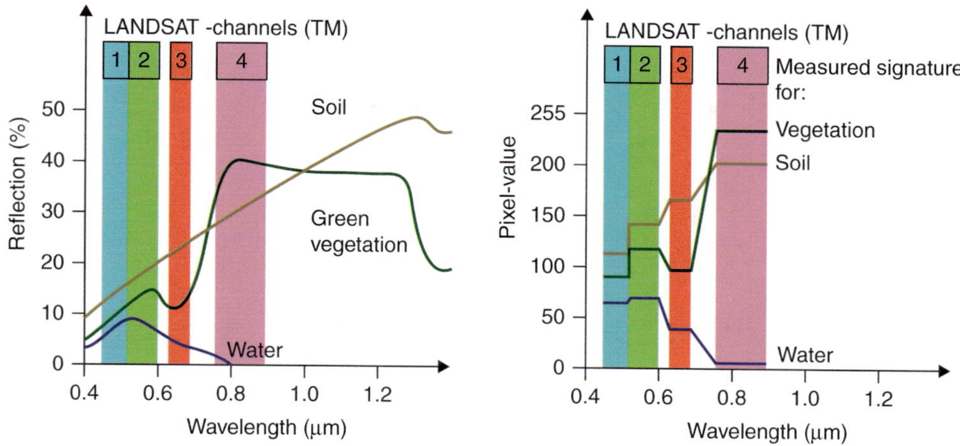

Figure 6.3. Spectral resolution is determined by the number and bandwidths of the channels used by a sensor. In this example, 4 Landsat-TM broadbands are super-imposed over the spectra of a typical soil, vegetation, and water. From: http://www .gisgeo.dk/course/demoeng/basic/tele1.htm

resolution is achieved by narrow bandwidths and a large number of bands, which collectively provide a better representation of an object's spectral signature. The number of bands on Earth observing sensors that are available for environmental studies can range from a couple of broad wavebands in the visible and infrared portion of the spectrum to 36 bands on the Moderate Resolution Imaging Spectroradiometer (MODIS), and as many as 220 bands (0.4 to 2.5 µm) on the EO-1 Hyperion sensor.

The spectral resolution properties of a sensor are related to its design objectives and intended applications and research use. For meteorological studies only three spectral bands may be needed: in the visible region, SWIR (atmosphere water vapour assessments) and thermal region (cloud temperature assessment), as in the case of the European Meteosat satellite. In contrast, for ecosystem studies, multiple bands in the visible, NIR, SWIR, and thermal are required for pigment, lignin, leaf structure and water assessments of the canopy.

6.2.3 Temporal Resolution

Temporal resolution refers to the frequency of sensor observation over a given area on Earth and is a function of the basic orbital characteristics of the sensor, a sensors repeat cycle, its field-of-view, pointing capabilities and the operational lifetime of the sensor. The larger the field-of-view the greater the frequency of coverage, relative to narrower field-of-view sensors which can only image small portions of the earth in any given orbital swath. With more frequent observations, one can composite the data to remove clouds and obtain gap-free temporal signatures of phenology for every pixel (Fig. 6.4).

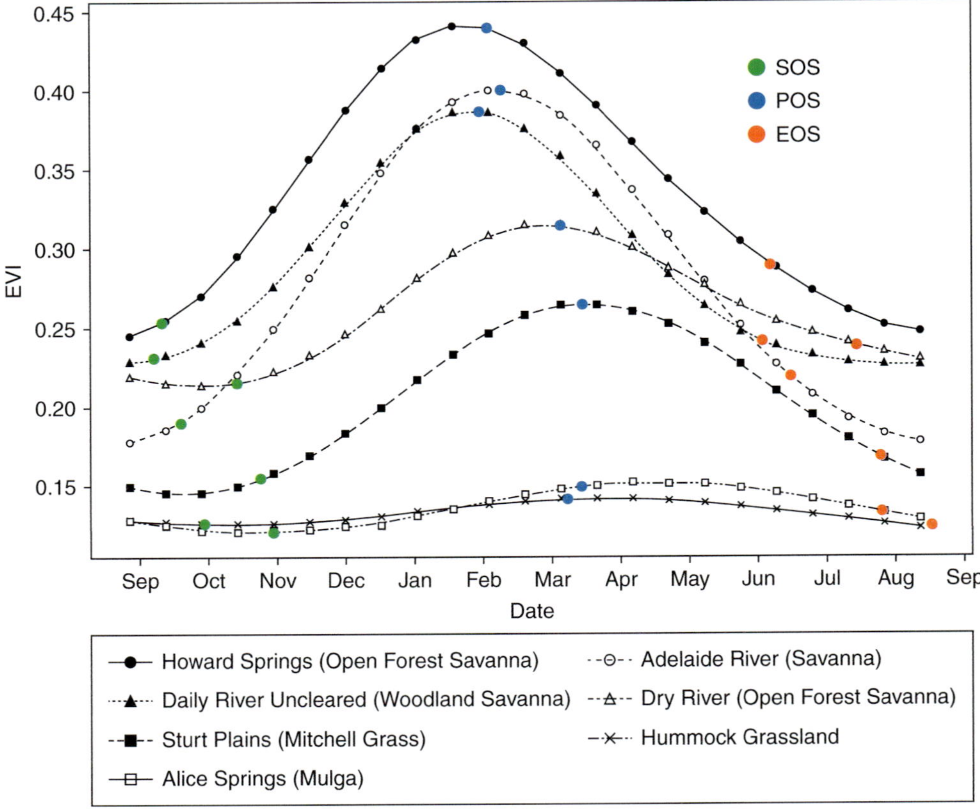

Figure 6.4. Temporal sequence of remotely-sensed 16-day enhanced vegetation index (EVI) values of seven sites in northern Australia as related to the temporal acquisition and processing of MODIS data. Redrawn from Ma et al. 2013.

Optimal temporal resolution is dependent on the desired objectives or project application. For ecological applications the best temporal resolution would depend on land cover dynamics, productivity patterns, or phenological features that need to be detected and retrieved. Temporal resolution is also very important for environmental monitoring applications, climate–vegetation studies, detection of disturbance events (floods, fire), and land cover change analysis. The temporal resolution of current orbiting sensors vary considerably with geostationary meteorological satellites providing repeat measurements at 10–15 minute intervals, while polar-orbiting meteorological satellites provide imagery every day. In contrast the moderate and finer spatial resolution sensors used in landscape studies have much coarser temporal resolutions, ranging from 16-day to monthly repeat observations.

Often the availability of satellite imagery may be considerably lower than the sensor repeat cycle. This is due to the presence of cloud cover which restricts observations and the acquisition of useful imagery. The Landsat sensor series have a repeat cycle of 16 days. However, if the Earth's surface is masked by clouds during an overpass time, it will be another 16 days before Landsat can image the same area, hence considerably lowering the 'effective' temporal resolution. Fortunately, some sensor systems have pointing capabilities allowing observations from adjacent orbits, thus increasing the effective temporal resolution. There may also be multiple copies of the same sensor with non-synchronous orbits (e.g., the MODIS sensor is present on the Aqua–and Terra–satellite platforms).

6.2.4 Radiometric Resolution

The radiometric resolution of a sensor refers to its measurent sensitivity in discriminating variations in reflected and emitted surface energy signals. The dynamic range (minimum and maximum detectable energy levels) of a sensor are divided into a number of grey levels, equivalent to the smallest detectable change in energy relative to the magnitude of electromagnetic energy. The practical limit to radiometric resolution is determined by the signal-to-noise ratio.

Radiometric resolution is important in image interpretation and distinguishing ground features differing only slightly in reflected or emitted energy. The radiometric resolution is expressed by the number of bits used for storing the image in binary format (e.g., 1 bit $= 2^1 = 2$). The human eye is only able to discern 64 grey levels (6 bit $= 2^6 = 64$). The Landsat ETM+ sensor is at 8 bit resolution, which is 256 discriminable grey levels (8 bit $= 2^8 = 256$), ranging from 0 (black) to 255 (white). The NOAA-AVHRR has 10 bit resolution or 1024 grey levels and MODIS is at 16 bit, or 65,536 grey levels. These grey levels, or digital counts, are readily converted into energy units by way of sensor calibration constants.

6.2.5 Optimum Sensor Resolution

Remote sensing systems measure spatial, spectral and temporal variations of electromagnetic energy emanating from ground surface features to varying extents of fidelity and performance. In the design of a sensor system, there are trade-offs made among spectral, spatial, temporal and radiometric resolutions (Fig. 6.5). For example, the narrower the bandwidth of a spectral filter, the less 'light' or energy that can pass through the filter and the lower the signal reaching a sensor detector. Similarly, a finer spatial resolution also results in less total reflected energy detectable due to the smaller area of the ground resolution cell (pixel). These examples both lead to reduced radiometric resolution or ability to detect fine energy differences.

To increase the amount of energy detected and improve the radiometric resolution, one has to widen and make coarser, the spectral waveband, thereby lowering the spectral resolution of the sensor. Conversely, a coarser spatial resolution (larger pixel sizes) allows improved radiometric and/or spectral resolution. Thus, the various types of resolution must be balanced against the desired capabilities and the spatial, spectral, and temporal objectives and ecological applications of the sensor. The resolutions necessary for an application further depends upon cost and data storage issues.

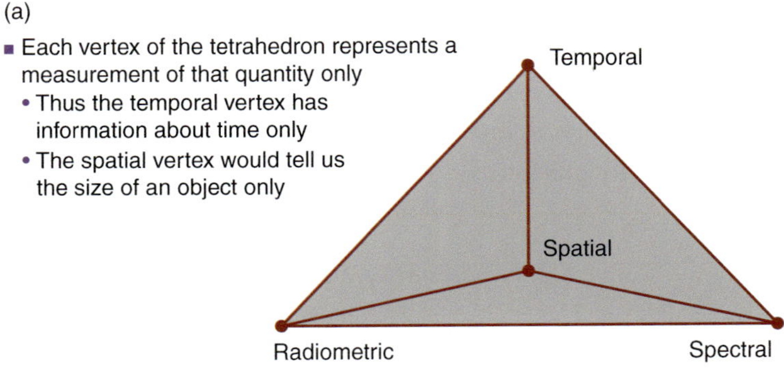

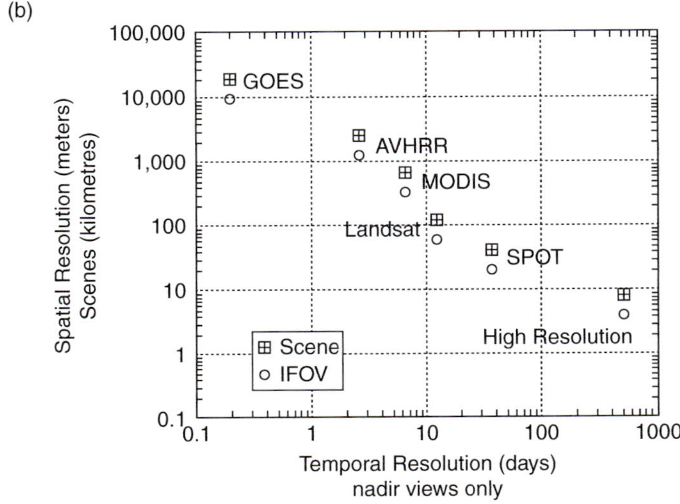

Figure 6.5. (a) Concept of sensor resolution trade-offs among spatial, radiometric, temporal and spectral resolutions in the design of a sensor (Courtesy K. Thome). Each vertex of the tetrahedron represents a sensor primarily designed for measurement of that surface property (size, spectrum, dynamics); (b) Example of spatial *versus* temporal resolution trade-offs across a range of orbiting sensors. Adapted from Goward and Williams (1997).

6.2.6 Angular Resolution and Properties

Angular resolution refers to a sensors capability to observe the same area on the ground across different viewing angles and sun–view geometries (Diner *et al.* 1999). Multi-angular observations improve the characterization of vegetation and soil structural properties, discrimination of land cover types and provide useful landscape hydrology and geomorphological information. The Multi-Angle Imaging Spectro-Radiometer (MISR) on board the *Terra* platform routinely provides imagery of the Earth through nine separate pushbroom CCD cameras with along-track angles (0, ±26.1°, ±45.6°, ±60°, ±70.5°) in four narrow spectral bands and a nine-day repeat cycle (Fig. 6.6). The change in reflectance at different view angles provides the means to distinguish different types of atmospheric particles (aerosols), cloud types, and

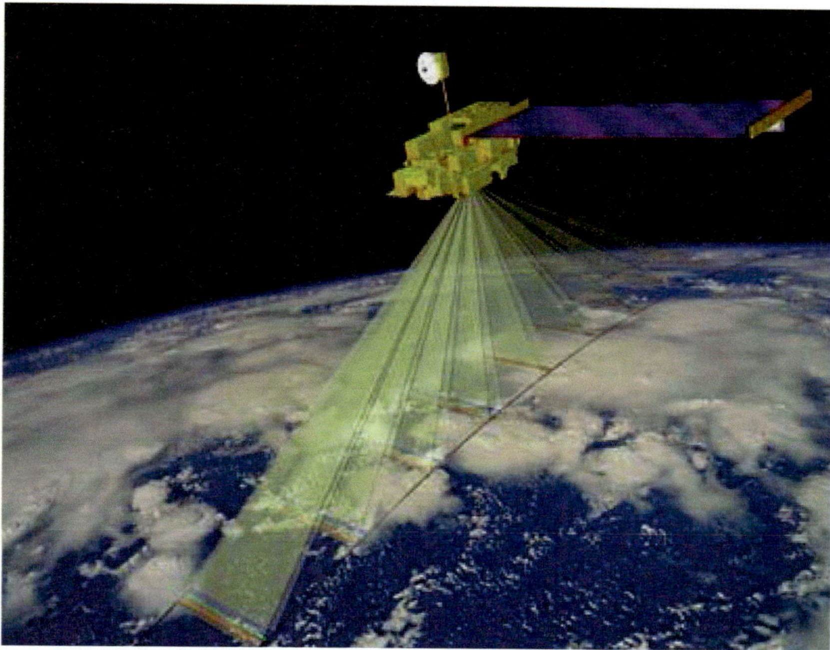

Figure 6.6. MISR imaging design involving nine imaging cameras at separate viewing geometries. NASA/JPL/Shigeru Suzuki and Eric M. De Jong, Solar System Visualization Project.

vegetation structures. Combined with stereoscopic techniques, this facilitates construction of 3-D models and estimation of the total amount of sunlight reflected (albedo) from the Earth.

Finer angular resolutions involve a greater number of sensor viewing angles and sun-view geometry configurations, as well as sensor pointing capabilities and narrower angular IFOVs. The High-Resolution Visible (HRV) sensor on board the French Systeme Probatoire pour l'Observation de la Terre (SPOT) platforms, has provided stereo-capabilities with a pointable (±27°) sensor since 1986. The Advanced Spaceborne Thermal Emission and Reflection Radiometer (ASTER), on board the Terra platform, also offers stereoscopic imagery with a backward-viewing telescope for same-orbit stereo data in a single, NIR channel at 15 m resolution. These sensors have greatly improved canopy structure and surface terrain characterization, involving drainage patterns, floodplains, terraces, relief, soil erosion potential, landscape stability, and surface roughness.

6.3 Orbital Systems

Satellite sensor systems may be classified by their Earth orbiting configuration. There are three types of orbits utilized in most satellite platforms, including geostationary, inclined, and polar-orbiting (Fig. 6.7).

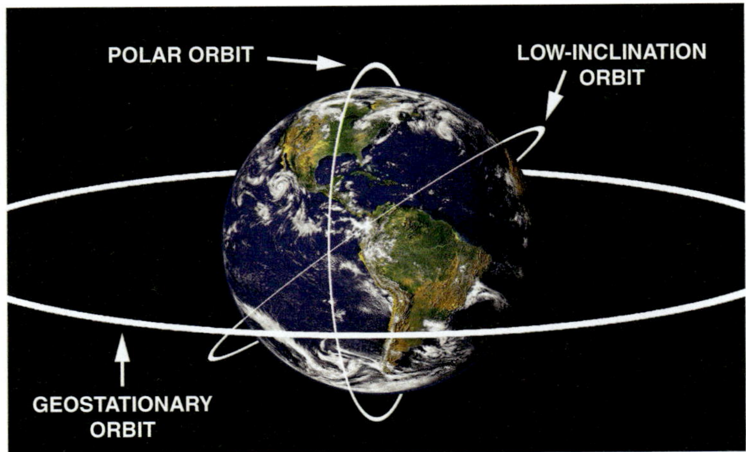

Figure 6.7. Three primary types of satellite orbiting platforms; Geostationary, polar orbiting, and low-inclination orbits. *Courtesy of NOAA.*

6.3.1 Geostationary Orbits

Geostationary sensors orbit the Earth on a path that is parallel with and at the same speed as the Earth's rotation. Hence the sensor appears positioned at a fixed point above the Earth throughout the day and night. These satellites observe an entire hemisphere of the Earth from large distances (36,000 to 41,000 km altitude), resulting in very coarse spatial resolutions, but very high temporal frequencies of 10–30 minutes. They are most frequently used to monitor the weather but are increasingly being utilized for terrestrial applications. The most common geostationary satellites are the Geostationary Operational Environmental Satellite (GOES) and Meteosat series. GOES has the Visible and Infrared Spin Scan Radiometer (VISSR) sensor with channels in the visible (0.55–0.70 μm) and thermal (10.5–12.6 μm) regions and has been operational since the mid–1970s. The VISSR on board the Meteosat has an additional band at 5.70–7.10 μm.

The Meteosat second generation (MSG) satellites (7, 8, 9, and 10) carry two instruments: the Spinning Enhanced Visible and InfraRed Imager (SEVIRI), which observes the Earth in 12 spectral channels and the Geostationary Earth Radiation Budget (GERB) instrument, a visible-infrared radiometer for Earth radiation budget studies (Fensholt *et al.* 2006). The MSG satellites operate over Europe and Africa and provides image data every 15 minutes. SEVIRI supports short-range weather forecasting capabilities over Europe and Africa and provide important observational inputs (rainfall and atmosphere water vapour) to numerical weather prediction and climate monitoring applications (Fig. 6.8). The Advanced Himawari Imager (AHI) on board Himawari 8/9 was launched in October 2014 and provides full hemisphere disk views over Asia and Oceania regions at 10-minute intervals in sixteen spectral bands. Future operational geostationary satellites are also planned over the Americas, including the GOES-R.

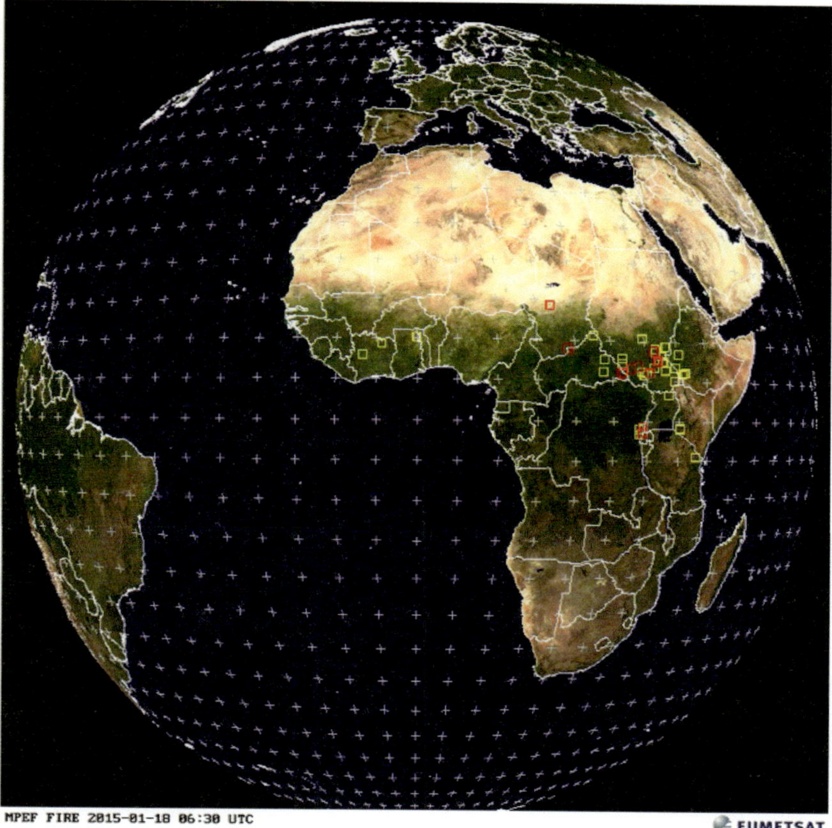

Figure 6.8. Full disk view and image of the Fire product from the EUMETSAT Spinning Enhanced Visible and InfraRed Imager (SEVIRI) geostationary satellite. The red coloured squares indicate hot spots of temperature anomalies. *Courtesy EUMETSAT.*

6.3.2 Low Inclined Orbits

Inclined orbits have an inclination between 0 degrees (equatorial orbit) and 90 degrees (polar orbit) and are aimed to provide increased coverage over specific areas of the Earth. The orbital altitude of these satellites is generally on the order of a few hundred km, so the orbital period is on the order of a few hours. The Tropical Rainfall Monitoring Mission (TRMM) is an example of a low inclination orbiting instrument to study the tropics and enable sensor microwave measurements of rainfall at 3-hour intervals between latitudes 40 N to 40 S. TRMM is widely used for extreme rainfall events and its ability to provide consistent values of precipitation, independent of the geographic distribution of ground-based meteorological stations.

6.3.3 Polar Orbiting (Sun-Synchronous)

Polar orbiting satellites orbit the Earth in successive orbits close to the poles, at altitudes of 600–950 km. These represent the majority of sensors used in terrestrial ecosystem studies and have a wide range of spatial, spectral, and temporal resolutions.

Coarse resolution sensors have pixel sizes ranging between 250 m to 100 km and have frequent temporal observations, ranging from daily to 4 days. They are designed for regional and global ecological applications, providing high frequency data sets depicting seasonal patterns of albedo, land and sea surface temperature, vegetation greening and snow and ice cover, useful in studies of biogeochemistry, hydrology, climate, ecology, erosion, land degradation, and fire detection (Anyamba and Tucker 2005).

Some widely used coarse resolution sensors include the NOAA–Advanced Very High Resolution Radiometer (AVHRR) sensor series, the Sea-Viewing Wide Field-of-View Sensor (SeaWiFS) (O'Reilly *et al.* 1998), the Moderate Resolution Imaging Spectroradiometer (MODIS) sensors on board the Terra and Aqua platforms, the Medium Resolution Imaging Radiometer (MERIS) on board the European Space Agency (ESA) Envisat platform, SPOT-VEGETATION, and the Visible Infrared Imaging Radiometer Suite (VIIRS) on board the Suomi National Polar-orbiting Partnership (NPP) platform.

The NOAA AVHRR sensor series contain a visible and near infrared channel and additional thermal channels. The AVHRR orbits at an altitude of 833 km with a wide field of view ($\pm 55.4°$) providing a repeat cycle of 12 hours (Cihlar *et al.* 1997). The SPOT VEGETATION series, launched in 1998, has a repeat cycle of once per day on a global basis at 1.15 km spatial resolution ($50.5°$ field-of view and 2200 km swath) and with spectral bands in the blue, red, NIR, and SWIR.

The Earth Observing System (EOS) began in 1999 with the EOS Terra platform, a sun synchronous polar orbit with a descending (southward) equatorial crossing in the morning. In 2002, EOS Aqua was launched into a sun synchronous polar orbit with an ascending (northward) equatorial crossing in the afternoon. The MODIS instrument was mounted on each of these two platforms and provides global coverage of the Earth in 36 bands with spatial resolutions that vary by band. Pixel sizes are 250 m for the red and NIR bands, 500 m for the blue, green, NIR2, and SWIR bands, and 1 km for the remaining bands, consisting of thermal channels, ocean colour bands, and atmosphere bands. With its wide ($\pm 55°$) field-of-view and large image swath (2300 km), MODIS is able to image most of Earth on a daily basis (Salomonson *et al.* 1989) (Fig. 6.9).

Satellite products derived from these sensors include vegetation indices, vegetation canopy moisture indices, higher-level canopy structure and chlorophyll products, land cover products, albedo and land surface temperature. Due to the high temporal frequency of coverage 1-2 day, coarse resolution measurements can be composited into weekly-to-monthly products to obtain cloud-free and higher quality products. This enables seasonal data sets to be generated, which are otherwise very difficult to acquire with finer spatial resolution sensors (Justice *et al.* 1998). The multi-angular MISR instrument, which is also on board the Terra platform, enables coarse-scale

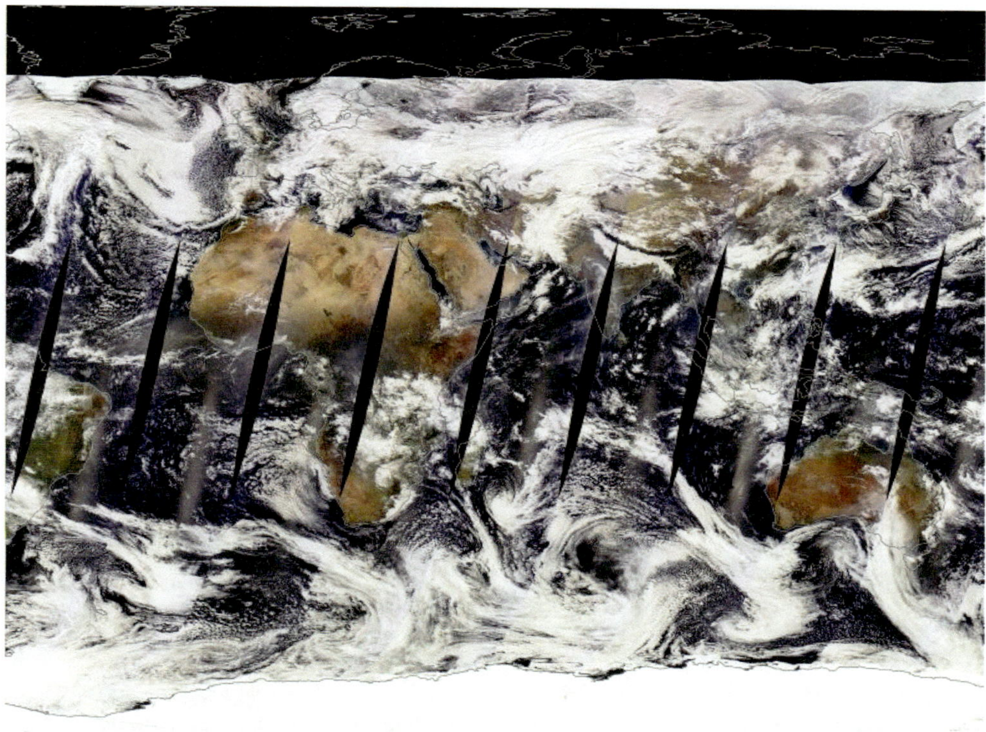

Figure 6.9. Terra–MODIS global view of the Earth on Jan. 20, 2015. Ten polar orbit swath widths are apparent across the day, with each swath encompassing 2300 km across. One can observe gaps along the equatorial zone indicating that MODIS is unable to achieve complete daily coverage of the Earth. *Courtesy NASA.*

geometric-optical studies for decoupling atmospheric effects from the surface and for characterizing surface BRDF properties. The integration of MODIS and MISR imagery further advances optical- geometric characterization studies of land surface structure variations.

MERIS is a programmable, medium-spectral resolution, imaging spectrometer operating in the solar reflective spectral range with 15 spectral bands that can be selected by ground command. The instrument's 68.5° field of view around nadir covers a swath width of 1150 km. VIIRS provides a bridge between EOS MODIS and the operational Joint Polar Satellite System (JPSS) planned for launch in 2016. VIIRS will extend many of the long term environmental data records initiated with the AVHRR and MODIS instruments (Miura *et al.* 2013).

Moderate resolution sensors, at spatial resolutions ranging from 10 m to 100 m pixels, acquire land surface information at near-nadir views for landscape and regional level ecosystem, soils, and earth resource mapping. The NASA Landsat series of satellites are the most well-known for delivering moderate resolution imagery. Landsat orbits the Earth in a near-polar, sun synchronous pattern with a mid-morning local solar equatorial crossing time. The image size or swath width of a Landsat scene is

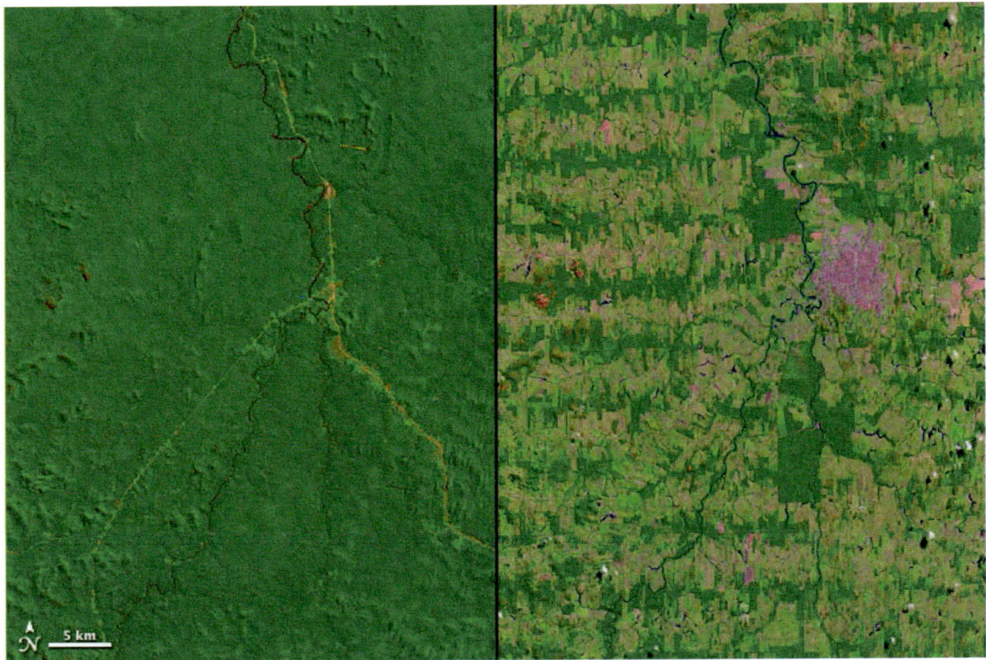

Figure 6.10. Deforestation in the Amazon rainforest as observed from Landsat in 1975 (left) and 2010 (right). In the Brazilian state of Rondônia, deforestation has the fishbone pattern with primary road-based clearings followed by secondary roads (*Courtesy U.S. Geological Survey, USGS*).

about 185 km across, covering broad areas of the Earth's surface with sufficient resolution to distinguish features like urban centres, agriculture, forests and other land uses (Fig. 6.10).

The earliest Landsat satellites contained the Multispectral Scanner (MSS) sensor consisting of 4 broad wavebands in the blue, green, red and NIR with a pixel size of approximately 80 m and repeat cycle of 18 days. The Landsat 4 and 5 satellites were launched in July 1982 and March 1984, respectively, and added the Thematic Mapper (TM) sensor with finer spatial, spectral, temporal, and radiometric resolutions. The TM sensor included 6 broad spectral bands at 30 m pixel size and a thermal waveband with 120 m pixel size. The newly added shortwave (SWIR) bands and thermal infrared (TIR) band provided sensing capabilities for moisture and drought studies. The repeat cycle of Landsat 4 and 5 further improved to 16 days.

Landsat 7 was launched in April 1999, following a failed Landsat-6 launch. Landsat 7 capabilities were improved with the Enhanced Thematic Mapper (ETM+) sensor that provided a 15 m panchromatic band and a 60 m thermal band. The most recent Landsat satellite, Landsat 8, was launched in February 2013 with the Operational Land Imager (OLI) and the Thermal Infrared Sensor (TIRS), providing 16-day coverage at spatial resolutions of 30 m (visible, NIR, SWIR); 100 m (thermal); and 15 m

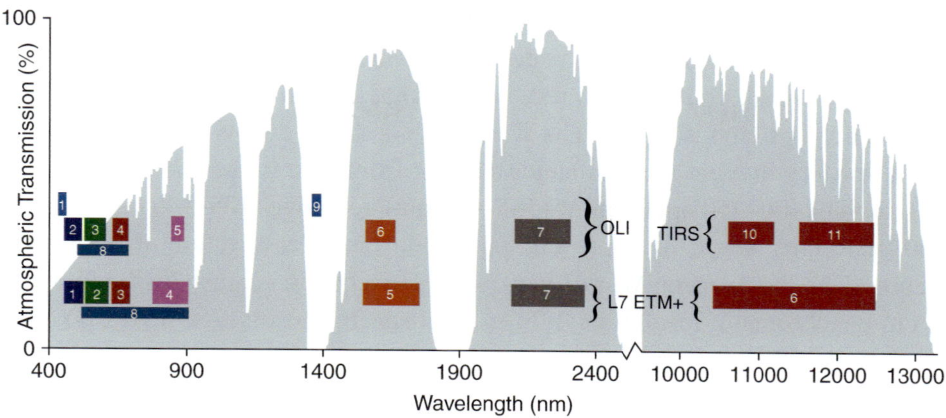

Figure 6.11. Comparisons of spectral bands, bandwidths, and their placements between Landsats 7 and 8. The vertical axis and shaded area are the atmosphere transmission spectra and windows (image from NASA).

(panchromatic). OLI also provides a new spectral band for detecting cirrus clouds and another new band for coastal zone observations (Irons *et al.* 2012) (Fig. 6.11). TIRS has 2 narrower spectral bands in the thermal region, formerly covered by one wide spectral band on Landsats 4–7.

The Landsat program has been in operation since July 1972 and constitutes the longest continuous set of Earth observation data of the land surface, a 40+ year record. Global mosaics of Landsat imagery are now freely available for detailed mapping of landscapes to global change mapping studies from the early 1970's (Tucker *et al.* 2004; Hermosilla *et al.* 2015).

Other examples of polar orbiting, moderate resolution sensors include SPOT-HRV and the ASTER sensors. The earlier SPOT satellites provided imagery in the green, red and NIR bands with 20 m spatial resolution, while more recent SPOT satellite (SPOT 4) have added a SWIR band and a panchromatic band with 10 m spatial resolution. The most recent SPOT 5 has 10 m spatial resolution for the red, green, NIR bands and a 5 m panchromatic band. ASTER, on board the EOS-Terra platform, has three spectral bands at 15 m pixel resolution in the visible and NIR; six bands in the spectral range 1 to 2.5 μm with 30 m pixel resolution; and five bands in the thermal infrared with 90 m resolution. Both SPOT and ASTER further provide stereoscopic imagery.

Fine resolution satellite sensors have the smallest pixel sizes, from 5 m to less than 1 m and generally are only commercially available. The very fine spatial resolutions provide a more complete representation of the landscape enabling one to discern individual trees and shrubs, gullies and ravines, and finer-scale surface features. The Space Imaging Ikonos and DigitalGlobe QuickBird instruments were some of the first commercial sensors to be launched in September 1999 and October 2001, respectively. Ikonos is pointable (±26°) and can image a given area of land at 1 meter (panchromatic) and 4 meter (multispectral) resolutions. QuickBird images the Earth in 4 spectral bands (blue, green, red, and NIR) at 2.44 m resolution and a panchromatic band at 61 cm resolution. The WorldView-2 satellite provides high spatial resolution imagery in eight spectral bands ranging from blue to the NIR, and includes a coastal blue (400–450 nm), yellow (585–625 nm), red-edge (705–745 nm) and NIR-2 (860–1040 nm) bands for improved applications and classification of Earth surface features. The most recent, WorldView-3, was launched in August 2014 and provides 31 cm panchromatic resolution, 1.24 m multispectral resolution, 3.7 m pixel resolution in the SWIR.

6.4 Hyperspectral Sensors

Hyperspectral imaging spectrometers acquire multispectral images in numerous, narrow and contiguous bands, such that a continuous spectrum of every pixel can be generated. The large number of bands carry valuable diagnostic information about the biogeochemistry of many Earth surface materials, related to mineralogy, liquid water, chlorophyll, cellulose, and lignin absorption features. Such fine resolution features are typically lost in coarser broadbands. Hyperspectral imaging has wide ranging

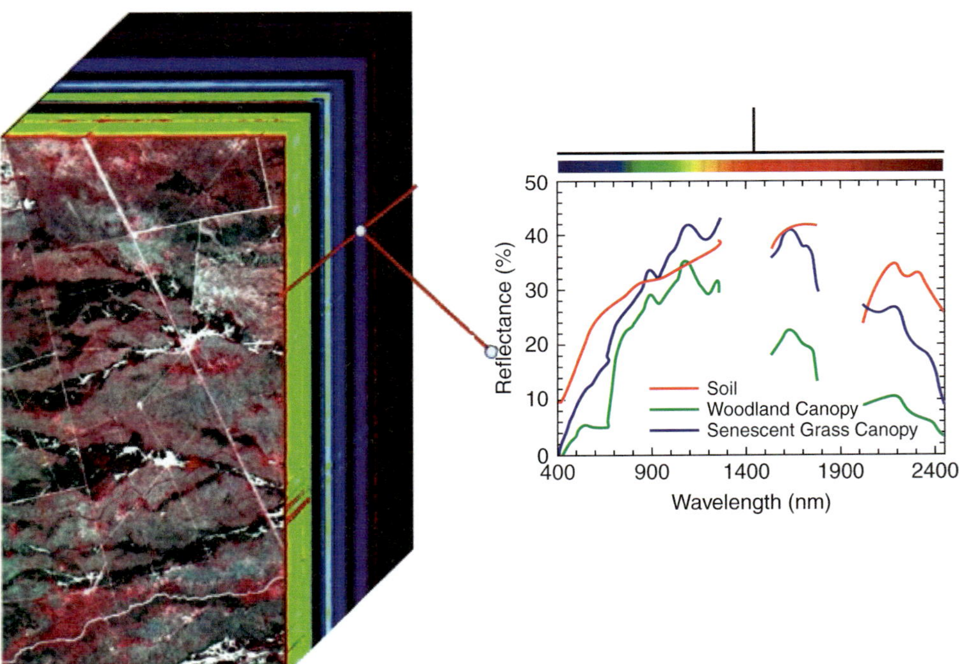

Figure 6.12. Example of EO-1 Hyperion data cube with continuous spectral signature extracts of soil, green vegetation, and woodland areas at Nacunan Reserve, Argentina (*Courtesy Jet propulsion Laboratory, NASA*).

applications in canopy chemistry analysis, forestry, agriculture, species diversity and species detection, and environmental management.

The first spaceborne Hyperspectral Imager, the Hyperion sensor was launched in November 2000 on board the Earth Observing-1 (EO-1) platform. Hyperion is a pushbroom sensor that measures Earth-reflected radiation in 220 spectral bands, from 400 nm to 2500 nm, at 10 nm nominal sampling intervals (Ungar *et al.* 2003) (Fig. 6.12). The instrument has 30 m spatial resolution and images a relatively small swath of ~7.7 km by 180 km land area per image. Besides Hyperion, the only other operational hyperspectral sensor is the ESA's Compact High Resolution Imaging Spectrometer (CHRIS) on Proba-1, which has a spatial resolution of 17 m in up to 62 bands.

6.5 Microwave Sensors

There are many passive and active microwave orbiting sensors used in Earth monitoring of landscapes. Passive microwave radiometers detect and image microwave radiation emanating from the Earth's surface, while radar (radio detection and ranging) represent 'active' sensors that generate their own source of energy and measure the reflected microwave signals that are returned from the surface.

The main factors influencing the microwave signal are the dielectric and roughness properties of the surface, such as vegetation roughness and moisture content, soil moisture content, soil roughness and topography. Microwave measurements are affected by the absorption and scattering properties of vegetation elements, which can be characterized by the total amount of biomass, its distribution among trunks, branches and leaves and vegetation moisture content (Ferrazzoli and Guerriero 1996). Moisture properties vary more dynamically than roughness enabling independent assessments of each.

Generally larger wavelengths (lower frequencies) will have greater penetration through vegetation, soil and ice surfaces than smaller wavelengths. Hence L band (23 cm) penetrates to a greater extent than C band (6 cm) and X band (3 cm).

The two types of imaging radars most commonly used are (a) real aperture radar, and (b) synthetic aperture radar (SAR). Real aperture radars are often called side-looking airborne radar (SLAR) with an illumination direction perpendicular to the flight line. These include Seasat (Band L), SIR-C (Bands L, C, and X); ERS-1,2 Envisat (Band C); Radarsat (Band C); and JERS (Band L). Frequencies that have been shown useful to monitor plant water status also include the X and Ka bands data (Pampaloni and Paloscia 1986).

Passive microwave satellite sensors will have coarser spatial resolutions and shorter revisit times compared to active microwave sensors. The Advanced Microwave Scanning Radiometer (AMSR-E) on board the Aqua polar orbiting satellite platform provides vertical (V) and horizontal (H) polarization data every 3 days at 0.25 degree pixels (Kawanishi *et al.* 2003). As the data are transparent to cloud conditions, the

3 day resolution is useful to assess soil and vegetation moisture variations during the growing season, which is particularly relevant for the detection of plant water stress.

Passive microwave vegetation indices (such as the Frequency Index (FI) and Emissivity Difference Vegetation Index (EDVI) (Min and Lin 2006) are sensitive to vegetation properties (Ferrazzoli and Guerriero 1996). Both FI and EDVI measures are sensitive to vegetation moisture content and vegetation structure and have been shown to potentially be sensitive to evapotranspiration fraction (Li *et al.* 2009). FI is calculated using the brightness temperatures at 37 GHz (Ka Band) and 10.6 GHz (X band) AMSR-E channels at vertical polarization. In forested and dense vegetation areas, soil moisture has a small effect on FI due to relatively high crown attenuation (X and Ka bands).

6.6 Solar-Induced Chlorophyll Fluorescence

Global space-based estimates of sun-induced chlorophyll fluorescence (SIF) have recently become available through the Japanese Greenhouse Gases Observing Satellite (GOSAT), where Fraunhofer lines are used to derive estimates of fluorescence emissions. Subsequently, global SIF data with better spatial and temporal sampling are now produced from the Global Ozone Monitoring Experiment-2 (GOME-2) instrument on board the Metop-A platform (Guanter *et al.* 2014) and the Orbiting Carbon Observatory-2 (OCO-2) launched in July 2014 (Frankenberg *et al.* 2014). Preparatory studies are also underway for the future European Fluorescence EXplorer (FLEX) satellite mission (Meroni *et al.* 2009) that will provide measurements characterizing the spectral shape of fluorescence emission and enable estimates of rates of photosynthesis under different vegetation stress conditions. In addition, the future Sentinel-5 Precursor Tropospheric Monitoring Instrument (TROPOMI) (Veefkind *et al.* 2012) satellite mission will provide advance spectrometer and fluorescence data with significantly finer spatial resolution.

6.7 LiDAR

Light Detection and Ranging (LiDAR) laser technology offers much potential for terrestrial carbon assessments, including assessments of standing wood biomass, forest disturbance, biomass loss and carbon accumulation through forest regrowth (Lefsky *et al.* 2002). LiDAR measures the physical structure of woody vegetation and can serve as a reliable replacement for field inventory plots in areas lacking field data. LiDAR integration with field inventory plots can provide calibrated LiDAR estimates of aboveground carbon stocks, which can then be up-scaled using satellite data on vegetation cover, topography and rainfall from satellite data to model carbon stocks (Asner *et al.* 2010).

6.8 GRACE

The Gravity Recovery and Climate Experiment (GRACE) satellites were launched in 2002 for the purpose of making detailed measurements of Earth's gravity field (Tapley *et al.* 2004). Variations in Earth's gravity are relatively constant over long time intervals, however, there are more dynamic gravity fields associated with land surface moisture, changes in the volume of ground water, sea ice, sea level rise and deep ocean currents. These can all be monitored with GRACE. Retrievals of groundwater and "unseen water reserves" from space are a significant new addition to hydrological studies that can substantially improve our knowledge of below–and aboveground water resources and associated changes to vegetation functioning and groundwater dependent ecosystems (GDEs) (Eamus *et al.* (2015). GRACE satellite data have been used to estimate groundwater depletion associated with severe droughts in Europe, the United States, China, and India (Leblanc *et al.* 2009; Rodell *et al.* 2009).

6.9 Airborne Sensors

Airborne sensor systems normally involve new, experimental sensors which are flown in research and development, to provide new insights and information regarding the quantitative assessment of Earth surface features. Such experiments include, although are not limited to (1) better placement of wavebands, both filter width and number of bands, to exploit the unique optical properties of the surface features of interest; (2) minimization of atmospheric contamination via better definition of atmospheric windows and narrower channels, for example, avoid water vapour absorption regions; (3) improved spectral stability and calibration accuracy of new sensor channels and systems; and (4) verification of satellite imagery, ground truthing and validation studies.

Aerial sensors extend locally-constrained ground-truth studies to kilometre length scales enabling sampling of landscape variability. Airborne sensors can also be flown at higher altitudes for large-area studies encompassing a range of terrestrial biome types, regional scaling and validation of satellite data.

Aircraft mounted sensors further provide greater flexibility in overpass times and repeat frequencies relative to satellite based systems. Examples of well-known airborne systems include the Airborne Visible/ Infrared Imaging Spectrometer (AVIRIS), a whiskbroom scanner first flown in 1987 on NASA's ER-2 at an altitude of 20 km, resulting in a swath width of 10 km (30° FOV) and a 20 m pixel size. Other commercially available airborne imaging spectrometers include the Compact Airborne Spectrographic Imager (CASI). CASI collects data in 288 channels between 0.4 and 0.9 µm at 1.8 nm intervals.

6.10 Continuity and Fusion of Data

There is much interest in maintaining long term and accurate satellite measures of land surface states and condition for land degradation and climate change assessments. The Intergovernmental Panel on Climate Change report (IPCC 2007) emphasized the need to assess climate change impacts on forest and agricultural resources through repeatable, consistent and long term observations. This necessitates a seamless fusion of satellite measures across multiple sensor systems for long term monitoring involving cross-sensor calibration methods that account for radiometric, spatial and spectral bandpass differences (Miura *et al.* 2013; Gobron *et al.* 2000).

Uncertainties in biophysical products derived from multiple sensors containing disparities in sensor and platform characteristics (e.g., the bandpass filter, pixel size, and the sun-target-sensor viewing geometry) are a major issue in biophysical parameter retrieval studies. Data continuity studies attempt to mitigate systematic biophysical retrieval errors generated by such sensor differences in order to preserve data continuity. Several studies have examined NDVI data continuity properties calculated across the AVHRR sensor series, MODIS on board Terra and Aqua satellite data, and the Visible Infrared Imager Radiometer Suite (VIIRS) data collected on board the Suomi National Polar-orbiting Partnership (Suomi NPP) (Miura *et al.* 2013).

6.10.1 Multi-Sensor Data Fusion

Multiple sensor systems with different combinations of spectral, spatial and temporal resolutions are often needed to effectively characterize ecosystem structure and function and capture the important spatiotemporal complexities of landscapes. Coarse spatial resolution sensors provide consistent and timely information of ecosystem health, functioning and large scale disturbance events, whereas species dynamics and more subtle land degradation processes, fragmentation and land use modifications are better resolved with finer spatial resolution satellite imagery. By realizing the spectral–spatial detail present in finer resolution data, one is able to fully interpret and characterize the spatial patterns hidden inside pixels of coarser spatial resolution satellite imagery. For example, Asner (2009) demonstrated that forest degradation and selective logging potentially contribute as much carbon loss as larger scale clear-cutting. Whereas coarse resolution satellites may detect large-scale clearings, the finer resolution data are needed for forest degradation assessments.

High temporal frequency MODIS satellite data are increasingly being blended with fine spatial resolution Landsat data for applications that require high resolution in both time and space. Two methods for generating time series of high spatial resolution imagery by fusion of MODIS and Landsat include the spatial and temporal adaptive reflectance fusion model (STARFM) algorithm (Gao *et al.* 2006) and the multi-temporal MODIS–Landsat data fusion method (Roy *et al.* 2008). Many benefits have also been found in fusing optical imagery with LiDAR data to improve monitoring of landscapes, land cover classification and assessment of vegetation conditions. The LiDAR provides 3-dimensional structural information while the spectral imagery provides surface biochemical and biophysical information for identification of surface features. Such methods to integrate multi-resolution satellite sensor data provide better resolution properties than the individual data sources and are vital to better understand interactions and processes that influence carbon stocks, water resources and land use activities.

6.11 References

Anyamba A and CJ Tucker, (2005). Analysis of Sahelian vegetation dynamics using NOAA-AVHRR NDVI data from 1981–2003. *Journal of Arid Environments* 63, 596–614.

Asner GP, (2009). Tropical forest carbon assessment: integrating satellite and airborne mapping approaches. *Environmental Research Letters* 4, 4003–4009.

Asner GP, GVN Powell, J Mascaro, DE Knapp, JK Clark, J Jacobson, T Kennedy-Bowdoin, A Balaji, G Paez-Acosta, E Victoria, L Secada, M Valqui and RF Hughes, (2010). High-resolution forest carbon stocks and emissions in the Amazon. *Proceedings of the National Academy of Sciences* 107, 16738–42.

Cihlar J, H Ly, Z Li, J Chen, H Pokrant and F Huang, (1997). Multi-temporal, multichannel AVHRR data sets for land biosphere studies—Artifacts and corrections. *Remote Sensing of Environment* 60, 35–57.

Diner DJ, GP Asner, R Davies,Y Knyazikhin, JP Muller, AW Noli and B Pinty B, (1999). New directions in earth observing: Scientific applications of multiangle remote sensing. *Bulletin of the American Meteorological Society* 80 (11), 2209–2228.

Eamus D, S Zolfaghar, R Villalobos-Vega, J Cleverly and AR Huete (2015). Groundwater-dependent ecosystems: a review of new techniques and an ecosystem-scale threshold response to differences in depth-to-groundwater. *Hydrological and Earth System Science* 19, 1–77.

Fensholt R, I Sandholt, S Stisen and C Tucker, (2006). Analysing NDVI for the African continent using the geostationary meteosat second generation SEVIRI sensor. *Remote Sensing of Environment* 101, 212–29.

Ferrazzoli P and L Guerriero, (1996). Passive microwave remote sensing of forests: a model investigation. *Geoscience and Remote Sensing, IEEE Transactions* 34, 433–43.

Frankenberg C, C O'Dell, J Berry, L Guanter, J Joiner, P Köhler, R Pollock and TE Taylor, (2014). Prospects for chlorophyll fluorescence remote sensing from the Orbiting Carbon Observatory-2. *Remote Sensing of Environment* 147, 1–12.

Gao F, J Masek, M Schwaller and F Hall, (2006). On the blending of the Landsat and MODIS surface reflectance: predicting daily Landsat surface reflectance. *Geoscience and Remote Sensing IEEE Transactions* 44, 2207–18.

Gobron N, B Pinty, NM Verstraete and JL Widlowski, (2000). Advanced vegetation indices optimized for up-coming sensors: Design, performance, and applications. *Geoscience and Remote Sensing IEEE Transactions* 38, 2489–505.

Guanter L, Y Zhang, M Jung, J Joiner, M Voigt, JA Berry, C Frankenberg, AR Huete, P Zarco-Tejada, J-E Lee, MS Moran, G Ponce-Campos, C Beer, G Camps-Valls, N Buchmann, D Gianelle, K Klumpp, A Cescatti, JM Baker and TJ Griffis, (2014). Global and time-resolved monitoring of crop photosynthesis with chlorophyll fluorescence. *Proceedings of the National Academy of Sciences* 111, E1327–E1333.

Hermosilla T, WA Wulder, JC White, NC Coops and GW Hobart, (2015). An integrated Landsat time series protocol for change detection and generation of annual gap-free surface reflectance composites. *Remote Sensing of Environment* 158, 220–324.

Hopkins, PF, AL Maclean and TM Lillesand, (1988). Assessment of Thematic Mapper imagery for forestry applications under Lake States conditions. *Photogrammetric Engineering and Remote Sensing* 54, 61–8.

IPCC 2007, Climate change (2007)-the physical science basis: Working group I contribution to the fourth assessment report of the IPCC, Solomon, S, D Qin, M Manning, Z Chen, M Marquis, KB. Averyt, M Tignor and HL Miller (eds.), vol. 4, Cambridge University Press, Cambridge, United Kingdom and New York, NY, USA.

Irons JR, JL Dwyer and JA Barsi, (2012). The next Landsat satellite: The Landsat Data Continuity Mission. *Remote Sensing of Environment* 122, 11–21.

Justice CO, E Vermote, JRG Townshend, R DeFries, DP Roy, DK Hall, VV Salomonson, JL Privette, G Riggs, A Strahler, W Lucht, RB Myneni, Y Knyazikhin, SW Running, RR Nemani, W Zhengming, AR Huete, W Van Leeuwen, RE Wolfe, L Giglio, JP Muller, P Lewis and MJ Barnsley, (1998). The Moderate Resolution Imaging Spectroradiometer (MODIS): land remote sensing for global change research. *Geoscience and Remote Sensing, IEEE Transactions* 36, 1228–49.

Kawanishi T, T Sezai, Y Ito, K Imaoka, T Takeshima, Y Ishido, A Shibata, M Miura, H Inahata and RW Spencer, (2003). 'The Advanced Microwave Scanning Radiometer for the Earth Observing System (AMSR-E), NASDA's contribution to the EOS for global energy and water cycle studies'. *Geoscience and Remote Sensing, IEEE Transactions* 41, 184–94.

Leblanc MJ, P Tregoning, G Ramillien, SO Tweed and A Fakes, (2009). Basin-scale, integrated observations of the early 21st century multiyear drought in southeast Australia. *Water Resources Research* 45, W04408.

Lefsky MA, WB Cohen, GG Parker and DJ Harding, (2002). Lidar remote sensing for ecosystem studies. *BioScience* 52, p. 19.

Li R, Q Min and B Lin, (2009). Estimation of evapotranspiration in a mid-latitude forest using the Microwave Emissivity Difference Vegetation Index (EDVI). *Remote Sensing of Environment* 113, 2011–2018.

Ma X, AR Huete, Q Yu, N Restrepo Coupe, K Davies, M Broich, P Ratana, J Beringer, LB Hutley, J Cleverly, N Boulain, D Eamus, (2013). Spatial patterns and temporal dynamics in savanna vegetation phenology across the North Australian Tropical Transect. *Remote Sensing of Environment* 139, 97–115.

Meroni M, M Rossini, L Guanter, L Alonso, U Rascher, R Colombo and J Moreno, (2009). Remote sensing of solar-induced chlorophyll fluorescence: Review of methods and applications. *Remote Sensing of Environment* 113, 2037–2051.

Min Q and B Lin, (2006). Remote sensing of evapotranspiration and carbon uptake at Harvard Forest. *Remote Sensing of Environment* 100, 379–87.

Miura T, JP Turner and & AR Huete (2013). Spectral Compatibility of the NDVI Across VIIRS, MODIS, and AVHRR: An Analysis of Atmospheric Effects Using EO-1 Hyperion. *Geoscience and Remote Sensing, IEEE Transactions* 51, 1349–1359.

O'Reilly JE, S Maritorena, BG Mitchell, DA Siegel, KL Carder, SA Garver, M Kahru, C McClain, (1998). Ocean color chlorophyll algorithms for SeaWiFS. *Journal of Geophysical Research* 103, 24937–24953.

Pampaloni P and S Paloscia, (1986). Microwave emission and plant water content: A Comparison between field measurements and theory. *Geoscience and Remote Sensing, IEEE Transactions* GE-24, 900–905.

Rodell M, I Velicogna and JS Famiglietti JS, (2009). Satellite-based estimates of groundwater depletion in India. *Nature* 460, 999–1002.

Roy DP, J Ju, P Lewis, C Schaaf, F Gao, M Hansen and E Lindquist, (2008). Multi-temporal MODIS–Landsat data fusion for relative radiometric normalization, gap filling, and prediction of Landsat data. *Remote Sensing of Environment* 112, 3112–3130.

Salomonson VV, W Barnes, PW Maymon, HE Montgomery and H Ostrow H, (1989). MODIS: advanced facility instrument for studies of the Earth as a system. *Geoscience and Remote Sensing, IEEE Transactions* 27, 145–53.

Tapley BD, S Bettadpur, M Watkins and C Reigber, (2004). The gravity recovery and climate experiment: Mission overview and early results. *Geophysical Research Letters* 31, L09607.

Townshend J and C Justice C, (1981). Information extraction from remotely sensed data. *International Journal of Remote Sensing* 2, 313–29.

Tucker CJ, DM Grant and JD Dykstra, (2004). NASA's global orthorectified Landsat data set. *Photogrammetric Engineering and Remote Sensing* 70, 313–22.

Ungar SG, JS Pearlman, JA Mendenhall and D Reute, (2003). Overview of the Earth Observing One (EO-1) mission. *Geoscience and Remote Sensing IEEE Transactions* 41, 1149–59.

Veefkind JP, I Aben, K McMullan, H Förster, J de Vries, G Otter, J Claas, HJ Eskes, JF de Haan, Q Kleipool, M van Weele, O Hasekamp, R Hoogeveen, J Landgraf, R Snel, P Tol, P Ingmann, R Voors, B Kruizinga, R Vink, H Visser and PF Levelt, (2012). 'TROPOMI on the ESA Sentinel-5 Precursor: A GMES mission for global observations of the atmospheric composition for climate, air quality and ozone lay applications'. *Remote Sensing of Environment* 120, 70–83.

7

Remote Sensing of Landscape Biophysical Properties

7.1 Introduction

The goal of this chapter is to discuss the basic optical properties of common landscape surface features, including plants, soils, water and detrital material. Variations of these optical properties over space and time provide the basis for remote sensing advancements on the detection and characterization of physical, chemical and biologic landscape properties across global terrestrial ecosystems. The remotely-sensed signal is a combination of the intrinsic optical properties of landscape components interacting with spectral, angular (sun–view geometries) and polarization radiation aspects of the incoming solar radiation. Information collected in spectral, spatial, angular and temporal optical domains are utilized in the development of methods for retrievals of ecological, hydrological and biogeochemical variables.

7.2 Spectral Signatures

Remotely-sensed information about landscape constituents is most often obtained through analyses of spectral reflectance signatures acquired across the reflected solar spectrum (400–2500 nm). Landscape spectral signatures are a function of the optical properties of the biogeochemical constituents, moisture condition and the size, shape and geometry of leaf elements or litter and soil particles (Fig. 7.1). There is considerable knowledge about vegetation, soil, litter and water optical signatures obtained through extensive field and laboratory spectroscopic analyses (Ben-Dor *et al.* 2008, Ustin *et al.* 2004).

7.2.1 Vegetation Optics

The absorption spectrum of a green leaf is related to its chemical (pigments, lignins and cellulose), moisture and physical properties (Fig. 7.2). Leaf optical properties in the visible wavelengths (400–700 nm) are controlled by the presence of biologically active pigments which generally result in low reflectance due to strong absorptions by

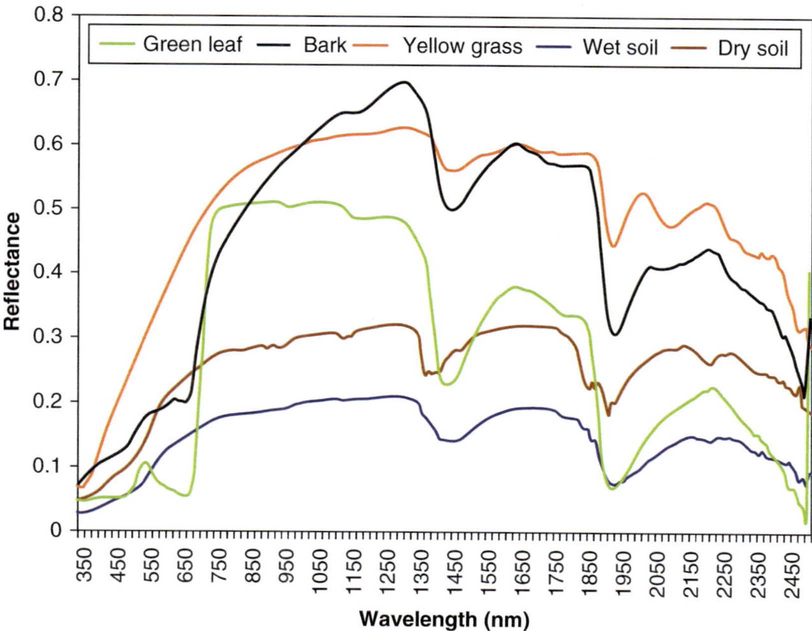

Figure 7.1. Examples of spectral reflectance signatures for field-measured landscape constituents.

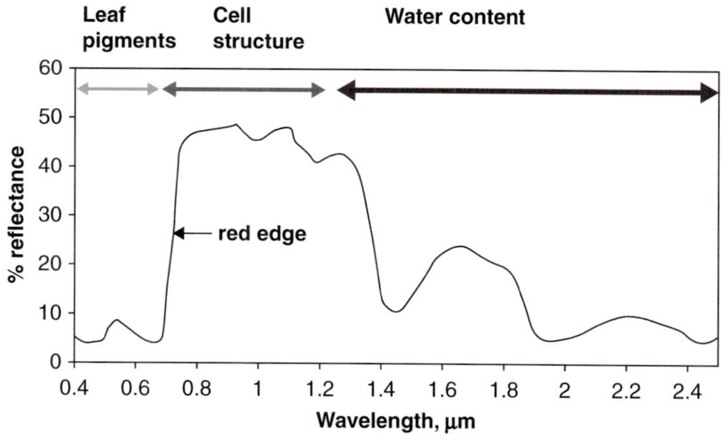

Figure 7.2. Leaf spectral reflectance signature depicting the three primary spectral regions and leaf controlling determinants.

chlorophylls, carotenoids, xanthophylls and anthocyanins (Fig. 7.3; Chapter 2; Carter and Knapp 2001, Blackburn 2007). Strong chlorophyll absorption features in the blue (450 nm) and red (680 nm) result in a 'green' appearance to healthy leaves as there is less-absorption and more reflection in the green.

Leaf pigment contents provide valuable information on the physiological performance of leaves. Plant disease and stress will alter pigment contents and hence the

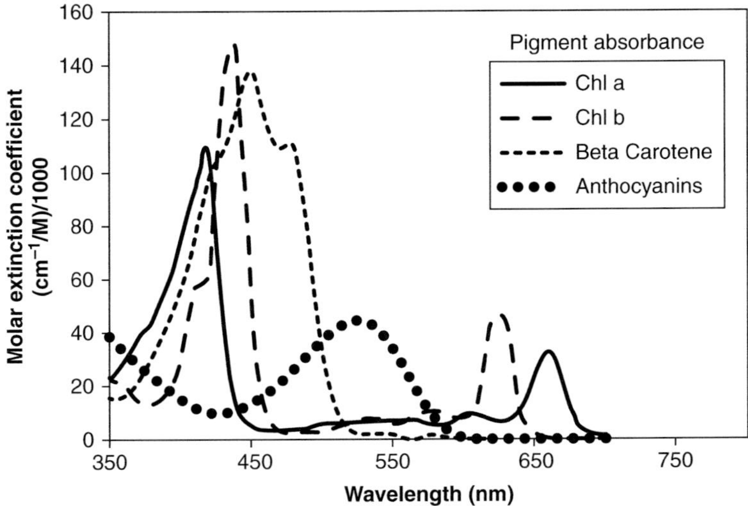

Figure 7.3. Absorption spectra of the major plant pigments (redrawn from Blackburn 2007).

visible spectrum (Knipling 1970). Mineral deficiencies affect chlorophyll content to varying extents with chlorosis directly affecting the visible spectra. Nitrogen deficiencies cause visible reflectance to increase and NIR reflectance to decrease by altering the number of cell layers and leaf necrosis results in spectral signatures comparable to that of senescent leaves (Gates *et al.* 1965).

Carotenoids function in processes of light absorption in plants as well as protecting plants from the harmful effects of high light conditions. The carotenoid reflectance index (CRI) is a sensitive measure of carotenoids in plant foliage (Gitelson *et al.* 2009):

$$CRI = (1/\rho_{510}) / (1/\rho_{550}) \qquad (7.1)$$

where ρ_{510} and ρ_{550} are reflectances at 510 and 550 nm, respectively. High CRI values are indicative of lower carotenoid/chlorophyll ratios and lower physiological stress (Fig. 7.4).

Carotenoid/chlorophyll ratios in green leaves as well as scaled light-use efficiencies (LUE) can be derived from spectral variations at 531 and 570 nm, commonly known as the photochemical reflectance index (PRI; Fig. 7.4) (Gamon *et al.* 1992, Sims and Gamon 2002):

$$PRI = (\rho_{531} - \rho_{570}) / (\rho_{531} + \rho_{570}) \qquad (7.2)$$

where ρ_{531} and ρ_{570} are reflectances at 531 and 570 nm respectively. Spectral variations at 531 nm (green spectral region) are closely associated with pigment changes in the xanthophyll cycle and their dissipation of excess light energy in order to protect the photosynthetic leaf apparatus (Ripullone *et al.* 2011).

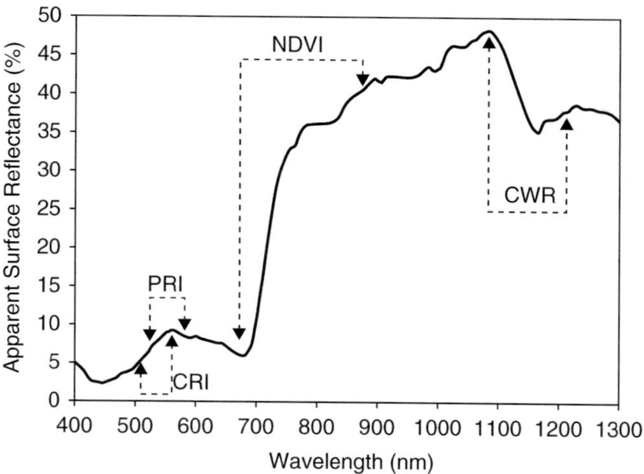

Figure 7.4. Canopy reflectance spectrum (400–1300 nm) of Hawaiian rain forest from the EO-1 Hyperion sensor showing spectral regions used in the derivation of the narrow-band normalised difference vegetation index (NDVI), photochemical reflectance index (PRI), carotenoid reflectance index (CRI) and the canopy water retrieval (CWR). Reproduced with permission from Asner *et al.* (2006).

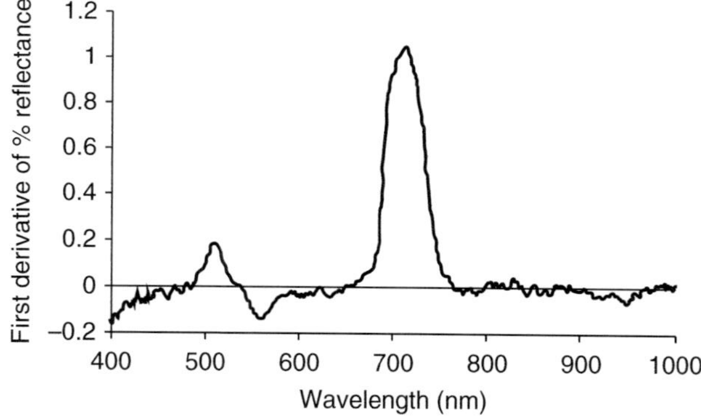

Figure 7.5. The first derivative of the reflectance spectrum of a typical leaf (Adapted from Blackburn 2007).

The inflection point signalling the transition from high absorption in the visible wavelengths to the high reflectance in near-infrared (NIR) wavelengths (above 740 nm) is termed the "red edge inflection point" (REIP) (Fig. 7.2). The REIP symbolises the boundary of the strong chlorophyll absorption feature and can shift to longer wavelengths with increasing chlorophyll content, resulting in a high correlation between REIP and leaf chlorophyll contents (Gitelson *et al.* 2003). To quantify the REIP, the first derivative of the spectral reflectance signature is often used (Fig. 7.5).

High leaf reflectance in the near-infrared (NIR) wavelengths (700–1300 nm) are due to internal leaf scattering with no absorption because nearly all of the NIR radiation is scattered (reflected and transmitted) in a manner dependent on leaf type,

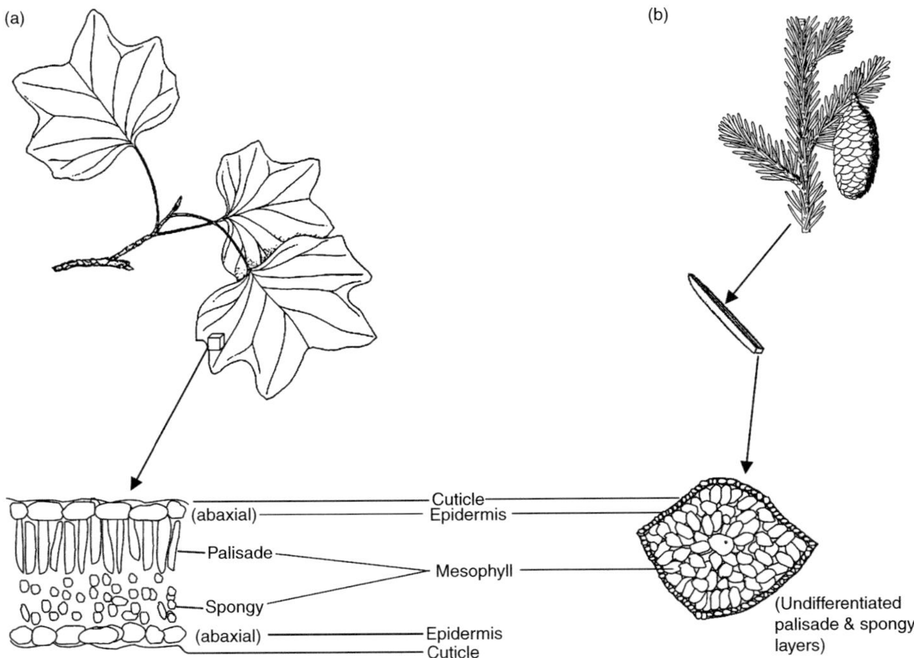

Figure 7.6. Schematic representation of leaves with (a) broadleaf (laminar structure); and (b) needle leaf structure and their general cell arrangements (Reproduced with permission from Ollinger 2010).

morphology and anatomical structure (Fig. 7.6). Leaf structure allows for regular contact between atmosphere and hydrated mesophyll cells and provides the optical environment for light capture for photosynthesis and NIR scattering at the air-cell interfaces within the leaf spongy mesophyll. NIR reflectance varies with the number of cell layers, the size of cells and the relative thickness of the spongy mesophyll layer Smith *et al.* (1997).

Leaves of dicotyledons have a larger reflectance than those of monocotyledons having the same thickness because their spongy mesophyll layer is more developed. Similarly, needle-leaved gymnosperms (conifers) have a more tightly packed cell structure with undifferentiated palisade and spongy mesophyll layers (Fig. 7.6). Drought adapted plants can also have a very high NIR reflectance that increase with number of cell layers and decrease with smaller cell size.

Leaf spectral properties are dominated by strong water absorption in the SWIR (shortwave infrared) spectral region (1300–2500 nm). Leaf water content not only controls leaf optics in the SWIR, but through its influence on cell turgor (Chapter 3), also has an indirect effect on the visible and NIR spectra. Strong water absorption in the SWIR obscures other biochemical features associated with cellulose, lignin and other carbon constituents. There are also secondary water absorption features in the NIR, at 970 and 1200 nm; these are used as measures of leaf moisture content (Gao 1996). The expression "equivalent water thickness, EWT" is sometimes used to

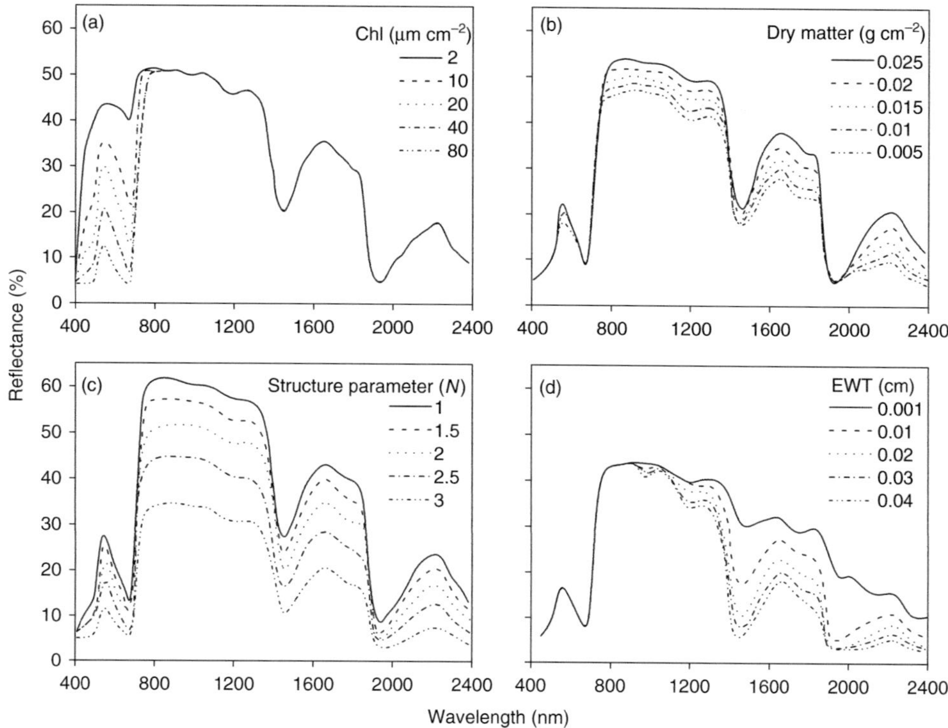

Figure 7.7. Leaf reflectance spectra simulated by the PROSPECT model, generated using a range of values for chlorophyll (Chl) concentration, dry matter content, equivalent water thickness (EWT) and the structure parameter N (Reproduced with permission from Ollinger 2010).

indicate the thickness of a sheet of water that can completely account for the absorption spectrum of a leaf in the 1.4 to 2.5 μm spectral range. In Figure 7.7, the PROSPECT leaf model is used to simulate variation in leaf EWT, along with chlorophyll concentrations, dry matter (cellulose and lignins) and a leaf layer structural parameter, N (Jacquemoud 1990).

7.2.2 Non-Photosynthetic Vegetation

Leaf optical properties change significantly during leaf senescence as chlorophyll pigments initially degrade and more persistent carotenoid and other brown pigments persist. In the visible spectrum, red reflectance increases, resulting in a yellow leaf colour and in certain environments (e.g. grasslands), there is no clear separation between a yellow *versus* green leaf, due to the gradual changes in pigment content and leaf structure (Fig. 7.8). The NIR reflectance of dehydrating leaves often increases as leaf volume decreases while micro-cavities remain between cell walls. The NIR reflectance eventually may decrease in the advanced stages of leaf senescence, resulting from the breakdown and deterioration of cell walls (Elvidge 1990).

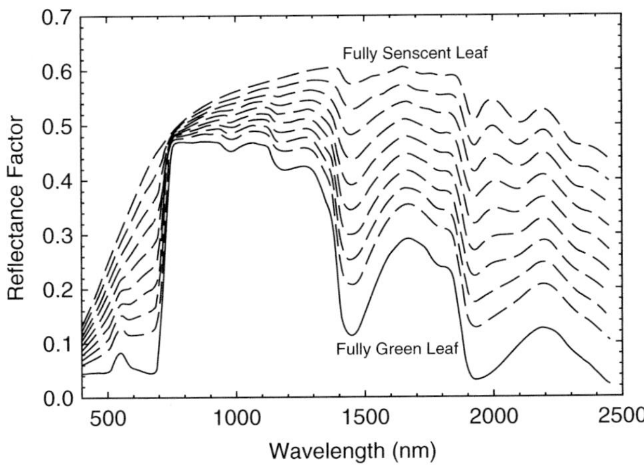

Figure 7.8. Contrast between spectral reflectance signatures of a green leaf and fully senescent leaf (Redrawn from Asner 1998).

In the SWIR, there are large increases in leaf reflectance associated with its drying and decrease of water content. This usually occurs after chlorophyll pigment degradation when the leaf is yellow. As plant leaves senesce or desiccate, the carbon features (cellulose and lignin) emerge in the SWIR region of the spectrum. Nagler *et al.* (2003) developed a cellulose absorption index (CAI) describing the average depth of the cellulose absorption feature at 2.1 μm:

$$CAI = 0.5\ (\rho_{2.0} + \rho_{2.2}) - \rho_{2.1} \qquad (7.3)$$

where $\rho_{2.0}$, $\rho_{2.2}$ and $\rho_{2.1}$ are reflectances at 2.0, 2.2 and 2.1 μm, respectively.

Nagler *et al.* (2003) found CAI useful for quantifying plant litter cover because it increased linearly with fraction of plant litter on the ground (Fig. 7.9). Woody plant parts, such as bark, can also exhibit spectral features that are similar to those of fully senescent foliage (Asner 1998).

7.2.3 Soil Spectral Signatures

Soils are three-dimensional living bodies, with spatially variable biologic, physical and chemical properties, that form the outer skin of the Earth's terrestrial surface. Soils exhibit great spatial variability as a result of the interactions of climate, topography, parent material and organisms acting on the soil body over time. Human activities and land use history also play important roles in determining soil properties and also contribute to heterogeneity in soil spatial patterns. Remote sensing, with synoptic, multi-scale and repetitive coverage of the land surface can provide quantitative measurements of such changes at the surface. These are needed to address a wide range of environmental and global change issues.

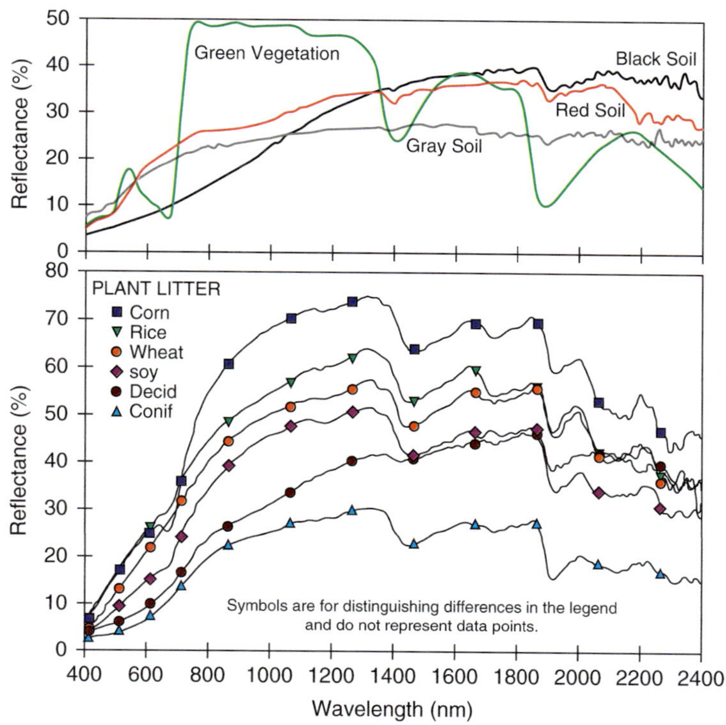

Figure 7.9. Variations in soil and plant litter spectral signatures (Reproduced with permission from Nagler *et al.* 2003).

Soil optical properties have been extensively studied using laboratory and field measurements and are primarily dependent on soil biogeochemical composition (minerals, salts and organic), soil surface geometrical-optical scattering (particle size, roughness) and surface moisture status, as reviewed in Baumgardner *et al.* (1985). However, in contrast to the strong absorption spectra of pure mineral samples, soil absorption features are only broadly defined as soils consist of aggregate mixtures of many mineral and organic constituents (Fig. 7.9). Many quantitative tools for extracting information on the physical and biochemical characteristics of soils have been developed and recognised as the field of soil spectroscopy. As an example, Ben-Dor and Bannin (1994) utilised a visible and near-infrared analysis scheme to predict a wide variety of soil chemical constituents, including $CaCO_3$, Fe_2O_3, Al_2O_3, SiO_2, free iron oxides and K_2O, from high-resolution spectra of arid and semiarid soils.

In field radiometric measurements of *in-situ* soils, there are strong optical-geometric interactions resulting from the angles at which the sun illuminates and sensors view the surface. Particle size distribution and size, shape and orientation of surface "roughness" elements are some of the most important factors influencing the reflectance properties of soils (Irons *et al.* 1992). There is a general decrease in reflectance with increasing surface roughness as coarse aggregates contain many inter-aggregate spaces and "light traps". The shortest wavelengths are more affected than longer

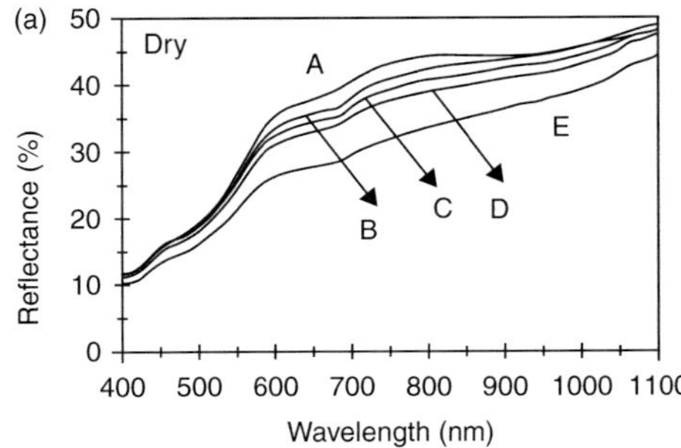

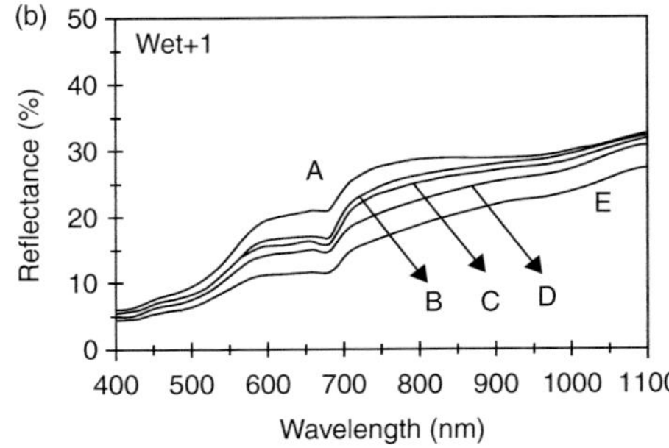

Figure 7.10. The influence of different microphytic communities on surface spectra properties over sand in the Negev desert, in order of increasing abundance from community A to E (Reproduced with permission from Karnieli *et al.* 1999).

wavelengths and hence such optical-geometric interactions alter a soil's spectral signature and the inferences made of soil biogeochemical properties such as soil colour and mineralogy.

Soil moisture also has a strong influence on the amount and composition of reflected energy from a soil across the entire spectrum, with a general decrease in reflectance, proportional to the thickness and energy status of the adsorbed water (Fig. 7.1). The decrease in reflectance appears similar to that caused by soil roughness, however, in the SWIR spectral region soil moisture will exert a much more pronounced reflectance drop relative to the decrease in the visible and NIR spectral regions. This is due to the greater sensitivity of SWIR wavelengths to water absorption.

Remote sensing only observes the immediate soil surface, which prevents the assessments of important sub-soil and root-zone soil properties. However, the soil surface is the most dynamic, biologic and hydrologic interface that responds immediately to climatic changes and human forcings. As an example, biogenic soil crusts can result in highly dynamic and complex soil spectral signatures in response to rainfall and soil wetting and drying cycles (Fig. 7.10).

In summary, soils exhibit large spatial and dynamic variations across a landscape due to variations in moisture content, biogeochemical composition, surface roughness elements and ongoing processes of land use, soil degradation, erosion and global change. Although not fully developed, remote sensing remains the only viable technique to map, monitor and manage the fragile soil resource; with most of the Earth's terrestrial surface having "open canopies," there is an appreciable soil component that can potentially be remotely sensed. Further advancements are needed to infer soil properties with depth through pedotransfer functions that enable the derivation of subsoil biogeophysical products from remotely sensed surface properties.

The presence of snow and water on the land surface can also yield highly dynamic spectral signatures in certain landscapes and hydrologic conditions (Fig. 7.11). Highly reflective snow, at visible wavelengths, contrasts greatly with the strong absorption properties of water and with the very strong absorptions in the NIR and SWIR.

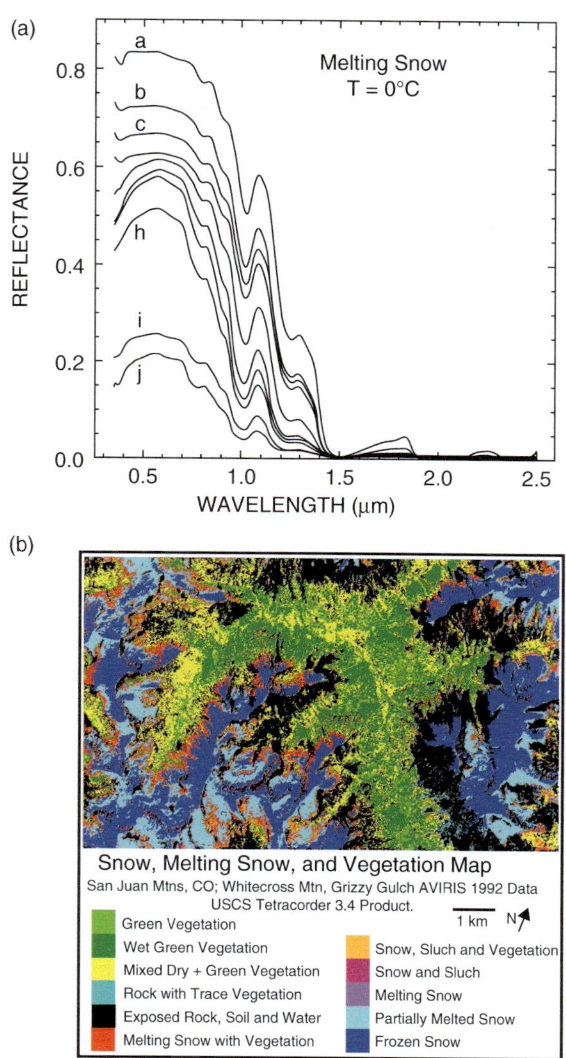

Figure 7.11. A series of reflectance spectra of melting snow (top), from snow (a) at 0°C with only a small amount of liquid water to the lowest, mostly water, spectrum (j); (bottom) an AVIRIS image of snow, melting snow, snow mixed with vegetation and wet and dry vegetation for a region in the San Juan Mountains of Colorado, August 1992. Spectra from Clark (1999) and image from USGS Speclab.

7.3 Landscape Optics

At the ecosystem scale, spectral measurements are considerably altered as individual leaf spectra are merged into plant canopy structural assemblages and mixed with soil, litter and non-photosynthetic vegetation (NPV) spectral signatures. Approximately 70 percent of the Earth's terrestrial surface is covered by partially vegetated canopies (deserts, grasslands, savannas, open woodlands and open forests) with vegetation overlying various proportions of exposed soil and litter background (Graetz 1990).

Although the basic properties of leaf spectra are well known, the interpretation of canopy-level spectra are hampered by a host of canopy-level traits (plant structures, leaf area, leaf angles), sun and shade leaves, leaf development and aging, soil background, topography, bidirectional reflectance distribution functions and the non-uniformity of incident solar radiation and shadows. Plant structures vary in canopy cover, crown sizes, clumping, spacing and canopy height that impart macro-scale shadowing between individual plants and land-cover types (Franklin *et al.* 1993). Satellite and airborne sensors collect integrative measurements of the various landscape constituents and their structural properties (Ollinger 2010).

Asner (1998) considered a set of dominant controls of the pixel-scale reflectance for arid and semi-arid ecosystems and described pixel reflectance in terms of the following scale-dependent variables:

$$R_{pixel} = f \text{(solar-view geometry, tissue optics, canopy structure,}$$
$$\text{landscape structure, soil optics)} \tag{7.4}$$

where solar-view geometry is the orientation of sun and sensor zenith and azimuth angles at the time of observation; tissue optics are the reflectance properties of green foliage, senescent foliage (litter) and woody material components; canopy structure is comprised of the amount of tissue present, usually expressed as leaf, wood and litter area index (m^2 tissue/m^2 ground area) and the architectural placement (angular distributions) of these tissues in the canopy (Myneni *et al.* 1989).

The optical properties of vegetation canopies vary greatly over time due to changing plant status, LAI, fractional vegetation cover (Fv) and phenological life form. As an herbaceous plant canopy develops, the soil contribution progressively decreases and the bare soil spectra are replaced by the developing plant spectra (Fig. 7.12). As the green vegetation cover becomes dense, spectral reflectance saturates. The LAI at which saturation occurs is dependent on the spectral region. In the visible and SWIR, with very low effective photon penetration depths from the top of the canopy, saturation is reached for LAI around 1–2. In contrast, NIR wavelengths can penetrate up to seven leaf layers resulting in an NIR enhancement effect in which much of the NIR energy scattered and transmitted from upper leaf layers can be reflected by lower leaves and retransmitted through upper leaves to enhance reflectance (Asner 1998). The value of LAI corresponding to saturation will also vary with plant geometry as higher LAI values are needed to reach saturation in plants having vertical leaves than with plants having horizontal leaves.

Overall, reflectance from a canopy will be lower than that from a single leaf because of general attenuation of radiation by variations in illumination angle, leaf orientation, shadows and non-foliage background surfaces such as soil. The relationship between leaf and canopy reflectance will be less sensitive to canopy structure within the visible region and more sensitive in the NIR due to the enhancement effect.

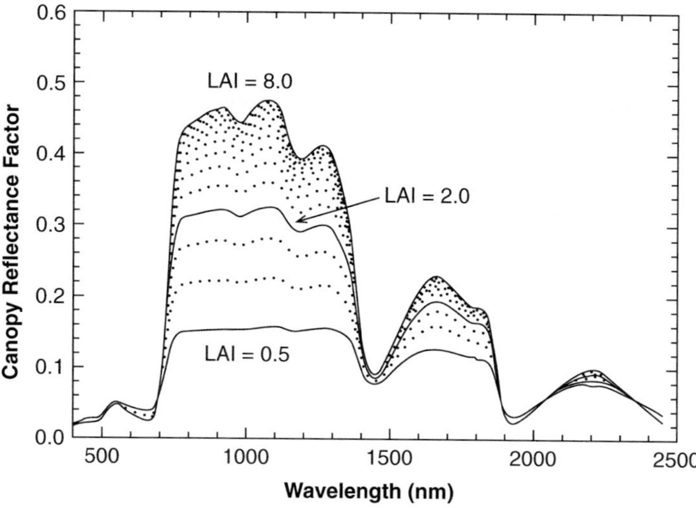

Figure 7.12. Spectral reflectance signatures for plant canopies with increasing LAI (Adapted from Asner 1998).

7.4 Canopy Biophysics

Various spectral functions, including empirical and semi-empirical indices and models, have been developed to translate optical information into biophysical information about a canopy. These include spectral vegetation indices (VIs) and satellite products of leaf area index (LAI), fraction of absorbed photosynthetically-active radiation (fAPAR), chlorophyll content, fractional vegetation cover (Fv), biomass and photosynthetic activity (Asrar *et al.* 1984, Baret and Guyot 1991). Most of these products utilise and combine the red, NIR and SWIR spectral regions to represent vegetation properties.

In general, vegetation is characterised by vegetation type (to distinguish trees from grass and shrubs), horizontal cover (as represented by green vegetation fraction, Fv) and vertical thickness/structure (as represented by leaf-area index, LAI and leaf angle distribution, LAD). Land-atmosphere exchanges of energy, water, carbon and other trace gases are all directly affected by these vegetation parameters. LAI is an indicator of photosynthetic capacity of a canopy, while fAPAR is an indicator of potential absorption of PAR. Both are coupled and may be used in Net Primary Production (NPP) and carbon models. Fractional vegetation cover is often used in erosion studies and climate models (Fig. 7.13).

7.4.1 Spectral Vegetation Indices

Vegetation indices (VIs) are spectral measures of the green foliage status of a canopy. They can be computed from reflectances derived by field spectroradiometers, tower-mounted sensors, airborne and spaceborne instruments. Due in part to their

                    Remote Sensing of Biophysical Properties

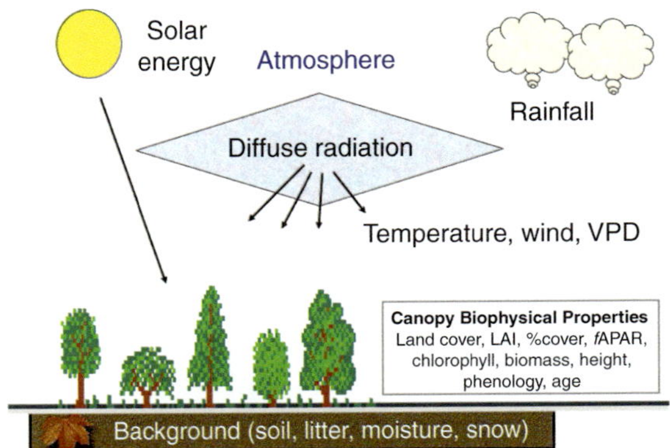

Figure 7.13. Remotely sensed vegetation canopy properties.

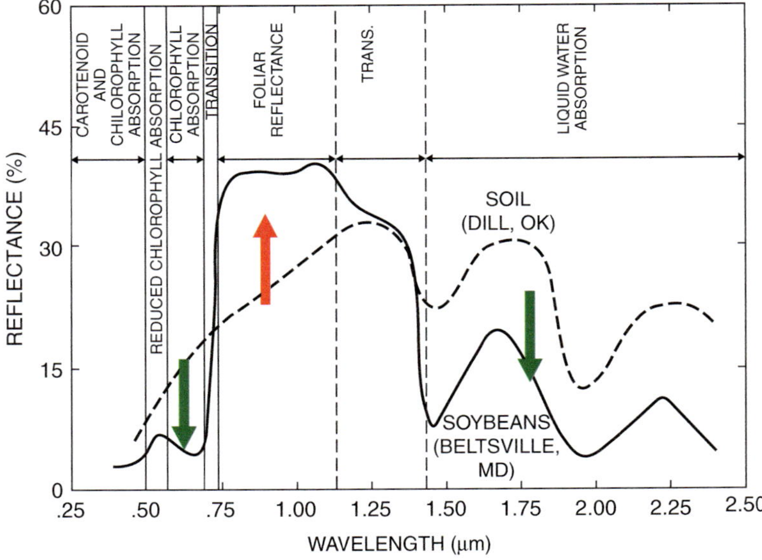

Figure 7.14. Contrasting spectral features between soil and green vegetation in the red, NIR and SWIR portions of the spectrum. Reproduced with permission from Tucker and Sellers (1986).

simplicity, they are the most widely used satellite products, providing key measurements in assessments of productivity, phenology, climate-vegetation interactions and biodiversity (Chapters 14–20). They have become indispensable tools across an array of ecohydrological, agricultural and natural resource management applications.

The theoretical basis for vegetation indices results from the contrast between a chlorophyll-absorbing leaf spectral feature with a non-absorbing one to capture plant biophysical phenomena (Fig. 7.14). Using the soil spectral signature as a baseline, increases in green vegetation will result in decreasing red reflectance simultaneous with increasing NIR reflectance and decreasing SWIR reflectance. The reflectance differences become larger under more densely vegetated conditions. Maximum red–NIR contrast occurs in densely foliated canopies with vigorous and healthy leaves; lower contrast is found in stressed or senescing canopies; the smallest contrast occurs in defoliated or sparse canopies.

Vegetation indices depict canopy *greenness*, an integrative measure of variations in leaf physiology, structure and area-averaged canopy photosynthetic activity. They are widely used as proxies in the assessment of canopy biophysical/ biochemical variables, including LAI, fAPAR, chlorophyll content, Fv, green biomass and canopy biophysical processes of evapotranspiration (ET) and gross primary production (GPP) (Field *et al.* 1995, Glenn *et al.* 2008). Several comprehensive reviews on the use of VIs to assess ecological properties are found in Kerr and Ostrovsky (2003), Pettorelli *et al.* (2005) and Huete and Glenn (2011).

There are a variety of ways in which two or more spectral bands can be combined to enhance the NIR–red reflectance contrast and provide measures of canopy greenness. This results in a multitude of VI formulae and variants that include two-band ratios and two-band differences, weighted differences and normalised differences (Tucker 1979), linear spectral band combinations, angle-based VIs (Jiang *et al.* 2006) and optimised spectral band combinations (Gobron *et al.* 2000, Huete 1988). VIs are positively related with plant canopy biophysical properties and, to a certain extent, many are functionally equivalent. Despite this, there are significant differences in how they depict vegetation foliage and multiple VIs can often provide a more complete characterization of complex canopies.

The Simple Ratio (SR) is the most basic measure of the red and NIR contrast of a pixel, computed as the ratio of NIR to red reflectance:

$$SR = \rho_{NIR} / \rho_{red}, \tag{7.5}$$

Normalised versions of these ratios are designed to constrain values between -1 and $+1$, of which the functionally equivalent normalised difference vegetation index (NDVI) is the most widely used (Tucker 1979):

$$NDVI = (SR - 1) / (SR + 1) = (\rho_{NIR} - \rho_{red}) / (\rho_{NIR} + \rho_{red}), \tag{7.6}$$

The SR and NDVI are successful as a vegetation measures in that they are sufficiently stable to permit meaningful comparisons of spatial, seasonal and inter-annual changes in vegetation growth and activity. As ratios, they reduce many forms of multiplicative noise (illumination differences, cloud shadows, atmospheric attenuation, certain topographic variations) present across multiple bands. However, they also exhibit certain disadvantages that relate to the inherent non-linearity of ratio-based indices and associated scaling problems. They are also subject to the influence of additive noise effects, such as atmospheric path radiances. Ratios are also very sensitive to variations in canopy background, with NDVI and SR values particularly high with darker soil backgrounds. The NDVI exhibits asymptotic (saturated) signals over

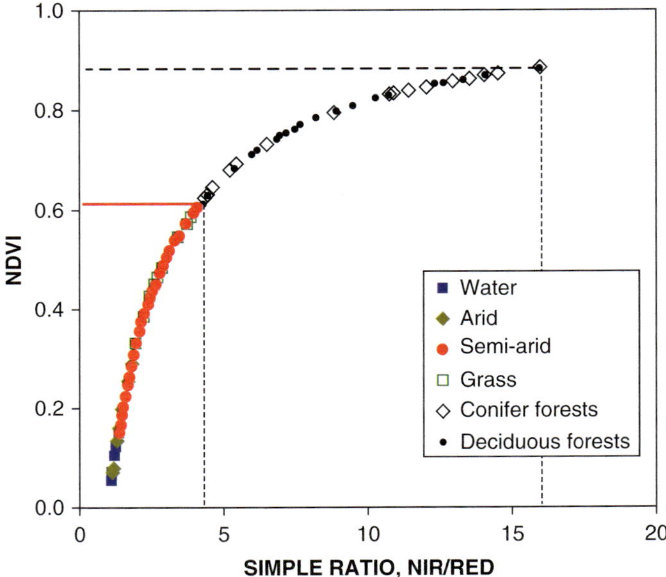

Figure 7.15. Functional equivalence in the simple ratio and NDVI equations across a range of global biome types.

high biomass conditions. One can note that the NDVI dynamic range is stretched in favour of low biomass conditions and compressed in high biomass, forested regions, relative to the simple ratio (Fig. 7.15). The opposite is true for the simple ratio, in which most of the dynamic range encompasses the high biomass, forests while very little variation is reserved for the lower biomass regions (grassland, semi-arid and arid biomes).

Optimised indices employ basic radiative transfer theory to account for and minimise atmosphere and/or soil-vegetation interactions. Verstraete and Pinty (1996) designed an AVHRR-based vegetation index, known as the Global Environment Vegetation Index (GEMI), to reduce the atmosphere effects in AVHRR time series data.

The soil-adjusted vegetation index (SAVI) incorporates a soil background coefficient, L, in order to remove the strong dependency of the NDVI on underlying soil background brightness:

$$\text{SAVI} = (1 + L) \, (\rho_{NIR} - \rho_{red}) \, / \, (L + \rho_{NIR} + \rho_{red}) \tag{7.7}$$

where ρ_{NIR} and ρ_{red} are reflectances in the NIR and red bands, respectively; L is the canopy background adjustment factor based upon differential red and NIR transmission through a canopy (Beer's law) (Huete 1988).

Figure 7.16 depicts the influence of soil on NDVI values derived from field measured canopy spectra of a broadleaf crop canopy (cotton). The NDVI values of a given amount of vegetation increase significantly in canopies with darker soil backgrounds and particularly at intermediate amounts of vegetation (Fig. 7.16a). The darker the

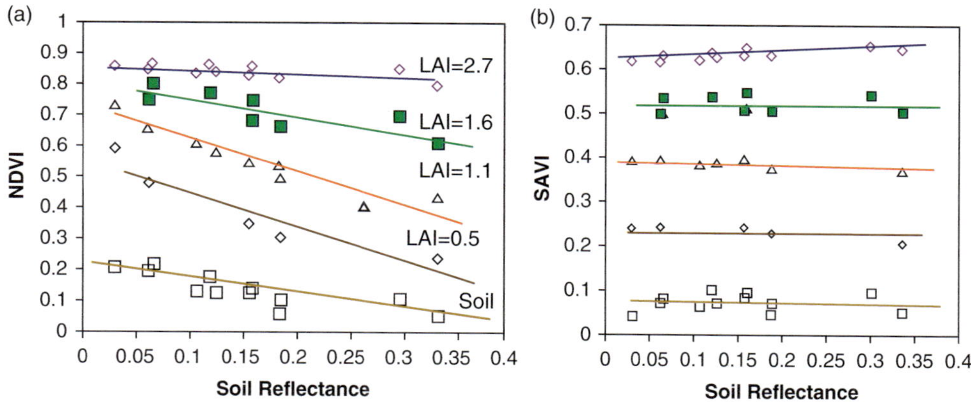

Figure 7.16. NDVI (a) and SAVI (b) relationships with green LAI for cotton canopies of varying LAI amounts and across a global range of soil types from dark organic soils to bright sandy soils (modified from Huete 1988).

canopy background, whether due to soil wetting, shadows, decomposed litter, roughness, or soil organic matter, the higher the NDVI value, for a constant amount of green vegetation. In contrast the soil-adjustment factor, *L*, in the SAVI equation achieves a general first order correction and removes the soil bias present in the NDVI, hence minimizing the need to assess soil reflectance (Fig. 7.16b).

Canopy background noise is more severe in open canopies at intermediate vegetation cover levels (30–70%), rather than in sparse canopies, since it is the 'coupled' influence of the canopy background and transmittance properties of the overlying canopy that determine the extent of noise in the VIs. The soil effect decreases in more dense canopies as the soil contribution weakens and also becomes weaker in sparser canopies as there is less vegetation to interact with the soil (Fig. 7.16). Removal or minimising this 'background' influence allows one to better interpret spatial and temporal variations associated with vegetation from variations associated with the canopy background.

The enhanced vegetation index (EVI) is another optimised index that derives its heritage from the SAVI and incorporates an aerosol resistance term developed in the atmosphere resistance vegetation index (ARVI; Kaufman and Tanré 1992), to stabilise aerosol influences. The EVI extracts canopy greenness, independent of the underlying soil background and atmospheric aerosol variations:

$$\text{EVI} = 2.5 \, (\rho_{\text{NIR}} - \rho_{\text{red}}) \, / \, (L + \rho_{\text{NIR}} + C1 \, \rho_{\text{red}} - C2 \, \rho_{\text{blue}}) \qquad (7.8)$$

where ρ_{NIR}, ρ_{red}, ρ_{blue} are reflectances in the NIR, red and blue bands, respectively; L is the canopy background adjustment factor; and C1 and C2 are aerosol resistance weights. The coefficients of the EVI equation, as used in MODIS and Landsat EVI are L = 1; C1 = 6 and C2 = 7.5 (Huete *et al.* 2002). Through a greater weighting on the NIR band, the EVI renders the VI signal more sensitive in high biomass areas relative to NDVI.

The "atmospheric resistance" concept is based on the wavelength dependency of aerosol effects, utilising the more atmosphere-sensitive blue band to correct the red band for aerosol influences. This is accomplished by finding a blue and red reflectance function that is stable against variations of atmospheric aerosol condition (Kaufman and Tanré 1992). Miura *et al.* (2001) and Xiao *et al.* (2003) showed the atmosphere-resistant VIs to successfully minimise aerosol effects in the Amazon during the biomass burning season and in Asia, respectively.

7.4.2 Linear Combination and Spectral Mixture Analysis

The spectral NIR and red bands may also be subtracted (NIR–red) or linearly combined ($a_{red} + b_{NIR}$) to depict and quantify vegetation status and green foliage amounts. Using hyperspetral data one can also take the first derivative of the 'red edge' to directly assess the slope of the red-NIR contrast. Linear vegetation indices include the perpendicular vegetation index (PVI; Richardson and Wiegand, 1977) and the Kauth and Thomas (1976) 'green vegetation index' transform (GVI).

The concept behind linear combination spectral indices and mixture models can be depicted through spectral band cross-plots, for example, the red-NIR cross-plot of Figure 7.17. A triangular cloud of points defined by vegetation and soil end-members characterise pixels of varying vegetation amounts and different canopy backgrounds representing a range of landscape surface conditions. The canopy background baseline is located close to the 1:1 line and represents the boundary condition of 'zero' vegetation, sometimes known as the "soil line". The end-members consist of dense green vegetation at the highest NIR and lowest red reflectances, bright canopy backgrounds (dry soil or snow) and dark backgrounds (wet soil, organic soil, or standing

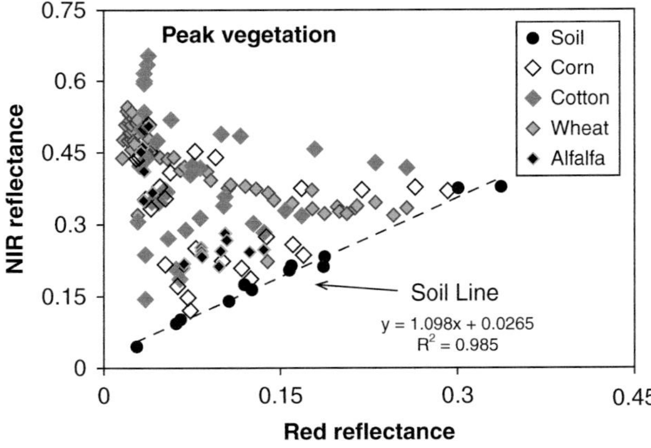

Figure 7.17. Cross-plot of red and NIR spectral reflectances for various crops, crop densities and soil backgrounds illustrating the triangular cloud of landscape features, end-members and component mixtures.

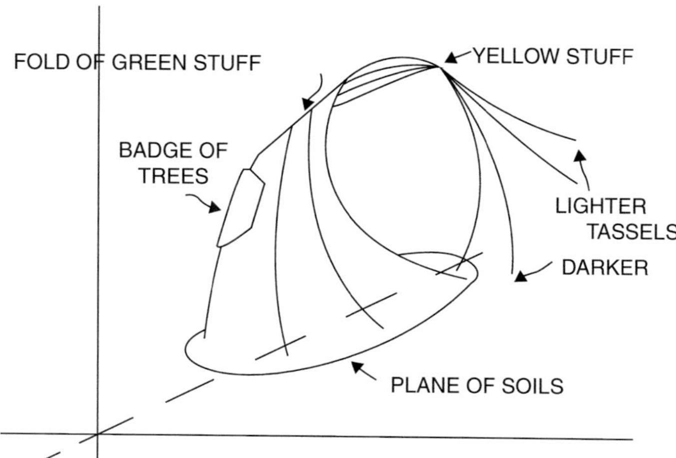

Figure 7.18. Tasseled Cap transform showing pixel trajectories as plants develop from bare soil through greening and subsequent senescing (yellowing) (adapted from Kauth and Thomas 1976).

water). The dense green apex exhibits the maximum red-NIR contrast while the soil baseline shows the smallest differences between red and NIR. Furthermore, all points inside the resulting triangular structure are mixed pixels composed of spectral signals from vegetation and canopy backgrounds. Partially vegetated pixels shift away from the lower baseline toward the apex of maximum NIR and lowest red reflectance in a manner dependent upon the optical and structural properties of the canopy and soil background type (Fig. 7.17). The larger the amount of green vegetation present in a pixel, the greater will be its red-NIR contrast and shift from the lower soil line.

Kauth and Thomas (1976) employed this concept in multi-dimensional spectral space utilizing all four bands available in Landsat MSS imagery and developed the Tasseled Cap Transform (Fig. 7.18) with a two-dimensional soil-plane base, a greenness axis representing the development of crops and a yellowness axis representing the drying and senescent phase of a crop's life cycle. Each axis is expressed by an index that linearly combines the four MSS spectral bands. The tasseled cap transform has more recently been applied to six-band Landsat ETM+ and MODIS imagery with additional axes, such as a Wetness feature that utilizes the SWIR bands (Crist and Cicone 1984).

Mixture models, by contrast, can provide information on vegetation, soil type and NPV properties for monitoring a range of landscape biophysical properties from multispectral broadband sensor systems such as Landsat Thematic Mapper, SPOT and MODIS. Hyperspectral sensors have improved the feasibility of unambiguously identifying numerous soil and plant absorption features, related to mineralogy, liquid water, chlorophyll, cellulose and lignin contents.

In spectral mixture analysis (SMA), the spectral responses from landscape surfaces are a function of the number and type of reflecting components, their

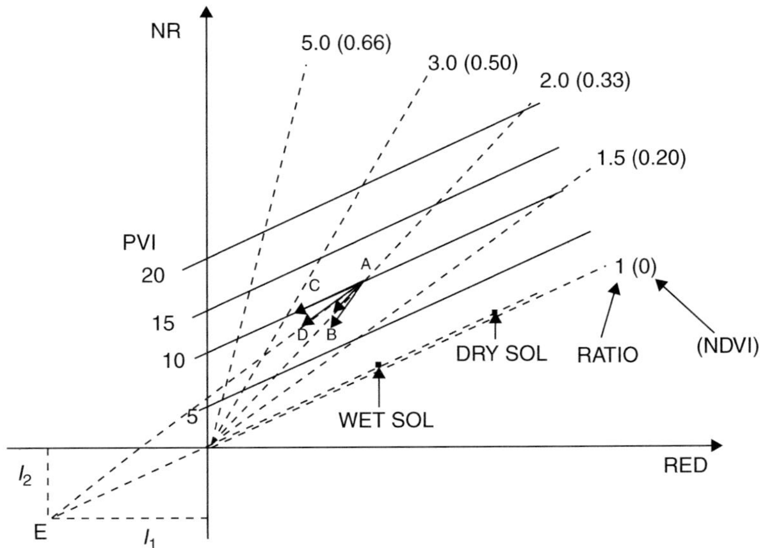

Figure 7.19. Conceptual diagram of vegetation index isolines (lines of constant VI value) in red–NIR space (Reproduced with permission from Huete 1988).

optical properties and their relative proportions. Thus, the weights that are used in linear combination indices become physically-based fractional (or concentration) amounts of a component feature within a pixel and the sum of all unmixed components add to one. The number of components is restricted to those with unique spectral signatures and by the number of bands available. Spectral mixture analysis has been widely used to unmix pixels into their respective soil, vegetation and NPV signal contributions, useful in soil and vegetation mapping, land degradation, soil erosion and land cover conversion studies. This has also been applied to hydrologic and biogeochemical landscape processes, to seasonally varying soil and vegetation cover and land cover conversions (Okin *et al.* 2001, Ustin *et al.* 2004).

The theoretical basis and defining characteristic of VIs (ratios, optimised, SMA and linear combinations) are in how they model the spectrally mixed pixels shown in Figure 7.17, their boundary conditions and associated variations in time and space. Linear combination and SMA indices model the triangular spectral cloud of pixels with parallel isolines (lines of equal VI values) of constant slope (≈ 1) and NIR intercepts that increase with increasing quantities of green vegetation (Fig. 7.19). This is in contrast to ratio-based indices such as the NDVI, in which VI isolines have a constant intercept (I = zero) and higher slopes with increasing amounts of vegetation. Optimised indices, on the other hand, involve the formulation of VI isolines to line up with actual biophysical isolines (e.g. lines of equal LAI), as in the SAVI in Figure 7.19. The SAVI has VI isolines with both varying intercepts and slopes in red and NIR cross-plots.

7.4.3 Vegetation Water Indices

Spectral vegetation indices may also be based on water absorption features instead of chlorophyll absorptions. Vegetation water indices are spectral measures of the moisture status of a canopy. They combine a water-absorbing leaf spectral feature in the shortwave infrared with the NIR to provide measures of canopy moisture content, as in the Moisture Stress Index (MSI) (Hunt and Rock 1989):

$$MSI = \rho_{NIR} / \rho_{SWIR} \qquad (7.9)$$

and the normalised difference water index (NDWI, Gao 1996):

$$NDWI = (\rho_{NIR} - \rho_{1240nm}) / (\rho_{NIR} + \rho_{1240nm}) \qquad (7.10)$$

Hardisky *et al.* (1983) developed the Normalised Difference Infrared Index (NDII), contrasting the NIR with SWIR (~1600 nm) wavelengths:

$$NDII = (\rho_{NIR} - \rho_{SWIR}) / (\rho_{NIR} + \rho_{SWIR}) \qquad (7.11)$$

This index was found to be strongly correlated with canopy water content. Xiao *et al.* (2006) used a similar normalised difference formulation between the NIR and SWIR bands (1580–1750 nm) from the SPOT-4 VEGETATION (VGT) and MODIS sensors and referred this as the Land Surface Water Index (LSWI). Fensholt and Sandholt (2003) also formulated the Shortwave Infrared Water Stress Index (SIWSI) from daily MODIS NIR and SWIR (1628–1652 nm) reflectances:

$$SIWSI = (\rho_{SWIR} - \rho_{NIR}) / (\rho_{SWIR} + \rho_{NIR}) \qquad (7.12)$$

and reported high correlations between SIWSI and soil moisture in the root zone in a Sahel vegetation study in Senegal (Fig. 7.20).

The global vegetation moisture index (GVMI) was developed for retrieval of equivalent water thickness (EWT, gH_2O/cm^2 leaf area) in a canopy (Ceccato *et al.* 2002):

$$GVMI = [(\rho_{NIR} + 0.1) - (\rho_{SWIR} + 0.02)] / [(\rho_{NIR} + 0.1) + (\rho_{SWIR} + 0.02)] \quad (7.13)$$

The use of two or more bands help minimise variations in leaf internal structure and leaf dry matter content as well as canopy geometry, shadowing and soil surface moisture that also influence SWIR reflectance.

7.4.4 Use of Multiple Vegetation Indices

The use of multiple VIs offers a more complete characterization of canopy properties. The advantages of the ratio indices (NDVI) is in their versatility under varying light conditions because the NDVI can reasonably be computed with airborne and field sensors under both cloudy and sunlit conditions. Linear combination indices and spectral mixtures have the advantage of simultaneous measurements of vegetation,

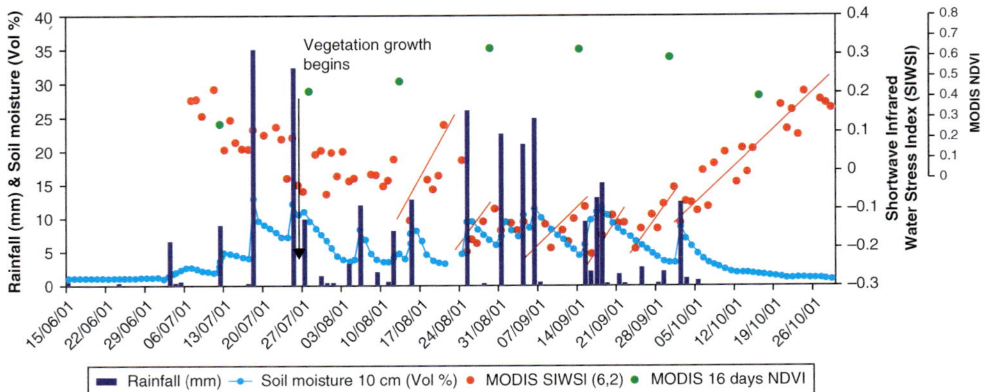

Figure 7.20. Daily measurements of rainfall (annual SUM=288 mm) and soil moisture, MODIS SIWSI(6,2) and MODIS 16-day composite NDVI in Dahra from June 15 to October 29, 2001 (Reproduced with permission from Fensholt and Sandholt, 2003).

soil and NPV. Optimised VIs incorporate physical theory, such as Beer's law, to best depict vegetation conditions. In addition, the PVI, EVI and TC greenness are more spectrally sensitive to NIR and will contain information from multiple canopy leaf layers due to the higher canopy optical penetration depths. In contrast, the NDVI has low optical depth penetration into canopies due to its heavier reliance on the red reflectance and thus will sense primarily the uppermost leaf layer.

In combination, chlorophyll-based VIs and vegetation water indices can add significant value in understanding canopy ecophysiological functioning, with model- and lab-based studies showing these indices to independently estimate canopy chlorophyll and water contents (Zarco-Tejada *et al.* 2003). However, water stress can only be detected in leaf scale optical measurements or under natural field conditions when severe water stress occurs (Bates and Hall 1981). Nevertheless, water indices have many applications and have been found useful in assessing canopy drying for fuel assessments and fire vulnerability (Caccamo *et al.* 2011).

7.4.5 LAI and Fractional Cover

Biophysical relationships of VIs with canopy properties are complicated by a lack of consensus on what VIs explicitly measure about a canopy and how to interpret a 'greenness' value. As a result VIs have been mostly used as empirical surrogates of canopy biophysical quantities. Vegetation indices respond to upper sunlit leaves to a larger extent than lower leaves, resulting in strongly non-linear relationships with field-derived LAI values. Generally VIs asymptotically approach saturation levels for LAI ranging from two to six, depending on which type of VI is used, canopy geometry and soil type. This relationship further differs in broadleaf *versus* needle leaf vegetation with NDVI sensitivity to LAI restricted to values below two or three and with differing correlations in broadleaf forests compared with coniferous stands (Fassnacht *et al.* 1997, Chen *et al.* 2005). NDVI–LAI relationships in a beech deciduous forest in Europe also varied significantly over different phenological periods, with poor correlations during periods of maximum LAI due to saturation of NDVI (Wang *et al.* 2005).

Linear combination and optimised indices offer extended LAI sensitivity as they are less prone to saturate in high-biomass areas (Fensholt *et al.* 2004). Houborg and Soegaard (2004) found MODIS EVI to accurately describe variations in green LAI up to five over agriculture areas in Denmark ($r^2 = 0.91$). Generally, poorer relationships are found when using coarser resolution VIs, such as MODIS and SPOT-VEGETATION (VGT), in comparison with relationships found with finer resolution Landsat satellite data. However, this may be more related to the added difficulties in validating disparate footprint size measurements.

A more sophisticated approach is to use physically based models that explicitly describe radiation transfer within canopies. These include geometrical–optical models and radiative transfer models (e.g. Myneni *et al.* 1989). Increasingly, satellite-based LAI retrieval algortihms are being developed.

The MODIS LAI/FPAR algorithm retrieves LAI and *f*APAR values based on comparisons of MODIS bi-directional surface reflectance and modelled reflectance stored in a Look-Up-Table (LUT) for an eight-biome classification map (Myneni *et al.* 2002; Chapter 1). The simulated reflectance values, per biome type, are generated using a 3-dimensional canopy radiative transfer model. There have been various global inter-comparisons of satellite-based LAI products across different sensor systems, along with their validation (Garrigues *et al.* 2008). A global review of LAI measurements derived with remote sensing is presented in Asner *et al.* (2003).

Fractional vegetation cover (F_v) is important for soil erosion studies and used in evapotranspiration and climate models. Vegetation Indices have been successfully correlated with F_v in Landsat data but with strong relationship dependencies with the extent of tree clumping, LAI and tree species (Smith *et al.* 2009). F_v can also be derived through spectral mixture analysis (SMA) and through seasonal time series measures. Donohue *et al.* (2008) used annual NDVI maximum and minimum values to compute seasonal foliage cover as:

$$Fv = (NDVI - NDVI_s) / (NDVI_g - NDVI_s),\qquad(7.14)$$

where $NDVI_s$ is the bare soil or minimum value of NDVI and $NDVI_g$ is the maximum NDVI value for dense vegetation.

7.4.6 fAPAR and Productivity

The absorbed photosynthetically active radiation (APAR), is the energy available for photosynthesis:

$$APAR = PAR_0 - R_c - T_c + T_c * R_{soil}\qquad(7.15)$$

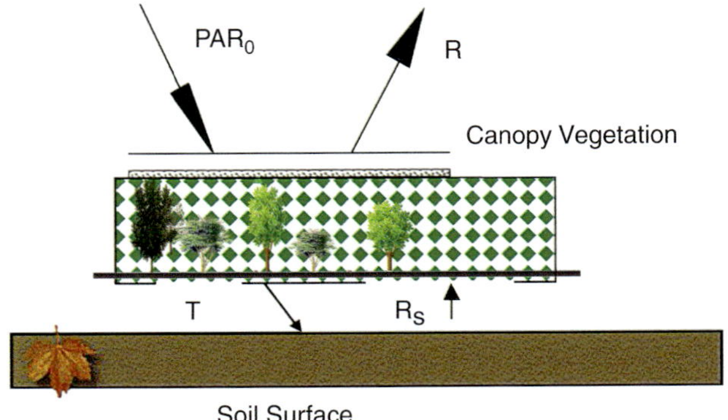

Figure 7.21. Diagram showing the components of photosynthetically active radiation (PAR) though an open canopy.

where PAR_o is the incident PAR; R_c is the hemispherical canopy-reflected PAR; T_c is the hemispherical canopy-transmitted PAR; and R_{soil} is soil reflectance.

The fraction of absorbed photosynthetically active radiation (fAPAR) is:

$$fAPAR = APAR / PAR_0, \text{ or} \qquad\qquad (7.16)$$

$$fAPAR = 1 - R_c - (1 - R_{soil}) * T_c \qquad\qquad (7.17)$$

The fraction absorbed PAR by the canopy (*f*APAR) can be linked experimentally to reflectance data and vegetation indices (Fig. 7.22; Asrar *et al.* 1984).

By theoretical analyses, Asrar *et al.* (1984) and Myneni and Williams (1994) showed *f*APAR to be linearly related with NDVI (Fig. 7.22) over a range of land cover types and leaf angle distributions (LAD) suggesting that covariance between *f*APAR and NDVI is insensitive to vegetative heterogeneity (Huemmrich *et al.* 2005, Fensholt *et al.* 2004).

Sims *et al.* (2006) empirically utilised a global-based, *f*APAR–NDVI relationship derived from multiple biomes:

$$fAPAR = 1.24 \times NDVI - 0.168 \qquad\qquad (7.18)$$

However, the type of soil background as well as non-green *f*APAR associated with NPV can alter NDVI–*f*APAR relationships and affect their linearity with important consequences to scaling (Jiang *et al.* 2006). Whereas NPV is present in only small amounts in annual herbaceous plants (forbs, grasses, crops), the NPV component may become significant in perennial grasslands and woody (shrub-tree) ecosystems. For example, the *f*APAR measured in a Kalahari woodland field campaign varied distinctly with NDVI during phenologic green-up and dry-down phenophases (Huemmrich *et al.* 2005).

Zhang *et al.* (2005) explored the differences between green and total *f*APAR and with use of a radiative transfer model combined with MODIS data, distinguished

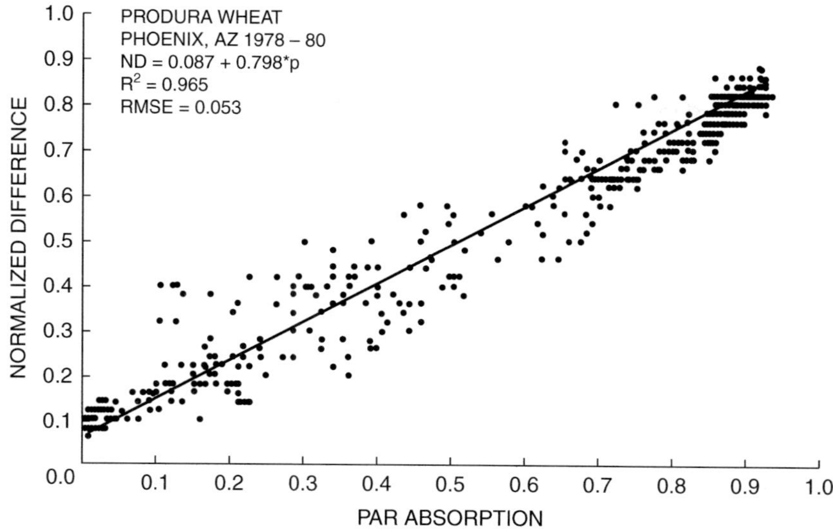

Figure 7.22. Linear relationship between NDVI and *f*APAR for wheat crop (Reproduced with permission from Asrar *et al.* 1984).

among chlorophyll-, leaf–and canopy-absorbed *f*APAR components. Large differences were found between the chlorophyll *f*APAR and canopy *f*APAR, noting that only the chlorophyll-absorbed *f*APAR is used in photosynthesis. They reported NDVI to be well correlated with total canopy *f*APAR, while EVI was better correlated with the chlorophyll *f*APAR, or green *f*APAR.

Chlorophyll spectral measures and indices have also been developed and investigated through extensive field measurements over a range of croplands (e.g. Gitelson *et al.* 2014). This has resulted in various chlorophyll indices (CI) for estimating total canopy chlorophyll content as:

$$CI \text{ (green)} = \rho_{NIR} / \rho_{green} - 1, \text{ and} \qquad (7.19)$$

$$CI \text{ (red edge)} = \rho_{NIR} / \rho_{red\,edge} - 1 \qquad (7.20)$$

Gitelson *et al.* (2014) found that spectral measures of total chlorophyll content related closely to green *f*APAR and significantly improved upon the estimation of gross primary production in croplands. But, as chlorophyll exceeds 2 g m^{-2}, the sensitivity of the relationship of green *f*APAR to chlorophyll content declined sharply.

A new satellite, chlorophyll-based product from the Medium Resolution Imaging Spectrometer (MERIS), known as the MERIS Total Chlorophyll Index (MTCI), is derived from inverse canopy reflectance modelling, at 300 m spatial resolution, by simultaneously mapping canopy chlorophyll content and LAI (Vuolo *et al.* 2012):

$$MTCI = (\rho_{754} - \rho_{709}) / (\rho_{709} - \rho_{681}) \qquad (7.21)$$

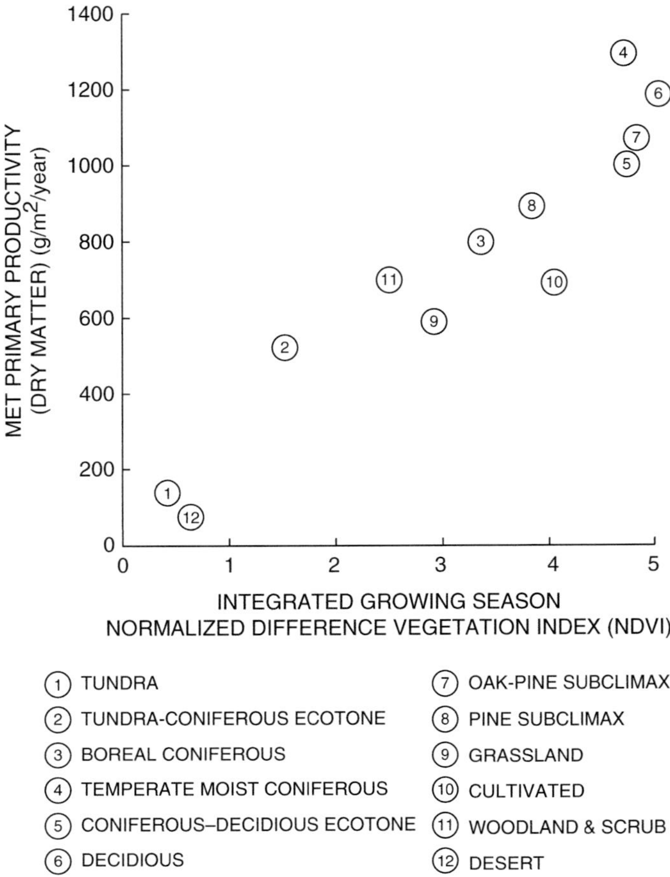

Figure 7.23. Integrated growing season NDVI values from the AVHRR instrument across major biome types of North America (Reproduced with permission from Goward *et al.* 1985).

where ρ_{754}, ρ_{709} and ρ_{681} are the normalised surface reflectance in the centre wavelengths of MERIS bands 8, 9 and 10 (754, 709 and 681 nm, respectively).

As the photosynthetic apparatus of a plant canopy transforms absorbed PAR energy into dry matter, the cumulative sum of *f*APAR over time can be used as a good estimate of net primary production (NPP). Goward *et al.* (1985) integrated AVHRR–NDVI values (iNDVI) over the growing season to estimate patterns in North American NPP across a wide range of biomes (Fig. 7.23). They found a good relationship of iNDVI, as a surrogate *f*APAR and thereby NPP for the growing season (Fig. 7.23).

Several subsequent studies suggest that annual integrals are a robust approximation of collective plant behaviour (Running *et al.* 2004). At seasonal time scales, there is a need to consider environmental stress factors and drivers of productivity. However, environmental stress forcings show up as reductions in NDVI via less chlorophyll and/or less foliage (LAI) and thereby such decreased productivity status

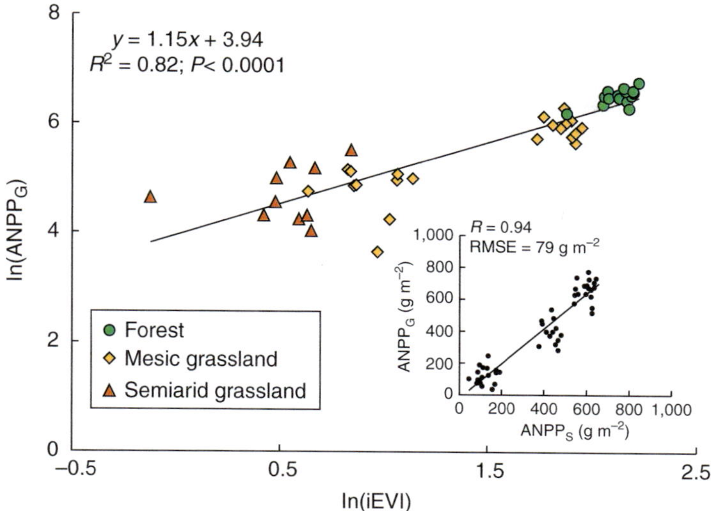

Figure 7.24. Relationship between *in situ* estimates of above-ground net primary production (ANPP$_G$) and annual integrated EVI (iEVI) derived from MODIS data (2000–2009) for ten sites across several biomes. The solid line represents the linear regression used to estimate ANPP from iEVI (ANPP$_s$). The inset shows the correlation between estimates of ANPP$_s$ and ANPP$_G$ for the ten sites over several years (Reproduced with permission from Ponce-Campos *et al.* 2013).

is captured by NDVI at these coarser time scales (weekly to monthly), dependent on biome type.

Annual integrated satellite observations of MODIS EVI have also been used as surrogates for annual above-ground NPP (ANPP) to quantify inter-annual vegetation dynamics across grassland to forest biomes (Zhang *et al.* 2013, Ponce-Campos *et al.* 2013). Ponce-Campos *et al.* (2013) calibrated *in-situ* estimates of ANPP made with conventional field methods between 2000 and 2009 for ten sites across the United States and related these with iEVI measurements for the same site and year. The resulting log–log regression equation showed iEVI to be a useful proxy for ANPP and this was subsequently used to estimate ANPP from iEVI values over a range of U.S. and Australian biomes (Fig. 7.24).

7.5 Conclusions

How plants interact with sunlight is central to the functioning of ecosystems and provides a basis for remote sensing of biogeochemical cycles of carbon and water fluxes at biome scales. The aim of airborne and satellite remote sensing of landscapes is to exploit and model the complex patterns of radiation interactions in order to better define, characterise and map, the composition, quantities, properties and health of vegetated land surfaces. This is accomplished through the use of multi-band spectral indices, spectral mixture analyses and physical radiation models to retrieve soil and vegetation information.

The accuracy and interpretation of retrievals of various landscape properties may be hampered by several external sources of noise and bias in satellite imagery that confound the landscape signals reflected from the surface. Potential sources of noise that may confuse landscape biophysical interpretations include: BRDF effects associated with systematic variations in sensor viewing and solar illumination geometries; topographic effects in which illumination conditions vary dramatically with slope and aspect of the surface; snow effects; and residual cloud and aerosol contamination. Snow and cloud masks are used to identify and filter contaminated pixels in satellite products, while BRDF models can be used to normalise sun angle conditions and viewing geometries to more accurately interpret changes in landscape signals.

Remote sensing can provide repeatable, frequent, long term and accurate satellite measures for global change and climate change impact studies of landscapes. Future advancements in remote sensing of the Earth's landscapes can be expected through more in-depth integration of remote sensing with ecological theory.

7.6 References

Asner GP, (1998). Biophysical and biochemical sources of variability in canopy reflectance. *Remote Sensing of Environment* 64, 234–253.

Asner GP, RE Martin, KM Carlson, U Rascher and PM Vitousek, (2006). Vegetation climate interaction among native and invasive species in Hawaiian rainforest. *Ecosystems* 9, 1106–1117.

Asner GP, JMO Scurlock and JA Hicke, (2003). Global synthesis of leaf area index observations: implications for ecological and remote sensing studies. *Global Ecology and Biogeography* 12, 191–205.

Asrar G, M Fuchs, ET Kanemaus and JL Hatfield, (1984). Estimating absorbed photosynthetic radiation and leaf area index from spectral reflectance in wheat. *Agronomy Journal* 76, 300–306.

Baret F and G Guyot (1991). Potentials and limits of vegetation indices for LAI and APAR assessment. *Remote Sensing of Environment* 35, 161–173.

Bates LM and AE Hall, (1981). Stomatal closure with soil water depletion not associated with changes in bulk leaf water status. *Oecologia* 50, 62–65.

Baumgardner MF, LF Silva, LL Biehl and ER Stoner, (1985). Reflectance properties of soils. *Advances in Agronomy* 38, 1–44.

Ben-Dor E and A Bannin, (1994). Visible and near infrared (0.4–1.1 μm) analysis of arid and semi-arid soils. *Remote Sensing of Environment* 48, 261–274.

Ben-Dor E, RG Taylor, J Hill, JAM Dematte, ML Whiting, S Chabrillat and S Sommer, (2008). Imaging spectrometry for soil applications. *Advances in Agronomy* 97, 321–392.

Blackburn GA, (2007). Hyperspectral remote sensing of plant pigments. *Journal of Experimental Botany* 58, 855–867. doi:10.1093/jxb/erl123.

Caccamo G, LA Chisholm, RA Bradstock and ML Puotinen, (2011). Assessing the sensitivity of MODIS to monitor drought in high biomass ecosystems. *Remote Sensing of Environment* 115, 2626–2639. Elsevier Inc. doi:10.1016/j.rse.2011.05.018

Carter GA and AK Knapp, (2001). Leaf optical properties in higher plants: Linking spectral characteristics to stress and chlorophyll concentration. *American Journal of Botany* 88, 67–84.

Ceccato P, N Gobron, S Flasse, B Pinty and S Tarantola, (2002). Designing a spectral index to estimate vegetation water content from remote sensing data: Part 1. Theoretical approach. *Remote Sensing of Environment* 82, 188–197.

Chen X, L Vierling, D Deering and A Conley, (2005). Monitoring boreal forest leaf area index across a Siberian burn chronosequence: a MODIS validation study. *International Journal of Remote Sensing* 26(24): 5433–5451.

Clark RN, (1999). Chapter 1: Spectroscopy of Rocks and Minerals and Principles of Spectroscopy, In: *Manual of Remote Sensing*, (Rencz AN, Ed.) John Wiley and Sons, New York, pp. 3–58.

Crist EP and RC Cicone, (1984). A physically based transformation of Thematic Mapper data–the TM Tasseled Cap. *IEEE Transactions on Geoscience and Remote Sensing* GE-22, 256–263.

Donohue RJ, ML Roderick and TR McVicar, (2008). Deriving consistent long-term vegetation information from AVHRR reflectance data using a cover-triangle-based framework. *Remote Sensing of Environment* 112, 2938–2949.

Elvidge CD, (1990). Visible and near infrared reflectance characteristics of dry plant materials. *International Journal of Remote Sensing* 11, 1775–1795.

Fassnacht KS, ST Gower, MD MacKenzie, EV Nordheim and TM Lillesand, (1997). Estimating the leaf area index of North Central Wisconsin forests using the Landsat Thematic Mapper. *Remote Sensing of Environment* 61(2):229–245.

Fensholt R and I Sandholt, (2003). Derivation of a shortwave infrared water stress index from MODIS near–and shortwave infrared data in a semiarid environment. *Remote Sensing of Environment* 87, 111–121.

Fensholt R, I Sandholt and MS Rasmussen, (2004). Evaluation of MODIS LAI, fAPAR and the relation between fAPAR and NDVI in a semi-arid environment using *in situ* measurements. *Remote Sensing of Environment* 91, 490–507.

Field CB, JT Randerson and CM Malmstrom (1995). Global net primary production: Combining ecology and remote sensing. *Remote Sensing of Environment* 51, 74–88.

Franklin J, J Duncan and D Turner, (1993). Reflectance of vegetation and soil in Chihuahuan desert plant communities from ground radiometry using SPOT wavebands. *Remote Sensing of Environment* 46, 291–304.

Gao BC, (1996). NDWI–A normalized difference water index for remote sensing of vegetation liquid water from space. *Remote Sensing of Environment* 58, 257–266.

Gamon JA, J Penuelas and CB Field, (1992). A narrow-waveband spectral index that tracks diurnal changes in photosynthetic efficiency. *Remote Sensing of Environment* 41, 35–44.

Garrigues S, R Lacaze, F Baret, JT Morisette, M Weiss, JE Nickeson, R Fernandes, S Plummer, NV Shabonov, RB Myneni, Y Knyazikhin and W Yang, (2008). Validation and inter-comparison of global leaf area index products derived from remote sensing data. *Journal of Geophysical Research*, 113:doi:10.1029/2007JG000635.

Gates DM, HJ Keegen, JC Schleter and VR Weidner, (1965). Spectral properties of plants. *Applied Optics* 4, 11–20.

Gitelson AA, Y Gritz and MN Merzlyak. (2003). Relationships between leaf chlorophyll content and spectral reflectance and algorithms for non-destructive chlorophyll assessment in higher plant leaves. *Journal of Plant Physiology* 160, 271–82.

Gitelson AA, OB Chivkunova and MN Merzlyak, (2009). Non-destructive estimation of anthocyanins and chlorophylls in anthocyanic leaves. *American Journal of Botany* 96, 1861–1868.

Gitelson AA, Y Peng, TJ Arkebauer and J Schepers, (2014). Relationships between gross primary production, green LAI and canopy chlorophyll content in maize: Implications for remote sensing of primary production. *Remote Sensing of Environment* 144, 65–72.

Glenn EP, AR Huete, PL Nagler and SG Nelson, (2008). Relationship between remotely sensed vegetation indices, canopy attributes and plant physiological processes: what vegetation indices can and cannot tell us about the landscape. *Sensors* 8, 2136–2160.

Gobron N, B Pinty, MM Verstraete and JL Widlowski, (2000). Advanced vegetation indices optimized for upcoming sensors: Design, performance and applications. *IEEE Transactions of Geosciences and Remote Sensing* 38, 2489–2505.

Goward SN, CJ Tucker and DG Dye. (1985). North American vegetation patterns observed with the NOAA-7 Advanced Very High Resolution Radiometer. *Vegetatio* 64, 3–14.

Graetz RD, (1990). Remote sensing of terrestrial ecosystem structure: an ecologist's pragmatic view, In: Hobbs RJ and HA Mooney, Eds., *Remote Sensing of Biosphere Functioning*. Springer-Verlag, New York, pp. 5–30.

Hardisky MA, RM Smart and V Klemas, (1983). Seasonal spectral characteristics and above ground biomass of the tidal marsh plant *Spartina alterniflora*. *Photogrametric Engineering and Remote Sensing* 49, 85–92.

Huemmrich KF, JL Privette, M Mukelabai, RB Myneni and Y Knyazikhin, (2005). Time series validation of MODIS land biophysical products in a Kalahari woodland, Africa. *International Journal of Remote Sensing* 26, 4381–4398.

Huete AR, (1988). A soil-adjusted vegetation index (SAVI). *Remote Sensing of Environment* 25, 295–309.

Huete AR, K Didan, T Miura, EP Rodriguez, X Gao and LG Ferreira, (2002). Overview of the radiometric and biophysical performance of the MODIS vegetation indices. *Remote Sensing of Environment* 83, 195–213.

Huete AR and EP Glenn, (2011). Recent advances in remote sensing of ecosystem structure and function, In: *Advances in Environmental Remote Sensing: Sensors, Algorithms and Applications* (Weng Q, Ed.), CRC Press, Taylor and Francis Group, pp. 291–319.

Hunt ER and BN Rock, (1989). Detection of changes in leaf water content using near–and middle–infrared reflectances. *Remote Sensing of Environment* 30, 43–54.

Irons JR, G Campbell, JM Norman, DW Graham and WM Kovalick, (1992). Prediction and measurement of soil bidirectional reflectance. *IEEE Transactions on Geoscience and Remote Sensing* GE-30, 249–260.

Jacquemoud S, (1990). PROSPECT: a model of leaf optical properties spectra. *Remote Sensing of Environment* 34, 75–91.

Jiang Z, AR Huete, J Li and Y Chen, (2006). An analysis of angle-based with ratio-based vegetation indices. *IEEE Transactions on Geoscience and Remote Sensing*, 44, 2506–2513.

Karnieli A, GJ Kidron, C Glaesser and E Ben-Dor, 1999. Spectral characteristics of cyanobacteria soil crust in semiarid environments. *Remote Sensing of Environment* 69, 67–75.

Kaufman Y and D Tanré, (1992). Atmospherically Resistant Vegetation Index (ARVI) for EOS-MODIS. *IEEE Transactions on Geoscience and Remote Sensing* 30, 261–270.

Kauth RJ and GS Thomas, (1976). The tasseled cap–a graphic description of the spectral-temporal development of agricultural crops as seen by Landsat. Proceedings of the Symposium on Machine Processing of Remotely Sensed Data, Purdue University, West Lafayette, Indiana USA, pp. 41–51.

Kerr JT and M Ostrovsky, (2003). From space to species: Ecological applications for remote sensing. *Trends in Ecology and Evolution* 18, 299–305.

Knipling EB, (1970). Physical and physiological basis for the reflectance of visible and near-infrared radiation from vegetation. *Remote Sensing of Environment* 1, 155–159.

Miura T, AR Huete, H Yoshioka and BN Holben, (2001). An error and sensitivity analysis of atmospheric resistant vegetation indices derived from dark target-based atmospheric correction. *Remote Sensing of Environment* 78, 284–298.

Myneni RB, Hoffman S, Knyazikhin Y, Privette JL, Glassy J, Tian Y and Wang Y, 2002. Global products of vegetation leaf area and fraction absorbed PAR from year one of MODIS data. *Remote Sensing of Environment* 83, 214–231.

Myneni RB, J Ross and G Asrar, (1989). A review on the theory of photon transport in leaf canopies. *Agricultural and Forest Meteorology* 45, 1–153.

Myneni RB and DL Williams, (1994). On the relationship between FAPAR and NDVI. *Remote Sensing of Environment* 49, 200–211.

Nagler PL, Y Inoue, EP Glenn, AL Russ and CST Daughtry, (2003). Cellulose absorption index (CAI) to quantify mixed soil-plant litter scenes. *Remote Sensing of Environment* 87, 310–325.

Okin GS, D Roberts, D, B Murray and WJ Okin, (2001). Practical limits on hyperspectral vegetation discrimination in arid and semiarid environments. *Remote Sensing of Environment* 77, 212–225.

Ollinger SV, (2010). Sources of variability in canopy reflectance and the convergent properties of plants. *New Phytologist* doi: 10.1111/j.1469-8137.2010.03536.x, pp. 1–20.

Pettorelli N, JO Vik, A Mysterus, J-M Gaillard, CJ Tucker and NC Stenseth, (2005). Using the satellite-derived NDVI to assess ecological responses to environmental change. *Trends in Ecology and Evolution* 20, 503–510.

Ponce-Campos GE, MS Moran, AR Huete, Y Zhang, C Bresloff, TE Huxman, D Eamus, DD Bosch, AR Buda, SA Gunter, TH Scalley, SG Kitchen, MP McClaran, WH McNab, DS Montoya, JA Morgan, DPC Peters, EJ Sadle, MS Seyfrie and PJ Starks, (2013). Ecosystem resilience under large-scale altered hydroclimatic condition. *Nature* 494, 349–352.

Richardson AJ and CL Wiegand, (1977). Distinguishing vegetation from soil background information. *Photogrammetric Engineering and Remote Sensing*, 43, 1541–1552.

Ripullone F, AR Rivelli, R Baraldi, R Guarini, R Guerrieri, F Magnani and S Raddi, (2011). Effectiveness of the photochemical reflectance index to track photosynthetic activity over a range of forest tree species and plant water statuses. *Functional Plant Biology* 38, 177–186.

Running SW, RS Nemani, FA Heinsch M Zhao, M Reeves and H Hashimoto, (2004). A continuous satellite-derived measure of global terrestrial primary production. *BioScience* 54, 547–560.

Sims DA, AF Rahman, VD Cordova, BZ El-Masri, DD Baldocchi, LB Flanagan, AH Goldstein, DY Hollinger, L Misson, RK Monson, WC Oechel, HP Schmid, SC Wofsy and L Xu, (2006). On the use of MODIS EVI to assess gross primary productivity of North American ecosystems. *Journal of Geophysical Research* 111, G04015, doi:04010 .01029/02006JG000162.

Sims DE and JA Gamon, (2002). Relationships between leaf pigment content and spectral reflectance across a wide range of species, leaf structures and developmental stage. *Remote Sensing of Environment* 81, 337–354.

Smith AMS., MJ Falkowski, AT Hudak, JS Evans, AP Robinson and CM Steele, (2009). A cross-comparison of field, spectral and lidar estimates of forest canopy cover. *Canadian Journal of Remote Sensing* 35, 447–459.

Smith WK., TC Vogelmann, EH Delucia, DT Bell, and KA Shepherd, (1997). Leaf form and photosynthesis: do leaf structure and orientation interact to regulate internal light and carbon dioxide? *BioScience* 47, 785–793.

Tucker CJ, (1979). Red and photographic infrared linear combinations for monitoring vegetation. *Remote Sensing of Environment* 8, 127–150.

Tucker CJ and PJ Sellers (1986). Satellite remote sensing of primary productivity, *International Journal of Remote Sensing* 7, 1395–1416.

Ustin SL, DA Roberts, JA Gamon, GP Asner and RO Green, (2004). Using imaging spectroscopy to study ecosystem processes and properties. *BioScience*, 54, 523–534.

Verstraete MM and B Pinty, (1996). Designing optimal spectral indexes for remote sensing applications. *IEEE Transactions of Geosciences and Remote Sensing* 34, 1254–1265.

Vuolo F, J Dash, PJ Curran, D Lajas and E Kwiatkowska, (2012). Methodologies and uncertainties in the use of the terrestrial chlorophyll index for the sentinel-3 mission. *Remote Sensing* 4, 1112–1133.

Wang Q, S Adiku, J Tenhunen and A Granier, (2005). On the relationship of NDVI with leaf area index in a deciduous forest site. *Remote Sensing of Environment* 94, 244–255.

Xiao X, B Braswell, Q Zhang, S Boles, S Frolking and I Moore, (2003). Sensitivity of vegetation indices to atmospheric aerosols: continental-scale observations in Northern Asia. *Remote Sensing of Environment* 84, 385–392.

Xiao X, SS Boles, C Frolking, JY Li, W Babu and B Moore III, (2006). Mapping paddy rice agriculture in south and southeast Asia using multi-temporal MODIS images. *Remote Sensing of Environment* 100, 95–113.

Zarco-Tejada PJ, CA Rueda and SL Ustin, (2003). Water content estimation in vegetation with MODIS reflectance data and model inversion methods. *Remote Sensing of Environment* 85, 109–124.

Zhang QY, XM Xiao, B Braswell, E Linder, F Baret and B Moore, (2005). Estimating light absorption by chlorophyll, leaf and canopy in a deciduous broadleaf forest using MODIS data and a radiative transfer model. *Remote Sensing of Environment* 99, 357–371.

Zhang Y, MS Moran, MA Nearing, G Ponce, AR Huete and HSG Kitchen, (2013). Extreme precipitation patterns reduced terrestrial ecosystem production across biomes. *Journal of Geophysical Research, Biogeosciences* 118, doi: 10.1029/2012JG002136.

Section Three

Modelling

8

An Introduction to Modelling in Plant Ecophysiology

8.1 Introduction

Knowledge of the exchange of water, carbon and energy above plant canopies provides a fundamental basis for understanding climate-vegetation interactions, land surface processes, carbon sequestration accounting, water-use-efficiency and water management. In this chapter, plant-environment interactions are discussed through insight of biophysical and biological processes (Chapters 2 and 3) within the soil-plant-atmosphere continuum (SPAC).

Theoretically, water and energy transfer in the SPAC provides the physical basis for understanding plant growth in relation to environmental variation, while plant physiological regulation is the biological basis for plant growth and environmental feedbacks. For example, increased diffuse radiation arising from cloud cover can increase canopy light-use-efficiency. Similarly increased atmospheric CO_2 concentration can lift photosynthetic rates and decrease transpiration. Changes in rates of carbon uptake and transpiration alter canopy water-use-efficiency (WUE), while decreased latent heat flux alters the radiation balance, reducing atmospheric and increasing leaf temperature. Soil water stress limits crop water-use, decreases rates of photosynthesis and alters canopy energy balances and crop yield. It is clear that these processes have significant interactions, including feedback and feed-forward loops. How to include the biophysical feedbacks and interactions is the subject matter of subsequent modelling chapters (Chapters 9–13).

8.2 Canopy Photosynthesis and Water Flow through the SPAC

8.2.1 Processes and Up-Scaling in Modelling

Biophysical and physiological processes within the SPAC can be divided into three components: (1) radiation exchange and energy balance; (2) hydrological architecture and water transfer; and (3) photosynthesis and physiological regulation of carbon gain. Vegetation responds to variation in environmental factors (e.g. light, rainfall, temperature, VPD) through changes in rates of photosynthesis and transpiration.

Feedbacks between canopies and local environmental factors determine canopy microclimate. The microclimate of vegetation is the growth environment in which plants exist at the local scale and which control radiation, heat, and water exchanges (Rosenberg *et al.* 1983). In the interaction of plant-and-environment, physical processes form the basis for change in microclimate, which vary at small-to-medium spatial and temporal scales.

Physiological responses to a changing environment have been intensively studied in the past three decades, from sub-cellular- and leaf-scales through to regional and global scales. Simulation models of physiological processes and their response to variation in environmental factors are the basis of the study of interactions between plant and environment, including the effects of global warming and elevated atmospheric CO_2 concentrations on vegetation. Such models also investigate the role that plant cover (LAI) plays in determining patterns and rates of climate change (that is, the feedbacks between plant cover and climate). Leaf-scale studies are the starting point of any examination of interactions between vegetation and atmosphere, but the understanding of feedbacks between leaf and atmosphere constitute the building blocks in modelling of processes at larger scales through up-scaling (Norman 1993, Jarvis 1995).

Up-scaling from leaf- to canopy-scale requires information about radiation interception through the canopy and turbulent transfer of water vapour, CO_2, and energy. Shading effect among leaves adds to the complexity of radiation transfer and this affects photosynthesis and transpiration. Fluxes of water and carbon are subject to a number of different resistances within the SPAC and these are summarised in Figure 8.1.

8.2.2 *Canopy Radiation, Turbulence and Water Exchange*

Solar radiation is the key input to leaf energy balances and this drives transpiration, in addition to being the energy source for photosynthesis. The net radiation balance of a leaf is partitioned into sensible heat, by heat conductance and turbulent transfer and into latent heat as evapotranspiration. Consideration of the net radiation rather than incoming shortwave solar radiation provides a better understanding of the steady-state of a canopy, because net radiation is a radiation balance; it is determined by changes in both sensible heat and latent heat exchanges. Leaf temperature, which is determined by the leaf energy balance, influences the biochemical reaction rates of all photosynthetic processes and thereby the photosynthetic rate. Leaf temperature also determines the leaf's emission of long wave radiation (Chapter 5) in addition to the saturated water vapour pressure in the sub-stomatal cavity, thereby influencing rates of transpiration. Consequently, leaf temperature influences stomatal conductance, transpiration and leaf water status.

Turbulent transfer of energy and mass within and above canopies influence rates of CO_2 and water vapour flux into and out of stomatal pores, respectively. The CO_2 flux

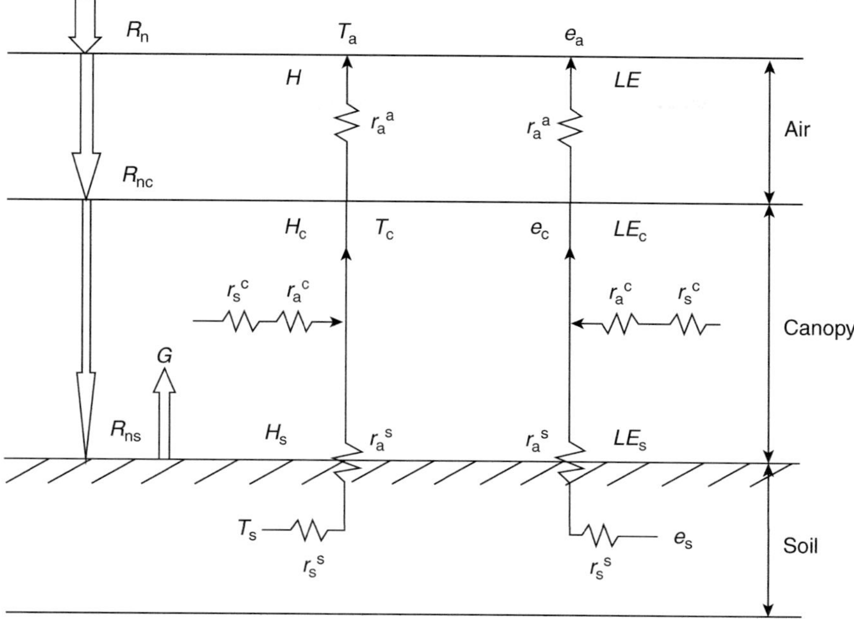

Figure 8.1. Simulation strategies for SPAC water and carbon processes.

and water vapour transfer are determined by stomatal conductance, boundary layer conductance, and differences of CO_2 and vapour pressure between the sub-stomatal cavity and ambient air (Chapter 2). In these processes there are complex interactions between photosynthesis, stomatal conductance and intercellular CO_2 concentration (C_i).

Water movement within the SPAC (Philip 1966) is driven by the gradient of water potential from high values in moist soil to lower values in leaves and even lower values in the atmosphere (Chapter 3). Feedbacks among soil and atmospheric water content and stomatal conductance, transpiration, and C_i are discussed in Chapters 2 and 3.

8.2.3 General Concepts of Water and Carbon Flux Modelling within the SPAC

The various modelling strategies used within the SPAC differ in the degree of simplification of representation of the SPAC (Wang 2006). Thus:

(1) Radiation transfer schemes vary from simple to complex representations. Welles and Norman (1991) divided competing canopy models into types according to the number of canopy layers: (a) big-leaf models in which the canopy is represented as a single horizontal "big-leaf", (b) two-leaf models in which the canopy is stratified into a sunlit and shaded fraction, to identify different light intensities for the two "leaves"; and (c) multi-layer models in which leaf energy balances and environmental gradients of temperature, light flux and VPD as well as foliar attributes (e.g. foliar nitrogen content), vary as a function of depth into the canopy.

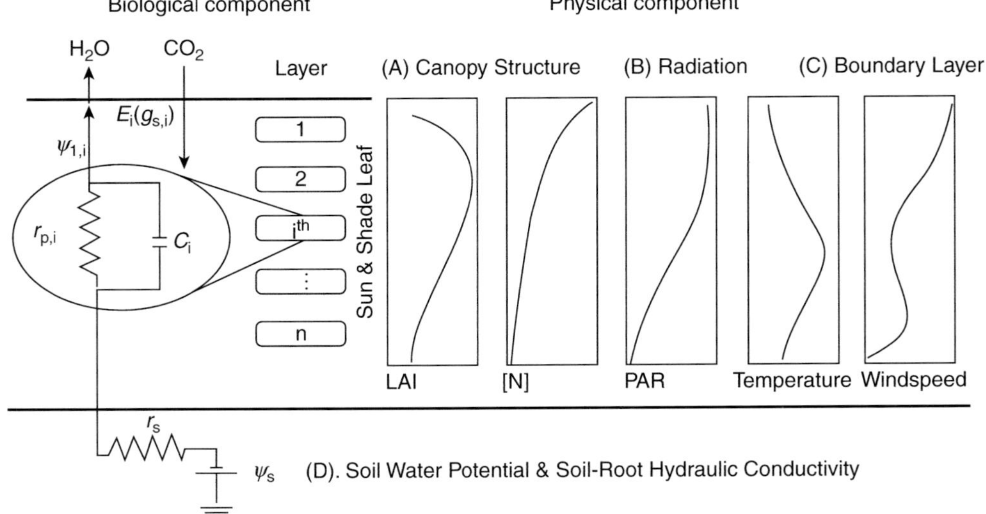

Figure 8.2. The structure the Soil-Plant-Atmosphere model (SPA, Williams *et al.* 1996).

The values of the light extinction coefficient and photosynthetic attributes differ for different plant functional types (PFTs; Chapter 1) within a multi-layered canopy (e.g. the C3 tree overstorey compared to the C4 grass understorey of a savanna; Chapter 16).

(2) Water flow within the SPAC has major impacts on energy balances and biogeochemical cycles. The key challenge to parameterising the water cycle is the determination of the values of a large number of resistances, arranged in parallel and series, along the pathway from soil to atmosphere. This needs integration of non-linear responses over coarse or fine spatial scales and many of these resistances vary in space and vary with plant size (Fig. 8.2).

(3) While modelling physiological regulation and responses to environmental variation is central to the development of an understanding of vegetation responses to climate variability and while multiple physiological responses can be detected directly at the leaf level, it is important to note that their feedbacks change with time.

Subsequent modelling chapters provide detailed examination of leaf-scale and canopy-scale modelling of carbon and water fluxes.

8.3 References

Jarvis PG, (1995). Scaling processes and problems. *Plant Cell and Environment* 18(10), 1079–1089.

Norman J, (1993). Scaling processes between leaf and canopy level. In: Ehleringer JR and CB Field. eds. *Scaling physiological processes*. Academic Press, London.

Philip JR, (1966). Plant water relations–some physical aspects. *Annual Review of Plant Physiology* 17, 245–268.

Rosenberg NJ, BL Blad and SB Verma, (1983). *Microclimate: The Biological Environment.* Wiley, pp. 528.

Welles JM and JM Norman, (1991). Instrument for indirect measurement of canopy architecture. *Agronomy Journal* 83(5), 818–825.

Wang J, (2006). Study on the experiment and simulation of crop growth and water, heat and CO_2 transfer in the agro-ecosystem. PhD Thesis, The Chinese Academy of Sciences.

Williams M, EB Rastetter, DN Fernandes, ML Goulden, SC Wofsy, GR Shaver, JM Melillo, JW Munger, S-M Fan and KJ Nadelhoffer, (1996). Modelling the soil-plant-atmosphere continuum in a *Quercus acer* stand at Harvard Forest: the regulation of stomatal conductance by light, nitrogen and soil/plant hydraulic properties. *Plant Cell and Environment* 19, 911–927.

9

Modelling Radiation Exchange and Energy
Balances of Leaves and Canopies

9.1 Introduction

Short wave solar radiation fluxes include direct beam and diffuse beam components (Chapter 1). This chapter introduces the calculation of global radiation, ground surface radiation, long wave radiation exchange and energy balances over land surfaces. It also introduces the calculation of albedo and light distribution within a canopy. To investigate ecosystem water, and biogeochemical cycles, researchers need to accurately establish the spatial distribution of solar radiation. However, compared to the large number of routine measurements of readily measureable meteorological variables such as sunshine hours, temperature and precipitation, few stations have recorded solar radiation directly because of the high cost of the necessary sensors and their maintenance (Liu *et al.* 2012). Consequently it is common practice to calculate solar radiation values through use of available meteorological variables (Chen *et al.* 2006, Yorukoglu and Celik 2006, Liu *et al.* 2012). Determination of light environment within plant canopies is essential for quantification of canopy photosynthesis, plant water-use, and energy balance.

9.2 Solar Radiation

9.2.1 Global Radiation (Ex-Terrestrial Radiation)

Global radiation is the solar radiation received at the outer edge of the earth's atmosphere. Therefore it is defined only by astronomic elements of both space and time including latitude (ψ), time of a day (determining hour angle, ω) and day of year (determining solar declination, δ). The global radiation level (Q_0) over a horizontal surface at time t can be calculated as per equation 9.1:

$$Q_0(t) = I_0 E_0 \sin h \tag{9.1}$$

where I_0 is the solar constant, E_0 is a correction factor for seasonal changes in solar-earth distance, h is solar altitude and, according to solar orbit equations $\sin h$ is determined as per equation 9.2:

$$\sin h = \sin \psi \sin \delta + \cos \psi \cos \delta \cos \omega \qquad (9.2)$$

At sunrise and sunset, $h=0$. The sunrise hour angle, ω_0, for equation 9.2 is obtained from equation 9.3:

$$\omega_0 = -\cos^{-1}\left(-\mathrm{tg}\,\psi\,\mathrm{tg}\,\delta\right) \qquad (9.3)$$

while the sunset hour angle, ω_0, is (eq. 9.4):

$$\omega_0 = \cos^{-1}\left(-\mathrm{tg}\,\psi\,\mathrm{tg}\,\delta\right) \qquad (9.4)$$

The daily total global radiation is integrated over ω from sunrise to sunset (eq. 9.5):

$$Q_0 = \frac{I_0 E_0 \tau}{\pi}\left(\omega_0 \sin \psi \sin \delta + \cos \psi \cos \delta \sin \omega_0\right) \qquad (9.5)$$

where τ is day length.

Global radiation and sunrise and sunset time are fundamental parameters for ground surface radiation. Solar declination, δ, can be calculated by equation 9.6:

$$\delta = 0.409 \sin\left(\frac{2\pi}{365}J - 1.39\right) \qquad (9.6)$$

where J is the number of the day in the year between 1 (1 January) and 365 or 366 (31 December).

When the time of sunrise and sunset is calculated, 24 hours in a day can be considered to correspond to 360 degrees in a circle, which means 15 degrees of rotation in an hour. The solar time of sunrise (T_1) and sunset (T_2) can therefore be described by equations 9.7 and 9.8 respectively:

$$T_1 = 12 - \omega_0 / 15 \qquad (9.7)$$

$$T_2 = 12 + \omega_0 / 15 \qquad (9.8)$$

9.2.2 Land Surface Solar Radiation

Solar radiation is diffused, absorbed and reflected by the atmosphere, clouds and objects on the earth's surface and consists of direct and diffuse beam radiation. If S and D are the daily total of direct and diffuse solar radiation at the ground surface, respectively and Q_0 is the daily global radiation, then the ratio of solar radiation to total hemi-spherical solar radiation incident on a horizontal surface at the outer edge of the earth's atmosphere is defined as the clearness index, which means the percentage of global solar radiation reaching the ground after passing through the atmosphere. Thus the clearness indices for direct (k_{Dir}) and diffuse (k_{Dif}) radiation, and for overall solar radiation (k_{cld}) are determined as(eq.9.9):

$$k_{\mathrm{Dir}} = \frac{S}{Q_0}, \quad k_{\mathrm{Dif}} = \frac{D}{Q_0}, \quad k_{\mathrm{cld}} = \frac{S+D}{Q_0} \qquad (9.9)$$

For modelling proposes, hourly solar radiation is required and daily values can be interpolated because of the known diurnal variation of global radiation. Assuming a zero cloud cover, solar radiation levels for every hour can be derived from the clearness index; a constant clearness index is often assumed for a site. The direct and diffuse solar radiation at time t can be expressed as follows (eq. 9.10):

$$D(t) = k_{\text{Dir}}Q_0(t), S(t) = k_{\text{Dif}}Q_0(t) \tag{9.10}$$

9.2.3 Derivation of Solar Radiation Over Land Surfaces from Meteorological Variables

Some dynamic models of ground level solar radiation such as METSTAT simulate radiation at time scales from seconds to decades. However, they are not widely used because of their complexity and the excessive input demand. In contrast simple empirical models are widely used. The Ångström model was established in 1924 (Ångström 1924) and modified in 1940 (Prescott 1940), in which solar radiation were estimated using sunshine hours. Thereafter, some similar empirical models were proposed for estimation of solar radiation using available sunshine hours (Chen *et al.* 2004, Menges *et al.* 2006, Bansouleh *et al.* 2009, Maghrabi 2009, Robaa 2009). A worldwide radiation model was developed by Bahel (1987) based on sunshine hours and radiation data from forty-eight stations, which was more complex than the Ångström model but has higher accuracy. Chen *et al.* (2004) developed a new scheme including effect of air temperature, which has proved to be more accurate than both the Ångström and Bahel models.

In addition to these sunshine-based models, other empirical models have been developed with inputs of routinely observed meteorological elements such as temperature and precipitation. As solar radiation is closely correlated to daily temperature variation between daily maximum and minimum air temperatures, Bristow and Compbell (1984), Hargreaves *et al.* (1985) and Allen (1997) proposed temperature-based models of solar radiation considering exponential and square root expressions of the daily temperature variation, respectively. Temperature, humidity and precipitation were all integrated to reformulated the Bristow and Compbell model by Thornton and Running (1999). It was found that the modified model performed better than the original over a wide range of climates. Wu *et al.* (2007) presented two new models adding rainfall and dew point to the Hargreaves model, and show significant improvement. Li *et al.* (2010) examined twelve models of the type based on the relationship between solar radiation and routinely-measured meteorological variables including maximum temperature, minimum temperature, rainfall, mean dew point temperature and fog days. It is shown that the model integrating all selected variables performed the best.

Models using more variables always produce a better performance (Thornton and Running 1999, Liu and Scott 2001, Chen *et al.* 2004, Wu *et al.* 2007, Li *et al.* 2010) than those with fewer variables but of course not all desired variables are available at any given site. Solar radiation can be directly estimated by the routinely observed meteorological variables while sunshine hours are not available. It is found models using only sunshine hours are superior to those using meteorological variables (Chen *et al.* 2004, Podesta *et al.* 2004, Wu *et al.* 2007). Liu *et al.* (2012) identified that the sunshine hours based model are much superior to those temperature-based ones, under Alpine regions on the western Tibetan Plateau.

The Ångström model, based on the linear relationship between radiation and sunshine hours (Ångström, 1924) was modified by Prescott (1940). This model (eq. 9.11) can be expressed as:

$$R_A = R_E \left[a + b \frac{S}{S_0} \right] \tag{9.11}$$

where R_A is actual daily solar radiation (MJ m^{-2} d^{-1}), R_E is daily extra–terrestrial solar radiation (MJ m^{-2} d^{-1}), S is sunshine hours and S_0 is potential sunshine hours.

Chen *et al.* (2004) found a new expression of the relationship between R_A/R_E and meteorological variables and suggested a new model (eq. 9.12) in which both sunshine hours and maximum and minimum temperatures (T_{min} and T_{max}) were included:

$$R_A = R_E \left[a + b \ln\left(T_{max} - T_{min}\right) + c \left(\frac{S}{S_0}\right)^d \right] \tag{9.12}$$

Chen *et al.* (2004) validated his model against data from China and found that the results were more accurate than those derived from the Ångström and Bahel models. Validation results of Wu *et al.* (2007) indicated that the Chen model performed the best among all selected models.

The modified Ångström model uses a non-linear relationship of S/S_0 with R_A (Bahel *et al.* 1987) or can be modified by introducing more meteorological variables, such as daily temperature range and rainfall (Chen *et al.* 2004).

Additional temperature-based models have also been developed (e.g. Li *et al.* 2010), but their input variables, for example, dew point temperate and fog, are not routinely observed and consequently these models are not widely used. The daily extra-terrestrial solar radiation is calculated according to the equations described by Allen *et al.* (1998) and the coefficient was fitted by numerical iteration methods (Hans and Francos 1999).

9.2.4 Incoming Atmospheric Long Wave Radiation

Downwelling long wave radiation is important for numerous applications requiring surface radiation and energy balance, including predicting evapotranspiration, snowmelt, surface temperature and the occurrence of frost. Meteorological stations rarely include long wave radiation sensors so measurements of incoming long wave radiation are generally not available. While surface radiative fluxes can be calculated with reasonable accuracy using complex radiative transfer models (Edwards and Slingo 1996; Pope *et al.* 2000), these models also require detailed measurements of the air column above a site, including cloud properties and vertical profiles of temperature, water vapour, trace gases and aerosols. Because these measurements are rarely available, downwelling long wave radiation is commonly estimated from algorithms that use more readily available meteorological observations, such as air temperature, humidity and solar radiation.

9.2.5 Parameterisation Schemes

Downwelling long wave radiation (L_d) is the thermal radiation emitted by the atmosphere downward to the ground surface. It is often computed from a general equation (eq. 9.13) following the form of the Stefan-Boltzmann equation:

$$L_d = \varepsilon_{eff} \sigma T_{eff}^{4} \tag{9.13}$$

where ε_{eff} and T_{eff} (K) are the effective emissivity and temperature of the atmosphere above the site, respectively and σ is the Stefan-Boltzmann constant. The temperature and emissivity are integrated quantities over an atmospheric column above the site and are typically estimated from ground-based meteorological observations only.

Typically, long wave parameterisations use near-surface air temperature and humidity measurements to calculate clear-sky fluxes, again based on the Stefan-Boltzmann equation:

$$L_{clr} = \varepsilon_{clr} \sigma T_o^{4} \tag{9.14}$$

where L_{clr} is incoming long-wave radiation for clear skies, ε_{clr} is clear-sky emissivity and T_o is near-surface air temperature (K).

Several parameterisations have been developed that produce estimates for downwelling long wave radiation (L_d) using synoptic observations only (Idso and Jackson 1969, Maykut and Church 1973, Jacobs 1978, Idso 1981, Aubinet 1994, Dilley and O'Brien 1998, Duarte *et al.* 2006, Lhomme *et al.* 2007). Thirteen such algorithms are given in Table 9.1 for estimating clear-sky downwelling long-wave flux (L_{clr}). The unit of the radiative flux is W m^{-2}, the unit of vapour pressure (e_o) is kPa and the unit of T_o is K.

Elevation can affect L_d because the air column above a site decreases with elevation. Deacon (1970) developed an elevation correction based on the Swinbank (1963) clear-sky equation. Iziomon *et al.* (2003) reported different parameter values for their algorithm based on sites at different elevations.

While numerous algorithms exist for predicting incident atmospheric long wave radiation under clear (L_{clr}) and cloudy skies, Flerchinger *et al.* (2009) made

Table 9.1. Algorithms for estimating clear-sky emmisivity following the form of the Stefan-Boltzmann equation or for estimating downwelling long-wave radiation directly

Source	Clear-sky algorithm
Ångström (1918)	$\varepsilon_{clr} = \left(0.83 - 0.18 \times 10^{-0.067 e_o}\right)$
Brunt (1932)	$\varepsilon_{clr} = \left(0.52 + 0.205\sqrt{e_o}\right)$
Brutsaert (1975)	$\varepsilon_{clr} = 1.723\left(\dfrac{e_o}{T_o}\right)^{1/7}$
Garratt (1992)	$\varepsilon_{clr} = 0.79 - 0.17\exp(-0.96 e_o)$
Idso and Jackson (1969); referred to as Idso-1	$\varepsilon_{clr} = 1 - 0.261\exp\left(-0.00077(T_o - 273.16)^2\right)$
Idso (1981); referred to as Idso-2	$\varepsilon_{clr} = 0.7 + 5.95 \times 10^{-4} e_o \exp\left(\dfrac{1500}{T_o}\right)$
Iziomon *et al.* (2003)	$\varepsilon_{clr} = 1 - X\exp\left(\dfrac{-Y e_o}{T_o}\right)$
Keding (1989)	$\varepsilon_{clr} = 0.92 - 0.7 \times 10^{-1.2 e_o}$
Niemela *et al.* (2001)	$\varepsilon_{clr} = \begin{cases} 0.72 + 0.09(e_o - 0.2) & for \quad e_o \geq 0.2 \\ 0.72 - 0.76(e_o - 0.2) & for \quad e_o < 0.2 \end{cases}$
Prata (1996)	$\varepsilon_{clr} = 1 - (1 + w)\exp\left(-(1.2 + 3w)^{1/2}\right)$
Satterlund (1979)	$\varepsilon_{clr} = 1.08\left[1 - \exp\left(-(10 e_o)^{T_o/2016}\right)\right]$
Swinbank (1963)	$L_{clr} = 5.31 \times 10^{-13} T_o^6$
Dilley and O'Brien (1998)	$L_{clr} = 59.38 + 113.7\left(\dfrac{T_o}{273.16}\right)^6 + 96.96\sqrt{w/2.5}$

Note: Coefficients are based on vapour pressure (e_o) in kPa, temperature (T_o) in K and rainfall water (w) in cm.
Source: Adapted from Flerchinger *et al.* (2009).

comparisons to assess the accuracy of the different algorithms. They concluded that clear-sky algorithms that excelled in predicting L_{clr} were the Dilley, Prata and Ångström algorithms (Table 9.1).

Clear-sky long wave radiation computed from algorithms in Table 9.1 is influenced by cloud cover. Most approaches adjust ε_{clr} for the fraction of cloud cover, c, to compute an effective atmospheric emissivity, ε_a. Cloudiness can be estimated from a clearness index (k_{clr}), equation 9.9. Campbell (1985) suggested that c can be linearly interpolated between $c = 1.0$ at a clearness index of 0.4 for complete cloud cover (k_{cld}) to $c = 0.0$ at a clearness index of 0.7 (k_{clr}). Others have used values of 0.35 and 0.6 for

k_{cld} and k_{clr}, respectively (Flerchinger 2000, Xiao *et al.* 2006); the optimum values are unknown and may depend on location.

9.2.6 Energy Balance Over Land Surface

The radiation balance equation (eq. 9.15) for a ground surface is given by:

$$R_n = (1 - \rho_a)Q - F_n \tag{9.15}$$

in which R_n is the net radiation above the canopy, ρ_a the albedo of canopy and soil to solar radiation, Q is the global radiation and F_n is surface effective radiation. The energy balance equation of land surface, neglecting the effects of advection, is given by (eq. 9.16):

$$R_n = H + \lambda E + G \tag{9.16}$$

where H, λE and G are sensible heat flux above the canopy, latent heat flux above the canopy and soil heat flux, respectively.

All surfaces receive short wave radiation during daylight and exchange long wave radiation continuously with the atmosphere. This exchange drives the surface energy balance, influences surface temperature and provides the energy for photosynthesis and plant growth. Accurate simulation of canopy microclimate and its influence on plant processes and water and CO_2 exchange, is contingent on simulation of the surface radiation balance. The net amount of radiation received by a surface is defined by equation 9.17:

$$R_n = S_b + S_d - S_u + L_d - L_u. \tag{9.17}$$

Here, S_b and S_d are direct and diffuse downward short wave radiation incident on the surface, S_u is short wave radiation reflected by the surface, L_d is downward long wave radiation emitted from the atmosphere and L_u is long wave radiation emitted from the surface. Measurements of S_d are typically input to simulation models and L_d is usually estimated from meteorological observations.

9.2.7 Radiation Exchange within Plant Canopies

The complexity of the canopy radiation models differs widely, which can be divided into single-layer (Monteith 1963), dual-source (Shuttleworth and Wallace 1985, Huntingford *et al.* 1995), or multiple-layer models (Norman 1979, Stockle 1992, Smith and Goltz 1994, Flerchinger *et al.* 1998, Zhao and Qualls 2005). The radiation distribution and microclimate profiles are simulated in multiple-layer models throughout the canopy, requiring calculation of the transmissivity, reflection and scattering of radiation within each canopy layer, which invariably requires assumptions regarding leaf orientation. Flerchinger and Yu (2007) presented relations for transmission and scattering of diffuse and direct radiation for ellipsoidal leaf angle distributions to facilitate modelling of radiation scattering without the need to divide the canopy into several inclination classes.

Multiple-layer canopy models must compute the energy balance for each layer within the canopy. This requires computation for downward direct radiation and upward and downward diffuse radiation being transmitted, reflected, scattered and absorbed, by each layer. Upward flux of diffuse short-wave radiation between canopy layer i and the next layer below, $i+1$, ($S_{u,i}$) can be estimated by equation 9.18:

$$S_{u,i} = \tau_{d,i+1}S_{u,i+1} + (\beta_l f_{d,\downarrow\uparrow} + \tau_l f_{d,\downarrow\downarrow})(1 - \tau_{d,i+1})S_{d,i}$$
$$+ (\beta_l f_{d,\uparrow\uparrow} + \tau_l f_{d,\downarrow\uparrow})(1 - \tau_{d,i})S_{u,i+1} + \beta_l f_{b,\downarrow\uparrow}(1 - \tau_{b,i+1})S_{b,i} \tag{9.18}$$

where $\tau_{d,l}$ is the transmissivity of canopy layer i to diffuse radiation (i.e. the fraction of diffuse radiation going through the canopy layer unimpeded), $\tau_{b,i}+1$ is the transmissivity of canopy layer $i+1$ to direct (or beam) radiation, β_l is the albedo of the canopy leaves, τ_l is the leaf transmissivity, $f_{d,\uparrow\downarrow}$ is the fraction of reflected downward diffuse radiation that is scattered upward, $f_{d,\downarrow\downarrow}$ is the fraction of reflected downward diffuse radiation that is scattered downward, $f_{b,\downarrow\uparrow}$ is the fraction of reflected downward direct radiation that is scattered upward and $S_{b,i}$ is direct radiation entering canopy layer $i+1$. It should be noted that $f_{d,\downarrow\uparrow} = f_{d,\uparrow\downarrow} = 1 - f_{d,\downarrow\downarrow} = 1 - f_{d,\uparrow\uparrow}$. Conversely, the fraction of downward diffuse radiation transmitted through the leaves that is scattered upward is equal to $1 - f_{d,\downarrow\uparrow}$ and that scatted downward is $1 - f_{d,\downarrow\downarrow}$. These same equalities hold true for downward direct radiation.

Common simplifications for leaf angle distribution are assumed to be horizontal, vertical, or spherical (Flerchinger 2000, Zhao and Qualls 2005). An ellipsoidal leaf angle density function is a more generalised approach (Campbell, 1986) and Campbell and Norman, 1998), which gives good approximation to leaf angle distribution of many real canopies (Barclay 2001, Falster and Westoby 2003) and can be extended to account for leaf clumping. However the ellipsoidal distribution needs complex computations required for scattering of reflected radiation and transmissivity of diffuse radiation. Computations for $f_{b,\downarrow\uparrow}, f_{d,\downarrow\uparrow}$, and τ_d in equation 9.17 require integration of complex functions for different combinations of leaf area index, leaf angle distribution and incident radiation angle and therefore must be re-computed at various time-steps within the model. This complexity has limited the application of ellipsoidal distribution method.

Transmission and absorption of long wave radiation is similar to solar radiation, with the exceptions that scattering of long wave radiation can be ignored and long wave emittance must be considered. Incoming long wave radiation from the atmosphere is computed using input air temperature and an average daily cloud cover estimated from observed solar radiation as described by Flerchinger *et al.* (1996). A long wave radiation balance is calculated for the soil surface and each residue layer and canopy layer based on the fluxes incident on and emitted by each side of the layer. For simplicity long wave emittance by a canopy layer is calculated using a leaf temperature for all plant species equal to air temperature within the layer. Thus emitted long wave radiation is biased by the difference between canopy air temperature and leaf temperature. However, these simplifications are not significant for most situations. Equations developed by Flerchinger and Yu (2007) for transmission and scattering of radiation for canopies with ellipsoidal leaf angle distributions make it unnecessary to divide the canopy into multiple leaf angle classes.

Short wave and long wave radiation exchange between canopy layers, residue layers and the snow or soil surface are computed by considering downward direct radiation and upward and downward diffuse radiation being transmitted, reflected and adsorbed by each layer. The upward flux of diffuse short-wave radiation above canopy layer i (numbered from the top of the canopy) is computed as (eq. 9.19):

$$S_{u,i} = \tau_{d,i}S_{u,i+1} + \alpha_i(1 - \tau_{d,i})S_{d,i} + \alpha_i(1 - \tau_{b,i})S_{b,i} \tag{9.19}$$

where $t_{d,I}$ is the transmissivity of canopy layer i to diffuse radiation, $t_{b,i}$ is the transmissivity of canopy layer i to direct radiation, α_i is the effective albedo of the canopy leaves in layer i (weighted by the leaf area and albedo of each plant type) and $S_{d,i}$ and $S_{b,i}$ are downward diffuse and direct radiation entering canopy layer i. A similar expression can be written for downward radiation at any point in the canopy. Transmissivity to direct radiation for each canopy layer is calculated from:

$$\tau_{b,i} = \exp\left(\sum_{j=1}^{NP}\Omega_j k_{b,j}L_{i,j}\right) \tag{9.20}$$

where $L_{i,j}$ and $k_{b,j}$ are leaf area index and extinction coefficient for direct radiation respectively for plant species j and canopy layer i, Ω_j a clumping factor to account for the fact that leaves are less efficient at intercepting radiation when clumped together (Campbell and Norman 1998) and NP is the number of plant species. This model is limited to vertical and spherical leaf orientations. Assuming a spherical leaf orientation, the extinction coefficient for direct radiation (k_b) can be computed from Campbell (1977) using equation 9.21:

$$k_b = \frac{1}{2\sin\beta} \tag{9.21}$$

where β is the solar elevation angle above the local slope.

9.3 Canopy Light Environment

9.3.1 Canopy Structure

In the Ross-Nilson model for leaf angle distribution, a canopy is defined as being composed of randomly distributed leaves. Each leaf position is defined by inclination

α and aspect β, which are continuously variable. In practice, modellers divide a canopy into a number of discrete fractions for α and β for ease of computation. Thus Ross and Nilson (1975) divided α into 6 fractions over 0–90 degrees from horizontal to vertical, while β was divided into 8 sections with 45 degree for each section. Therefore 48 compartments are identified (6 × 8).

9.3.2 Canopy Light Extinction Coefficient and Albedo

The values of extinction coefficients for direct beam visible radiation and direct beam infrared radiation in a canopy can be determined using equations 9.22 and 9.23, respectively (Goudriaan 1977):

$$k_v(i) = 0.0353 + 0.94623 \times k_{hv} \times k_b(i) \tag{9.22}$$

$$k_i(i) = 0.0353 + 0.94623 \times k_{hi} \times k_b(i) \tag{9.23}$$

where i is from 1 to 9, which denotes the inclination angles of 0–10°, …, 81–90°, respectively, k_b is the extinction coefficient for a canopy with black leaves, k_{hv} and k_{hi} are extinction coefficients for a canopy with horizontal leaves for visible radiation and infrared radiation, respectively. The coefficient for diffuse radiation is described by equation 9.24:

$$\exp(-k_d L) = \sum_{i=1}^{9} B_u(i)\exp(k_f(i)L) \tag{9.24}$$

where k_d is the extinction coefficient for diffuse radiation in a canopy with leaves that absorb all incoming short wave radiation and B_u is the normal distribution function of diffuse radiation within a canopy. Thus, extinction coefficients for diffuse visible radiation and diffuse infrared radiation of a canopy are described in equations 9.25 and 9.26, respectively:

$$k_{dv} = -\ln\left(\sum_{i=1}^{9} B_u(i)e^{k_v(i)L}\right)\bigg/ L \tag{9.25}$$

and

$$k_{di} = -\ln\left(\sum_{i=1}^{9} B_u(i)e^{k_i(i)L}\right)\bigg/ L \tag{9.26}$$

Because the scattering coefficient s is close to zero for the attenuation of long wave radiation, the attenuation of long wave radiation is:

$$\exp(-k_l L) = \sum_{i=1}^{9} B_u(i)\exp(k_b(i)L) \tag{9.27}$$

The extinction coefficient for long wave radiation (k_l) of a canopy is:

$$k_l = -\ln\left(\sum_{i=1}^{9} B_u(i)e^{k_b(i)L}\right)\bigg/ L \tag{9.28}$$

The albedo of a canopy with horizontal leaves for direct visible radiation (ρ_{fv}) and direct infrared radiation (ρ_{fi}) is defined by equations 9.29 and 9.30, respectively:

$$\rho_{fv} = 1 - e^{-2\rho_{hv}\left(1+\frac{L\times k_{dv}}{1+k_{dv}}\right)\times\frac{k_b}{1+k_b}} \tag{9.29}$$

$$\rho_{fi} = 1 - e^{-2\rho_{hi}(1+\frac{L\times k_{di}}{1+k_{di}})\times\frac{k_b}{1+k_b}} \tag{9.30}$$

The albedo of canopy for direct visible radiation and direct infrared radiation is given by equations 9.31 and 9.32, respectively:

$$\rho_v(i) = -0.0111544 + 1.117 \times \rho_{fv}(i) \tag{9.31}$$

$$\rho_i(i) = -0.0111544 + 1.117 \times \rho_{fi}(i) \tag{9.32}$$

The albedo of canopy for diffuse visible radiation and diffuse infrared radiation is:

$$\rho_{dv} = \sum_{i=1}^{9} B_u(i)\rho_v(i) \tag{9.33}$$

$$\rho_{di} = \sum_{i=1}^{9} B_u(i)\rho_i(i) \tag{9.34}$$

The energy stored through canopy photosynthesis is ignored on hourly and daily time steps.

9.3.3 The Distribution of Direct Beam Radiation within a Canopy

Solar radiation above a canopy can be divided into direct visible radiation (S_{bv}), direct infrared radiation (S_{bi}), diffuse visible radiation (S_{dv}) and diffuse infrared radiation (S_{di}). Light intensity (I) inside a canopy decreases exponentially with increasing leaf area index according to Beer's law (eq. 9.35). That is:

$$I(L) = I_c e^{-kL} \tag{9.35}$$

where I_c is the light intensity at the top of canopy, k is the extinction coefficient and L is the leaf area index. Because the extinction coefficient and albedo of a canopy for direct radiation and diffuse radiation are different, they are calculated separately. Leaves in a plant community are assumed randomly distributed and their aspects are evenly distributed. The direct solar radiation at level i, can be calculated as (eq. 9.36):

$$S(i) = S_0 \exp(-k_i L_i) \tag{9.36}$$

where $L_{(i-1)}$ is the depth of leaf area index for the layer (i−1), which is accumulated from canopy top. The direct beam extinction coefficient k_i depends on the geometrical structure of leaves and solar altitude and aspect, such that ki is described in equation 9.37 (Ross, 1981):

$$k_{\mathrm{i}} = G / \sin h \tag{9.37}$$

in which

$$G = \int_0^{\pi/2} A(\alpha, h)\, g(\alpha)\sin\alpha\, d\alpha \tag{9.38}$$

and $A(\alpha, h)$ can be calculated by

$$A(\alpha, h) = \begin{cases} \cos\alpha\sin h & \alpha \le h \\ \cos\alpha\sin h + \arcsin(\tan\alpha\cot h) + 2\sin h\sin\alpha(1 - \tan^2\alpha\cot^2 h)^{1/2}/\pi & \alpha > h \end{cases} \tag{9.39}$$

In equation 9.38, $g(\alpha)$ is a distribution function for leaf declination, which is ratio of the leaf area with a declination of α and the total leaf area in the canopy and which satisfies equation 9.40:

$$\int_0^{\pi/2} \sin\alpha\, d\alpha = 1 \tag{9.40}$$

Direct radiation over a leaf surface at a specific time is given by:

$$S(\alpha, \beta, h, A) = I_0 E_0 \left| \cos\theta(\alpha, \beta, h, A) \right| \tag{9.41}$$

where I_0 is light intensity at the canopy surface perpendicular to the incoming radiation, θ is the included angle of solar beam and the leaf surface is normal, which can be expressed as:

$$\theta(\alpha, \beta, h, A) = \cos^{-1}[\sin h\cos\alpha + \cos h\sin\alpha\cos(A - B)] \tag{9.42}$$

which can be written as:

$$\begin{aligned}
\theta(\alpha, \beta, \delta, \omega, \psi) = \cos^{-1}[&\sin\alpha\sin\beta\cos\delta\sin\omega + \sin\psi\cos\alpha\sin\delta \\
&+ \sin\psi\sin\alpha\cos\beta\cos\delta\cos\omega \\
&- \cos\psi\sin\alpha\cos\beta\sin\delta + \cos\psi\cos\alpha\cos\delta\cos\omega]
\end{aligned} \tag{9.43}$$

in which α, β are leaf declination angle and leaf aspect. Direct solar light will illuminate on the other side of leaf when $\theta > 90°$ (Ross, 1981).

9.3.4 Diffuse Radiation Distribution

Sky diffuse beam radiation can be simplified as being defined as identical for all directions. Therefore, the extinction coefficient for diffuse radiation (k_{d}) is not considered to vary with solar position and can also be assumed to not change with canopy structure. The coefficient, k_{d}, is 0.68 for a canopy with vertical leaves and is 1.00 for horizontally distributed leaves, while others range in between (Wang and Jarvis 1990).

The penetration rate τ_d of diffuse radiation for the 'i-th' layer is given by:

$$\tau_d = 2\int_0^{\pi/2} \exp[-L_{i-1}G(\alpha)/\cos\alpha]\cos\alpha\sin\alpha \, d\alpha \tag{9.44}$$

$$G(\alpha,\beta) = \frac{1}{2\pi}\int_0^{2\pi}\int_0^{\pi/2} g(\alpha,\beta)|\cos\theta|\sin\alpha \, d\alpha \, d\beta \tag{9.45}$$

The diffuse radiation at the i-th layer is:

$$D(i) = D_0\tau_d \tag{9.46}$$

Both sides of a leaf are illuminated by diffuse radiation. The leaf surface diffuse radiation is actually equal to horizontal surface radiation:

$$D_{li}(\alpha) = D(i) \tag{9.47}$$

The total solar radiation at each layer in the canopy can be calculated by equations 9.36 (direct radiation) and 9.47 (diffuse radiation).

From this chapter we have established how we can model total solar radiation arriving at the top of a canopy and model the distribution of radiation within a canopy. These are fundamental requirements for models of carbon and water fluxes of canopies and land surfaces, which is the subject of the remaining modelling chapters.

9.4 References

Allen RG, 1997. Self-Calibrating Method for Estimating Solar Radiation from Air Temperature. *Journal of Hydrologic Engineering* 2(2), 56–67.

Allen RG, LS Pereira and D Raes, 1998. Crop evapotranspiration–guidelines for computing crop water requirements, FAO Irrigation and Drainage. Food and Agricultural Organization of the United Nations, pp. 56.

Ångström A, 1918. A study of the radiation of the atmosphere. *Smithsonian Miscellaneous Collections* 65, 1–159.

Ångström A, 1924. Solar and terrestrial radiation. *Quarterly Journal of the Royal Meteorological Society* 50, 121–125.

Aubinet M, 1994. Sky Radiation Parametrizations. *Solar Energy* 53(2), 147–154.

Bahel V, H Bakhsh and R Srinivasan, 1987. A Correlation for Estimation of Global Solar-Radiation. *Energy* 12(2), 131–135.

Bansouleh BF, MA Sharifi and H Van Keulen, 2009 Sensitivity analysis of performance of crop growth simulation models to daily solar radiation estimation methods in Iran. *Energy Conversion and Management* 50(11), 2826–2836.

Barclay HJ, 2001. Distribution of leaf orientations in six conifer species. *Canadian Journal of Botany, Revue Canadienne De Botanique* 79(4), 389–397.

Bristow KL and GS Campbell, 1984. On the Relationship between Incoming Solar-Radiation and Daily Maximum and Minimum Temperature. *Agricultural and Forest Meteorology* 31(2), 159–166.

Brunt D, 1932. Notes on radiation in the atmosphere. *Quarterly Journal of the Royal Meteorological Society* 58, 389–420.

Brutsaert W, 1975. Derivable Formula for Long-Wave Radiation from Clear Skies. *Water Resources Research* 11(5), 742–744.

Campbell GS, 1977. *An Introduction to Environmental Biophysics.* Springer-Verlag, New York.

Campbell GS, 1985. *Soil Physics with Basic: Transport Models for Soil-plant Systems.* Elsevier, New York, New York.

Campbell GS, 1986. Extinction Coefficients for Radiation in Plant Canopies Calculated Using an Ellipsoidal Inclination Angle Distribution. *Agricultural and Forest Meteorology* 36(4), 317–321.

Campbell GS and JM Norman, 1998. *An Introduction to Environmental Biophysics,* second edition. Springer, New York.

Chen RS, K Ersi, JP Yang, SH Lu and WZ Zhao, 2004 Validation of five global radiation models with measured daily data in China. *Energy Conversion and Management* 45(11–12), 1759–1769.

Chen RS, SH Lu, ES Kang, JP Yang and XB Ji, 2006 Estimating daily global radiation using two types of revised models in China. *Energy Conversion and Management* 47(7–8), 865–878.

Deacon EL, 1970. Derivation of Swinbanks long-wave radiation formula. *Quarterly Journal of the Royal Meteorological Society* 96(408), 313–319.

Dilley AC and DM O'Brien, 1998 Estimating downward clear sky long-wave irradiance at the surface from screen temperature and precipitable water. *Quarterly Journal of the Royal Meteorological Society* 124(549), 1391–1401.

Duarte HF, NL Dias and SR Maggiotto, 2006 Assessing daytime downward radiation estimates for clear and cloudy skies in Southern Brazil. *Agricultural and Forest Meteorology* 139(3–4), 171–181.

Edwards JM and A Slingo, 1996. Studies with a flexible new radiation code .1. Choosing a configuration for a large-scale model. *Quarterly Journal of the Royal Meteorological Society* 122(531), 689–719.

Espana ML, F Baret, F Aries, M Chelle, B Andrieu and L Prevot, 1999 Modeling maize canopy 3D architecture–Application to reflectance simulation. *Ecological Modelling* 122(1–2), 25–43.

Falster DS and M Westoby, 2003. Leaf size and angle vary widely across species: what consequences for light interception? *New Phytologist* 158(3), 509–525.

Flerchinger GN, 2000. The Simultaneous Heat and Water (SHAW) Model: Technical Documentation, USDA-ARS, Northwest Watershed Research Center, Boise, Idaho.

Flerchinger GN, CL Hanson and JR Wight, 1996 Modeling evapotranspiration and surface energy budgets across a watershed. *Water Resources Research* 32(8), 2539–2548.

Flerchinger GN, WP Kustas and MA Weltz, 1998. Simulating surface energy fluxes and radiometric surface temperatures for two arid vegetation communities using the SHAW model. *Journal of Applied Meteorology* 37(5), 449–460.

Flerchinger GN, W Xaio, D Marks, TJ Sauer and Q Yu, 2009. Comparison of algorithms for incoming atmospheric long-wave radiation. *Water Resources Research* 45(3).

Flerchinger GN and Q Yu, 2007. Simplified expressions for radiation scattering in canopies with ellipsoidal leaf angle distributions. *Agricultural and Forest Meteorology* 144(3–4), 230–235.

Garratt JR, 1992 Extreme Maximum Land Surface Temperatures. *Journal of Applied Meteorology* 31(9), 1096–1105.

Goudriaan J, 1977. *Crop micrometeorology: a simulation study. Center for Agricultural Publishing and Documentation,* Wageningen, the Netherlands.

Hans VS and WZ Francos, 1999. *Statistical analysis in climate research.* Cambridge University Press, Cambridge.

Hargreaves GL, GH Hargreaves and JP Riley, 1985. Irrigation Water Requirements for Senegal River Basin. *Journal of Irrigation and Drainage Engineering, ASCE* 111(3), 265–275.

Huntingford C, SJ Allen and RJ Harding, 1995. An Intercomparison of Single and Dual-Source Vegetation-Atmosphere Transfer Models Applied to Transpiration from Sahelian Savanna. *Boundary-Layer Meteorology* 74(4), 397–418.

Idso SB, 1981. A Set of Equations for Full Spectrum and 8-Mu-M to 14-Mu-M and 10.5-Mu-M to 12.5-Mu-M Thermal-Radiation from Cloudless Skies. *Water Resources Research* 17(2), 295–304.

Idso SB and RD Jackson, 1969. Thermal Radiation from Atmosphere. *Journal of Geophysical Research* 74(23), 5397–5403.

Iziomon MG, H Mayer and A Matzarakis, 2003. Downward atmospheric irradiance under clear and cloudy skies: Measurement and parameterization. *Journal of Atmospheric and Solar-Terrestrial Physics* 65(10), 1107–1116.

Jacobs JD, 1978. Radiation climate of Broughton Island, in Energy Budget Studies in Relation to Fast-Ice Breakup Processes in Davis Strait (Occas. Pap. 26), pp. 105–120, Univ. of Colorado, Inst. of Arctic and Alp. Res., Boulder, Colorado.

Keding I, 1989. Klimatologische Untersuchung über die atmosphärische Gegenstrahlung und Vergleich von Berechnungsverfahren anhand langjähriger Messungen im Oberrheintal. Ber Deutscher Wetterd 178, 72 pp.

Lhomme JP, JJ Vacher and A Rocheteau, 2007. Estimating downward long-wave radiation on the Andean Altiplano. *Agricultural and Forest Meteorology* 145(3–4), 139–148.

Li MF, HB Liu, PT Guo and W Wu, 2010. Estimation of daily solar radiation from routinely observed meteorological data in Chongqing, China. *Energy Conversion and Management* 51(12), 2575–2579.

Liu DL and BJ Scott, 2001. Estimation of solar radiation in Australia from rainfall and temperature observations. *Agricultural and Forest Meteorology* 106(1), 41–59.

Liu JD, JM Liu, HW Linderholm, DL Chen, Q Yu, DR Wu and S Haginoya, 2012. Observation and calculation of the solar radiation on the Tibetan Plateau. *Energy Conversion and Management* 57, 23–32.

Maghrabi AH, 2009. Parameterization of a simple model to estimate monthly global solar radiation based on meteorological variables and evaluation of existing solar radiation models for Tabouk, Saudi Arabia. *Energy Conversion and Management* 50(11), 2754–2760.

Maykut GA and PF Church, 1973. Radiation climate of Barrow, Alaska, 1962–66. *Journal of Applied Meteorology* 12, 620–628.

Menges HO, C Ertekin and MH Sonmete, 2006. Evaluation of global solar radiation models for Konya, Turkey. *Energy Conversion and Management* 47(18–19), 3149–3173.

Monteith JL, 1963. Gas exchange in plant communities. In: LT Evans (Editor), *Environmental Control of Plant Growth*. Springer, New York, pp. 95–112.

Niemela S, P Raisanen and H Savijarvi, 2001 Comparison of surface radiative flux parameterizations–Part I: radiation. *Atmospheric Research* 58(1), 1–18.

Norman JM, 1979. Modeling the complete crop canopy. In: BJ Barfield and JF Gerber (Editors), *Modification of the Aerial Environment of Plants. American Society of Agricultural Engineers*, St. Joseph, Michigan, pp. 249–277.

Podesta GP, L Nunez, CA Villanueva and MA Skansi MA, 2004. Estimating daily solar radiation in the Argentine Pampas. *Agricultural and Forest Meteorology* 123(1–2), 41–53.

Pope VD, ML Gallani, PR Rowntree and RA Stratton, 2000. The impact of new physical parametrizations in the Hadley Centre climate model: HadAM3. *Climate Dynamics* 16(2–3), 123–146.

Prata AJ, 1996. A new long-wave formula for estimating downward clear-sky radiation at the surface. *Quarterly Journal of the Royal Meteorological Society* 122(533), 1127–1151.

Prescott JA, 1940. Evaporation from a water surface in relation to solar radiation. *Transactions of the Royal Society of South Australia* 64: 114–118.

Robaa SM, 2009. Validation of the existing models for estimating global solar radiation over Egypt. *Energy Conversion and Management* 50(1), 184–193.

Ross J, 1981, *The Radiation Regime and Architecture of Plant Stands*, Dr W Junk Publisher, The Hague, pp. 127–226.

Ross J and T Nilson, 1975. Radiation exchange in plant canopies, In: DA de Vries and HH Afgan (Editors), *Heat and Mass Transfer in the Biophere*, Scripta, Washington, DC, pp. 327–336.

Satterlund DR, 1979. Improved Equation for Estimating Long-Wave-Radiation from the Atmosphere. *Water Resources Research* 15(6), 1649–1650.

Shawcroft RW, ER Lemon, LH Allen, DW Stewart and SE Jensen, 1974. Soil-Plant-Atmosphere model and some of its predictions. *Agricultural Meteorology* 14(1–2), 287–307.

Shuttleworth WJ and JS Wallace, 1985. Evaporation from Sparse Crops–an Energy Combination Theory. *Quarterly Journal of the Royal Meteorological Society* 111(469), 839–855.

Smith JA and SM Goltz, 1994. A thermal excitance and energy-balance model for forest canopies. *Transactions on Geoscience and Remote Sensing* 32(5), 1060–1066.

Stockle CO, 1992. Canopy Photosynthesis and Transpiration Estimates Using Radiation Interception Models with Different Levels of Detail. *Ecological Modelling* 60(1), 31–44.

Swinbank WC, 1963. Long-wave radiation from clear skies. *Quarterly Journal of the Royal Meteorological Society* 89(381), 339–348.

Thornton PE and SW Running, 1999. An improved algorithm for estimating incident daily solar radiation from measurements of temperature, humidity and precipitation. *Agricultural and Forest Meteorology* 93(4), 211–228.

Wang YP and PG Jarvis, 1990. Influence of crown structural-properties on PAR absorption, photosynthesis and transpiration in Sitka Spruce–application of a model (Maestro). *Tree Physiology* 7(1–4), 297–316.

Wu GF, YL Liu and TJ Wang, 2007. Methods and strategy for modeling daily global solar radiation with measured meteorological data–A case study in Nanchang station, China. *Energy Conversion and Management* 48(9), 2447–2452.

Xiao W, GN Flerchinger, Q Yu and YF Zheng, 2006. Evaluation of the SHAW model in simulating the components of net all-wave radiation. *Transactions of the Asabe* 49(5), 1351–1360.

Yorukoglu M and AN Celik, 2006. A critical review on the estimation of daily global solar radiation from sunshine duration. *Energy Conversion and Management* 47(15–16), 2441–2450.

Zhao WG and RJ Qualls, 2005. A multiple-layer canopy scattering model to simulate short wave radiation distribution within a homogeneous plant canopy. *Water Resources Research* 41(8), W08409.

10

Modelling Leaf and Canopy Photosynthesis

10.1 Introduction

Physiological responses to a changing environment have been intensely studied in the past decade at all spatial levels, from sub-cellular and leaf-scales through to regional and global scales. Simulation models of physiological processes in response to environmental factors are the basis of the study of the interactions between plant and environment, such as the effects of global warming and elevated atmospheric CO_2 concentrations on vegetation. Such models also investigate the role that plant cover plays in determining patterns and rates of climate change (i.e. the feedbacks between plant cover and climate). Leaf-scale studies are the starting point of any examination of interactions between vegetation and atmosphere but the understanding of the feedbacks between leaf and atmosphere constitute the building blocks in modelling processes at larger scales through up-scaling (Norman 1993, Jarvis 1995).

The biophysical and physiological processes at leaf-scales may be summarised as radiation exchange, heat and water transfer, physiological regulation and biochemical reactions (Fig. 10.1). Plants can alters these processes in response to changes in environmental drivers, including solar radiation, temperature, humidity and soil water and nutrient contents, as well as CO_2 concentration. Thus:

(1) **Energy exchange**: solar radiation is the key input of leaf energy balances which drive transpiration, in addition to it being the energy source for photosynthesis. The net radiation balance of a leaf is partitioned into sensible heat by heat conductance and turbulent transfer and latent heat by transpiration. Leaf temperature, as determined by leaf energy balance, influences the biochemical reaction rates of all photosynthetic processes and thereby the photosynthetic rate. Leaf temperature determines the leaf's emission of long-wave radiation and also determines the saturated water vapour pressure in the sub-stomatal cavity (Chapter 2). Consequently leaf temperature influences stomatal conductance and transpiration.

(2) **Mass flow**: CO_2 and water vapour pass into and out of stomatal pores respectively. The CO_2 flux and water vapour transfer are determined by stomatal conductance,

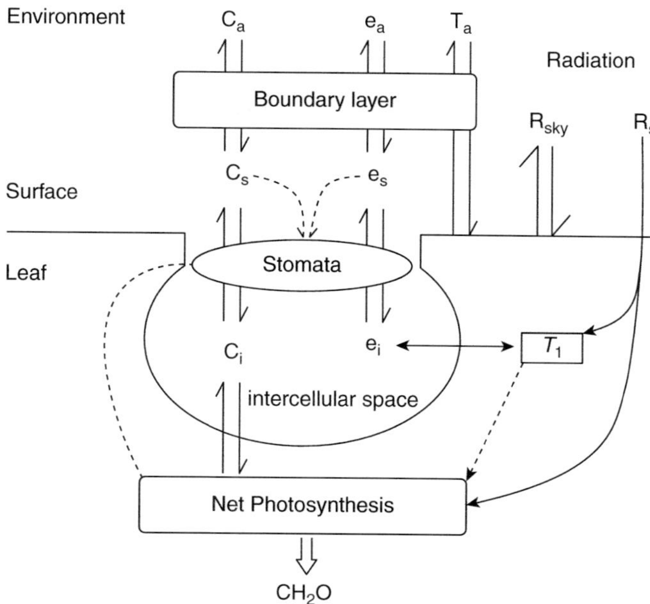

Figure 10.1. A scheme showing the mass and energy fluxes and the interactions between physiological processes and environmental variables, as considered in the combined model. Adapted from Collatz *et al.* 1991.

boundary layer conductance and differences of CO_2 and vapour pressure between the sub-stomatal cavity and ambient air. In these processes, there are complex interactions between photosynthesis, stomatal conductance and intercellular CO_2 concentration.

(3) **Physiological regulation** and **biochemical reactions**: Stomatal regulation is based on complex physiological processes that are regulated by light and water availability, CO_2 concentration, as well as the hormone ABA. Photosynthesis is influenced by solar radiation, leaf temperature and intercellular CO_2 concentration (C_i). Although stomatal conductance influences photosynthesis by limiting transport of CO_2 from air outside the leaf to mesophyll chloroplasts, photosynthesis itself influences stomatal conductance through its influence on C_i and through the formation of ATP which drives ion transport into guard cells, as previously discussed in Chapter 2. Cowan (1977) proposed that the adaptation of plants to environmental variation maximises leaf-scale water-use-efficiency, as discussed in Chapter 3.

10.2 Models of Leaf-Scale Photosynthesis

Plant photosynthesis (P) is responsive to environmental variables such as solar radiation, temperature, water and CO_2 concentration. It is common to describe responses of photosynthesis to a single specified variable while all others are kept constant.

This can identify characteristics of plant sensitivity to changes in these variables. For example, the response of photosynthesis to light intensity can be examined by determining a light response curve whereby temperature, vapour pressure deficit (*VPD*) and water supply, are all optimised experimentally and only light supply is varied. The influence of variation in other factors can then be assessed by quantifying the response of the two principle features of a light response curve, namely, the initial slope (α) of the linear response of P to increasing light supply (I) and the maximum photosynthesis rate of light saturated point (P_{max}). There are many formulations for the mathematical description of a light response curve and these are now discussed.

10.2.1 Light Response Curves

Light is the most significant limiting factor for plant growth as it can vary from darkness to supra-optimal values over a twenty-four-hour period. The light response curve can be described as a Michaelis-Menten equation (eq. 10.1), which is a rectangular hyperbola:

$$P(I) = \frac{\alpha I P_{max}}{\alpha I + P_{max}} \tag{10.1}$$

where P is gross photosynthetic rate and I the light intensity (Thornley 1976).

There are two parameters in the equation. The first, α, is quantum yield and is the slope of the rate of change in P as I increases for low values of I (typically in the range 30–100 μmol m^{-2} s^{-1}). The second is P_{max}, the maximum photosynthetic rate. Therefore, the net photosynthetic rate (P_n) can be written as:

$$P_n(I) = \frac{\alpha I P_{max}}{\alpha I + P_{max}} - R_d \tag{10.2}$$

where R_d is the rate of dark respiration. Its value is much lower than P_n (typically 5–10% of P_n) and is often taken as a constant in fitting the light response curve.

The photosynthetic light response curve can also be expressed as exponential equations:

$$P(I) = P_{max}[1 - e^{-(\alpha I / P_{max})}] \tag{10.3}$$

or

$$P(I) = \alpha I[1 - e^{-(\alpha I / P_{max})}] \tag{10.4}$$

After these two equations are expressed by a Taylor series expansion, these two equations can be approximated by a rectangular hyperbola as per equation 10.1.

Although a rectangular hyperbolic equation (eq. 10.1) can fit the general trend of light response curves, it fails in describing curvature of the response. Therefore a non-rectangular hyperbolic model has been introduced to include one more parameter to give a flexible bending of the response curve. The leaf gross photosynthetic rate can be calculated using equation 10.5 (Thornley 1976):

$$\theta P(I)^2 - (\alpha I + P_{max})P(I) + \alpha I P_{max} = 0 \tag{10.5}$$

where θ is the convexity, when θ equals zero, equation 10.5 will degrade to the rectangular hyperbolic model (eq. 10.1). This means equation 10.1 is a typical form of equation 10.5.

A reasonable solution is shown in the following equation (eq. 10.6):

$$P(I) = \frac{\alpha I + P_{max} - \sqrt{(\alpha I + P_{max})^2 - 4\theta(\alpha I P_{max})}}{2\theta} \tag{10.6}$$

Therefore $P_n = P(I) - R_d$, in which R_d is dark respiration. The parameters α, θ and P_{max}, relate to biochemical processes that are influenced by environmental factors and vary for specific plants. P_{max} is considered to be the maximum photosynthetic rate expressed in the field but it is actually the maximum rate when light intensity goes to infinity (∞) and this value can be much higher than its maximum rate observed in the field, where I does not reach values close to ∞. Consequently an unrealistic value of P_{max} is estimated if observed photosynthetic rates do not reach an asymptote in the field. The obvious deficiency of these curves is, therefore, that they cannot provide the saturation point of photosynthesis.

In many field studies it has been shown that photosynthetic rates can decrease above a maximum rate when I becomes supra-saturating. These observations, which arise because of photo-inhibition, can be integrated with a standard rectangular equation (Yu *et al.* 2002). Their empirical modification assumes a linear or exponential decline of P_{max} with light intensity, such as:

$$P_{max} = P_{max0} - bI \tag{10.7}$$

This modification to the hyperbolic response function of photosynthesis can describe the P_n declines with increasing light intensity once I is larger than that which causes light saturation of photosynthesis, especially under high light environments.

A new leaf photosynthesis model (eq. 10.8) was proposed by Ye and his colleagues, based on the photosynthetic electron transport of PS II (Ye 2007, Ye and Yu 2008, Ye *et al.* 2013):

$$P(I) = \alpha' I \frac{1 - \beta I}{1 + \gamma I} \tag{10.8}$$

in which, α', β, and γ are coefficients which are independent of the light intensity. It can also be found the following relationships with equation 10.7 hold:

$$P_{max0} = \frac{\alpha'}{\gamma + \beta} \tag{10.9}$$

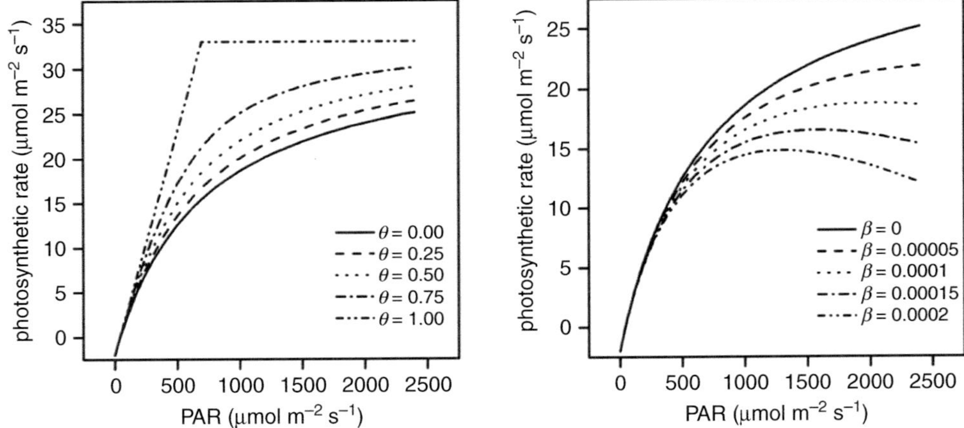

Figure 10.2. Light response curves of non-rectangular hyperbola model (left) and modified rectangular hyperbola model (right) at $\alpha = 0.05$, $P_{max} = 35$ µmol m^{-2} s^{-1} and $R_d = 2$ µmol m^{-2} s^{-1}.

$$b = \frac{\alpha'\beta}{\gamma+\beta} \tag{10.10}$$

Because net photosynthetic rate is zero at the light compensation point and the gross rate of photosynthesis is equal to the rate of respiration, the following equation (eq. 10.11) is obtained:

$$P_n(I) = \alpha'(I - I_c)\frac{1-\beta I}{1+\gamma I} \tag{10.11}$$

where I_c is the light compensation point.

Comparisons of light response curves are shown in Figure 10.2. The modified equation can give a maximum photosynthetic rate which agrees well with measured values, while P_{max} in rectangular and non-rectangular equations is the limit of the photosynthetic rate when $I \to \infty$. This value can be much higher (and hence unrealistic) than that observed in the field.

From equation 10.11, the rate of dark respiration R_d is given as follows for $I = 0$ (eq. 10.12):

$$R_d = -P(I = 0) = -\alpha' I_c \tag{10.12}$$

Where R_d is only dependent on the coefficient α and the light compensation point I_c.

Differentiating equation 10.11, $P'(I)$ is given by (eq. 10.13):

$$P'(I) = \alpha'\frac{1-2\beta I - \beta\gamma I^2 + (\gamma+\beta)I_c}{(1+\gamma I)^2} \tag{10.13}$$

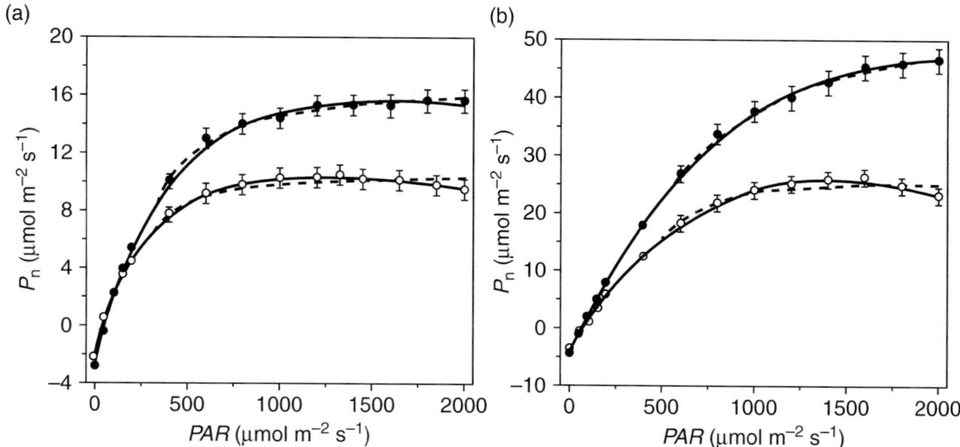

Figure 10.3. The light response curves fitted by rectangular and non-rectangular equations (eq. 10.2, 10.6) and the Ye's equation (eq. 10.8). Redrawn from Ye and Yu (2008).

For $I = 0$ the quantum yield, α, can be obtained at this point (eq. 10.14):

$$\alpha = P'(I = 0) = \alpha'[1 + (\gamma + \beta)I_c]$$ (10.14)

The saturation light intensity I_m is obtained by setting $P'(I) = 0$ in equation 10.13:

$$I_m = \frac{\sqrt{(\beta + \gamma)(1 + \gamma I_c)/\beta} - 1}{\gamma}$$ (10.15)

The maximum photosynthetic rate, $P(I_m)$, is given by (eq. 10.16):

$$P(I_m) = \alpha' \frac{1 - \beta I_m}{1 + \gamma I_m}(I_m - I_c)$$ (10.16)

Figure 10.3 shows examples of this model fitted against observed light response data (Ye and Yu 2008). Three photosynthesis models fit the data well giving coefficients of determination of 0.95–0.99. However, the parameter values fitted by the three models are significantly different from each other. The new photosynthesis model, however, provided reasonable estimates of maximum rates of photosynthesis and the light saturation point, which has three parameters, same as the non-rectangular hyperbolic model.

10.2.2 CO₂ Response Model of Photosynthesis

Photosynthetic responses to changes in atmospheric CO_2 concentration are similar in shape to those observed for light response curves, which can be expressed either as a rectangular equation or non-rectangular equation (Thornley 1976). Thus:

$$P = \frac{\alpha C_a P_{max}}{\alpha C_a + P_{max}}$$ (10.17)

or

$$\theta P^2 - P(\alpha I + \tau C_a) + P_{max}(\alpha I + \tau C_a) = 0 \qquad (10.18)$$

where C_a is atmospheric CO_2 concentration.

We can consider parameters in the light response equations are a function of CO_2 concentration and assume an absolute maximum on photosynthetic rate, P_0, when CO_2 approaches infinity. Thus:

$$P_{max} = \frac{\eta C_a P_0}{\eta C_a + P_0} \qquad (10.19)$$

where η is a constant. Combining equations 10.17 or 10.18 and 10.19, gives equation 10.20:

$$P = \frac{P_{max} \alpha I \eta C_a}{P_{max} \alpha I + P_{max} \eta C_a + \alpha I \eta C_a} \qquad (10.20)$$

Similarly, there is a non-rectangular expression for the CO_2 response; see the detailed derivation in Thornley (1976). These equations have the same deficiency as noted for light response curves in that no saturation point can be found and Pmax can be predicted to be much higher than the observed maximum value. The modification of the photosynthesis model used in equation 10.8, allows estimation of the CO_2 saturation point and compensation point (Ye and Yu 2008).

The initial quantum yield (α) is greatly influenced by CO_2 concentration, which can be described as in equation 10.21:

$$\alpha = \alpha_0 (C_i - \Gamma) / (C_i + 2\Gamma) \qquad (10.21)$$

where α_0 is the initial quantum efficiency of CO_2 assimilation and C_i is intercellular CO_2 concentration. If the influence of variation in light or CO_2 supply on photosynthetic rates are measured in a controlled environment with stable temperature and humidity, smooth response curves can be obtained. However, the parameter values (e.g. P_{max} and α) in each response curve can be functions of temperature, humidity and leaf water content. Thus, more comprehensive models to represent how these factors influence photosynthesis should be included.

10.3 Modelling the Biochemistry of Photosynthesis

Why do we need a biochemical model for prediction of plant response to variable climate and soil conditions? There are many models describing changes in the rate of photosynthesis in response to changes in light, CO_2 and temperature. These response curves consider the response of photosynthesis to single variables while others are kept constant. Of course in reality photosynthesis responds to multiple environmental variables simultaneously. Plant nutrient and water status also influence photosynthetic capacity. Consequently a process-based model can be useful for predicting rates of photosynthesis in the field where multiple factors simultaneously act to constrain carbon (C) gain. For example, when light levels increase after dawn, not only light supply increases, but temperature and *VPD* also increase. Furthermore, changes in stomatal conductance (g_s) occur, partly independently of photosynthesis and partly dependent on photosynthesis (via C_i). These diverse physiological responses can be depicted by process-based biochemical models.

10.3.1 Biochemical Models

Biochemical models of photosynthesis describe the light and dark reactions of photosynthesis (Chapter 2) occurring within chloroplasts and therefore do not include gaseous transfer between leaf and air. The input variables of the model are light intensity, temperature, *VPD*, CO_2 concentration and atmospheric pressure. Farquhar *et al.* (1980) and von Caemmerer and Farquhar (1981) generalised the main biochemical features of photosynthesis whereby photosynthesis is expressed as a function of intercellular CO_2 concentration (C_i), photosynthetic photon flux density (I) and leaf temperature (T_l). In this biochemical model (described later), C_i is taken as an input variable, so it is a biochemical model reflecting reactions in chloroplasts and excluding explicit consideration of diffusion through stomata. The input factor, C_i, depends on the rates of photosynthesis and stomatal conductance. The stomatal conductance and photosynthesis sub-models are interdependent, so a combination of them is needed to predict physiological responses to environmental factors. Collatz *et al.* (1991) and Leuning (1995) revised the original von Caemmerer and Farquhar model to simulate the coupling of photosynthesis and transpiration.

In the biochemical model of von Caemmerer and Farquhar, leaf gross photosynthesis (P) may be described as (eq. 10.22):

$$P = \min\left\{J_e, J_c, J_s\right\} \tag{10.22}$$

where J_c and J_e are the gross rates of photosynthesis limited by carboxylation reactions catalysed by ribulose-1,5-bisphosphate carboxylase/oxygenase and by the rate of ribulose-1,5-bisphosphate (RuP2) regeneration limited by the rate of electron transport supported by radiant energy received, respectively (Chapter 2). The term "min $\{J_e, J_c, J_s\}$" represents the minimum of the three values within the brackets; J_e depends not only on C_i and temperature, but also on irradiance. J_c depends on intercellular CO_2 concentration (C_i) and temperature. J_s is the capacity for the export or utilisation of photosynthate. Some simplified models (e.g. equation 10.23) neglect J_s and provided similar photosynthetic response to environmental factors:

$$P_n = \min\{J_c, J_e\} - R_d \tag{10.23}$$

where J_c and J_e are the rates of photosynthesis limited by Rubisco activity and the rate of RuP2 regeneration through electron transport, respectively. Rates of respiration (R) depend mainly on leaf temperature and is proportional to the maximum rate of carboxylation at an ambient oxygen concentration of 21 percent (V_m; Collatz *et al.* 1991), such that:

$$R_d = rV_m \tag{10.24}$$

where r is a proportionality constant.

RuBP regeneration is controlled by the rate of electron transport driven photo-phosphorylation which is given by equation 10.25 (Collatz *et al.* 1991):

$$J_e = a \cdot \alpha \cdot I \frac{(P_i - \Gamma)}{(P_i + 2\Gamma)} \tag{10.25}$$

where a is leaf absorbance of photosynthetically active radiation, α is the intrinsic quantum efficiency for CO_2 uptake, I is the leaf absorbed photosynthetically active radiation and P_i is the partial pressures of CO_2 in the intercellular air space of the leaf. Γ, the CO_2 compensation point, is defined by the equation:

$$\Gamma = \frac{[O_2]}{2\tau} \tag{10.26}$$

where τ is a ratio of kinetic parameters describing the partitioning of RuBP to the carboxylase or oxygenase reactions of Rubisco. τ can be determined experimentally from gas exchange experiments on intact leaves (see Brooks and Farquhar 1985) or from enzymatic analysis *in vitro* (see Jordan and Ogren 1981). $[O_2]$ is assumed constant (209 mmol mol^{-1}), P_i is given by $P_i = P_a \times C_i$ *and* c_i is determined using equation 10.27:

$$c_i = c_a - P_n \frac{1.56g_s + 1.37g_b}{2.14g_s g_b} \tag{10.27}$$

Note that C_a, C_i and diffusion gradients are in mole fraction, while the kinetic expressions for biochemical reactions require partial pressure and P_a is the atmospheric pressure.

The Rubisco-limited rate of photosynthesis is defined as per equation 10.28 by Collatz *et al.* (1991):

$$J_c = \frac{V_m(P_i - \Gamma)}{P_i + K_c(1 + O / K_o)} \tag{10.28}$$

where V_m is the maximum rate of carboxylation at an ambient oxygen concentration of 21 percent, O is the partial pressures of the O_2 in the intercellular air space of the leaf and K_c and K_o are the Michaelis–Menten kinetic parameters for CO_2 and

O_2, which are calculated from equations 10.29 and 10.30, respectively (Nikolov *et al.* 1995):

$$K_c = P_a K_{c25} \exp[32.462 - 80470 / (RT_1)] \tag{10.29}$$

$$K_o = P_a K_{o25} \exp[5.854 - 14510 / (RT_1)] \tag{10.30}$$

where K_{c25} and K_{o25} are the corresponding parameter values at 25°C. V_m can be determined from measurements of the slope of the response of P_n to P_i at $P_i = \Gamma$ (eq. 10.31; see Collatz *et al.* 1990):

$$V_m = \frac{dP_n}{dP_i}\left[\Gamma + K_c\left(1 + \frac{[O_2]}{K_o}\right)\right] \tag{10.31}$$

The gross assimilation rate limited by the rate of triose phosphate utilisation is defined by equation 10.32 (Collatz *et al.* 1991):

$$J_s = \frac{V_m}{2} \tag{10.32}$$

Because the transition from the limitation by one factor to the limitation by another appears to be gradual, to allow for some co-limitation between J_c, J_e and J_s, Collatz *et al.* (1991) solved the following quadratics (eq. 10.33 and eq. 10.34) for their smaller roots:

$$\beta_1 J_p^2 - J_p(J_e + J_c) + J_e J_c = 0 \tag{10.33}$$

$$\beta_2 P^2 - P(J_p + J_s) + J_p J_s = 0 \tag{10.34}$$

where P is the gross rate of CO_2 uptake, J_p an intermediate variable which represents the smaller of J_c and J_E and β_1 and β_2 are convexity coefficients describing the transition between limitations and are close to 1.0.

The combined temperature response of P_n described above does not account for the well-known thermal inhibition at temperatures exceeding 35°C (Berry and Raison 1981). In order to simulate realistic rates at high temperature a gradual temperature inhibition of the V_m is introduced to the model, as described by equation 10.35:

$$V_m = V_{m0}\left\{1 + \exp\left[\frac{(-a_1 + a_2 T_1)}{RT_1}\right]\right\}^{-1} \tag{10.35}$$

Where

$$V_{m0} = V_{m25}Q_{10}^{(T_1 - 25)/10} \tag{10.36}$$

and a_1 and a_2 are parameters, R is the ideal gas constant and V_{m0} is an intermediate variable. V_{m25} is V_{m0} at 25°C. R_d is also inhibited at high temperature and for this response the following form is used (eq. 10.37):

$$R_d = R_{d0}\left\{1 + \exp\left[1.3\left(T_1 - 55\right)\right]\right\}^{-1} \qquad (10.37)$$

which predicts an abrupt collapse in R_d as T_1 approaches 55°C (Bjorkman *et al.* 1980).

Generally, the temperature response of plant photosynthesis is characterised by the three cardinal temperatures, i.e. the minimum and maximum temperatures and the optimum temperature. Plant net photosynthetic rate is determined by photosynthesis and respiration, both of which are influenced by temperature. Consequently leaves do not maintain a constant rate because temperature is not constant (Fitter and Hay 1987). When temperature varies over a wide range, for example, in boreal forest (Chapter 14), there is a clear temperature envelope and obvious temperature limitations to photosynthesis will be apparent. The temperature response can commonly be represented by an asymmetric bell-shaped curve (Pisek *et al.* 1973, Fitter and Hay 1987). Plant temperature optima for photosynthesis vary with species, developmental stage and current growth conditions.

10.3.2 Simulation of Dynamic Change in Photosystems: Photoinhibition

Irradiation over leaves in excess of what can be used in photosynthesis may result in photo–inhibition. This is manifested as a decline in maximum quantum efficiency of photosynthesis. There have been few attempts to integrate experimental results using mathematical tools, whilst much work has examined photo–inhibition using biophysical, biochemical and physiological methods (Powles 1984, Demmig-Adams and Adams 1992 and Long *et al.* 1994). Apparent quantum use efficiency (α) is an essential parameter in many photosynthesis models, but is introduced as a constant (Hall 1979, Johnson and Thornley 1984, Harley *et al.* 1992), or a function of only leaf temperature and CO_2 concentration (Goudriaan *et al.* 1985, Harley and Tenhunen 1991, Cannell and Thornley 1998). However α may decrease significantly on clear days with no other stress due to photo-inhibition. The diurnal courses in α and the convexity (θ) in the non-rectangular hyperbola model (eq. 10.5) used to describe light response curves of photosynthesis, can be estimated from the extent of decrease of F_v/F_m as a function of irradiance. It is estimated that daily rate of net photosynthesis can be decreased by 13 percent as a result of photo-inhibition, even with no other stress (Ogren and Sjostrom 1990). However in the field, photo-inhibition is often accompanied by other stresses such as dry and hot weather (Xu and Shen 1997). If effect of photo-inhibition is not included in the models, the net photosynthetic rate may be considerably over-estimated.

The the photosynthesis-transpiration-stomatal conductance model can be extended to include the photo-inhibition sub-model to evaluate midday depression of P_n and transpiration (E) due to stomatal and non-stomatal limitations under variable environmental conditions. The primary site of photo-inhibition is photo-system II (PS2) (Powles 1984). When photo-inhibition occurs, maximum quantum use efficiency of PS2 always decreases. The extent of photo-inhibition increases with photosynthetic photon flux density (I). Therefore, assuming that the rate of change in the apparent maximum quantum use efficiency α_t with time (t) is proportional to the amount of radiant energy absorbed, the following equation (eq. 10.38) holds (Yu *et al.* 2001):

$$\mathrm{d}\alpha_t / \mathrm{d}t = -K_i I \tag{10.38}$$

in which $K_i (>0)$ is the photo-inhibition coefficient.

There is a linear relationship between F_v/F_m and α (Greer *et al.* 1986, Demmig and Bjorkman 1987, Demmig-Adams *et al.* 1989, Kao and Forseth 1992, Edwards and Baker 1993). Fluorescence ratio (F_v/F_m) can be directly measured, therefore the change in F_v/F_m in leaves is an index of photo-inhibition. The recovery rate of photo-inhibition can be derived from the formulae (eq. 10.39) for a pseudo first-order process of F_v/F_m (Greer and Laing 1988a, 1988b):

$$\mathrm{d}\alpha_t / \mathrm{d}t = K_r \left(\alpha_n - \alpha_t \right) \tag{10.39}$$

where K_r is the recovery coefficient. As leaf photo-inhibition and its recovery occur simultaneously, the two components of inhibition and recovery in equations 10.38 and 10.39, can be combined into one equation (eq. 10.40) (Yu *et al.* 2001):

$$\mathrm{d}\alpha_t / \mathrm{d}t = K_t \left(\alpha_n - \alpha_t \right) - K_i I \tag{10.40}$$

where α_t is α at time t and α_n is the maximal value of α after recovery.

The recovery of photosynthesis from photo-inhibition is temperature-dependent with negligible recovery occurring below 15°C and maximum recovery around 30°C for certain plant species (Greer *et al.* 1986, Greer and Laing 1988a, Greer and Laing 1988b). A one-peaked equation (eq. 10.41) can be used to simulate the temperature response curve (Greer *et al.* 1986):

$$K_t = \frac{K_0 \exp\left[\left(H_k / RT_0 \right)\left(1 - T_0 / T \right) \right]}{1 + \exp\left(RT \left[S_k T - H_d \right] \right)} \tag{10.41}$$

where the parameters are: $K_0 = 0.0001$ s^{-1}, $H_k = 79500$ J mol^{-1}, $S_k = 650$ J mol^{-1} K^{-1} and $H_d = 199$ KJ mol^{-1}. R is the universal gas constant, T_0 is the reference temperature (293.2 K).

To solve equation 10.40 the initial value of α when $t = 0$ is needed which requires the parameters α_0, K_i and K_r. In the simulations shown in Figure 10.4, the parameters were set as follows: $\alpha_0 = 0.08$ mol(CO_2)/mol(quantum) and $K_i = 0.21 \times 10^{-8}$ µmol m^{-2}.

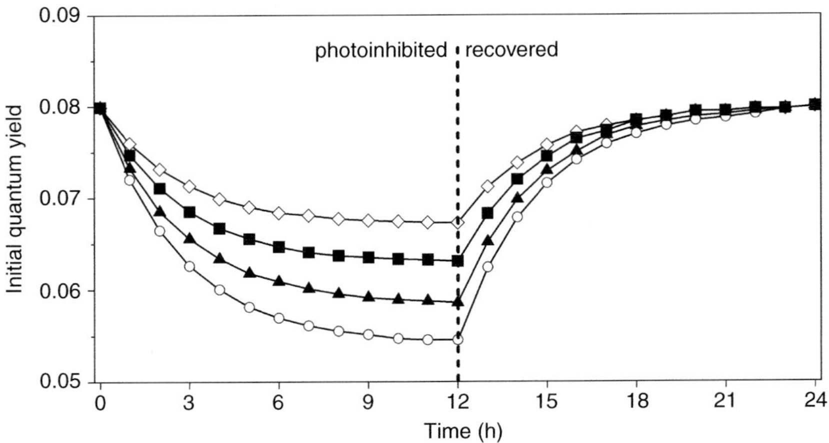

Figure 10.4. Simulation of time course of photo-inhibition of initial quantum yield under different irradiances (in the first 12 h) and its recovery in dark condition (the second 12 h). Photosynthetic photon flux density increases from 600 μmol m^{-2} s^{-1} ($\Diamond$) to 1200 μmol m^{-2} s^{-1} ($\circ$) in 200 μmol m^{-2} s^{-1} intervals. Adapted from Yu *et al.* (2001).

10.4 Parameter Optimisation for Photosynthesis Models

10.4.1 Fitting Model equations

Experimental data can be used for model fitting to estimate parameter values. The values of the model parameters characterise photosynthetic capacity. It is difficult to compare photosynthetic rates directly across different experiments because it can vary widely as a function of soil moisture content, temperature, species and many other factors. There are two methods to measure rates of leaf-scale photosynthesis and transpiration. Observations can be made under natural field conditions. Alternatively controlled conditions are applied experimentally and only one factor is altered systematically (e.g. increasing the supply of light to the leaf). From this second approach, individual light, temperature and CO$_2$ response curves are derived.

Determining leaf response curves is a convenient way to get model parameter values. A deficiency of this approach, however, is that it does not readily include interactive responses of photosynthesis and transpiration to different combinations of environmental variables. Therefore the response curve data are suitable for single model fitting but data measured in the field with natural diurnal variation of multiple factors simultaneously can be used along with combined models of leaf photosynthesis, transpiration and stomatal conductance.

As described in Chapter 9, model fitting is based on a least-squares approach. The objective function of the least-square method is described in equation 10.42:

$$\delta = \sum_{i=1}^{n} \left(P_{n,i} - \left[\frac{\alpha I_i P_{max}}{\alpha I_i + P_{max}} - R_d \right] \right)^2 \tag{10.42}$$

where n is the total number of observations and i refers to the i-th observation. The method involves searching for the best fit in the least-squares sense which minimises the sum of squared residuals (δ) between model predictions and observations. Typically commercially available statistical software is used to do this, such as Origin, SigmaPlot, Mathematica and SPSS, for example, to achieve the goal. Modellers can input data to the software and choose one existing equation or edit an equation with a curve best corresponding to the data. The software can fit the model iteratively and output model parameter values, coefficients of determination, significance test and statistical errors.

10.4.2 Parameterization for a Set of Combined equations of Photosynthesis

The rate of photosynthesis is a key indicator of the effect the local environment has on an individual plant. This section describes how to estimate model parameters using nonlinear regression techniques combined with a Monte Carlo method for estimating confidence intervals for individual parameters. The first objective function takes the standard form for least squares optimisation (eq. 10.43):

$$F_1(\beta) = \sum_{i=1}^{n} \left(P_n(x_i, \beta) - \hat{P}_{n,i} \right)^2 \tag{10.43}$$

where $x_i = (I, T, C_s, VPD)$ at the i-th observation, β denotes the vector of parameter values to be determined, P_n is given by photosynthesis models (e.g. Yu *et al.* 2002) and $\hat{P}_{n,j}$ is the observed net photosynthetic rate at the i-th observation. The value of β which minimises the value of the objective function is sought.

Mathematica can be chosen as the development platform for implementing the first formulation, due to its high-level functionality. One of the main inbuilt functions used is "Find Minimum" which searches for a local minimum using the interior point method. When a model has a multiple outputs, for example P_n, g_s and E, it has to fit multiple objective constraints of the least-square (e.g. Chew *et al.* 2012).

Understanding modelling of leaf-scale photosynthesis is required before up-scaling to canopy photosynthesis can be undertaken because photosynthesis is primarily a leaf-scale process and all the biochemistry of photosynthesis occurs within leaves. However, up-scaling from leaves to canopies comes with its own problems and this is the topic of consideration for the rest of this chapter.

10.5 Modelling Canopy Photosynthesis

Models of canopy-scale photosynthesis differ from leaf-scale models of photosynthesis through consideration of three additional features of canopies: (1) light distribution through the canopy; (2) canopy structure in terms of leaf angle and leaf area index; and, (3) vertical variation of microclimate. Up-scaling from leaf models can be undertaken using a range of simplifying assumptions to produce big-leaf, two-leaf and multi-layer models. Within a reasonably dense plant canopy, direct beam light creates regions of illumination and regions of umbra (shadows arising from leaves and branches within the canopy). The leaves of a canopy can then be represented as sunlit and shaded fractions. The light intensities for sunlit and shaded fractions are then used to drive a leaf-scale photosynthesis model applied to the two fractions. Two types of leaf photosynthesis models can be used to calculate canopy photosynthesis: (1) a light response curve with temperature modification of P_{max}; or (2) the biochemical model developed by Farquhar *et al.* (1980). However, adoption of a light-use-efficiency models is perhaps the simplest way to quantify canopy photosynthesis. This is now discussed.

10.5.1 General Form of Canopy Photosynthetic Rate

Canopy photosynthetic rate can calculated from leaf models according to the light distribution within a canopy for specific populations of leaves of specific aspects and declination angles and these are then summed to the whole canopy per unit ground area (eq. 10.44):

$$P_c(h,I) = \int_0^L \int_0^{2\pi} \int_0^{\pi/2} P_l(I,\alpha,\beta)p(\alpha)\mathrm{d}\alpha\,\mathrm{d}\beta\,\mathrm{d}L \tag{10.44}$$

where α the leaf inclination angle, β leaf aspect, $p(\alpha)$ the leaf area distribution density, $P_l(I,\alpha,\beta)$ is leaf photosynthetic rate and I is irradiance in the photosynthetic active radiation waveband which is a function of canopy structure and position of within the canopy, that is $I = I(L_i, \alpha, \beta)$. The discrete form can be written as follows (eq. 10.45):

$$P_c(h,S) = \sum_{L_i=1}^{L} \sum_{\beta_i=1}^{8} \sum_{\alpha_i=1}^{6} P(I,\alpha,\beta)p(\alpha_i) \tag{10.45}$$

And where:

$$p(\alpha_i) = g(\alpha_i) / \sum_{i=1}^{6} g(\alpha_i) \tag{10.46}$$

These two equations assume a canopy with eight aspects and six leaf inclinations although the user can modify these values to suit their own needs.

10.5.2 Daily Canopy Photosynthesis (P_{cd})

Daily photosynthesis is summed from instantaneous canopy photosynthetic rates over time from sunrise to sunset as in equation 10.47:

$$P_{cd} = \int_{-\omega_0}^{\omega_0} P_c\,\mathrm{d}\omega \tag{10.47}$$

Often P_{cd} is calculated at 30-minute or hourly time-steps and then integrated over the day. From integral trapezoid formula, the model can be written as in equation 10.48:

$$P_{cd}(I) = H \sum_{i=1}^{n-1} P_c(t) + DT_1 \cdot P_c(0) + DT_2 \cdot P_c(n) \tag{10.48}$$

in which, DT_1 is the time from sunrise to the first integral point, DT_2 is time from the last integral point to sunset and H is set to 3600 seconds for hourly time step as the unit of P_c is in second. The parameters DT_1 and DT_2 are determined by equations 10.49 and 10.50, respectively:

$$DT_1 = INT(T_1) + 1 - T_1 \tag{10.49}$$

$$DT_2 = T_2 - INT(T_2) \tag{10.50}$$

where T_1 is time of sunrise and T_2 is time of sunset, which can be obtained from equations 9.7 and 9.8, respectively. INT indicates the integer hours across time from sunrise to sunset, for example 6:40am as 6.

10.5.3 Big-Leaf Photosynthesis Models

Big-leaf models simplify the representation of a canopy as a single giant leaf and therefore the single leaf model can be directly applied to a canopy, assuming there is no error introduced in ignoring the shaded fraction of the canopy. The strategy of up-scaling single leaf models to entire canopies has two benefits: (1) being able to include explicit biochemical and biophysical processes; and, (2) simplification of calculation. However, leaf nitrogen content is not homogeneous throughout the depth of a canopy and this will alter photosynthetic capacity through the canopy. Similarly the micro-environment differs within a canopy, especially light, temperature and *VPD* and big-leaf models tend to neglect such within-canopy differences. However big-leaf models are convenient and have fewer parameters than multiple layered models. Consequently they are widely applied in site-specific, regional and global models.

Sellers *et al.* (1992) proposed the use of leaf biochemical photosynthesis models to calculate canopy photosynthesis. In these models electron transfer and Rubisco content are assumed to have linear correlations with leaf N content and canopy N content is distributed in accordance with average PAR distribution. Amthor (1994) developed a big-leaf model for canopy photosynthesis, including major variables such CO_2 concentration and absorbed PAR. Sands (1995) scaled a leaf-scale photosynthesis model to a canopy and proposed a daily photosynthesis model, while Friend *et al.* (1997) incorporated a big-leaf model within the terrestrial ecosystem model 'Hybrid'.

Big-leaf models neglect variations in plant water status and microclimate and this deficiency is significant for dense canopies. It may overestimate canopy photosynthesis using average light intensities for leaves with different inclinations and aspects at each layer (Wang and Leuning, 1998).

Light intensities within a canopy can be calculated using a negative exponential function with a canopy extinction coefficient (k). Canopy gross photosynthetic rate (P_{cg}) is integrated over the canopy using a leaf area index (L) such that (eq. 10.51):

$$P_{cg} = \frac{P_{max}}{k} \ln \frac{\alpha I + P_{max}}{\alpha I e^{-kL} + P_{max}} \tag{10.51}$$

where P_{max} is a function of temperature, soil or leaf water potential and CO_2 concentration (Collatz *et al.* 1991, Yu *et al.* 2002). The difference between leaf-scale and canopy big-leaf models is that the later includes consideration of LAI.

By adopting Ye's leaf model (eq. 10.8) and summing for an entire canopy, a new canopy big-leaf model (eq. 10.52) can be derived using the method similar to derivation of equation 10.51 above:

$$P_{n,c} = \int_0^L \alpha I \frac{1 - \beta I}{1 + \gamma I} dL \tag{10.52}$$

By combining the light intensity within a canopy, $I = I_0 e^{-kL}$, equation 10.52 becomes equation 10.53:

$$P_{n,c} = \int_0^L \alpha I_c e^{-kL} \frac{1 - \beta I_c e^{-kL}}{1 + \gamma I_E e^{-kL}} dL = -\frac{\alpha}{k} \int_0^L \frac{1 - \beta I_c e^{-kL}}{1 + \gamma I_c e^{-kL}} dI_c e^{-kL} \tag{10.53}$$

By making $x = I_c e^{-kL}$,

$$P_{n,c} = -\frac{\alpha}{\eta} \int_{I_c}^{I_c e^{-kL}} \frac{1 - \beta x}{1 + \gamma x} dx \tag{10.54}$$

$$\frac{1 - \beta x}{1 + \gamma x} = -\frac{\beta}{\gamma} + \left(\frac{\beta}{\gamma} + 1 \right) \frac{1}{1 + \gamma x} \tag{10.55}$$

$$P_{n,c} = -\frac{\alpha}{k} \int_{I_c}^{I_c e^{-kL}} \left[-\frac{\beta}{\gamma} + \left(\frac{\beta}{\gamma} + 1 \right) \frac{1}{1 + \gamma x} \right] dx$$

$$= -\frac{\alpha}{k} \left[\left(-\frac{\beta}{\gamma} x + \left(\frac{\beta + \gamma}{\gamma^2} \right) \ln|1 + \gamma x| + C^* \right) \right]_{I_c}^{I_c e^{-kL}} \tag{10.56}$$

$$= \left[\frac{\alpha \beta}{k \gamma} x - \frac{\alpha(\beta + \gamma)}{k \gamma^2} \ln|1 + \gamma x| + C \right]_{I_c}^{I_c e^{-kL}}$$

So, equation 10.52 can descripted as:

$$P_{n,c} = \int_0^L \alpha I \frac{1-\beta I}{1+\gamma I} \mathrm{d}L$$

$$= \frac{\alpha \beta I_c}{k\gamma}\left(e^{-kL}-1\right) + \frac{\alpha(\beta+\gamma)}{k\gamma^2}\ln\left|\frac{1+\gamma I_c}{1+\gamma I_c e^{-kL}}\right| \tag{10.57}$$

From this equation, canopy photosynthetic rate can be direcly calculated from light intensity above the canopy I_0 and leaf area index (and hence light distribution) through the canopy.

10.5.4 Two-Leaf Photosynthesis Models

Two-leaf models sum the rates of photosynthesis calculated separately for sunlit ($P_{nc.n}$) and shaded ($P_{nc.e}$) leaves, using the absorptance of the two "leaves" and their respective leaf area indices (eq. 10.58):

$$P_{nc} = P_{nc,n} + P_{nc,e} \tag{10.58}$$

Two-leaf models are intermediate between big-leaf models and multi-layer models in their complexity. Big-leaf models cannot represent processes within a canopy but multi-layer models have far more parameters making it more difficult to accurately parameterise the model. Wang and Leuning (1998) proposed methods to calculate canopy conductance, photosynthesis and energy partitioning, including a simple and precise radiation model, an improved photosynthesis-stomatal conductance model and a new parameterisation scheme for the leaf energy balance.

The advantage of two-leaf models is that they are more realistic than single big-leaf models but they are also computationally simpler than multi-layer models.

10.5.5 Multi-Layer Photosynthesis Models

Theoretically, multi-layer photosynthesis models are necessary for precise description of biophysical and biochemical processes within plant canopies. There are significant differences between upper and lower leaf layers in terms of microclimate, such as light intensities, temperature, wind speed, humidity and CO_2 concentration, as well as leaf attributes including leaf nitrogen and specific leaf area. Multi-layer models of photosynthesis generally adopt one of the following schemes:

(1) Variations in light level within the canopy, leaf inclination and leaf aspect are considered (e.g. Baldocchi 1992)
(2) Leaf N distribution as a function of canopy position is a primary feature. Leuning (1995) included N content attenuation from top canopy along a negative potential curve in a tempo-spatial integrative model of canopy photosynthesis

(3) Microclimate differences are included within sub-models of micrometeorological variables (Flerchinger *et al.* 1996) and CO_2 distribution within canopy (Gu *et al.* 1999).

Once the canopy has been divided into multiple layers, canopy photosynthesis is calculated separately for each layer. Normally, the canopy can be divided into n layers and light intensity declines with depth into the canopy as a function of leaf area index. Other microclimate variables (temperature, CO_2 concentration, humidity) can be assumed as a constant or derived from multi-layer water and energy balance models. In each layer, leaves are assumed to be horizontal.

10.6 References

Amthor JS, (1994). Scaling CO_2-photosynthesis relationships from the leaf to the canopy. *Photosynthesis Research* 39(3), 321–350.

Baldocchi D, (1992). A lagrangian random-walk model for simulating water-vapor, CO_2 and sensible heat-flux densities and scalar profiles over and within a soybean canopy. *Boundary-Layer Meteorology* 61(1–2), 113–144.

Berry J and JJ Raison, (1981). Responses of macrophytes to temperature. In: Lange OL, PS Nobel, CB Osmond, H Ziegher. (eds.) *Physiological plant ecology I. Responses to the physical environment*, Vol. 12A. Springer-Verlag, Berlin.

Bjorkman O, MR Badger and PA Armond, (1980). *Responses and adaptation to high temperature*. In: NC Turner and PJ Kramer (eds.), *Adaptation of plants to water and high temperature stress*. Wiley, New York, pp. 233–249.

Brooks A and GD Farquhar, (1985). Effect of temperature on the CO_2/O_2 specificity of Ribulose-1,5-Bisphosphate carboxylase oxygenase and the rate of respiration in the light–estimates from gas-exchange measurements on Spinach. *Planta* 165(3), 397–406.

Cannell MGR and JHM Thornley, (1998). Temperature and CO_2 responses of leaf and canopy photosynthesis: A clarification using the non-rectangular hyperbola model of photosynthesis. *Annals of Botany* 82(6), 883–892.

Chew KL, T Langtry, Y Zinder, Q Yu and LH Li, (2012). Estimation of biochemical parameters from leaf photosynthesis. *ANZIAM Journal* 53, C218–C235.

Collatz GJ, JT Ball, C Grivet and JA Berry, (1991). Physiological and environmental-regulation of stomatal conductance, photosynthesis and transpiration–a model that includes a laminar boundary-layer. *Agricultural and Forest Meteorology* 54(2–4), 107–136.

Collatz GJ, JA Berry, GD Farquhar and J Pierce, (1990). The relationship between the Rubisco reaction-mechanism and models of photosynthesis. *Plant Cell and Environment* 13(3), 219–225.

Cowan IR, (1977). Stomatal behaviour and environment. *Advances in Botanical Research* 4, 117–228.

Demmig-Adams B and WW Adams, (1992). Photoprotection and other responses of plants to high light stress. *Annual Review of Plant Physiology and Plant Molecular Biology* 43, 599–626.

Demmig-Adams B, K Winter, A Kruger and FC Czygan, (1989). Zeaxanthin synthesis, energy dissipation and photoprotection of photosystem II at chilling temperatures. *Plant Physiology* 90(3), 894–898.

Demmig B and O Bjorkman, (1987). Comparison of the effect of excessive light on chlorophyll fluorescence (77k) and photon yield of O_2 evolution in leaves of higher-plants. *Planta* 171(2), 171–184.

Edwards GE and NR Baker, (1993). Can CO_2 assimilation in maize leaves be predicted accurately from chlorophyll fluorescence analysis. *Photosynthesis Research* 37(2), 89–102.

Farquhar GD, SV Caemmerer and JA Berry, (1980). A biochemical-model of photosynthetic CO_2 assimilation in leaves of C-3 species. *Planta* 149(1), 78–90.

Fitter AH and RKM Hay, (1987). *Environmental physiology of plants*. Academic Press.

Flerchinger GN, JM Baker and EJA Spaans, (1996). A test of the radiative energy balance of the SHAW model for snowcover. *Hydrological Processes* 10(10), 1359–1367.

Friend AD, AK Stevens, RG Knox and MGR Cannell, (1997). A process-based, terrestrial biosphere model of ecosystem dynamics (Hybrid v3.0). *Ecological Modelling* 95(2–3), 249–287.

Goudriaan J, HH van Laar, H van Keulen and W Louwerse, (1985). *Photosynthesis, CO_2 and plant production*. In: W Day and RK Atkin (ed.), *Wheat Growth and Modelling*. Plenum Press, New York.

Greer DH, JA Berry and O Bjorkman, (1986). Photoinhibition of photosynthesis in intact bean-leaves–role of light and temperature and requirement for chloroplast-protein synthesis during recovery. *Planta* 168(2), 253–260.

Greer DH and WA Laing, (1988a). Photoinhibition of photosynthesis in intact Kiwifruit (Actinidia-Deliciosa) leaves–effect of light during growth on photoinhibition and recovery. *Planta* 175(3), 355–363.

Greer DH and WA Laing, (1988b). Photoinhibition of photosynthesis in intact Kiwifruit (*Actinidia deliciosa*) leaves–recovery and its dependence on temperature. *Planta* 174(2), 159–165.

Gu LH, JD Fuentes, HH Shugart, RM Staebler and TA Black, (1999). Responses of net ecosystem exchanges of carbon dioxide to changes in cloudiness: Results from two North American deciduous forests. *Journal of Geophysical Research-Atmospheres* 104(D24), 31421–31434.

Hall AE, (1979). Model of leaf photosynthesis and respiration for predicting carbon-dioxide assimilation in different environments. *Oecologia* 43(3), 299–316.

Harley PC and JD Tenhunen, (1991). *Modeling the photosynthetic response of C3 leaves to environmental factors*. In: KJ Boote and RS Loomis (eds.), *Modeling Crop Photosynthesis: from Biochemistry to Canopy. Special publication of the American Society of Agronomy*, Madison, WI, pp. 17–39.

Harley PC, RB Thomas, JF Reynolds and BR Strain, (1992). Modeling photosynthesis of cotton grown in elevated CO_2. *Plant Cell and Environment* 15(3), 271–282.

Jarvis PG, (1995). Scaling processes and problems. *Plant Cell and Environment* 18(10), 1079–1089.

Johnson IR and JHM Thornley, (1984). A model of instantaneous and daily canopy photosynthesis. *Journal of Theoretical Biology* 107(4), 531–545.

Jordan DB and WL Ogren, (1981). A sensitive assay procedure for simultaneous determination of Ribulose-1,5-Bisphosphate carboxylase and oxygenase a. *Plant Physiology* 67(2), 237–245.

Kao WY and IN Forseth, (1992). Responses of gas-exchange and phototropic leaf orientation in soybean to soil-water availability, leaf water potential, air-temperature and photosynthetic photon flux. *Environmental and Experimental Botany* 32(2), 153–161.

Leuning R, (1995). A critical-appraisal of a combined stomatal-photosynthesis model for C-3 plants. *Plant Cell and Environment* 18(4), 339–355.

Long SP, S Humphries and PG Falkowski, (1994). Photoinhibition of photosynthesis in nature. *Annual Review of Plant Physiology and Plant Molecular Biology* 45, 633–662.

Nikolov NT, WJ Massman and AW Schoettle, (1995). Coupling biochemical and biophysical processes at the leaf level–an equilibrium photosynthesis model for leaves of C-3 plants. *Ecological Modelling* 80(2–3), 205–235.

Norman J, (1993). *Scaling processes between leaf and canopy level*. In: JR Ehleringer and CB Field (eds.) *Scaling physiological processes. Scaling Physiological Processes*. Academic Press, London.

Ogren E and M Sjostrom, (1990). Estimation of the effect of photoinhibition on the carbon gain in leaves of a willow canopy. *Planta* 181(4), 560–567.

Pisek A, W Larcher, A Veges and K Nap-Zin, (1973). *The normal temperature range.* In: H Precht, J Christopherson, H Hensel and W Larcher (eds.), *Temperature and life.* Springer-Verlag, Berlin, pp. 102–194.

Powles SB, (1984). Photoinhibition of photosynthesis induced by visible-light. *Annual Review of Plant Physiology and Plant Molecular Biology* 35, 15–44.

Sands PJ, (1995). Modeling canopy production: 2. from single-leaf photosynthetic parameters to daily canopy photosynthesis. *Australian Journal of Plant Physiology* 22(4), 603–614.

Sellers PJ, JA Berry, GJ Collatz, CB Field and FG Hall, (1992). Canopy reflectance, photosynthesis and transpiration. A reanalysis using improved leaf models and a new canopy integration scheme. *Remote Sensing of Environment* 42(3), 187–216.

Thornley JHM, (1976). *Mathematical models in plant physiology.* London: Academic Press, 86–110.

Von Caemmerer S and GD Farquhar, (1981). Some relationships between the biochemistry of photosynthesis and the gas-exchange of leaves. *Planta* 153(4), 376–387.

Wang YP and R Leuning, (1998). A two-leaf model for canopy conductance, photosynthesis and partitioning of available energy I: Model description and comparison with a multi-layered model. *Agricultural and Forest Meteorology* 91(1–2), 89–111.

Xu DQ and YG Shen, (1997) Diurnal variations in the photosynthetic efficiency in plants. *Acta Phytophysiologica Sinica* 23(4), 410–416.

Ye Z, JD Suggett, P Robakowski, HJ Kang, (2013). A mechanistic model for the photosynthesis-light response based on the photosynthetic electron transport of PS II in C_3 and C_4 species. *New Phytologist*, 199(1), 110–120.

Ye Z and Q Yu, (2008). A coupled model of stomatal conductance and photosynthesis for winter wheat. *Photosynthetica* 46(4), 637–640.

Ye ZP, (2007). A new model for relationship between irradiance and the rate of photosynthesis in *Oryza sativa*. *Photosynthetica* 45(4), 637–640.

Yu Q, J Goudriaan and TD Wang, (2001). Modelling diurnal courses of photosynthesis and transpiration of leaves on the basis of stomatal and non-stomatal responses, including photoinhibition. *Photosynthetica* 39(1), 43–51.

Yu Q, YF Liu, JD Liu and TD Wang, (2002). Simulation of leaf photosynthesis of winter wheat on Tibetan Plateau and in North China Plain. *Ecological Modelling* 155(2–3), 205–216.

11

Modelling Stomatal and Canopy Conductance

11.1 Introduction

Stomata are located at the critical interface between a leaf's internal and external environment. Stomata are the point of coupling between the flux of CO_2 for photosynthesis and water for transpiration. Modelling stomatal conductance (g_s) has a long history but mostly it has been empirical rather than mechanistic. Recently, however, this has changed and more mechanistic models are emerging. In this chapter various models that are currently used to describe stomatal and canopy conductance are discussed. The physiological basis of stomatal movements underpinning these models is discussed extensively in Chapter 2 while examples of stomatal behaviour in the field are presented in Chapters 14–18.

Stomata respond to environmental factors in such a manner that it not only protects the plant from excessive water loss, but also optimises water-use-efficiency (Cowan 1977, Chapter 2). Regulation of g_s is a key element in mass flow through the soil-plant-atmosphere continuum (SPAC) with respect to water loss and carbon gain. Though much is known about stomatal behaviour in the field, a mechanistic model of field behaviour of stomatal conductance has been elusive. Mathematical simulation models of stomatal conductance are mostly semi-empirical, with a notable recent exception which is described more fully in Section 11.2 (Medlyn *et al.* 2011). Early work by Upadhyaya *et al.* (1983), and Fu and Wang (1994) are also noteworthy in their attempts to construct mechanistic models which include leaf water potential, photosynthetic electron transport, and ion fluxes, into or out of the guard cells. These models, though more analytical, are rarely applied today because they contain variables which are either not readily measurable or they are rarely measured by ecophysiologists.

Although many studies simulate stomatal behaviour and there are multiple combined photosynthesis-stomatal conductance models, systematic analysis of physiological responses to environmental variation has rarely been undertaken. This is despite the primary importance in model building of the need to construct a sound mechanistic framework so that observed phenomena can be simulated under widely different environments and circumstances.

During numerical analysis of physiological responses it is important to study the sensitivity of outputs to variation in inputs. Models of physiological responses to environmental factors have a wide range of use in the modelling of crop yield, forest growth, and dynamic global vegetation models and land surface exchange models.

11.2 Semi-Empirical Models of Stomatal Conductance

11.2.1 The Jarvis Model

Steady-state models of stomatal function, which are more popular than dynamic ones, may be classified into three types: (1) multiple correlation models; (2) phenomeno-logical models; and, (3) models based on a range of mechanisms that may regulate stomatal behaviour. Multiple correlation models, expressing stomatal conductance as a linear multi-regression function of light, temperature, humidity and water potential, are of little value in producing an understanding of the mechanisms underpinning observed behaviours. Phenomenological models, a representative one of which was proposed by Jarvis (1976) (and further developed in the Jarvis-Stewart formulation and then modified by Whitley *et al.* [2009]), has been widely adopted by modellers because it fits a wide range of experimental data across a large range of ecosystems. Models based on a range of mechanisms that may regulate stomatal behaviour include Ball-Woodrow-Berry type models and models based upon optimisation theory. These are discussed in Sections 11.1.2 and 11.2. See Sections 2.4.5 and 2.4.9 for discussion of the ecophysiological aspects of stomatal responses to abiotic factors and Whitley's modified Jarvis-Stewart approach.

There are five main environmental (abiotic) factors affecting stomatal conductance (g_s) under natural conditions (Chapter 2): (1) solar radiation; (2) air temperature; (3) humidity; (4) CO_2 concentration; and, (5) soil water potential. The g_s observed in the field can be estimated from the maximum conductance (g_{max}) under suitable conditions modified by a series of correction coefficients (Jarvis 1976), one for each of the key abiotic variables that limit g_s. Thus (eq. 4.1):

$$g_s = g_{max} f(I) f(T_a) f(C_a) f(VPD) f(\psi) \tag{11.1}$$

where I is absorbed photosynthetic photon flux density, T_a is air temperature, C_a is CO_2 concentration, VPD is vapour pressure deficit and ψ is leaf water potential. In the most commonly applied form of the Jarvis model, functions for C_a are generally omitted because C_a doesn't vary much on a daily or seasonal time-scale, and soil moisture content (θ) frequently replaces Ψ because θ is measured more frequently than soil water potential. A detailed treatment of the modified Jarvis-Stewart model and its applications, is given in Chapter 2.

11.2.2 Ball-Woodrow-Berry Types Models

Wong *et al.* (1979) showed that stomata regulate C_i at a relatively constant value (at the value where electron transport and Rubisco activity co-limit photosynthesis; see Chapter 10 on modelling leaf-scale photosynthesis and Chapter 2 for the physiology of photosynthesis). Under steady state conditions there is a linear relation between stomatal conductance and CO_2 (g_{sc}) and photosynthesis (P_n) when the concentration of CO_2 and humidity at the leaf surface (C_s and h_s respectively) are not changing. From this premise, the Ball-Woodrow-Berry (BWB) model for g_{sc} can be written (eq. 11.2):

$$g_{sc} = m\frac{P_n h_s}{C_s} + b \tag{11.2}$$

where P_n is net photosynthetic rate ($\mu mol\ m^{-2}\ s^{-1}$); h_s and C_s are relative humidity and CO_2 concentration at the leaf surface, respectively; m and b are parameters, the latter being the intercept on the coordinate g_{sc} near zero. $P_n h_s / C_s$ is called "a stomatal conductance index". Since equation 4.2 is not applicable at low CO_2 concentrations, Leuning (1990) modified the BWB model and used $C_s\Gamma$, instead of C_s, where Γ is the CO_2 compensation concentration, to give a better fit. Recent experiments revealed that stomata respond to evaporative demand (D) more than to relative humidity of air (Aphalo and Jarvis 1991, Mott and Parkhurst 1991; Chapter 2). By adopting these modifications, Leuning (1995) proposed a revised form of the BWB model (eq. 11.3):

$$g_{sc} = m\frac{P_n}{(C_s - \Gamma)(1 + VPD_s / VPD_0)} + g_{s0} \tag{11.3}$$

where g_{s0}, the intercept, is equivalent to b in equation 11.2, and VPD_0 is a parameter.

There is a close relationship between rates of photosynthesis and stomatal conductance. Stomatal conductance regulates photosynthesis by influencing the intercellular CO_2 concentration and, thereby, biochemical processes in the chloroplast. The extent of stomatal opening is jointly determined by light intensity, and by plant, atmospheric and soil water status (Chapter 2). The former involves both the action of light receptors and the response to intercellular CO_2 concentration, and the latter is determined by the water balance of leaf, root, and stem tissues, and the loss of water from the guard and epidermal cells to the immediate aerial environment (Chapter 2).

The stomatal models used in almost all ecological models are semi-empirical, among which the one proposed by Ball *et al.* (1987) (the BWB model) is widely accepted (Collatz *et al.* 1991, Hatton *et al.* 1992, McMurtrie *et al.* 1992, Sellers *et al.* 1996). These stomatal models have generally been constructed around laboratory data.

Under natural field conditions however, feedbacks among the interactions between physiological and physical processes are far more important than in lab studies. For example, an increase in solar radiation will induce an increase in leaf temperature,

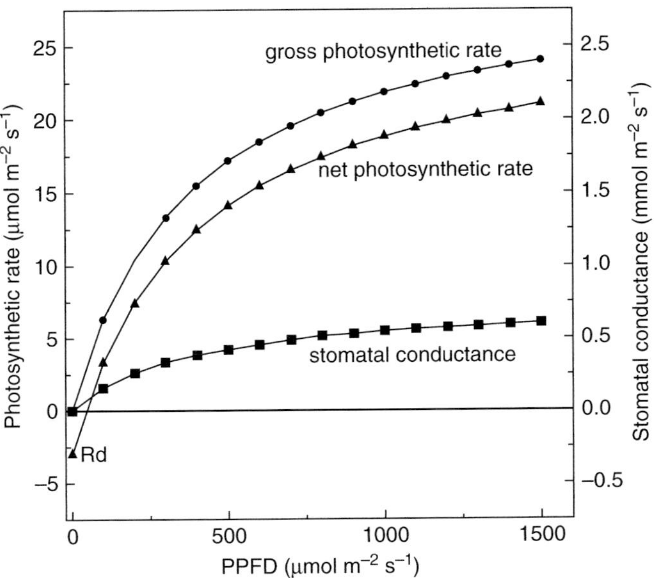

Figure 11.1. Illustration of the relationship between stomatal conductance, net and gross photosynthetic rates, and light intensity. Adapted from Yu *et al.* (2004).

photosynthesis, and leaf-to-air vapour pressure difference. In contrast, when light levels are manipulated in the laboratory, leaf temperature is controlled artificially and consequently responses observed in the field may be different from those observed in the lab. The validity of stomatal models under natural conditions is often taken for granted and seldom examined critically. In the following discussion we make a theoretical analysis of several stomatal models based on Ball-Woodrow-Berry's original model and evaluate the models by fitting experimental data obtained under field conditions.

Fu and Wang (1994) found that the relationship between g_s and an index of g_s ($P_n h_s/C_s$) changes when boundary layer conductance (g_b) changes. This implies that the BWB model cannot interpret the impact of changes in g_b on g_s, despite the BWB model including consideration of the impacts of light, temperature, CO_2 and humidity, on g_s (Fig. 11.1). It was therefore suggested that there may be a better index reflecting the relationship between leaf conductance and its responses to ambient CO_2 concentration and humidity. However, as noted in Chapter 2, stomata respond to transpiration rate rather than relative humidity (Monteith 1995). Thus (eq. 11.4):

$$\frac{g_s}{g_{sm}} = 1 - E / E_m \tag{11.4}$$

where g_{sm} and E_m are characteristic parameters for stomatal conductance and transpiration.

Dewar (1995) combined equation 11.4 with the BWB model and obtained equation 11.5:

$$g_s = \frac{mP_n}{C_s - \Gamma}(1 - E/E_m) + g_0 \tag{11.5}$$

where the parameter g_0 is always a small value near zero.

Mott and Parkhurst (1991) and Eamus *et al.* (2008) observed that stomatal aperture, which determines g_s, is more closely correlated with transpiration (the rate of water loss) rather than with D. Therefore equation 11.4 gives the relation between g_s and transpiration. Using a combined photosynthesis-transpiration-g_s model in which g_s is a hyperbolic function of VPD_s (the vapour pressure deficit at the leaf surface), Yu and Wang (1998) showed that g_s declines linearly with transpiration under changing air humidity. This result suggests an implicit relation between equation 11.3 and 11.5, as given in equation 11.6:

$$E = g_s \cdot VPD_s \tag{11.6}$$

Substituting equation 11.6 into equation 11.4 gives equation 11.7:

$$\frac{g_s}{g_{sm}} = 1 - \frac{g_s \cdot VPD_s}{E_m} \tag{11.7}$$

After rearrangement this becomes equation 11.8:

$$g_s = \frac{1}{1/g_{sm} + VPD_s/E_m} \tag{11.8}$$

Given that $VPD_0 = E_m/g_{sm}$, equation 11.8 can be written as equation 11.9:

$$g_s = g_{sm}\frac{1}{1 + VPD_s/VPD_0} \tag{11.9}$$

The correlation between g_s and transpiration in equation 11.4 accords with that between g_s and VPD_s, which means that a linear relation between g_s and transpiration, as proposed by Monteith, is equivalent to a hyperbolic relation between g_s and VPD_s, as shown by equation 1.9 and as is frequently observed (Chapter 2).

As g_s begins to increase immediately with increasing light, even below the light compensation point, Yu *et al.* (2001) proposed that the gross assimilation rate should be used instead of net assimilation, and correspondingly, $(C_s - \Gamma)$ should be replaced by C_s in equation 11.3 to give equation 11.10:

$$g_s = m\frac{P}{C_s(1 + VPD_s/VPD_0)} \tag{11.10}$$

where P is the gross assimilation rate and C_s is the CO_2 concentration at the leaf surface. In this expression parameter g_0 in equation 11.3 is taken as zero because P_g and g_s tends to zero in the dark.

The rate of photosynthesis (P) is a function of many environmental variables including light supply, soil water content, and g_s. A revision to take account of the limitation of photosynthesis by g_s can be written as in equation 11.11 (Yu *et al.* 2004):

$$P = \frac{P_{max}\,\alpha I\,\eta C_a}{P_{max}\,\alpha I + P_{max}\,\eta C_a + \alpha I\,\eta C_a}\,\frac{g_s}{g_s + g_{int}} \tag{11.11}$$

where P_{max} is the maximum catalytic capacity of Rubisco per unit leaf area, α is initial photochemical efficiency, and η is the initial slope of the CO_2 response curve (μmol m^{-2} s^{-1}/μmol mol^{-1}).g_{int} is a parameter. If $g_{int} \rightarrow 0$, $g_s/(g_s + g_{int})$ is equal to 1 the equation shortens to the original light and CO_2 response curves (Chapter 10).

P_{max} is mainly determined by the activity of Rubisco, and this is very much determined by temperature and CO_2 partial pressure (P_i; Collatz *et al.* 1991):

$$P_{max} = V_m\left(P_i - \Gamma\right)/\left(P_i + 2\Gamma\right) \tag{11.12}$$

where V_m is a function of temperature with a maximum as given by Collatz *et al.* (1991) (eq. 11.13):

$$V_m = V_0\,\frac{Q_{10}^{(T_a - 25)/10}}{1 + \exp\{[-a_1 + b_1(T_a + 273)]/[R(T_a + 273)]\}} \tag{11.13}$$

where a_1 and b_1 are parameters and $V_m = V_0$ at $T_a = 25\,^\circ$C and R is the universal gas constant.

It is assumed that $g_s/(g_s + g_{int})$ is determined chiefly by leaf or soil water status for a particular plant. Therefore, equation 11.11 can be converted into the following two equations (eq. 11.14 and eq. 11.15):

$$P = \frac{V_m\,\alpha I\,\eta C_a}{V_m\,\alpha I + V_m\,\eta C_a + \alpha I\,\eta C_a}\,f(\psi) \tag{11.14}$$

The water-stress coefficient, $f(\psi)$, is most simply characterized by a linear relation from the water potential at wilting point (ψ_0) to water potential at field capacity (ψ_m), that is, relative extractable water (Lagergren and Lindroth 2002). Therefore, by combining equations 11.10 and 11.14, g_s can be expressed as a function of environmental variables in the following form (eq. 11.15):

$$g_s = m\,\frac{P_{max}\,\alpha I\,\eta}{(P_{max}\,\alpha I + P_{max}\,\eta C_a + \alpha I\eta C_a)}\,\frac{1}{(1 + VPD_s\,/\,VPD_0)}\,\frac{\psi - \psi_0}{(\psi_m - \psi_0)} \tag{11.15}$$

Stomata generally close in the dark so that g_s is zero in darkness (but see the discussion of nocturnal transpiration in Chapter 3). Thus modellers assume that g_s is zero in dark and this is satisfied by equation 4.6. Boundary conditions of stomatal response to light, VPD_s and water potential are also satisfied by equation 11.15. The relationship expressed in equation 11.15 has recently been revised by replacing relative humidity with leaf water potential (Ψ_1) to include influences of both atmospheric *VPD* and soil water content (Tuzet *et al.* 2003), to yield the following relationship (eq. 11.16):

$$g_s = m \frac{P_n}{(C_s - \Gamma)} f(\psi_1) + g_0 \tag{11.16}$$

where Γ is the CO_2 compensation point, g_0 is cuticular conductance, and m is an empirical parameter. The function $f(\psi_1)$ is a correction factor which ranges from 0 to 1. An empirical logistic function is used to describe the dependence of stomatal conductance on ψ_1 as per equation 11.17:

$$f(\psi_1) = \frac{1 + \exp(s_1 \psi_f)}{1 + \exp[s_1(\psi_f - \psi_1)]} \tag{11.17}$$

where ψ_f is a reference potential and s_1 is a sensitivity parameter.

11.3 Models Based on Conservative Water-Use-Efficiency

Photosynthesis is modelled using our mechanistic understanding of the processes of CO_2 fixation and electron transport and their responses to environmental factors (especially CO_2 concentration, light, and temperature; Chapter 10). In contrast transpiration is modelled using an empirical representation of g_s. Although g_s is correlated with the rate of photosynthesis (P), the ratio P/g_s varies with *VPD* and an understanding of the cellular (mechanistic) basis of stomatal responses to *VPD*, concentration of CO_2 and light remains elusive and contested. Consequently, empirical models of g_s (such as those by Ball *et al.* 1987 and Leuning 1995) are widespread. This is because parameters can be estimated from site-specific data and the models are simple. However, because these models are empirical their parameters are essentially meaningless and site-specific.

Mechanistic models of g_s do exist (Dewar 2002, Gao *et al.* 2002, Buckley *et al.* 2003) but they tend to be too complex to implement at scales larger than plot-scale. They require complex numerical, iterative solutions, and values for their parameters are unobtainable at regional or global scales.

In further contrast to complex mechanistic models of g_s, there is a clear and simple theory of optimal stomatal behaviour (Chapter 2) that applies to a wide range of scales and environments (Cowan and Farquhar 1977). This theory is based on the idea that stomatal aperture is such that it maximises photosynthesis (P_n) while minimising transpiration (E). Thus optimal stomatal behaviour is one that minimises the integrated sum of:

$$E - \lambda P_n \tag{11.18}$$

where λ (mol H_2O mol^{-1} C) is a parameter representing the marginal water cost of plant carbon gain.

In a recently published model, Medlyn *et al.* (2011) derive a simple model for g_s based upon this optimisation approach. Most importantly, a theoretical interpretation exists for the model parameters and the model is applicable across biomes, climates, and spatial scales. By combining the optimal stomatal conductance model with the RuBP regeneration-limited model of photosynthesis Medlyn *et al.* (2011) derived the following approximation for the optimal stomatal conductance (g_s^*; eq. 11.19; Fig. 11.2):

$$g_s^* = g_0 + 1.6\left(1 + g_1 / \sqrt{VPD}\right)\left(\frac{P}{C_a}\right) \tag{11.19}$$

The analytical expression in this model is analogous to the empirical models of Leuning (1995) and Ball *et al.* (1987). This unified stomatal model is derived from the optimal model but similar in form to the empirical model. Most importantly the slope parameter g_1 can be directly obtained by fitting to experimental (lab or field) data as is done with empirical models. However a theoretical interpretation for g_1 is available. The slope parameter increases linearly with the combination of terms:

$$g_1 \infty \sqrt{\Gamma \lambda} \tag{11.20}$$

Thus g_1 should increase with the marginal water cost of carbon (λ) and with the CO_2 compensation point (Γ). g_1 is proportional to both the CO_2 compensation point and the marginal water cost of carbon gain. The CO_2 compensation point depends on temperature and the relationship between g_1 and temperature can be assumed to be constant across

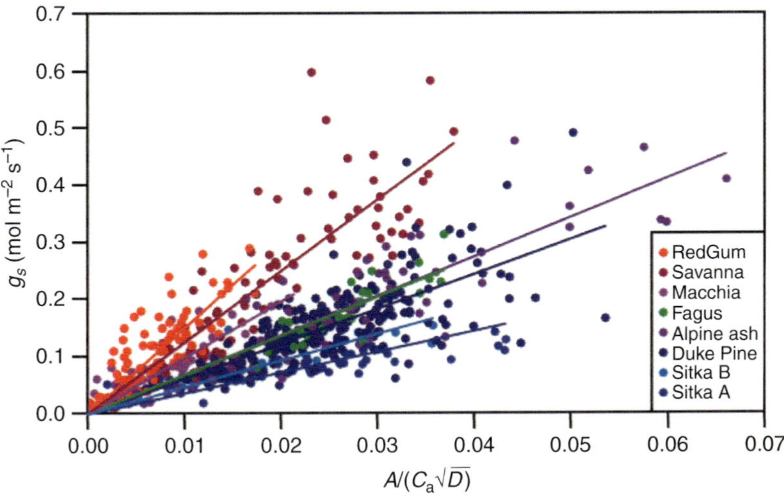

Figure 11.2. Visualization of the unified stomatal model equation 4.19 fitted to eight datasets from contrasting forest ecosystems. Reproduced with permission from Medlyn *et al.* (2011).

C3 species (Bernacchi *et al.* 2001). For a given species g_1 increases with growth temperature and that rate of increase follows the square root of the temperature-dependence of the CO_2 compensation point, as shown by Medlyn *et al.* (2011).

This unified model of g_s makes a significant improvement over earlier models because it is the first to correctly capture the response of g_s to C_a, namely a declining g_s as C_a increases.

11.4 Canopy Conductance/Resistance Models

Canopy resistance to water vapour, the reciprocal of canopy conductance, is the sum of leaf conductance and boundary conductance of the canopy. Canopies are often assumed to be a random array of leaves in big-leaf models, and canopy resistance (r_c) consists of the stomatal resistance of each layer in the canopy ($i = 1, \ldots, m$) ($r_{s,i} = 1/g_s$) (Lhomme 1991), thereby giving equation 11.21:

$$\frac{1}{r_c} = \sum_{i=1}^{m} \frac{1}{r_{s,i}}. \tag{11.21}$$

To up-scale the stomatal model from leaf-scale to canopy-scale, the relationship between canopy resistance, canopy net photosynthesis (P_{cn}), atmospheric CO_2 (C_s), and ψ_l, is given by equation 11.22:

$$\frac{1}{r_c} = a_1 \frac{P_{n,c}}{(C_s - \Gamma)} f(\psi_l) + \frac{1}{r_{c0}} \tag{11.22}$$

where r_{c0} is cuticular resistance and $1/r_{c0}$ is a summation of total g_0 of each leaf. Thus (eq. 11.23):

$$1/r_{c0} = g_0 L \tag{11.23}$$

The following sections (11.3.1 and 11.3.2) provide details of the multiple applications of methods to calculate canopy conductance, using different approaches as summarised in Table 11.1.

11.4.1 Canopy Conductance Based On a Two-Leaf Canopy Model

In two-leaf models the canopy is considered to consist of two fractions: a sunlit fraction and a shaded fraction. Consideration of these two sets of leaves is made in the following contrasting formulations (Wang, 2006):

(1) Average conductance of sunlit and shaded leaves (Bailey and Davies 1981) is derived using equation 11.24:

$$g_c = [(g_{s,n} + g_{s,e})/2] \times L \tag{11.24}$$

where $g_{s,n}$ and $g_{s,e}$ are sunlit and shaded leaf conductance, respectively.

(2) Alternatively, the average conductance of sunlit leaves in the upper layers of the canopy is derived using the following (Whitehead *et al.* 1981):

$$g_c = g_{s,n,T} L \tag{11.25}$$

where $g_{s,n,T}$ is average leaf conductance for the 25 percent leaf area index of the uppermost layers of the canopy, and T indicates top layers. Because the sunlit fraction is assumed to dominate the behaviour of the entire canopy, no consideration of the shade fraction is made.

(3) Alternatively, a weighting of conductances of sunlit and shaded leaves can be applied as per equation 11.26 (Duncan *et al.* 1967):

$$g_c = g_{s,n} L_n + g_{s,e} L_e \tag{11.26}$$

in which L_n and L_e are leaf area indices for sunlit and shaded leaves, respectively.

(4) Alternatively, stomatal conductance for the two leaf area indices (the sunlit and shaded fractional LAI) is computed as per equation 4.27 (Szeicz and Long 1969):

$$g_c = [(g_{s,n} + g_{s,e})/2] \times L_f \tag{11.27}$$

where effective leaf area index (L_f) is expressed as $L_f = L$ if $L \leq L_x/2$ or as $L_f = L_x/2$ if $L > L_x/2$, and L_x is the maximum leaf area index.

11.4.2 Canopy Conductance Based on Multi-Layered Canopies

In contrast to the two-leaf model, multiple-layered formulations consider the canopy to consist of more than two fractions. Four examples are presented here:

1) Canopy conductance for layered canopies is given by (Sellers *et al.* 1986):

$$g_c = \sum_{i=1}^{3} (L_{n,i} \times g_{s,n,i} + L_{e,i} \times g_{s,e,i}) \tag{11.28}$$

2) Canopy conductance for multiple leaf angles is given by (Sellers *et al.* 1986):

$$g_c = \sum_{i=1}^{3} \left[\sum_{j=1}^{5} (g_{s,n,i,j} \times L_{n,i,j}) + g_{s,e,i} \times L_{e,i} \right] \tag{11.29}$$

where *i* stands for sequence of layers, and *j* is the sections for leaf declination.

3) The canopy is divided into sunlit, umbral, and penumbral fractions, and integration of leaf stomatal conductance is made for the entire canopy:

$$g_c = \int_0^L \left[g_s(I_s, T_{l,s}, VPD_s) P_s(L') + g_s(I_u, T_{l,u}, VPD_u) P_u(L') + g_s(I_p, T_{l,p}, VPD_p) P_p(L') \right] dL'$$

$$\tag{11.30}$$

Table 11.1. Up-scaling strategies for canopy conductance from leaf conductance (Wang, 2006).

Model description	LAI	Two-leaf	Multi-layer	Nitrogen content	References
$g_c = g_{sun,t} \times L$	√				Whitehead *et al.* 1981
$g_c = [(g_{sun} + g_{shade})/2] \times L$	√	√			Bailey & Davies 1981
$g_c = [(g_{sun} + g_{shade})/2] \times L_{eff}$	√	√			Szeicz & Long, 1969
$g_c = g_{sun} \times L_{sun} + g_{shade} \times L_{shade}$	√	√			Duncan *et al.* 1967 Irmak *et al.* 2008 Zhang *et al.* 2011 Blonquist *et al.* 2009
$g_c = \sum_{i=1}^{3} (L_{sun,i} \times g_{sun,i} + L_{shade,i} \times g_{shade.i})$	√	√	√		
$g_c = \sum_{i=1}^{3} [\sum_{j=1}^{5} (g_{sun,i,j} \times L_{sun,i,j}) + g_{shade,i} \times L_{shade,i}]$	√	√	√		Sellers *et al.* 1986
$g_c = \int_{0}^{L} g_s (Q_{ps}, T_{ls}, D_s) P_s(L') +$ $g_s(Q_{pu}, T_{lu}, D_u) P_u(L') + g_s (Q_{pp}, T_{lp}, D_p) P_p(L') dL'$	√	√	√		Baldocchi *et al.* 1991
$g_c = g_s \dfrac{1 - \exp(-kLAI)}{k}$	√			√	Sellers *et al.* 1992
$A_{n,c} = A_{sun} L_{sun} + A_{shade} L_{shade}$ $g_c = g_1 \dfrac{A_{n,c} p}{C_s (1 + D/D_0)} + g_0 L$	√	√		√	Wang and Leuning 1998 Arora 2003
$g_c = \sum_{1}^{i=n} g_{s,i} L_i$			√		Furon *et al.* 2007

where I_s, I_u, and I_p are solar radiation over sunlit, umbral, and penumbral leaves, respectively, $T_{1,s}$, $T_{1,u}$, and $T_{1,p}$ are leaf temperatures, and VPD_s, VPD_u, and VPD_p the leaf surface water pressure deficits, accordingly. P_s, P_u, and P_p are leaf area fractions in sunlit, umbral, and penumbral fractions, respectively. This equation may need a radiation transfer model and an energy balance model to be resolved. The umbra is the innermost and darkest part of a shadow and occurs where the sun is entirely blocked by the canopy; the penumbra is the part of the shadow where only part of the sun is blocked by the canopy and is therefore less shaded than the umbral region.

4) Integration of stomatal conductance is made for the entire canopy. Thus Sellers *et al.* (1992) and Cox *et al.* (1998) assume that plants optimally use resources and that leaf nitrogen concentration declines as a negative exponential with canopy depth. From this assumption it is possible to calculate the maximum carboxylation rate (V_{max}), which corresponds well with light intensity, thus:

$$V_{max}(L) = V_{max}(0)e^{-kL} \qquad (11.31)$$

where $V_{max}(L)$ and $V_{max}(0)$ are the maximum rates of carboxylation at the top of the canopy and within the canopy, respectively, with an LAI of L, k is the extinction coefficient of light through the canopy.

Canopy photosynthesis can be expressed as an integration of leaf photosynthesis over the entire canopy leaf area index assuming a proportional relationship between photosynthesis and light intensity:

$$P_c = \int_0^L P e^{-kL} dL = P \frac{1 - e^{-kL}}{k} \qquad (11.32)$$

Thus, canopy conductance is as follows if the ratio of photosynthetic rate and stomatal conductance is a constant:

$$g_c = g_s \frac{1 - e^{-kL}}{k} \qquad (11.33)$$

Similarly Arora (2003) proposed the following model of canopy conductance as a function of canopy photosynthesis ($P_{n,c}$) and leaf area index (L) based on the revised BWB model (Leuning 1995):

$$g_c = g_1 \frac{P_{n,c}}{C_s \left(1 + VPD / VPD_0\right)} + g_0 L \qquad (11.34)$$

In general, canopy resistance can be expressed as follows for n layers of a canopy (Furon *et al.* 2007):

$$r_c = \left[\sum_1^{i=n} g_{s,i} L_i \right]^{-1} \qquad (11.35)$$

11.5 References

Aphalo PJ and PG Jarvis, (1991). Do stomata respond to relative-humidity. *Plant, Cell and Environment* 14(1), 127–132.

Arora VK, (2003). Simulating energy and carbon fluxes over winter wheat using coupled land surface and terrestrial ecosystem models. *Agricultural and Forest Meteorology* 118(1–2), 21–47.

Bailey WG and JA Davies, (1981). Bulk Stomatal-Resistance Control on Evaporation. *Boundary-Layer Meteorology* 20(4), 401–415.

Baldocchi DD, BB Hicks and TP Meyers, (1988). Measuring biosphere-atmosphere exchanges of biologically related gases with micrometeorological methods. *Ecology* 69(5), 1331–1340.

Baldocchi DD, RJ Luxmoore and JL Hatfield, (1991). Discerning the forest from the trees–an essay on scaling canopy stomatal conductance. *Agricultural and Forest Meteorology* 54(2–4), 197–226.

Ball JT, IE Woodrow and JA Berry,(1987). A model predicting stomatal conductance and its contribution to the control of photosynthesis under different environmental conditions. In: J. Biggins (Editor), *Progress in Photosynthesis Research*. Martinus Nijhoff Publishers, Dordrecht–Boston–Lancaster, pp. 221–224.

Bernacchi CJ, EL Singsaas, C Pimentel, AR Portis and SP Long, (2001). Improved temperature response functions for models of Rubisco-limited photosynthesis. *Plant, Cell and Environment* 24(2), 253–259.

Blonquist Jr, JM, JM Norman and B Bugbee, (2009). Automated measurement of canopy stomatal conductance based on infrared temperature. *Agricultural and Forest Meteorology*, 149(12): 2183–2197.

Buckley TN, KA Mott and GD Farquhar, (2003). A hydromechanical and biochemical model of stomatal conductance. *Plant, Cell and Environment* 26(10), 1767–1785.

Collatz GJ, JT Ball, C Grivet and JA Berry, (1991). Physiological and environmental-regulation of stomatal conductance, photosynthesis and transpiration–a model that includes a laminar boundary-layer. *Agricultural and Forest Meteorology* 54(2–4), 107–136.

Cowan IR, (1977). Stomatal behaviour and environment. *Advances in Botanical Research* 4, 117–228.

Cowan IR and GD Farquhar, (1977). Stomatal function in relation to leaf metabolism and environment. *Symposium of the Society for Experimental Biology* 31, 471–505.

Cox PM, C Huntingford and RJ Harding, (1998). A canopy conductance and photosynthesis model for use in a GCM land surface scheme. *Journal of Hydrology* 211–213(79–94).

Dewar RC, (1995). Interpretation of an empirical-model for stomatal conductance in terms of guard cell function. *Plant Cell and Environment* 18(4), 365–372.

Dewar RC, (2002). The Ball-Berry-Leuning and Tardieu-Davies stomatal models: synthesis and extension within a spatially aggregated picture of guard cell function. *Plant Cell and Environment* 25(11), 1383–1398.

Duncan WG, RS Loomis, WA Williams and R Hanau, (1967). A model for simulating photosynthesis in plant communities. *Hilgardia* 38(4), 181–205.

Eamus D, DT Taylor, CMO Macinnis-Ng, S Shanahan and L De Silva, (2008). Comparing model predictions and experimental data for the response of stomatal conductance and guard cell turgor to manipulations of cuticular conductance, leaf-to-air vapour pressure difference and temperature: feedback mechanisms are able to account for all observations. *Plant, Cell and Environment* 31(3), 269–277.

Fu W and TD Wang,(1994). A mechanistic model of stomatal responses to environmental factors. *Acta Phytophysiol Sin* 20, 277–284 (in Chinese).

Furon J, JS Warland and W Wagner-Riddle, (2007). Analysis of scaling-up resistances from leaf to canopy using numerical simulations. *Agronomy Journal* 99, 1483–1491.

Gao Q, P Zhao, X Zeng, X Cai and W Shen, (2002). A model of stomatal conductance to quantify the relationship between leaf transpiration, microclimate and soil water stress. *Plant, Cell and Environment* 25(11), 1373–1381.

Hatton TJ, J Walker, WR Dawes and FX Dunin, (1992). Simulations of hydroecological responses to elevated CO_2 at the catchment scale. *Australian Journal of Botany* 40(4–5), 679–696.

Irmak, S, D Mutiibwa, A Irmak, TJ Arkebauer, A Weiss, DL Martin and DE Eisenhauer, (2008). On the scaling up leaf stomatal resistance to canopy resistance using photosynthetic photon flux density. *Agricultural and Forest Meteorology,* 148(6–7): 1034–1044.

Jarvis PG, (1976). Interpretation of variations in leaf water potential and stomatal conductance found in canopies in field. *Philosophical Transactions of the Royal Society of London Series B-Biological Sciences* 273(927), 593–610.

Jarvis PJ and KG McNaughton, (1986). Stomatal control of transpiration: scaling up from leaf to region. In: A. MacJadyen and R.D. Ford (Editors), *Advances in Ecological Research.* Academic Press, London, pp. 205–265.

Lagergren F and A Lindroth, (2002). Transpiration response to soil moisture in pine and spruce trees in Sweden. *Agricultural and Forest Meteorology* 112(2), 67–85.

Leuning R, (1990). Modeling stomatal behavior and photosynthesis of *Eucalyptus grandis. Australian Journal of Plant Physiology* 17(2), 159–175.

Leuning R, (1995). A critical-appraisal of a combined stomatal-photosynthesis model for C-3 plants. *Plant, Cell and Environment* 18(4), 339–355.

Lhomme JP, (1991). The concept of canopy resistance: Historical survey and comparison of different approaches. *Agricultural and Forest Meteorology* 54(2–4), 227–240.

McMurtrie RE, HN Comins, MUF Kirschbaum and YP Wang, (1992). Modifying existing forest growth-models to take account of effects of elevated CO_2. *Australian Journal of Botany* 40, 657–677.

Medlyn BE, RA Duursma, D Eamus, DS Ellsworth, IC Prentice, CVM Barton, KY Crous, P de Angelis, M Freeman and L Wingate, (2011). Reconciling the optimal and empirical approaches to modelling stomatal conductance. *Global Change Biology* 17(6), 2134–2144.

Monteith JL, (1995). A reinterpretation of stomatal responses to humidity. *Plant, Cell and Environment* 18(4), 357–364.

Mott KA and DF Parkhurst, (1991). Stomatal responses to humidity in air and helox. *Plant, Cell and Environment* 14(5), 509–515.

Sellers PJ, JA Berry, GJ Collatz, CB Field and FG Hall, (1992). Canopy reflectance, photosynthesis, and transpiration. A re-analysis using improved leaf models and a new canopy integration scheme. *Remote Sensing of Environment* 42(3), 187–216.

Sellers PJ, Y Mintz, YC Sud and A Dalcher, (1986). A Simple Biosphere Model (Sib) for use within general circulation models. *Journal of the Atmospheric Sciences* 43(6), 505–531.

Sellers PJ, DA Randall, GJ Collatz, JA Berry, CB Field, DA Dazlich, C Zhang, GD Collelo and L Bounoua, (1996). A revised land surface parameterization (SiB2) for atmospheric GCMs .1. Model formulation. *Journal of Climate* 9(4), 676–705.

Szeicz G and IF Long, (1969). Surface resistance of crop canopies. *Water Resources Research* 5(3), 622–633.

Tuzet A, A Perrier and R Leuning, (2003). A coupled model of stomatal conductance, photosynthesis and transpiration. *Plant, Cell and Environment* 26(7), 1097–1116.

Upadhyaya SK, RH Rand and JR Cooke, (1983). A mathematical-model of the effects of CO_2 on stomatal dynamics. *Journal of Theoretical Biology* 101(3), 415–440.

Wang J, (2006). Study on the experiment and simulation of crop growth and water,heat and CO_2 transfer in the agro-ecosystem. PhD Thesis, The Chinese Academy of Sciences.

Wang YP and R Leuning, (1998). A two-leaf model for canopy conductance, photosynthesis and partitioning of available energy I: Model description and comparison with a multi-layered model. *Agricultural and Forest Meteorology* 91(1–2), 89–111.

Whitehead D, DUU Okali and FE Fasehun, (1981). Stomatal response to environmental variables in 2 tropical forest species during the dry season in Nigeria. *Journal of Applied Ecology* 18(2), 571–587.

Whitley R, B Medlyn, M Zeppel, C Macinnis-Ng and D Eamus, (2009). Comparing the Penman Monteith equation and a modified Jarvis-Stewart model with an artificial neural network to estimate stand-scale transpiration and canopy conductance. *Journal of Hydrology* 373, 256–266.

Wong SC, IR Cowan and GD Farquhar, (1979). Stomatal conductance correlates with photosynthetic capacity. *Nature* 282(5737), 424–426.

Yu Q, J Goudriaan and TD Wang, (2001). Modelling diurnal courses of photosynthesis and transpiration of leaves on the basis of stomatal and non-stomatal responses, including photoinhibition. *Photosynthetica* 39(1), 43–51.

Yu Q and TD Wang, (1998). Simulation of the physiological responses of C-3 plant leaves to environmental factors by a model which combines stomatal conductance, photosynthesis and transpiration. *Acta Botanica Sinica* 40(8), 740–754.

Yu Q, YG Zhang, YF Liu and PL Shi, (2004). Simulation of the stomatal conductance of winter wheat in response to light, temperature and CO_2 changes. *Annals of Botany* 93(4), 435–441.

Zhang B, Y Liu, D Xu, J Cai and F Li, (2011). Evapotranspiraton estimation based on scaling up from leaf stomatal conductance to canopy conductance. *Agricultural and Forest Meteorology,* 151(8): 1086–1095.

12

Modelling Leaf and Canopy Transpiration and the Soil-Plant-Atmosphere Continuum

12.1 Introduction

The concept of the soil-plant-atmosphere continuum (SPAC) was first proposed by Phillip in 1966. In this conceptualisation, water flows from soil to the atmosphere via a plant's hydraulic pathways (Chapter 3). The flow is driven by the gradient of water potential from high to low, arising principally from differences in solute and turgor potentials along the SPAC (Chapter 3).

The biophysical and physiological processes occurring through the soil-plant-atmosphere continuum (SPAC) can be divided into three components: (1) radiation absorption and energy balance; (2) water transfer; and, (3) photosynthesis and physiological regulation of carbon and water fluxes. This chapter focuses on the second process, that is, water transfer, while Chapter 9 deals with the first and Chapter 10 deals with the third process.

There are two significant differences between canopy-scale and leaf-scale transpiration which need to be considered explicitly in models. These are: (a) radiation interception; and, (b) turbulent transfer. It is the shading and multiple layering within canopies that increases the complexity of radiation interception and turbulent transfer. A brief consideration of these is given prior to a more complete description of leaf and canopy transpiration.

12.2 Canopy Radiation Exchange

12.2.1 Radiation Interception

Solar radiation is the key input to leaf energy balances and this drives transpiration, in addition to it being the energy source for photosynthesis.

The net radiation balance of a leaf is partitioned into sensible heat by heat conductance and turbulent transfer and into latent heat by transpiration. Because net radiation is a radiation balance between changes in sensible and latent heat exchanges, net radiation represents a steady-state, at any given moment. Leaf temperature, as determined by a leaf's energy balance, influences the biochemical reaction rates of all photosynthetic processes. Leaf temperature determines the leaf's emission of long wave radiation and also determines the saturated water vapour pressure in the sub-stomatal cavity (Chapter 2). Consequently leaf temperature influences stomatal conductance and transpiration.

12.2.2 Turbulent Transfer of Energy and Mass Within and Above a Canopy

CO_2 and water vapour pass into and out of stomatal pores, respectively (Chapter 2). The rate of CO_2 flux and water vapour transfer are determined by stomatal conductance, boundary layer conductance and differences in CO_2 concentration and vapour pressure between the sub-stomatal cavity and ambient air. In these processes there are complex interactions between photosynthesis, stomatal conductance and intercellular CO_2 concentration. These are discussed later in this chapter.

12.2.3 Physiological Responses and Feedbacks

Leaf gas exchange, as discussed in the previous chapter, provides feedbacks on plant microclimate. All of these gas exchange processes interact with each other and either aid or constrain water, carbon and energy, exchanges through all components of the SPAC. The modelling strategies used to model water flow through the SPAC are of two main types:

(a) radiation transfer schemes–which can be categorised into big-leaf, two-leaf, multi-layer types; and,
(b) the use of water flows to define energy balances–the key challenges to parameterising models of water flux is the difficulty in measuring the series of resistances along the pathway from soil to atmosphere through the plant and the integration of non-linear responses over both coarse and fine spatial scales. Furthermore these resistances (or, more usually, their conductances) are not static and vary over time (Chapters 2 and 3), adding a further difficulty to modelling the SPAC.

12.3 Transpiration at the Leaf-Scale

Stomata are sensitive to both CO_2 concentration and water vapour concentrations; their aperture enlarges as CO_2 concentration or *VPD* decrease (Chapters 2 and 11). In early formulations of combined models of stomatal conductance (combining assimilation rate, transpiration rate, *VPD* and CO_2 concentration within a single model; Chapter 10), *VPD* is taken as an environmental factor. When boundary layer conductance is large enough to be non-limiting, a situation always satisfied in a well-stirred leaf chamber, the relation described in Ball's model of g_s (Chapter 11) holds. However, under natural field conditions, the influence of changes in boundary layer conductance (g_b), which depends on wind speed and leaf size, always needs to be taken into account. When g_b is low, the humidity difference between ambient air and leaf surface may not be negligible and a similar situation also exists with respect to CO_2 concentration. Taking this into consideration, Aphalo and Jarvis (1993a) derived an expression for the difference in vapour pressure between intercellular air spaces and the air layer just above the leaf surface (VPD_s), as a function of the vapour pressure difference between intercellular and ambient air, g_{sw} and $g_{bw,}$ as follows, assuming that leaf temperature equals air temperature ($T_l = T_a$):

$$VPD_s = VPD_a(1 - g_{tw} / g_{bw}) \tag{12.1}$$

where VPD_a is the water vapour pressure difference between intercellular and ambient air; g_{tw} and g_{bw} are total conductance and boundary layer conductance to water vapour (mol m^{-2} s^{-1}), respectively. In fact, equation 12.1 is derived from the equation for transpiration, in which VPD_s represents the driving force for transpiration. Thus transpiration can be calculated from the following (eq. 12.2):

$$E = g_{sw} \cdot VPD_s \tag{12.2}$$

where E is transpiration rate, g_{sw} is stomatal conductance to water vapour and VPD_s is as previously defined. The relation between g_{sw} with g_{sc} and total stomatal conductance to CO_2 (g_{tc}) can be expressed in terms of g_{sc} and g_{bw} as follows:

$$g_{sw} = 1.56 g_{sc} \tag{12.3}$$

$$g_{tc} = (1 / g_{sc} + 1.37 / g_{bw})^{-1} \tag{12.4}$$

$$g_{tw} = (1 / g_{sw} + 1 / g_{bw})^{-1} \tag{12.5}$$

in the steady state, by applying the general diffusion equation, the following relations hold:

$$C_s = C_a - P_n / g_{bc} \tag{12.6}$$

$$C_i = C_a - P_n / g_{tc} \tag{12.7}$$

where g_{bc} is boundary layer conductance to CO_2. The saturated vapour pressure at any leaf temperature is calculated by the Goff-Gorech equation.

When leaf and air temperatures are not identical, the following expression (eq. 12.8) is used to replace equation 12.1 (Aphalo and Jarvis 1993b):

$$VPD_s = \left[VPD + s(T_l - T_a) \right](1 - g_{tw} / g_{bw}) \tag{12.8}$$

where g_{tw} is total conductance ($g_{sw} + g_{bw}$) to water vapour (mol m^{-2} s^{-1}), T_l and T_a are temperatures of leaf and air, respectively and s is the slope of the saturated water vapour pressure/leaf temperature curve. The calculation of saturated vapour pressure and *s* can be found in the previously described energy balance sub-model of Chapter 9.

12.3.1 Leaf Energy Balance

In order to calculate canopy-scale fluxes of water (that is, transpiration), some models use canopy temperature while Penman-Monteith type models eliminate the need for canopy temperature. The leaf's energy balance (Chapter 9) can be assumed to be in steady state with a balance between energy into and out of the leaf; the energy used by photosynthesis is a negligible fraction of total energy input to the leaf. Similarly, leaf heat capacity is small enough compared with energy input for it to be ignored. This means leaf temperature is dependent on current environmental conditions but not previous conditions. The change in leaf temperature with time is driven by solar radiation, transpiration rates, air temperature, humidity and wind speed. Resolving steady state equations is much simpler than non-steady state models. Thus leaf energy balance equations can be considered as a balance between input and outputs of energy, with storage set to zero at a daily time-scale.

A leaf receives solar radiation from only one side but absorbs and emits longwave radiation from both sides. Therefore, the longwave radiation terms are doubled. For simplification, however, shortwave and longwave radiation exchange over the sunlit side of a leaf is described and longwave radiation exchange of the shaded side is neglected. This exchange is assumed negligible because emitted longwave radiation from the shaded side may simply offset its longwave absorption from surrounding leaves. Therefore, an energy balance equation from one leaf side can be established, which includes absorbed solar radiation, longwave radiation from the sky and longwave emission of the leaf (Chapter 9).

The one-side leaf energy balance expression can also be extended to include energy exchange of both sides. The equation is also applicable for a 'big-leaf' model of canopies (Chapter 10) to calculate canopy temperature as a whole, assuming that a canopy can be treated as a one-sided giant leaf interacting with the atmosphere for radiation and sensible and latent energy exchanges. See Chapter 13 for details about resolution of leaf temperature from the leaf energy balance equation.

12.4 Water Flow through the Soil-Plant-Atmosphere Continuum and Evapotranspiration Models

Plant water-use (i.e. transpiration, E) can be limited principally by two variables: soil water content and atmospheric evaporative demand. Classically these are termed water-limited and energy-limited conditions. Water flow through the soil-plant-atmosphere continuum (SPAC) is driven by gradients of water potential between soil, root, leaf and atmosphere and limited by serial resistances (Chapter 3). Canopy evapotranspiration (ET) models may include sub-models of solar radiation interception and absorption, thermal radiation exchange, single or multi-layered water and heat transfer and schemes of resistances. These are now discussed more extensively.

12.4.1 A Framework for Model Consideration

Water flow through the SPAC is constrained by the balance between loss to the atmosphere and supply by soil. These, in turn, are regulated by resistances (or conductances; Chapter 3) within roots, stems and the canopy. Water flow within the SPAC can be usefully divided into two, generally equal, parts. First is transpiration from a canopy, as determined by (a) leaf water status (leaf water potential and stomatal conductance) and (b) atmospheric evaporative demand. Second is soil water uptake by roots, as determined by gradients in water potential between soil and root xylem. Whole plant resistance influences the rate of flow between root surface and canopy (Chapter 3; Yu and Flerchinger 2008).

Water flow can be described using an electrical resistance model as an analogy. Transpiration can be calculated at any point within a plant from the following equation (eq. 12.9), assuming there is equilibrium flow and no change in plant water storage (that is, no capacitance in the "circuit"):

$$E_{c,j} = \sum_{k=1}^{NS} \frac{\psi_k - \psi_{x,j}}{r_{r,k,j}} = \sum_{i=1}^{NC} \frac{\psi_{x,j} - \psi_{l,i,j}}{r_{l,i,j}} = \sum_{i=1}^{NC} \frac{\rho_{vs,i,j} - \rho_{v,i}}{r_{s,i} + r_{h,i,j}} \tag{12.9}$$

where $E_{c,j}$ is transpiration rate for plant species j within the plant canopy; ψ_k, $\psi_{x,j}$ and $\psi_{l,i,j}$ are water potential in layer k of the soil, in the plant xylem and in the leaves of canopy layer i, respectively; $r_{r,k}$ and $r_{l,i,j}$ are the resistance to water flow through the roots of layer k and the leaves of layer i, respectively, for plant species j; $\psi_{vs,i,j}$ and $\psi_{v,i}$ are the vapour density within the stomatal pores and of the air within the canopy layer, respectively; $r_{s,i,j}$ is stomatal resistance per unit of leaf area index; and NS and NC are the number of soil and canopy layers, respectively (Flerchinger, 2000).

Root resistance for each plant species within each soil layer is allocated by dividing the total root resistance for the plant by its fraction of roots within each layer. Leaf resistance for each plant species within each canopy layer is determined by stomatal models and its leaf area index within each canopy layer (see Chapter 11). Transpiration from the leaves of each plant species within each canopy layer, $E_{c,i,j}$, is computed from the last term in equation 12.9.

Soil drying leads to a decrease in soil water potential and to a large increase in the hydraulic resistance at the soil-root interface. The water potential of leaves will decrease and xylem water potentials also decrease in response to soil drying. Therefore cavitation and emboli (Chapter 3) develop and the hydraulic conductivity of the distal parts of the tree will decrease (Cruiziat *et al.* 2002). This is a generic scheme for water flow through the SPAC and is also a core component of land surface modelling and ecohydrological modelling.

12.4.2 Evapotranspiration Models

ET can be calculated by static or dynamic models; the former is driven by meteorological and soil variables and do not include time as a variable. The latter may consider dynamic processes within the soil and vegetation. These models also integrate physiological regulation, that is, canopy conductance or resistance and change in hydrological architecture. Some statistical models of ET use empirical relationship between ET and temperature, solar radiation. These statistical models have two merits: ease of application and simple formulae. However, because they are empirical it is difficult to extend beyond their site-specific calibrations. Furthermore they do not have a solid physical basis and cannot be used for prediction and are not discussed further. The following provides a brief description of several static models (Wang, 2006).

The single layer or big-leaf model considers soil and vegetation as a whole and assumes canopy-atmosphere exchange occurs from one layer. Such models include canopy solar radiation absorption (Chapter 9) and its conversion into plant latent and sensible heat fluxes. These models are applicable for dense vegetation and soil heat flux is neglected because the soil surface receives little solar radiation. These models have fewer parameters and are easier to apply than multi-layer models. A few examples of big leaf models are now described.

The Penman Equation

Penman (1948) combined a convection equation and an energy balance equation for sensible and latent heat fluxes (λE) and derived the Penman equation. This equation can be used to calculate evaporation from a uniform wet surface such as a short grass canopy with an ample water supply:

$$\lambda E = \frac{\Delta(R_{\mathrm{n}} - G) + \rho C_{\mathrm{p}} VPD / r_{\mathrm{a}}}{\Delta + \gamma} \tag{12.10}$$

where Δ is the slope of the curve of saturated vapour pressure against temperature, ρ air density, C_{p} specific heat, VPD the water vapour pressure deficit, r_{a} is aerodynamic resistance and γ is the psychrometric coefficient.

The Penman equation is a simple equation and provides the basis for calculation of transpiration and evapotranspiration but it does not include a vegetation element and cannot be used to evaluate actual ET over vegetated landscape. For this to occur, Monteith had to add a consideration of plant resistances into the Penman equation, as now discussed.

The Penman-Monteith Equation

Monteith introduced the concept of canopy resistance (r_c) into the Penman equation and thereby extended the applicability of the Penman equation from free-water surfaces or ideal plant conditions to more diverse vegetated surfaces. The Penman-Monteith equation (eq. 12.11) has received global attention and is the basis of very many models of ET (Monteith 1965, Monteith 1973):

$$\lambda E = \frac{\Delta(R_n - G) + \rho c_p D / r_a}{\Delta + \gamma\left(1 + \dfrac{r_c}{r_a}\right)}$$

(12.11)

(Terms as defined for equation 12.10.) Canopy resistance is the sole element in the Penman-Monteith equation that reflects the influence of vegetation. Consequently this variable has to include the impact of a large number of factors that are known to influence transpiration, such as leaf area index, soil water content and stomatal conductance (which itself is influenced by photosynthetic rate). This classic equation is widely applied in calculation of vegetation evapotranspiration (Doorenbos and Pruitt 1977, Brutsaert and Stricker 1979, Brutsaert 1982, Katul and Parlange 1992, Parlange and Katul 1992, Konzelmann *et al.* 1997, Domingo *et al.* 1999).

The Priestley-Taylor equation

Priestley and Taylor (1972) introduced the concept of equilibrium evaporation and simplified the Penman equation thus:

$$\lambda E = \frac{\alpha\Delta(R_n - G)}{\Delta + \gamma}$$

(12.12)

where α is a coefficient and normally set to 1.26. Debruin and Holtslag (1982) developed equations to calculate water flux for short crops under ample water supply based on the Priestley-Taylor equation:

$$\lambda E = \alpha' \frac{\Delta}{\Delta + \gamma}(R_n - G) + \beta$$

(12.13)

$$H = \frac{(1 - \alpha')\Delta}{\Delta + \gamma}(R_n - G) - \beta$$

(12.14)

where α' and β are empirical constants. This model is applicable for humid and semi-humid regions, but may have limitations in arid and semi-arid regions.

The Crop Coefficient

The rate of evapotranspiration (ET) is determined by evaporative demand and soil water supply. The crop coefficient (K_c) is the ratio of actual to potential (ET_p)

Table 12.1. Indicative mid-season crop coefficients, (K_c) for well-watered crops in a sub-humid climate ($RH_{min}>$ 45%, wind speed 2 m s^{-1})

Species	Mid-season K_c
Rice	1.2
Coniferous forest	1.0
Sweet corn (*Zea mays*)	1.15
Cotton	1.15–1.2
Strawberries	0.85
Sisal	0.4–0.7

Source: From Allen *et al.* (1998).

evapotranspiration, the former is limited by both meteorological variables and soil water content; the latter is a function only of atmospheric conditions. Crop coefficients are determined by crop coverage and soil and climate factors during a growing season. Thus:

$$ET = K_c ET_p \tag{12.15}$$

Potential evapotranspiration can be calculated by the Penman-Monteith equation or the Priestley-Taylor equation. Doorenbos and Pruitt (1977) provide values of crop coefficients for diverse climate conditions and different growth stages of crops which allow easy calculation of crop water-use for regions without data. Table 12.1 provides some indicative values of the crop coefficient for a range of crops/species. Crop coefficients vary as a function of LAI and hence differ across the early-to-mid-to later growing season.

12.4.3 Two-Source Models

Single layer models or big-leaf models are simple in expression and easy in application due to their reliance on few parameters. They are applicable for vegetation with a large LAI. In sparse vegetation, for example in arid and semi-arid regions, their application is limited because soil plays a significant role in evapotranspiration. Consequently it is necessary to separate these two sources (plant and soil) of ET. Single layer models can be extended to two-source or two-layer models. Thus Menenti (1984) extended the Penman equation to two layers and applied remote sensing methods to calculate evapotranspiration in deserts.

The energy balance equations of canopy and soil are expressed by:

$$R_{nc} = H_c + \lambda E_c \tag{12.16}$$

$$R_{ns} = H_s + \lambda E_s + G \tag{12.17}$$

where H_c and H_s are the sensible heat fluxes of canopy and soil, respectively and E_c and E_s are the latent heat fluxes of canopy and soil, respectively. The typical two-source model was first proposed by Shuttleworth and Wallace (1985), which considered both soil and canopy layers and calculated the energy balance for each layer. The Shuttleworth and Wallace (S-W) equation can be expressed as:

$$\lambda E = C_c PM_c + C_s PM_s \tag{12.18}$$

where PM_c is plant transpiration with total coverage of plant (no soil evaporation), PM_s is bare soil evaporation (no transpiration) and C_c and C_s are resistance coefficients. PM_c and PM_s are terms similar to those in the Penman-Monteith equation and are calculated from the following equations:

$$PM_c = \frac{\Delta(R_n - G) + [\rho c_p D - \Delta r_a^c (R_{ns} - G)] / (r_a^a + r_a^c)}{\Delta + \gamma\left(1 + \dfrac{r_s^c}{r_a^a + r_a^c}\right)} \tag{12.19}$$

and

$$PM_s = \frac{\Delta(R_n - G) + [\rho c_p D - \Delta r_a^s (R_n - R_{ns})] / (r_a^a + r_a^s)}{\Delta + \gamma\left(1 + \dfrac{r_s^s}{r_a^a + r_a^s}\right)} \tag{12.20}$$

To solve the S-W model C_c and C_s are calculated from the following:

$$C_c = \frac{1}{1 + R_c R_a / \left[R_s (R_c + R_a)\right]} \tag{12.21}$$

and

$$C_s = \frac{1}{1 + R_s R_a / \left[R_c (R_s + R_a)\right]} \tag{12.22}$$

To solve equations 12.21 and 12.22 R_a, R_c and R_s are calculated from the following three equations:

$$R_a = (\Delta + \gamma) r_a^a \tag{12.23}$$

$$R_c = (\Delta + \gamma) r_a^c + \gamma r_s^c \tag{12.24}$$

$$R_s = (\Delta + \gamma) r_a^s + \gamma r_s^s \tag{12.25}$$

Vapour pressure deficit at the canopy source height (D_0) is determined by the latent heat flux from the following:

$$D_0 = D_a + [\Delta(R_n - G) - (\Delta + \gamma)\lambda E] r_a^a / \rho c_p \tag{12.26}$$

Soil evaporation and crop transpiration are therefore given by:

$$\lambda E_{\mathrm{c}} = \frac{\Delta(R_{\mathrm{n}} - R_{\mathrm{ns}}) + \rho C_{\mathrm{p}} D_0 / r_{\mathrm{a}}^{\mathrm{c}}}{\Delta + \gamma\left(1 + \dfrac{r_{\mathrm{s}}^{\mathrm{c}}}{r_{\mathrm{a}}^{\ \mathrm{c}}}\right)} \tag{12.27}$$

$$\lambda E_{\mathrm{s}} = \frac{\Delta(R_{\mathrm{ns}} - G) + \rho c_{\mathrm{p}} D_0 / r_{\mathrm{a}}^{\mathrm{s}}}{\Delta + \gamma\left(1 + \dfrac{r_{\mathrm{s}}^{\mathrm{s}}}{r_{\mathrm{a}}^{\ \mathrm{s}}}\right)} \tag{12.28}$$

and sensible heat fluxes from canopy and soil are estimated by the following expressions:

$$H_{\mathrm{c}} = \rho c_{\mathrm{p}} \frac{T_{\mathrm{c}} - T_{\mathrm{a}}}{r_{\mathrm{a}}^{\ \mathrm{a}} + r_{\mathrm{a}}^{\mathrm{c}}} \tag{12.29}$$

$$H_{\mathrm{s}} = \rho C_{\mathrm{p}} \frac{T_{\mathrm{s}} - T_{\mathrm{a}}}{r_{\mathrm{a}}^{\ \mathrm{a}} + r_{\mathrm{a}}^{\mathrm{s}}} \tag{12.30}$$

where λ is the latent heat of vaporization (J kg^{-1}); D_{a} the vapour pressure deficit at reference height; T_{a}, T_{c} and T_{s} is the air temperature at reference height, canopy temperature and soil surface temperature, respectively; $r_{\mathrm{a}}^{\ \mathrm{a}}$ is the aerodynamic resistance between the canopy source height and reference height; $r_{\mathrm{a}}^{\mathrm{c}}$ is the boundary layer resistance of canopy; $r_{\mathrm{a}}^{\mathrm{s}}$ is the aerodynamic resistance between the substrate and canopy source height; $r_{\mathrm{s}}^{\mathrm{c}}$ is the canopy resistance; and, $r_{\mathrm{s}}^{\mathrm{s}}$ is soil resistance.

The S-W equation has been widely applied in farmland, grassland and savannas (Stannard 1993, Farahani 1994). This two-source model separates water and energy exchanges of soil and plant and this is more realistic than big-leaf models of a landscape where soil fluxes are ignored. However, the largest challenge to two-source models is the determination of the serial resistances.

Choudhury and Monteith (1988) proposed a model of four-level water and heat transfer, which is based on Menenti (1984) and Shuttleworth and Wallace (1985) models for homogenous land surfaces. Vegetated land surfaces are divided into two layers: the plant canopy and the soil and their energy balances are calculated using the S-W equation. However, in the Choudhury-Monteith model the soil is divided into dry and wet layers.

12.4.4 Multi-Source Models

When the land surface is covered by diverse plants, such as a mixture of shrubs, grass and trees (for example, savannas include C3 and C4 grasses, upper storey trees and mid-storey shrubs), single layer and two source models cannot represent the contribution of these different layers to total heat and water fluxes. It is inappropriate to use one value of the bulk canopy conductance parameter to represent vegetation containing multiple plant types with different attributes. Dolman (1993) developed a multi-source land surface energy balance model, which was employed in tropical rainforest, savannas and crops. Brenner and Incoll (1997) further developed the Dolman model by dividing the land surface into vegetated and non-vegetated parts.

Multi-layer models divide the canopy and soil into several compartments horizontally and calculate vertical heat and water transfer. In multi-layer models sensible and latent heat fluxes and photosynthetic C fluxes are calculated using micro-meteorological variables for each layer, i.e. the profiles of temperature, humidity, radiation and wind speed, within a canopy. Some models also include CO_2 concentration distribution. The total fluxes are summed and the dynamics of variation in plant processes, such as photosynthesis, transpiration and energy balance and their feedbacks, are often included. Generally, these models only have numerical solutions, as they include plant feedbacks and need numerical iterative calculation to yield results (Flerchinger 2000). Multi-layer models incorporate more processes or sub-models than big-leaf or multi-source models and therefore include more parameters, many of which are difficult to obtain. Multi-layer models may have large uncertainty of model prediction which often limits their applicability.

12.5 Microclimate Within and Over Canopies

Turbulence is an important feature of atmospheric dynamics. Turbulence allows mixing and influences the fluxes of sensible and latent heat because turbulence carries heat and water vapour away (on average) from the land (vegetation) surface. It is the correlations among wind velocity, CO_2 concentration, water vapour concentration and wind direction that allow the eddy covariance technique to quantify the exchange of water and CO_2 between vegetation and the atmosphere (Chapters 2, 17 and 18). The importance of turbulence lies in the fact that in the absence of turbulence, photosynthetic fixation of C by a forest would deplete all of the CO_2 contained in a 30 m layer of air adjacent to the canopy within a single day (Chapter 2).

There are three methods of parameterisation of turbulence transfer over the atmospheric boundary layer: (i) Gradient-dispersion theory (K-theory; where transport occurs by diffusion down a concentration gradient); (ii) the higher order closure method; and, (iii) the Lagrangian equation. K-theory is the most popular and simplest method. It is difficult to determine turbulence transfer and this is the limiting aspect in the application of multi-layer models. Field observations have reported anti-gradient transfer of gases and this is counter to K-theory (Denmead and Bradley 1987, Thurtell 1989), because scales of eddies within many plant canopies are similar to canopy height and K-theory is then not applicable. Some models adopt higher-order closure methods (Meyers *et al.* 1988, Wilson 1989), or Lagrangian theory (Dolman and Wallace 1991, Hsieh *et al.* 1997) to simulate eddy transfer. However, this requires detailed information about canopy structure and this information is often absent. Wu *et al.* (2001) compared K-theory and Lagrangian methods within single and two-leaf models and only found small differences in results (Wang 2006).

Heat and vapour fluxes within a canopy are determined by computing rates of transfer between layers of the canopy and considering the source terms for heat ($H_{c,i,j}$) and transpiration ($E_{c,i,j}$) from canopy leaves of each plant species (j) within each canopy layer (i). Gradient-driven transport, or K-theory, is generally used for transfer within the canopy although Lagrangian approaches for modelling turbulent transport processes within the canopy have been successfully applied. Nevertheless, many have found K-theory particularly useful for simulating transport processes within the canopy (Goudriaan 1989, Van de griend and Van boxel 1989, Nichols 1992, Huntingford *et al.* 1995) while other studies indicate relatively small differences in flux predictions between Lagrangian and K-theory (Dolman and Wallace 1991, Wilson *et al.* 2003). Heat flux and temperature within the air space of the canopy is described by equation 12.31:

$$\frac{\partial}{\partial z}\left(\rho_a c_a k_e \frac{\partial T}{\partial z}\right) + H_c = \rho_a c_a \frac{\partial T}{\partial t} \tag{12.31}$$

where z is height above the soil surface (m), k_e is a transfer coefficient within the canopy ($m^2\ s^{-1}$), T is temperature within the canopy, t is time (s) and H_c is heat transferred from leaves to the air space within the canopy ($W\ m^{-3}$). Vapour flux through the canopy is described similarly to heat flux (eq. 12.32):

$$\frac{\partial}{\partial z}\left(k_e \frac{\partial \rho_v}{\partial z}\right) + E_c = \frac{\partial \rho_v}{\partial t} \tag{12.32}$$

where E_c is transpiration from the leaves within the canopy ($kg\ s^{-1}\ m^{-2}$) and ρ_v is vapour density within the canopy. Transfer of heat and vapour within the canopy is dependent upon location within the canopy and several approaches for computing the transfer coefficient, k_e, have been developed. Flerchinger and Pierson (1991) used the following expression (eq. 12.33) above the zero plane displacement (d):

$$k_e = \kappa u_* (z - d + z_h) / \varphi_h \tag{12.33}$$

and for heights less than d:

$$k_e = \kappa u_* z_h / \varphi_h \tag{12.34}$$

where κ is von Karman's constant, u_* is the friction velocity ($m\ s^{-1}$), z is height above the residue or soil surface (m), z_h is the thermal surface roughness parameter (m) and φ_h is a diabatic correction factor dependent on the Richardson's number computed from sensible heat flux above the canopy.

Heat transfer from the vegetation elements (leaves) to the air space within a canopy layer for plant species j (W m^{-2}) is computed from:

$$H_{c,i,j} = -\rho_a c_a L_{i,j} \frac{(T_{l,i,j} - T_{a,i})}{r_{h,i}} \qquad (12.35)$$

where, $L_{i,j}$ and $T_{l,i,j}$ are leaf area index and leaf temperature of plant species j respectively and $T_{a,i}$ is air temperature within canopy layer i. Resistance to convective transfer from the canopy leaves per unit leaf area index, $r_{h,i}$, is computed from (Campbell 1977; eq. 12.36):

$$r_{h,i} = 307 \left(\frac{d_l}{u_i} \right)^{1/2} \qquad (12.36)$$

where d_l is leaf dimension (m), u_i is wind speed within the canopy layer (m s^{-1}) and 307 is a coefficient (s$^{1/2}$ m^{-1}) for the thermal diffusivity and viscosity of air. Leaf temperature of plant j for each layer within the canopy ($T_{l,i,j}$) is determined from a leaf energy balance of the canopy layer assuming the leaves within the canopy have negligible heat capacity:

$$R_{n,i,j} + H_{c,i,j} + \lambda E_{c,i,j} = 0 \qquad (12.37)$$

where $R_{n,i,j}$ is net all-wave radiation for the leaf surfaces within canopy layer i (W m^{-2}) and $E_{c,i,j}$ is transpiration within each canopy layer.

Water uptake, transpiration and leaf temperature are coupled through the energy balance of the leaf, which is calculated for each plant species within each canopy layer. The leaf energy balance is computed iteratively with heat and water vapour transfer equations (Eqs. 12.31 and 12.32) and transpiration within the canopy.

12.5.1 Aerodynamic Resistance

The Penman-Monteith equation contains an aerodynamic factor and a canopy resistance factor. The canopy resistance describes the resistance to vapour flow through stomata, cuticle and the boundary layer around leaves and lying close to the soil surface. Aerodynamic resistances are the resistances experienced as air flows over any surfaces, such as leaves and branches in canopies or the ground surface. The canopy and aerodynamic resistances are in series.

Aerodynamic resistance (r_a) under near neutral conditions is a function of wind speed and can be calculated (Thom *et al.* 1975) from equation 12.38:

$$r_a = \ln^2[(z_r - d) / z_0] / (\kappa^2 u) \qquad (12.38)$$

where z_r is the reference height and u the wind speed. Aerodynamic resistances decline with increasing wind speed (Fig. 12.1).

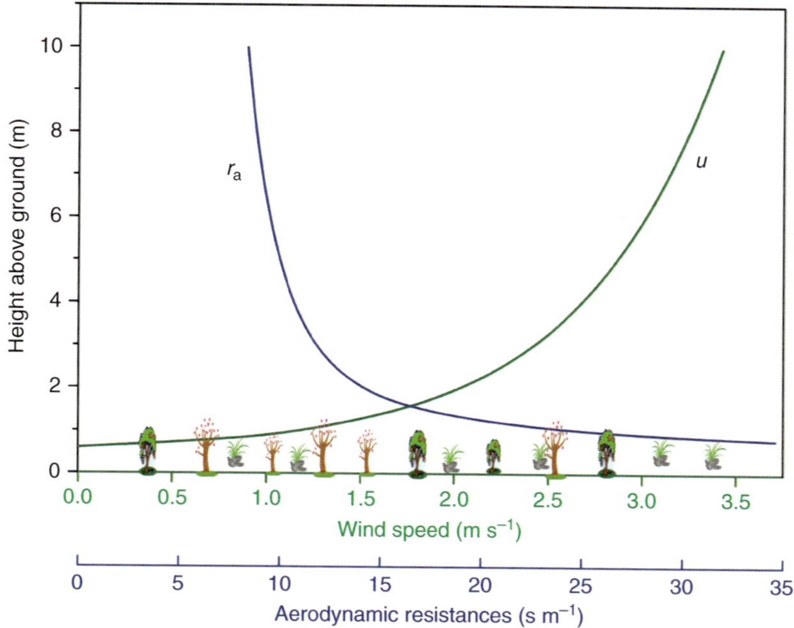

Figure 12.1. Aerodynamic resistance (r_a) and wind speed (u) over ground under neutral atmospheric conditions.

The vegetation and soil are considered separately, the characteristics of which are expressed by the values of zero plane displacement (d) and roughness length (z_0), given by the expressions (Benmehrez *et al.* 1992; equations 12.39 and 12.40):

$$d = 0.63\sigma_\alpha h \tag{12.39}$$

$$z_0 = (1 - \sigma_\alpha)z_b + \frac{\sigma_\alpha(h - d)}{3} \tag{12.40}$$

where h is the crop height, z_b the roughness length of bare soil, usually given the value 0.01 m (Vanbavel and Hillel 1976) and σ_α is the momentum partition coefficient which depends on leaf area index (L) expressed as equation 12.41 (Shaw and Pereira 1982):

$$\sigma_\alpha = 1 - \left(\frac{0.5}{0.5 + L}\right)\exp\left(-\frac{L^2}{8}\right) \tag{12.41}$$

r_a^c, r_a^a and r_a^s can be expressed as fractions of the overall aerodynamic resistance for momentum transfer in the bulk soil-vegetation aerodynamic resistance term (r_a) such that (Eqs. 12.42, 12.43 and 12.44; Anadranistakis *et al.* 1999):

$$r_a^c = \frac{u_h}{\sigma_\alpha u_*^2} = \frac{u_E}{\sigma_\alpha u_r}r_a \tag{12.42}$$

$$r_a^{\ a} = \frac{u_r - u_h}{u_*^2} = \frac{u_r - u_h}{u_r} r_a \tag{12.43}$$

$$r_a^{\ s} = \frac{u_h}{(1 - \sigma_\alpha) u_*^2} = \frac{u_h}{(1 - \sigma_\alpha) u_r} r_a \tag{12.44}$$

where u_r is the wind speed at the reference height and r_a is a function of atmospheric stability and can be expressed by (eq. 12.45):

$$r_a = \frac{1}{\kappa^2 u(z)} \left[\ln \frac{z-d}{z_0} - \psi_M \right] \Big/ \left[\ln \frac{z-d}{z'_0} - \psi_H \right] \tag{12.45}$$

where $z'_0 = Z_0/7$ (Garratt, 1978).

If the atmosphere is under neutral stability conditions, $\psi_M = \psi_H = 0$ and wind speed at mean canopy height is described by equation 12.46:

$$u_h = 0.83 u(z) \sigma_\alpha + (1 - \sigma_\alpha) u_r \tag{12.46}$$

12.6 Soil Water and Heat Dynamics

Soils can be divided into n layers with variable depth. Movement of water in soils can be described by the Richards equation (eq. 12.47):

$$c(h_s, i) \frac{\partial h_s}{\partial t} = \frac{\partial}{\partial i} \left(K(h_s, i) \frac{\partial h_s}{\partial i} \right) - \frac{\partial K(h_s, i)}{\partial i} - S(i, t) \tag{12.47}$$

where $c(h_s, i)$ is the specific water capacity of soil in the i-th layer, h_s is soil-water potential, $K(h_s, i)$ is the hydraulic conductivity of soil at h_s in the i-th layer and $S(i,t)$ is the root uptake rate (Wang *et al.* 2006). The Richards equation is a corner-stone of any model describing soil water fluxes.

The discrete form describing vertical water movement in soil is described by:

$$D_1 \frac{dw_c}{dt} = P + I - E_s - Q_{1,2} - S_1 \tag{12.48}$$

$$D_i \frac{dw_{c,i}}{dt} = Q_{i-1,i} - Q_{i,i+1} - S_i \tag{12.49}$$

$$D_n \frac{dw_{c,n}}{dt} = Q_{n-1,n} - Q_n - S_n \tag{12.50}$$

where i is the layer number ($i = 2, \ldots, n-1$), D is layer thickness, w_c the soil volumetric water content, Q the water flux through the i-th layer and $(i+1)$-th layer interface between two adjacent compartments, with downward flux taken as positive and S is the absorption rate of roots. P and I are intensities of precipitation and irrigation, respectively. E_s is the soil evaporation rate and t is time.

12.6.1 Root Water Uptake

The water absorption by roots can be simulated by taking account of a weighting factor of soil water potential (Feddes and Zaradny, 1978)). The water absorption rate (S) can be expressed by the function of the transpiration rate, root length and soil water potential:

$$S_i = \frac{\alpha(\psi_i)}{\int_0^{Lr} \alpha(\psi_i)\,dz} E_c \tag{12.51}$$

where ψ_i is the soil water potential at layer i, L_r the root length, E_c the transpiration rate and $\alpha(\psi_i)$ is the weighting factor of soil water potential and is defined as (e.g., Wang *et al.* 2006):

$$\alpha(\psi) = \begin{cases} \dfrac{\psi}{\psi_1} & \psi_1 \le \psi \le 0 \\[2mm] 1 & \psi_2 \le \psi < \psi_1 \\[2mm] \dfrac{\psi - \psi_3}{\psi_2 - \psi_3} & \psi_3 \le \psi < \psi_2 \\[2mm] 0 & \psi < \psi_3 \end{cases} \tag{12.52}$$

where ψ_1, ψ_2 and ψ_3 are thresholds of soil water potential for the three ranges. This equation is suitable for root absorption $\alpha(\psi) = 1$ when its potential ranges from ψ_1 to ψ_2 and $\alpha(\psi)$ is lower than 1 when water potentials are lower than ψ_2 or higher than ψ_3 due to drying or excess water and low aeration. Soil water is not available for roots when its potential is lower than ψ_3.

The rate of root water uptake at each layer can be represented as:

$$S(i,t) = \frac{\psi_{s,i} - \psi_1}{r_{s,i} + r_{r,i} + r_{x,i}} \tag{12.53}$$

where $r_{s,i}$, $r_{r,i}$ and $r_{x,i}$ are the hydraulic resistance of the soil, the root radial resistance and root axial resistance, respectively, at the i-th layer.

The hydraulic resistance of the soil is calculated according to single-root radial flow theory (Wang *et al.* 2002):

$$r_{s,i} = \ln(b_i / a_r) / (2\pi K_i L_{d,i} D_i) \tag{12.54}$$

where a_r is fine root radius, $L_{d,i}$ is root length density and b_i is the path length for water uptake. The root radial resistance is expressed as:

$$r_{r,i} = r_r' / (L_{d,i} D_i),$$ (12.55)

where r_r' is root radial resistivity. The root xylem resistance is estimated from:

$$r_{x,i} = r_x' z_{d,i} / (0.5 f L_{d,i})$$ (12.56)

where r_x' is root axial resistivity, $z_{d,i}$ is the depth of the midpoint of the soil layer i and f represents the fraction of the number of roots that connect to the shoot compared with the total number of roots.

12.6.2 Soil Heat Diffusion and Soil Water Potential

Soil heat fluxes are routinely measured when the eddy covariance technique is used to measure water and CO_2 fluxes above vegetation. The absence of soil heat flux measurements introduces errors when attempting to close the energy balance of a site. The heat diffusion equation for heterogeneous soil can be expressed by equation 12.57:

$$c(z)\frac{\partial z}{\partial t} = \frac{\partial}{\partial z}\left(K(z)\frac{\partial T}{\partial z}\right)$$ (12.57)

where $c(z)$ is the specific heat of soil at depth z and $K(z)$ is the soil heat conductivity at depth z. The heat diffusion equation can be solved by implicit difference scheme of integration. Soil heat flux at depth z is described by eq. 12.58:

$$G = -K(z)\frac{\partial T}{\partial z} + \int_0^z \frac{\partial T}{\partial t} c(z)\mathrm{d}z$$ (12.58)

Under many conditions the effect of $\int_0^z \frac{\partial T}{\partial t} c(z)\mathrm{d}z$ for soil heat flux can be neglected (Leyton, 1975) and then soil heat flux through the ground surface can be calculated by the difference method:

$$G = -K(z)[(T_g - T_s) / \Delta z]$$ (12.59)

where T_g is the soil surface temperature and T_s the average temperature from surface ground to depth z.

Changes in soil water potential have large impacts on rates of root water uptake (Chapter 3). Therefore modelling soil water potential is central to modelling within the SPAC. Soil water potential can be calculated from equation 12.60:

$$\psi_s = \psi_0 \left(\frac{w_c}{w_{c,sat}}\right)^{-b}$$ (12.60)

where ψ_s, ψ_0 is soil water potential and suction at saturation (MPa), respectively; w_c and $w_{c,sat}$ is soil water content and saturation water (v/v), respectively; and, b is Campbell's pore-size distribution index.

12.6.3 Soil Water Conductivity

Determination of the resistances to flux (of heat or gases) is required to solve the energy balance equation. Aerodynamic resistance, canopy boundary layer resistances, canopy resistance and soil resistance all need to be calculated. Soil resistance (the inverse of conductance) is calculated with an empirical function dependent on surface soil water content and is described in the model by Lin and Sun (1983) in equation 12.61:

$$r_s^s = b_1 \left(\frac{w_{c,sat}}{w_c} \right)^{b_2} + b_3 \tag{12.61}$$

where b_1, b_2 and b_3, are empirical constants.

12.7 Dynamic Water and Heat Exchanges Across the SPAC

The CO_2 and water vapour fluxes of vegetation canopies are two important components of water-use efficiency when soil respiration is relatively low. Therefore their measurement and simulation are valuable for revealing mechanisms of mass and energy transfer and controls of crop productivity (Powell and Thorpe 1977; Chen and Coughenour 1994; Sellers *et al.* 1996; Wang and Leuning 1998).

Many models assume static plant conditions for which model parameters are also invariant. For example, plant-water status (leaf and xylem water potentials, stem storage capacitance), the maximum photosynthetic rate (P_{max}) and apparent quantum yield (α), are often held as constants through the day, despite the fact that they change over time (Yu *et al.* 2001).

Non-steady-state water transfer in the SPAC has been reported since the 1970s (Lhomme 2001). Stomatal closure can occur frequently around midday and closure also begins as leaf-water potential (or stem water potential; Chapter 3) declines beyond a set-point (Chapter 2) as soils dry (Jensen *et al.* 2000). Part of this decline in stomatal conductance is the result of abscisic acid transport from roots to shoots (Davies and Zhang 1991, Ali *et al.* 1998, Liu *et al.* 2005) and models have been developed to relate the rate of synthesis of ABA in response to soil drying (Lhomme 2001). Furthermore, a decline in soil and root hydraulic conductance due to soil drying are an important control mechanism of stomatal closure (Yao *et al.* 2001) and there are several models which simulate water transport in the soil-root-canopy pathway (Grant 2001, Tuzet *et al.* 2003). The issue of midday stomatal closure is discussed in Chapter 3.

Recently a model of leaf-water potential changes was incorporated into a SPAC model of water and heat transfer and CO_2 assimilation. This model can replicate the known influences of water potential on CO_2 and water-vapour fluxes over wheat fields. This leaf water potential model can be integrated with leaf photosynthesis and stomatal models and up-scaled to canopy scales to calculate CO_2 and H_2O fluxes (Yu and Flerchinger 2008).

Leaf water status, such as water content, and potential, are determined by the balance between water loss through transpiration and supply from soil through root absorption. Changes in plant water content can be computed from water absorption and water loss based on mass conservation (eq. 12.62):

$$c\frac{d(\psi_s - \psi_l)}{dt} = T_r - \frac{\psi_s - \psi_l}{R_{sl}} \tag{12.62}$$

where c is specific water capacitance, defined as the incremental changes in leaf water content (W) with ψ_l ($dW/d\psi_l$), ψ_s is the average soil water potential in the root zone, ψ_l is the leaf water potential and, R_{sl} is the resistance of water transfer from soil to leaf. The water potential of plants with lower water capacitance is more sensitive to changes in water content. R_{sl} changes with soil types and ψ_l. The difference between the water potential of soil and plant increases when soil becomes dry and transpiration decreases (Chapter 3). Therefore, the total resistance between roots and soil increases with decreased soil water potential (Cowan 1965). Consequently ψ_l is computed from equation 12.63:

$$-c\frac{d\psi_l}{dt} = T_r - \frac{\psi_s - \psi_l}{R_{sl}} \tag{12.63}$$

This equation was used to simulate h_l based on transpiration rate. The ψ_l value determined the stomatal conductance and transpiration. The model, as developed, was applied for single plants and expressed per unit of ground area (Yu *et al.* 2007).

Changes in leaf water potential with time can be calculated thus (Yu *et al.* 2007; eq. 12.64):

$$\frac{d\psi_{l,j}}{dt} = \frac{\psi_s - \psi_{l0,j}}{r_{sl0} \cdot e^{k(\psi_s - \psi_{l0,j})} \cdot c} - \frac{T_{r,j}}{c} \tag{12.64}$$

$\Psi_{l,j}$ and $\Psi_{l0,j}$ are leaf potential in current and previous time steps, respectively; $T_{r,j}$ is leaf transpiration rate ($m^3\ s^{-1}$) (j = 1, sun leaf and j = 2, shade leaf); c is water capacitance (M^3/MPa); and, r_{sl0} and k are parameters.

A key requirement is to be able to predict the limitation imposed on stomatal conductance by declining leaf water potentials. This can be modelled by equations 12.65 and 12.66:

$$g_s = m\frac{P_n}{(C_s - \Gamma)(1 + VPD_s / VPD_0)}f(\psi_l) + g_0 \tag{12.65}$$

$$f(\psi_1) = \frac{1 + e^{s\psi_f}}{1 + e^{s(\psi_f - \psi_1)}} \tag{12.66}$$

where ψ_f is a reference potential and s is a sensitivity parameter (Tuzet *et al.* 2003). This model includes leaf and canopy photosynthesis sub-models, a soil-water dynamics sub-model and an evapotranspiration sub-model. The photosynthesis model is a version of the simplified biochemical model of Farquhar *et al.* (1980), which combines the photosynthesis-stomatal conductance sub-model (Collatz *et al.* 1991) with a version of the Ball-Berry model revised by Leuning (1995; Chapter 11) and a newly proposed dynamic model of leaf-water potential. Water-vapour flux is simulated by the Shuttleworth-Wallace model in which parameterisation of canopy resistance was up-scaled from leaf to canopy.

As a dynamic process, transpiration is calculated from predawn when the water potential gradient from soil to plant is smallest, through to sundown. For predawn periods, leaf water potential is assumed equal to soil water potential and transpiration rate is calculated. Then the leaf water potential is calculated, intercellular CO_2 is calculated and the photosynthetic rate and the energy balance and leaf temperature are derived. As a gradient in soil-to-leaf water potential develops, this is used to recalculate transpiration rates.

12.8 Model Solution

Models of water and C exchange within the SPAC consist of an evapotranspiration module, a canopy photosynthesis module and a soil water and heat transfer module. Meteorological variables serve as driving factors of these modules. These three modules are coupled by using rates of canopy photosynthesis from which to calculate stomatal/canopy conductance which are used directly in the evapotranspiration module. Then soil water content and fluxes are calculated using the evapotranspiration module, as well as for calculation of soil and canopy temperatures through solution of energy balance equations for canopy and ground surface. Canopy temperatures are then used to calculate canopy photosynthesis and soil respiration. Simultaneously, the canopy evapotranspiration module is coupled with the soil water and heat transfer modules for the solution of energy balance equations. Site specific values of parameters for soil physics and plant physiological traits are provided in "look-up" tables and used as inputs before numerical solution.

Similar to the numerical methods of leaf-scale models described in Chapter 10, canopy photosynthesis influences transpiration and energy balance through its correlation with stomatal and canopy conductances. Photosynthesis models are incorporated within models to solve leaf water potential, stomatal conductance and leaf temperature, using leaf energy balance equations.

For soil heat and water dynamic equations, the solution for each time step involves alternating back and forth between a Newton-Raphson iteration for the heat flux

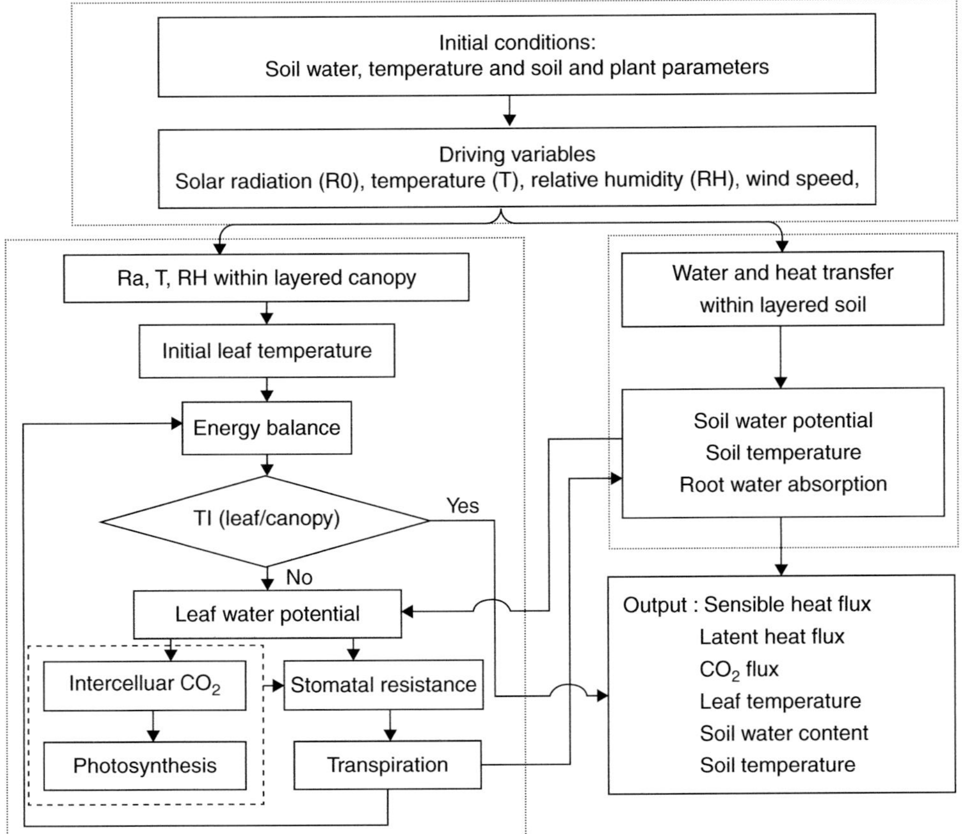

Figure 12.2. Feedbacks and calculation scheme in the integrated SHAW and photo-synthesis models (SHAW-Pn; redrawn from Yu and Flerchinger 2008).

equations and one for the water flux equations. An iteration is conducted for the heat flux equations and temperature estimates (water content in the case of melting snow) and at the end of each time step these new values are up-dated. This is followed by another iteration for the water flux equations, where updated vapour density within the canopy and matric potential in soil layers are determined. Upon completion of this iteration for the water flux equations the solution reverts back to an iteration for the heat flux equations, with the updated values. This cycle of iterations is repeated until all subsequent iterations of both heat and water flux equations for each layer are within a prescribed tolerance. Thus the heat and water flux equations are solved simultaneously, maintaining a correct balance between the two coupled equations (Flerchinger 2000). This process is summarised in Figure 12.2.

The rates of solute transport are computed using liquid fluxes from the water balance calculations after convergence for the heat and water flux equations has been achieved. If more-than-one iteration is required for energy and water balance convergence, it is likely that there is sufficient moisture movement to affect solute concentrations and the newly-calculated solute concentrations will be significantly different from those used in the energy and water balance calculations. In this case, the program will use the new solute concentrations to calculate the energy and water balance iteratively until convergence is met (Flerchinger, 2000).

Soil surface temperature and canopy temperature are calculated by iteration in models. The initial values of canopy and soil surface temperatures are given as inputs for the start of the model run and then sensible heat fluxes for canopy and soil are calculated. The values thus obtained are used in the energy balance equation and the iteration is continued until the difference between the sum of sensible and latent heat fluxes of canopy and net radiation of the canopy is <0.1 W m^{-2}, for example. Similarly, iterations are repeated until the difference between the sum of sensible and latent heat fluxes of soil and net radiation absorbed by soil is <0.1 W m^{-2}. Then sensible heat fluxes of the canopy and soil are obtained from canopy temperature and soil temperature and latent heat fluxes of canopy and soil are obtained.

12.9 References

Ali M, Jensen CR and Mogensen VO, (1998). Early signals in field grown wheat in response to shallow soil drying. *Australian Journal of Plant Physiology* 25(8), 871–882.

Allen RG, LS Pereria, D Raes and M Smith, (1998). Crop evapotranspiration–guidelines for computing crop water requirements. FAO Irrigation and drainage paper 56. Food and Agriculture Organisation of the UN, Rome.

Anadranistakis M, Kerkides P, Liakatas A, Alexandris S and Poulovasilis A, (1999). How significant is the usual assumption of neutral stability in evapotranspiration estimating models? *Meteorological Applications* 6(2), 155–158.

Aphalo PJ and Jarvis PG, (1993a). An analysis of Balls empirical-model of stomatal conductance. *Annals of Botany* 72(4), 321–327.

Aphalo PJ and PG Jarvis, (1993b). The boundary layer and the apparent responses of stomatal conductance to wind speed and to the mole fraction of CO_2 and water vapor in the air. *Plant Cell Environment*, 1993, 16, 771–783.

Benmehrez M, Taconet O, Vidalmadjar D and Valencogne C, (1992). Estimation of stomatal-resistance and canopy evaporation during the Hapex-Mobilhy experiment. *Agricultural and Forest Meteorology* 58(3–4), 285–313.

Brenner AJ and LD Incoll, (1997). The effect of clumping and stomatal response on evaporation from sparsely vegetated shrublands. *Agricultural and Forest Meteorology* 84(3–4), 187–205.

Brutsaert W, (1982). Some exact-solutions for non-linear desorptive diffusion. *Zeitschrift Fur Angewandte Mathematik Und Physik* 33(4), 540–546.

Brutsaert W and H Stricker, (1979). Advection-aridity approach to estimate actual regional evapotranspiration. *Water Resources Research* 15(2), 443–450.

Campbell GS, (1977). *An Introduction to Environmental Biophysics, Spring-Cerlag*, Berlin. pp. 223–233.

Chen DX and MB Coughenour,(1994). Gemtm–a general-model for energy and mass-transfer of land surfaces and its application at the five sites. *Agricultural and Forest eteorology* 68 (3–4), 145–171.

Choudhury BJ and JL Monteith, (1988). A 4-layer model for the heat-budget of homogeneous land surfaces. *Quarterly Journal of the Royal Meteorological Society* 114(480), 373–98.

Collatz GJ, JT Ball, C Grivet and JA Berry, (1991). Physiological and environmental-regulation of stomatal conductance, photosynthesis and transpiration–a model that includes a laminar boundary-layer. *Agricultural and Forest Meteorology* 54(2–4), 107–136.

Cowan IR, (1965). Transport of water in the soil-plant-atmosphere system. *Journal of Applied Ecology* 2, 221–239.

Cruiziat P, H Cochard and T Ameglio, (2002). Hydraulic architecture of trees: main concepts and results. *Annals of Forest Science* 59(7), 723–752.

Davies WJ and JH Zhang, (1991). Root signals and the regulation of growth and development of plants in drying soil. *Annual Review of Plant Physiology and Plant Molecular Biology* 2, 55–76.

Debruin HAR and AAM Holtslag, (1982). A simple parameterization of the surface fluxes of sensible and latent-heat during daytime compared with the Penman-Monteith concept. *Journal of Applied Meteorology* 21(11), 1610–1621.

Denmead OT and EF Bradley, (1987). On scalar transport in plant canopies. *Irrigation Science* 8 (2), 131–149.

Dolman AJ, (1993). A multiple-source land-surface energy-balance model for use in general-circulation models. *Agricultural and Forest Meteorology* 65 (1–2), 21–45.

Dolman AJ and JS Wallace, (1991). Lagrangian and K-Theory approaches in modeling evaporation from sparse canopies. *Quarterly Journal of the Royal Meteorological Society* 117 (502), 1325–1340.

Domingo F, L Villagarcia, AJ Brenner and J Puigdefabregas, (1999). Evapotranspiration model for semi-arid shrub-lands tested against data from SE Spain. *Agricultural and Forest Meteorology* 95(2), 67–84.

Doorenbos J and WO Pruitt, (1977). Guidelines for predicting crop water requirements. FAO Irrigation and Drainage Paper, No.24, Rome.

Farahani HJ, (1994). Evaporation and transpiration processes, root zone water quality model Tech Rep.No.2. USDA-ARS GPSR, Fort Clooins, CO.

Farquhar GD, SV Caemmerer and JA Berry, (1980). A biochemical-model of photosynthetic CO_2 assimilation in leaves of C-3 species. *Planta* 149(1), 78–90.

Feddes RA and H Zaradny, (1978). Model for simulating soil-water content considering evapotranspiration–comments. *Journal of Hydrology* 37(3–4), 393–397.

Flerchinger GN, (2000). The Simultaneous Heat and Water (SHAW) Model: Technical Documentation, USDA-ARS, Northwest Watershed Research Center, Boise, Idaho.

Flerchinger GN and FB Pierson, (1991). Modeling plant canopy effects on variability of soil-temperature and water. *Agricultural and Forest Meteorology* 56(3–4), 227–246.

Garratt JR, (1978). Transfer characteristics for a heterogeneous surface of large aerodynamic roughness. *Quarterly Journal of the Royal Meteorological Society* 104(440), 491–502.

Goudriaan J, (1989). Simulation of micrometeorology of crops, some methods and their problems and a few results. *Agricultural and Forest Meteorology* 47(2–4), 239–258.

Grant RF, (2001). *A review of the Canadian ecosystem model ECOSYS.* In: S. M (Editor), *Modeling carbon and nitrogen dynamics for soil management.* CRC Press, Boca Raton, FL, pp. 175–264.

Hsieh CI, GG Katul, J Schieldge, JT Sigmon and KK Knoerr, (1997). The Lagrangian stochastic model for fetch and latent heat flux estimation above uniform and nonuniform terrain. *Water Resources Research* 33(3), 427–438.

Huntingford C, SJ Allen and RJ Harding, (1995). An intercomparison of single and dual-sourcevegetation-atmosphere transfer models applied to transpiration from Sahelian savanna. *Boundary-Layer Meteorology* 74(4), 397–418.

Jensen CR, SE Jacobsen, MN Andersen, N Nunez, SD Andersen, L Rasmussen and VO Mogensen, (2000). Leaf gas exchange and water relation characteristics of field quinoa (*Chenopodium quinoa* Willd.) during soil drying. *European Journal of Agronomy* 13(1), 11–25.

Katul GG and MB Parlange, (1992). A Penman-Brutsaert model for wet surface evaporation. *Water Resources Research* 28(1), 121–126.

Konzelmann T, P Calanca, G Muller, L Menzel and H Lang, (1997). Energy balance and evapotranspiration in a high mountain area during summer. *Journal of Applied Meteorology* 36(7), 966–973.

Leuning R, (1995). A critical-appraisal of a combined stomatal-photosynthesis model for C-3 plants. *Plant Cell and Environment* 18(4), 339–355.

Leyton L, (1975). *Fluid Behaviour in Biological Systems*. Clarendon Press, Oxford, pp. 167–175.

Lhomme JP, (2001). Stomatal control of transpiration: Examination of the Jarvis-type representation of canopy resistance in relation to humidity. *Water Resources Research* 37(3), 689–699.

Lin JD and SF Sun, (1983). Moisture and heat flow in soil and theirs effects on bare soil evaporation. Trans. *Water Conservancy* 7, 1–7 (in Chinese).

Liu FL, MN Andersen, SE Jacobsen and CR Jensen, (2005). Stomatal control and water use efficiency of soybean (Glycine max L. Merr.) during progressive soil drying. *Environmental and Experimental Botany* 54(1), 33–40.

Menenti M, (1984). Physical Aspects and Determination of Evaporation in Deserts Applying Remote Sensing Techniques, Instituut von Cultuurtechnik en Waterhuishounding. Wageningen, The Netherlands, pp. 202.

Meyers TP, DD Baldocchi and BB Hicks, (1988). Processes promoting exchange between the surface and the atmosphere. *Abstracts of Papers of the American Chemical Society* 196, 136-AGRO.

Monteith JL, (1965). Light distribution and photosynthesis in field crops. *Annals of Botany* 29(113), 17–37.

Monteith JL, (1973). *Principles of Environmental Physics*. Edward Arnold, London.

Nichols DJ, (1992). Plants at the K/T Boundary. *Nature* 356(6367), 295–295.

Parlange MB and GG Katul, (1992). Estimation of the diurnal variation of potential evaporation from a wet bare soil surface. *Journal of Hydrology* 132(1–4), 71–89.

Penman HL, (1948). Natural evaporation from open water, bare soil and grass. *Proceedings of the Royal Society of London Series a-Mathematical and Physical Sciences* 193(1032), 120–145.

Philip JR, (1966). Plant water relations: some physical aspects. *Annual Review of Plant Physiology* 17, 45–268.

Powell DB and B Thorpe, (1977). *Dynamic aspects of plant-water relations in environmental effects on crop physiology*. Academic Press, London, pp. 159–279.

Priestley CHB and RJ Taylor, (1972). Assessment of surface heat-flux and evaporation using large-scale parameters. *Monthly Weather Review* 100(2), 81–92.

Sellers PJ, DA Randall, GJ Collatz, JA Berry, CB Field, DA Dazlich, C Zhang, GD Collelo and L Bounoua, (1996). A revised land surface parameterization (SiB2) for atmospheric GCMs. 1. Model formulation. *Journal of Climate* 9(4), 676–705.

Shaw RH and AR Pereira, (1982). Aerodynamic roughness of a plant canopy–a numerical experiment. *Agricultural Meteorology* 26(1), 51–65.

Shuttleworth WJ and JS Wallace, (1985). Evaporation from sparse crops–an energy combination theory. *Quarterly Journal of the Royal Meteorological Society* 111(469), 839–855.

Stannard DI, (1993). Comparison of Penman-Monteith, Shuttleworth-Wallace and modified Priestley-Taylor evapotranspiration models for wildland vegetation in semi-arid rangeland. *Water Resources Research* 29(5), 1379–1392.

Thom AS, JB Stewart, HR Oliver and JHC Gash, (1975). Comparison of aerodynamic and energy budget estimates of fluxes over a Pine forest. *Quarterly Journal of the Royal Meteorological Society* 101(427), 93–105.

Thurtell GW, (1989). Comments on using K-theory within and above the plant canopy to model diffusion processes. *Estimation of Areal Evaotranspiration* 177, 81–85.

Tuzet A, A Perrier and R Leuning, (2003). A coupled model of stomatal conductance, photosynthesis and transpiration. *Plant Cell and Environment* 26(7), 1097–1116.

Van de griend AA and JH Van Boxel, (1989). Water and surface-energy balance model with a multilayer canopy representation for remote-sensing purposes. *Water Resources Research* 25(5), 949–971.

Vanbavel CHM and DI Hillel, (1976). Calculating potential and actual evaporation from a bare soil surface by simulation of concurrent flow of water and heat. *Agricultural Meteorology* 17(6), 453–476.

Wang J, (2006). Study on the experiment and simulation of crop growth and water heat and CO_2 transfer in the Agro-ecosystem. Doctor Thesis, Insititute of Geographic Sciences and Natural Resources Research, Chinese Academy of Sciences, 211 pp.

Wang J, Q Yu, J Li, LH Li, XG Li, GR Yu and XM Sun, (2006). Simulation of diurnal variations of CO_2, water and heat fluxes over winter wheat with a model coupled photosynthesis and transpiration. *Agricultural and Forest Meteorology* 137(3–4), 194–219.

Wang SS, RF Grant, DL Verseghy and TA Black, (2002). Modelling carbon-coupled energy and water dynamics of a boreal aspen forest in a general circulation model land surface scheme. *International Journal of Climatology* 22(10), 1249–1265.

Wang YP and R Leuning, (1998). A two-leaf model for canopy conductance, photosynthesis and partitioning of available energy I: Model description and comparison with a multi-layered model. *Agricultural and Forest Meteorology* 91(1–2), 89–111.

Wilson JD, (1989). Turbulent transport within the plant canopy. In E*stimation of Areal Evapotranspiration*, Black TA, DL Spittlehouse, MD Novak and DT Price (Eds.), IAHS Publication 17, pp. 43–80.

Wilson TB, JM Norman, WL Bland and CJ Kucharik, (2003). Evaluation of the importance of Lagrangian canopy turbulence formulations in a soil-plant-atmosphere model. *Agricultural and Forest Meteorology* 115(1–2), 51–69.

Wu A, A Black, DL Verseghy and WG Bailey, (2001). Comparison of two-layer and single-layer canopy models with Lagrangian and K-theory approaches in modelling evaporation from forests. *International Journal of Climatology* 21(14), 1821–1839.

Yao C, S Moreshet and B Aloni, (2001). Water relations and hydraulic control of stomatal behaviour in bell pepper plant in partial soil drying. *Plant Cell and Environment* 24(2), 227–235.

Yu Q and GN Flerchinger, 2008. Extending Simullttaneous Heat and Water (SHAW) model to simulate carbon dioxide and water fluxes over wheat canopy. In *Resonse of crops to limited water: understanding and modeling water stress effects on plant growth processes*, Ahuja LR, VR Reddy, SA Saseendran and Q Yu (Eds.), Advances in Agricultural Systems Modeling 1. SSSA-ASA-CSSA Publication, USA.

Yu Q, J Goudriaan and TD Wang, 2001. Modelling diurnal courses of photosynthesis and transpiration of leaves on the basis of stomatal and non-stomatal responses, including photoinhibition. *Photosynthetica* 39(1), 43–51.

Yu Q, SH Xu, J Wang and XH Lee, 2007. Influence of leaf water potential on diurnal changes in CO_2 and water vapour fluxes. *Boundary-Layer Meteorology* 124(2), 161–181.

Yu Q and GN Flerchinger, 2008. Combining Simultaneous Heat and Water (SHAW) with photosynthesis model to simulate water and CO_2 fluxes over wheat canopy. In *Response of crops to limited water: Understanding and modeling water stress effects on plant growth processes*. Advances in Agricultural Systems Modeling Series 1. Ed: L.R. Ahuja, V.R. Reddy, S.A. Saseendran, and Q Yu (SSSA-ASA-CSSA publication), pp. 191–214.

13

Coupling Models of Photosynthesis, Transpiration and Stomatal Conductance and Environmental Controls of Leaf Function

13.1 Introduction

To simulate physiological responses of leaves to the environment, a semi-empirical model (the BWB model; Chapter 11) proposed by Ball *et al.* (1987) summarised the relationship between stomatal conductance (g_s) and an index relating net photosynthesis, relative humidity and CO_2 concentration, with a single linear equation. Since g_s and net photosynthesis are interdependent, the solution of the BWB model needs a photosynthesis sub-model and a biochemical photosynthesis model also needs a stomatal model to provide stomatal regulation of intercellular CO_2 concentration (C_i). Therefore Leuning (1990) proposed a method to solve a combined photosynthesis-g_s model. By combining the BWB stomatal model with Farquhar's intercellular biochemical model of photosynthesis, an integrated photosynthesis, transpiration and stomatal conductance, model was developed (Collatz *et al.* 1991, 1992). Aphalo and Jarvis (1993) studied the effects of boundary layer conductance (g_b) on leaf gas exchange and constructed a mathematical relation among vapour pressure difference between the sub-stomatal cavity and the leaf surface (VPDs), g_b and g_s, by using a gaseous diffusion equation. By incorporating the equation of Aphalo and Jarvis (1993) into the photosynthesis-g_s model (Collatz *et al.* 1991, Leuning 1995), Yu and Wang (1998) simulated stomatal responses to changes in g_b. The coupled model has been widely used in the analysis of many kinds of ecophysiological problems across different scales, for example in modelling evapotranspiration of a plant canopy (Vogel *et al.* 1995). In this chapter the coupling of photosynthesis, transpiration and stomatal conductance into single models is discussed. Such coupling allows application to real-world modelling problems.

13.2 Leaf Temperature

Leaf temperature is an important variable in models of leaf function, which is determined by radiation and heat exchanges, and reflects coupling of biophysical processes. Leaf temperature influences all biochemical processes, influences biophysical

321

processes such as the conversion of liquid water to water vapour and also influences the saturated water vapour pressure within leaves and thus influences the gradient in water vapour pressure between the leaf sub-stomatal cavity and the boundary layer.

Leaf temperature can be estimated using the energy budget equation:

$$R_\text{n} = \frac{\rho C_p (T_\text{s} - T_\text{a})}{r_\text{a}} + \frac{\rho C_p \left[e_\text{s}(T_\text{s}) - e_\text{a} \right]}{\gamma (r_\text{s} + r_\text{a})} + \varepsilon \sigma T^4 \tag{13.1}$$

Where $e_\text{s}(T_\text{s})$ is the saturated water vapour pressure, e_a is ambient water vapour pressure, γ is the psychrometric constant and r_a and r_s are boundary and stomatal resistances, respectively.

Equation 13.1 may be solved using linear or quadratic Taylor approximations, with the linear version being common (Paw U 1987):

$$T_\text{s} = T_\text{a} + \left\{ R_\text{i} - \varepsilon \sigma T_\text{a}^4 - h_\text{e} \left[e_\text{s}(T_\text{a}) - e_\text{a} \right] \right\} / \left(4 \varepsilon \sigma T_\text{a}^3 + h_\text{t} + h_\text{e} S \right) \tag{13.2}$$

where S is the slope of the saturated vapour pressure curve ($= \text{d}e_\text{s}/\text{d}T$); h_e is the transfer coefficient for water vapour defined as:

$$h_\text{e} = \rho c_\text{p} / \left[\gamma (r_\text{a} + r_\text{s}) \right] \tag{13.3}$$

And h_t is the transfer coefficient for heat, defined as:

$$h_\text{t} = \rho c_\text{p} / r_\text{a} \tag{13.4}$$

A more accurate approximation may be obtained using a second-order Taylor approximation, leading to a quadratic solution:

$$T_\text{s} = T_\text{a} + \left(-b + \sqrt{b^2 - 4ac} \right) / (2a) \tag{13.5}$$

where

$$a = 6 \varepsilon \sigma T_\text{a}^2 + \left(\text{d}^2 e_\text{s} / \text{d}T^2 \right) h_\text{e} / 2 \tag{13.6}$$

$$b = 4 \varepsilon \sigma T_\text{a}^3 + h_\text{t} + h_\text{e} S \tag{13.7}$$

$$c = -R_\text{i} + \varepsilon \sigma T_\text{a}^4 + h_\text{e} e_\text{a} \tag{13.8}$$

where $\text{d}^2 e_\text{s} / \text{d}T^2$ is the derivative of the slope S and the second derivative of saturated vapour pressure function e_s with respect to temperature.

Iterative solutions are also used to solve the operation temperature equation. The most widely used method is the Newton-Raphson method. In each iteration a new value for the variable is calculated using the previous iteration value, a function of the previous value and, the derivative of the function (equation 13.9):

$$T_\text{i+1} = T_\text{i} + f(T_\text{i}) / f'(T_\text{i}) \tag{13.9}$$

where T_{i+1} is the previous iteration value of T, T_i is the last iteration value of T, $f(T)$ is the energy budget residual function, defined as:

$$f(T_i) = R_i - \varepsilon \sigma T_i^4 - \rho c_p (T_i - T_a) / r_a \tag{13.10}$$

and $f'(T_i)$ is the derivative of f:

$$f'(T_i) = -4\varepsilon \sigma T_i^3 - \rho c_p / r_a \tag{13.11}$$

The first iteration value for T_s is usually the linearised solution to T_s (Tracy *et al.* 1984, Paw U 1987). Previously iterative solutions have been used to solve for the surface temperature of wet organisms. The residual equation (eq. 6.9) can be again used with the residual functions f and f' now being:

$$f(T_i) = -R_i + \varepsilon \sigma T_i^4 + h_t (T_i - T_a) + h_e \left[e_s (T_i) - e_a \right] \tag{13.12}$$

$$f'(T_i) = 4\varepsilon \sigma T_i^3 + h_t + h_e S \tag{13.13}$$

A quartic solution method for the energy budget is plausible except for the non-linearity of the saturated vapour pressure term $e_s(T_s)$. A solution is possible if the saturated vapour pressure function is approximated as a fourth- or lower-order polynomial. Then the temperature terms arising from the saturated vapour pressure function may be combined with those from the energy budget for a single quartic equation in T_a (Paw U 1987).

13.3 Numerical Solutions for Combined Leaf Models

When model parameter values are available either from published sources, experimental measurement, or from model fitting, the model can be ready to run by supplying the driving variables, normally the most active variables of meteorology, that is, solar radiation, temperature, humidity, wind speed, atmospheric CO_2 concentration and soil water content.

There is rarely an analytic solution for a set of non-linear equations. The combined equations can be solved using an iterative method to get a numerical solution. Figure 13.1 summarises the approach used to solve a set of non-linear equations. For the combined leaf model, C_i can be set first as this is usually a fairly conservative value (i.e. does not show very large variation during the day when stomata are open) and an iteration occurs to seek for convergence of C_i. Once this has been done, estimated values of g_s are used to calculate the leaf energy balance and to get convergence of leaf temperature. A numerical approach similar to that of Collatz *et al.* (1991) can be used to couple the sub-models. In finding the solution, the value of C_i is given first and substituted into the biochemical photosynthesis model to get P_n. Then stomatal conductance is calculated from the stomatal model (Chapter 11)

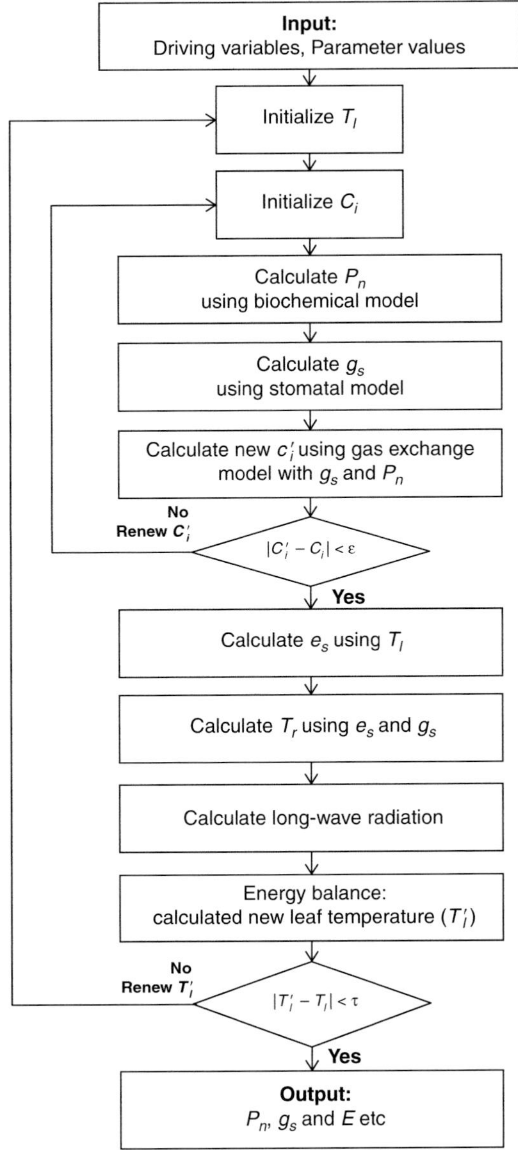

Figure 13.1. Scheme for combined photosynthesis-transpiration-stomatal conductance models and leaf energy balance, as used by Yu and Flerchinger (2008). The ε and τ represent tolerances in the iterative calculations.

and a new C_i is found by the diffusion model. These steps are repeated until the change in C_i is less than a certain small (user-defined) allowed value, the "tolerance". All of the physiological variables are solved simultaneously. The negative physical and physiological feedback processes in the model guarantee the convergence of the set of equations used in respect of the range of changes of the various environmental factors.

The models for leaf gas exchange are generally based on steady state conditions, which can be driven by environmental variables. This may assume constant parameters over a period of time (for example, a day or an entire growing season).

If growth is considered as a response, to changes in environmental variables such as rainfall, changes in photosynthetic parameters should be considered and, consequently, more dynamic equations should be included in the model. These models are time-dependent, with variable parameter values whereby their value is a function of that of the previous time step.

13.4 Uncoupling of Integrated Photosynthesis-Transpiration-Stomatal Conductance Models

The stomatal conductance model describes g_s as a function of P_n, where P_n is in turn a function of g_s. This is a steady state model while physiological relationships are kept stable and no driving-response or feedbacks were included in the model. The model derives g_s purely as a function of P_n and *vice versa*. To resolve this model it is necessary to use interactive methods to get model results.

It is predicted that g_s is regulated such as to keep a conservative C_i, which means C_i does not vary much. A direct expression of C_i can be obtained from two environmental variables, i.e. CO_2 and humidity by simplifying the stomatal model (Yu *et al.* 2002).

From flux-gradient relationships:

$$C_i = C_s - P_n / g_s \tag{13.14}$$

where C_i is the intercellular CO_2 concentration. Under weak light, P_n is close to 0 and g_s is also approximately 0. Because g_0 is a constant near 0, if $g_0 = 0$, the following relation holds:

$$C_i = C_s - (C_s - \Gamma)(1 + VPD_a / VPD_0) / m \tag{13.15}$$

This simplified equation is adopted to calculate intercellular CO_2 concentration (C_i) from atmospheric CO_2 concentration and humidity.

The photosynthetic rate in equation 13.14 can be obtained directly. First, C_i is obtained from eq.13.15, α from equation 10.1, P_{max} from equations 10.7–10.10 and R_d from equation 10.12 and then the values of these variables can be used to substitute for these in equation 10.2 and solve this to get the photosynthetic rate.

In conclusion, this simplified photosynthesis model can be divided into two parts: (1) based on a stomatal model (Chapter 11), the relation between intercellular CO_2 concentration (C_i) and atmospheric CO_2 concentration and atmospheric humidity is derived, that is, $C_i = f(C_s, VPD)$; and, (2) according to the biochemical model, the relation between photosynthetic rate and light intensity, temperature and C_i is developed, that is, $P_n = f(PPFD, T_a, C_i)$ (Yu *et al.* 2002).

13.5 A Modelling Perspective of Physiological Responses to Environmental Variables

From the previous section, a combined photosynthesis-transpiration-stomatal conductance model was proposed (Collatz *et al.* 1991, 1992). Yu and Wang (1998) integrated Farquhar's photosynthesis biochemical model (Farquhar *et al.* 1980), Leuning's equation of BWB stomatal model (Chapter 11) and gaseous transfer model proposed by Aphalo and Jarvis (1993). Numerical analysis can be conducted to show physiological responses to changes in environmental variables, which then provide insight to plant response to environment change and also demonstrate the model's performance.

13.5.1 The Influence of Boundary Layer Conductance on Leaf Function

In Figure 13.2, it can be seen that when boundary layer conductance to CO_2 (g_{bc}) is initially small, a small decrease in boundary layer thickness caused by wind will induce a sharp increase in total conductance and thereby increases in C_i, assimilation and transpiration rates. Since stomata respond to CO_2 concentration, an increase in g_{bc} may elevate C_s (CO_2 concentration at the leaf surface) and enlarge VPD_s, both of which have the effect of decreasing g_{sc} and, therefore, diminish the extent of the increase in P_n caused by wind, but not totally abolishing the increase. As there is an additional mesophyll resistance for photosynthesis, the reduction in g_{bc} will have a smaller effect on P_n than on E. Consequently water-use–efficiency (WUE) will increase with decreasing g_{bc}. Thus any reduction in g_{bc} has a more severe effect on water loss than on photosynthesis. Model simulations also show that in warm environments with low solar radiation inputs (e.g. in the understorey of rainforests), large leaves with thick boundary layers have an advantage in WUE through reduced transpiration.

While photosynthesis and transpiration are very sensitive to change in g_b when it has a very small value, the responses of P_n and E will no longer increase with any further increase in g_b when g_b reaches about 0.5 mol m^{-2} s^{-1}. Consequently, the extent to which the concentration of water vapour and CO_2 at the leaf surface differs from that in ambient air is very small. Thus the effect of g_b can be neglected when g_b is much larger than g_{sc}.

Collatz *et al.* (1991) first proposed the combined model of photosynthesis-transpiration-stomatal conductance and included a leaf energy balance sub-model. This model simulated the frequently observed midday depression of transpiration (Chapter 3) arising from stomatal closure. In the Collatz model midday closure of stomata was a result of a positive feedback mediated by boundary layer conductance, for high or low values of g_b and this led to excessive drying or heating, respectively, of the air at the leaf surface (Schuepp 1993). Such simulations show that the dependence of leaf transpiration on boundary layer conductance is such

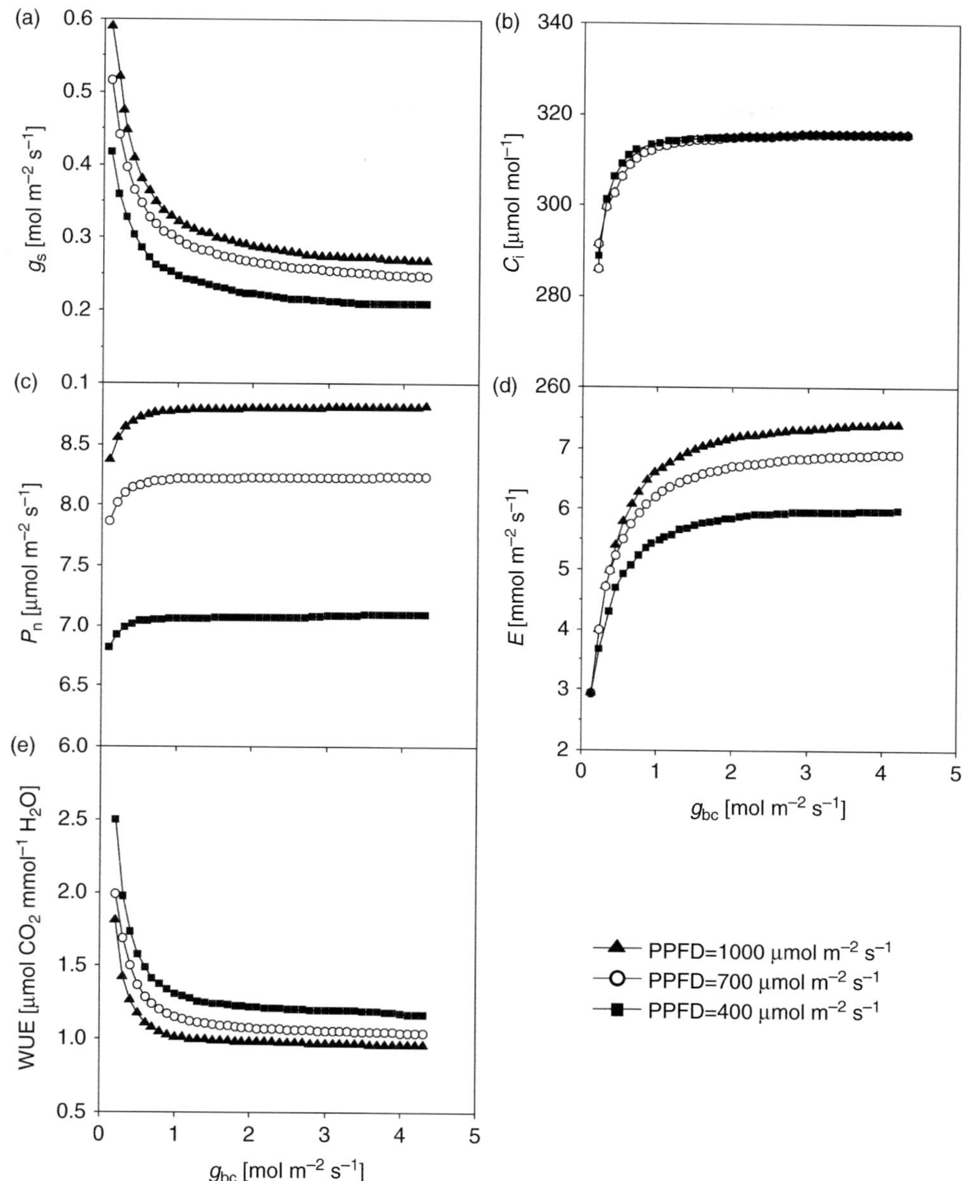

Figure 13.2. Simulated physiological response to boundary layer conductance at different irradiances. e_a is 1200 P_a; C_a is 350 μmol mol^{-1}; T_l is 25°C; P_n is net assimilation; E is transpiration; g_s is stomatal conductance to CO_2; g_{bc} is boundary layer conductance to CO_2; C_i is intercellular CO_2 concentration; and WUE is water-use-efficiency. Adapted from Yu and Wang (1998).

that maximum transpiration and photosynthesis occur at intermediate values of g_b. The model outlined above, which replaces the BWB stomatal model with Leuning's model and including Aphalo and Jarvis's gaseous exchange model, cannot simulate P_n and E decline at high g_b (Yu and Wang 1998).

Figure 13.2 considered the effect of g_b under laboratory conditions, at a given constant leaf temperature. Under these conditions excessive drying does not occur because even when g_b is infinite, the water vapour pressure over a leaf surface (e_s) can only be as low as that in ambient air (e_a) and cannot be lower. The effect of g_b on leaf function is likely to be minimal when g_b exceeds 0.5 mol m^{-2} s^{-1} and differences in CO_2 concentration and water vapour between leaf surface and ambient air are very small.

13.5.2 The Influence of Vapour Pressure Deficit (VPD) on Leaf Function

In model simulations of responses of photosynthesis, transpiration and stomatal conductance to air humidity, VPD is varied by changing the value of ambient water vapour pressure (e_a). In Figure 13.3 it can be seen that g_s declines hyperbolically with an increase in VPD, as observed in many studies (Chapter 2). The response of stomata to a change in *VPD* has a strong influence on rates of transpiration. C_i declines almost linearly with an increase in VPD, because of stomatal closure and P_n declines with increasing *VPD* by a weakly accelerating rate as a result of declining C_i, which becomes progressively more limiting to photosynthesis. Such a response has been observed experimentally by Comstock and Ehleringer (1993) and C_i responds to *VPD* more strongly than to variation in other factors, such as light flux. Since small changes in humidity can affect stomata without directly affecting assimilation rate in many species (Schulze *et al.* 1987), C_i varies strongly with humidity (Morison 1987). In Leunings' revised model, there is an almost linear correlation between C_i and VPD_a as shown in Figure 13.3b. Because of the differential effect of VPD on P_n and E, WUE strongly decreases with increasing *VPD* as observed in leaf and canopy-scale observations (Comstock and Ehleringer 1993).

The difference in water vapour pressure between leaf air spaces and ambient air (*LAVPD*) is the driving force for transpiration. Transpiration is also affected by g_s and g_s responds to *LAVPD*, or more precisely, g_s responds to E (Chapter 2). There is thus a feedback loop between g_s and E. That is, an increase in the rate of transpiration can cause a decrease in g_s which in turn will decrease the rate of transpiration to some extent. Consequently, E can increase with increasing *LAVPD* but the increase is smaller than would occur in the absence of a decline in g_s.

A nearly linear decrease of g_s with increasing E is found from the model simulation (Fig. 13.4). In a seminal review published in 1995, Monteith showed across fifty-two sets of measurements that g_s responds to E rather than *LAVPD*. This point is extensively discussed in Chapter 2.

13.5.3 The Influence of Leaf Temperature on Leaf Function

When leaf temperature is used as a model input, the energy balance module is not included in calculations. This means that changes in transpiration will not alter leaf temperature. Temperature affects photosynthesis and g_s in two different ways. First,

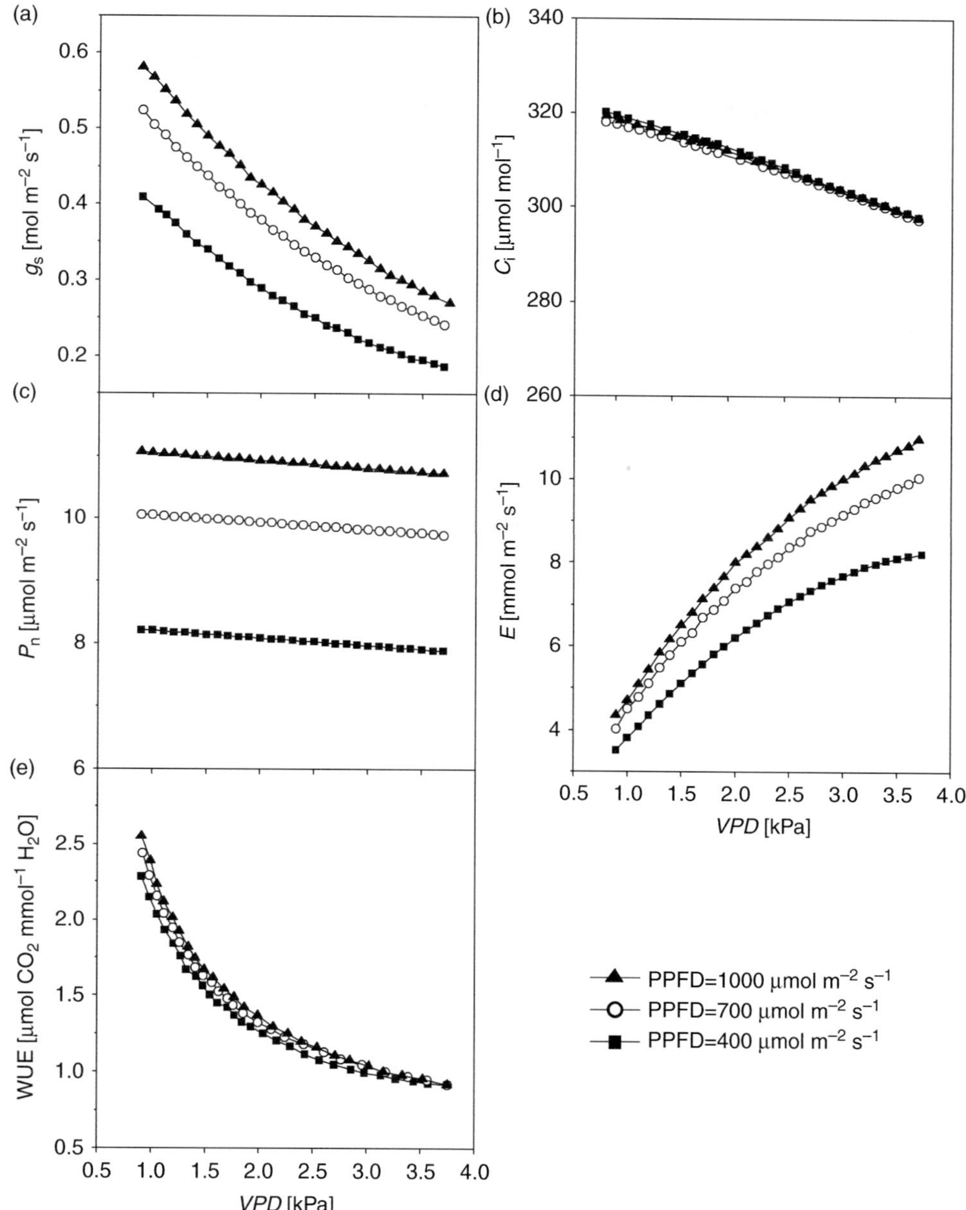

Figure 13.3. Simulated physiological response to vapour pressure deficit (VPD) at 3 different irradiances. The conditions and symbols are the same as in Figure 13.2, except for boundary layer conductance, which is 1.0 mol CO$_2$ m^{-2} s^{-1}. Adapted from Yu and Wang (1998).

temperature affects the biochemistry of photosynthesis; second, it affects *LAVPD* through its effect on the saturated intercellular vapour pressure within leaves. A declining P_n in response to increasing temperatures can arise because of a declining g_s arising from increasing *LAVPD*.

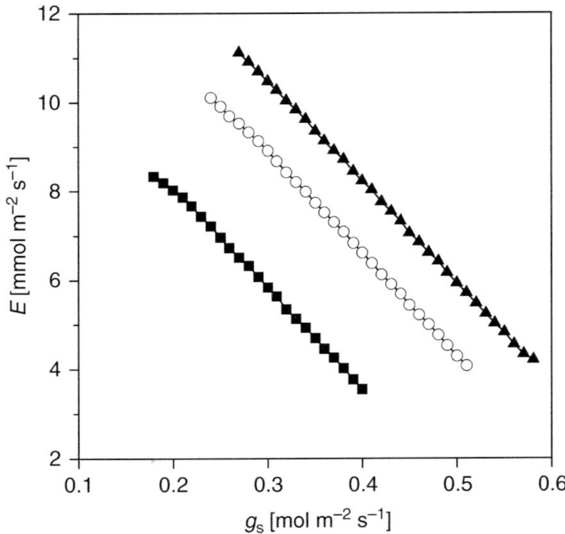

Figure 13.4. Transpiration rate (E) *versus* stomatal conductance to CO_2 (g_s) under varying values of vapour pressure deficit (*VPD*). Values of VPD increase in moving right-to-left up each line. Each line presents a single PPFD. Data and symbols are the same as in Figure 13.2, except for boundary layer conductance, which is 1.0 mol CO_2 m^{-2} s^{-1}. Adapted from Yu and Wang (1998).

Figure 13.5 shows that assimilation, transpiration and stomatal conductance vary with temperature as one-peaked curves. Measurements of the response of g_s to temperature give similar response curves (Jarvis 1980). The optimum temperature of g_s is lower than that of assimilation. This is because although g_s is proportional to assimilation in controlled experimental leaf-chamber studies (Wong *et al.* 1985a, b, c), increases in temperature cause exponential increases in *LAVPD* which make g_s decline hyperbolically. Consequently, the magnitude of the decline in g_s is larger than those observed in P_n and E. When leaf temperature is lower than the optimum temperature for g_s, E increases sharply with increasing T_l. Both *LAVPD* and g_s, the two factors determining transpiration rate, increase with increasing temperature. When leaf temperature exceeds slightly its optimum value for g_s, E continues to increase before the decrease of g_s is large enough to counter the increase of *LAVPD*. With further increases in T_l, a point is reached where stomatal closure is large enough to cause a decrease in E despite *LAVPD* increasing (the third phase of the three phase response discussed in Chapter 2).

WUE declines monotonically with an increase in temperature because the increase in *LAVPD* is much steeper than that in CO_2 concentration deficit. From the analysis made above it follows that C_i/C_a is determined mainly by the changes in g_s which are caused by changes in *LAVPD*, so an increase in temperature causes a monotonous decrease in C_i.

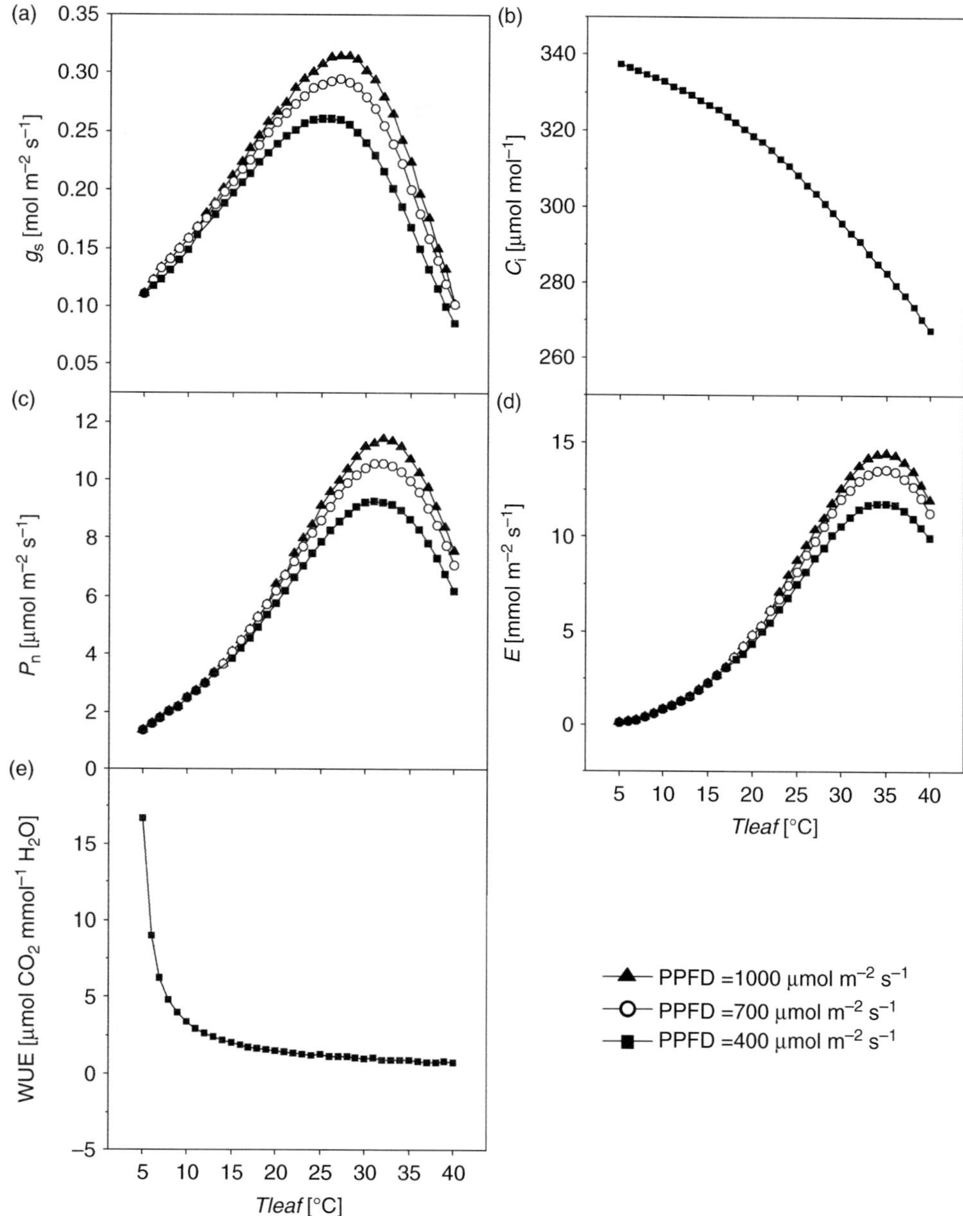

Figure 13.5. Simulated physiological response to leaf temperature at 3 different irradiances. The conditions and symbols are the same as in Figure 13.2, except for boundary layer conductance, which is 1.0 mol CO_2 m^{-2} s^{-1}. Adapted from Yu and Wang (1998).

13.5.4 The Influence of Light Intensity on Leaf Function

Light irradiance affects physiological processes in two ways: (i) providing the energy for photosynthesis (Chapter 2); and, (ii) physically affecting leaf energy balance

(Chapter 9), which determines leaf temperature. The former also includes direct and indirect responses of g_s which include guard cell responses to C_i and the photoreceptors of guard cells (Chapter 2). Under field conditions, both aspects require consideration. Here the former is taken into account, omitting impacts of solar radiation on leaf energy budgets; thus leaf temperature is assumed constant so that model predictions can be compared with experimental results under controlled environmental conditions.

The response of g_s, P_n, C_i, E and WUE, to increasing *PPFD* are shown in Figure 13.6. Almost all of these responses are mediated through changes in rates of photosynthesis and stomatal conductance. As irradiance increases, P_n increases along a Michaelis-Menten-type curve. Increasing activation of Rubisco and electron transport cause increased rates of C fixation. The intersection of the curves in Figure 6.6a shows the influence of temperature on both photosynthesis and respiration. C_i decreases with increasing irradiance, but the magnitude of the change is relatively small over a wide range of irradiances and is mostly evident when irradiance is low (Fig. 13.6b). An increase in irradiance causes an increase in photosynthesis, leading to a decrease in C_i and stomata open in response to this decrease in C_i (Sharkey 1987). Similarly Wong *et al.* (1985a) experimentally found that P_n and g_s change in the same proportion, so that C_i changes very little over a wide range of irradiances.

When *LAVPD* is kept constant transpiration is determined mainly by g_s. Photosynthesis and transpiration both increase with increased irradiance as stomatal aperture increases, but their ratio, WUE, increases with irradiance (Fig. 13.6e); although its rate of increase is smaller than that of either P_n or E. Increasing WUE is the result of the fact that increases in irradiance cause larger increases in photosynthesis than transpiration. This is in contrast to the effect of increased boundary layer conductance which has a larger impact on transpiration than photosynthesis (Fig. 13.3e). When irradiance exceeds about 600 μmol m^{-2} s^{-1}, the increase in the WUE is not significant, as has been observed experimentally (Shi and Wang 1994).

13.5.5 The Influence of Ambient CO$_2$ Concentration on Leaf Function

Physiological responses of plants to elevated atmospheric CO_2 concentration can be divided into two types. Short-term responses do not include change in photosynthetic potential, that is, no acclimation occurs (Eamus and Jarvis 1989). In contrast long-term responses usually include an acclimation of photosynthetic potential to elevated CO_2 (Eamus and Jarvis 1989, Mott 1990). Here only short term responses under steady state conditions are considered.

In Figure 13.7, an almost linear increase in C_i with an increase in ambient CO_2 is generated by the model. It shows that the intercept is near zero, so C_i/C_a, or more precisely C_i/C_s, is almost constant, as has been observed experimentally (Wong *et al.* 1979, 1985a, Sharkey and Raschke 1981, Morison and Gifford 1983).

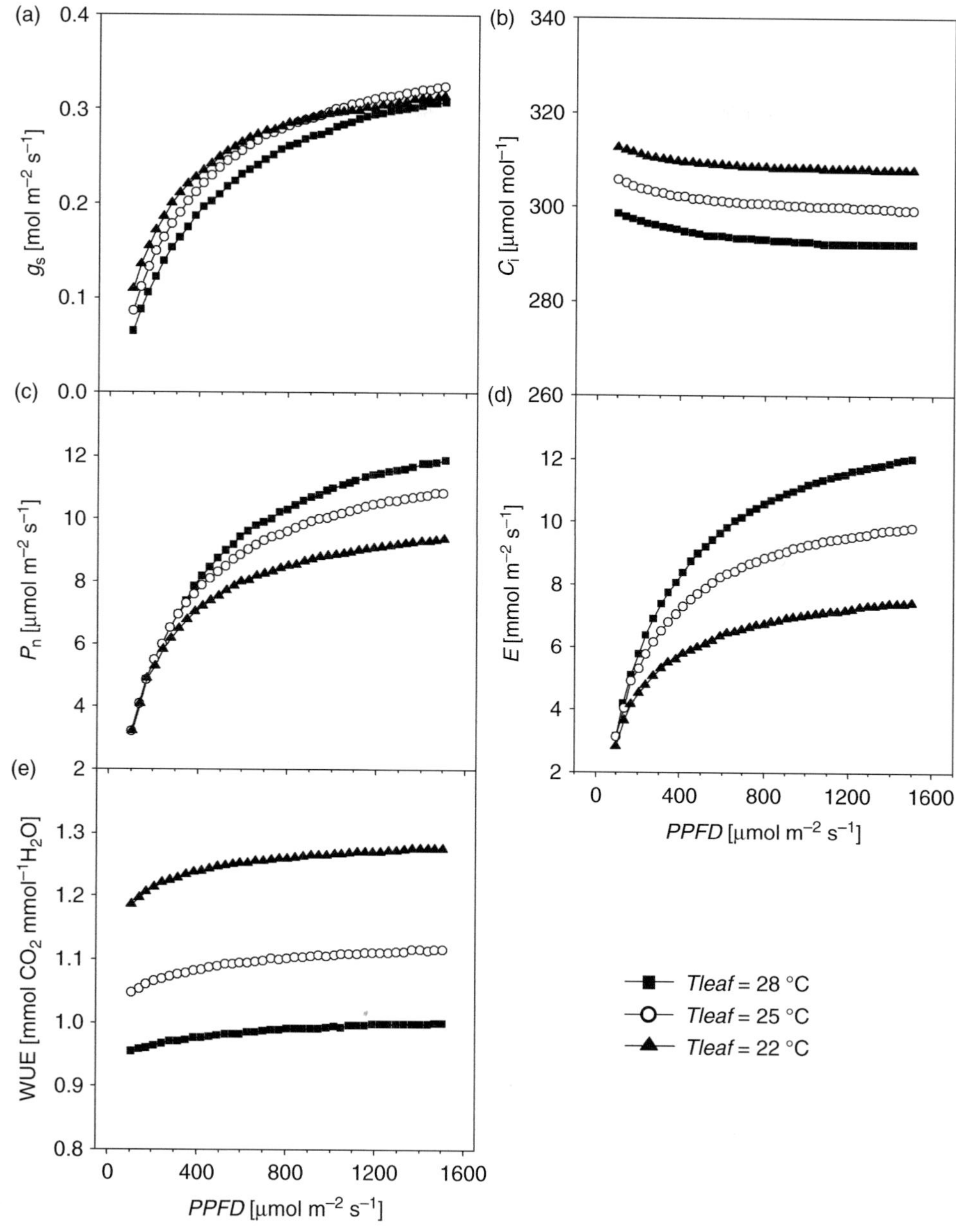

Figure 13.6. Simulated physiological response to photosynthetic photon flux density (irradiance) at different leaf temperatures. The conditions and symbols are the same as in Figure 13.2, except for the boundary layer conductance, which is 1.0 mol CO_2 m^{-2} s^{-1}. Adapted from Yu and Wang (1998).

Because there is a linear relation between C_a and C_i (Fig. 13.7b), the relation between P_n and C_a, shown in Fig. 13.7c can also be taken as the relation between P_N and C_i. The relation between P_n and C_a obeys the Michaelis-Menten equation. As P_n increases and g_s decreases with an increase in C_i, stomatal sensing of C_i occurs.

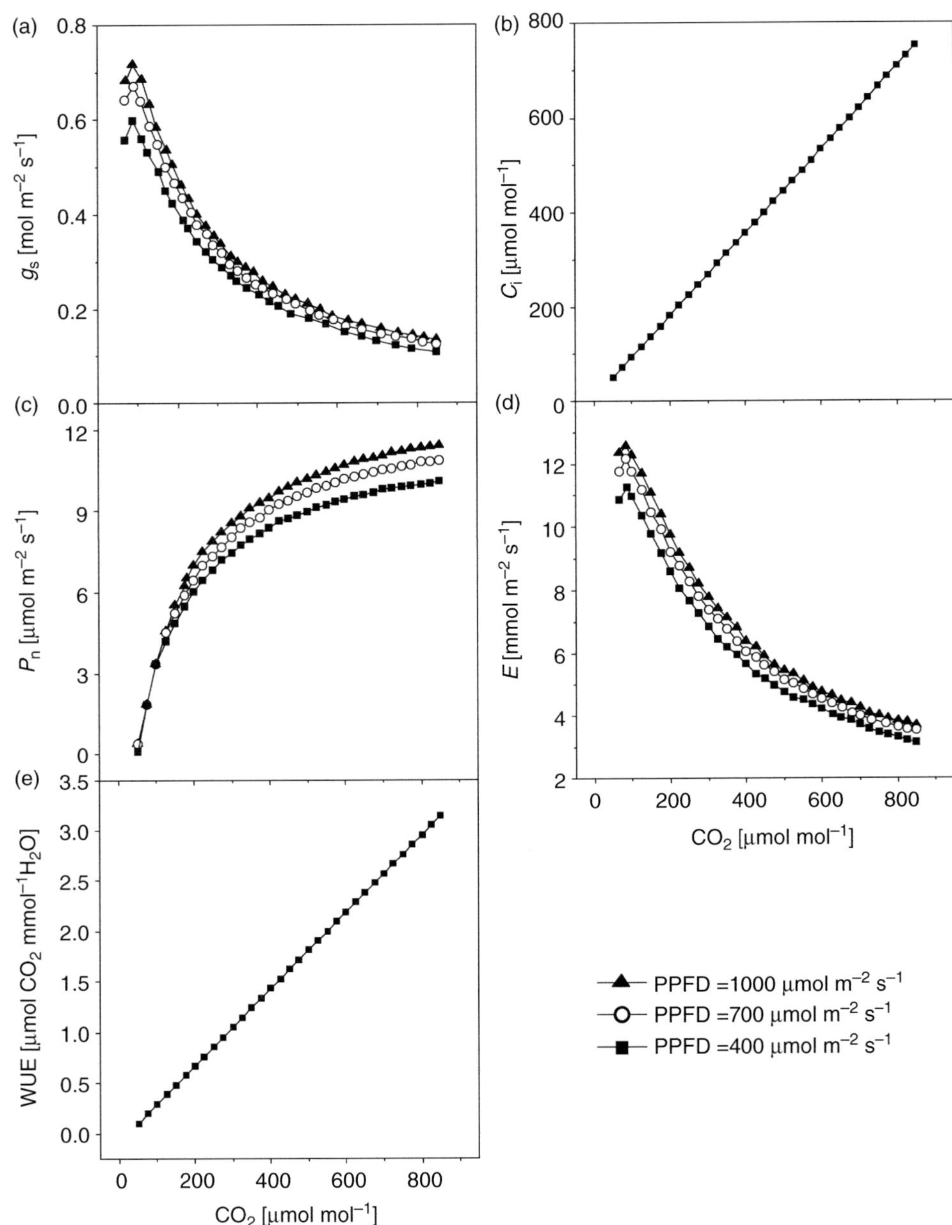

Figure 13.7. Simulated physiological response to ambient CO_2 concentration at 3 different irradiances. The conditions and symbols are the same as in Figure 13.2, except for the boundary layer conductance, which is 1.0 mol CO_2 m^{-2} s^{-1}. Adapted from Yu and Wang (1998).

Almost universally, g_s declines curvilinearly with increasing CO_2 concentration (Fig. 13.7a), with larger changes in g_s occurring at concentrations below 300 μmol mol^{-1} than at higher concentrations. Only recently has this behaviour been replicated in a model of stomatal conductance (Chapter 11). When $P_n \to 0$ (when irradiance

< 50 μmol m^{-2} s^{-1} or C_s < 100 μmol mol^{-1}) the use of C_s-Γ instead of C_s in Ball's model will fit better at low CO_2 concentrations (Leuning 1990). In some reports, maximum g_s occurs at an ambient CO_2 of about 100 μmol mol^{-1} (Raschke 1976, Dubbe *et al.* 1978); in most other studies, it occurs at 0 μmol mol^{-1}. The nature of the g_s response at low CO_2 concentrations may correlate closely with the CO_2 compensation point in the simulation. The peak of g_s at high Γ is obvious and occurs at higher CO_2 concentrations than at small Γ.

Transpiration also declines with increased CO_2 concentration, in the same pattern as g_s when *LAVPD* is nearly constant and so WUE increases approximately linearly with increasing CO_2. The extent to which WUE increases is about 100 percent under doubled CO_2 concentration, as shown in Fig. 13.7e. Morison and Gifford (1983) found that under doubled CO_2 concentration, a 67 percent increase in WUE on average was observed for young plants. Many experiments found photosynthesis and WUE increases with increasing CO_2 concentration (Morison and Gifford 1984; Eamus 1991; Harley *et al.* 1992; Barton *et al.* 2012).

The integrated model discussed in this section is capable of simulating physiological responses to boundary layer conductance, whereas previous models cannot. The main difference of this model from others is the consideration of gaseous transfer into and out of leaves via stomata. Ball's model is based on the fact that g_s varies in proportion with P_n when C_s and h_s are held constant, as observed by Wong *et al.* (1979). This is equivalent to holding C_i/C_s constant on the condition that humidity is not changing. In this kind of model C_i is determined by C_s and *LAVPD* and therefore if C_s and *LAVPD* are given as input factors, C_i can be obtained directly. Consequently in the biochemical model of photosynthesis, P_N can be determined when T_l, irradiance and C_i are known and g_s can be calculated by the stomatal model.

In this calculation, after g_b is incorporated into the combined model C_s and e_s are no longer taken as constant and are therefore not used as constant inputs into the model. Rather they are treated as variables controlled by physiological and ecological factors. For example, when irradiance changes the change in g_s will induce changes in C_s and e_s. Consequently C_i is not solely dependent on *LAVPD* as inferred from previous models but changes also with irradiance. The combined model also assumes that C_i/C_s is proportional to *LAVPD* by taking g_b into account and *LAVPD* will change with other environmental factors, such as irradiance. C_i also varies with irradiance, as is regularly found experimentally when other factors are kept constant. This significantly improves the performance of the model.

In determining the mechanisms of stomatal responses (Chapters 2 and 11), it is advisable to construct a dynamic model with experiments on a time-scale of minutes. The time courses of changes in P_n, C_i and g_s are different under different levels of CO_2 concentration, irradiance and *LAVPD*, because other physiological processes act as feedbacks.

13.6 Environmental Controls of Diurnal Variation of Photosynthesis, Transpiration and Stomatal Conductance

In the previous section, the response of P_n, g_s and E, to changes in single factors such as irradiance, temperature, or *LAVPD*, were simulated. This provides a sound understanding of plant responses to specific factors. In this section modelling field behaviours of P_n, g_s and E, in response to interactive effects of solar radiation, temperature and humidity, over the daily time-course are discussed. For convenience, the dependence of P_n, E and g_s, on diurnal courses of meteorological variables under a steady state are simulated and plant leaf status is kept unchanged.

13.6.1 Physiological Responses to Diurnal Variation of Environmental Factors

Figure 13.8 shows the diurnal course of changes in solar radiation and air temperature under a typical cloudless day at middle latitudes. Diurnal changes in solar radiation are the same as those of global radiation, being symmetric with respect to solar noon. Air temperature increases with solar radiation after sunrise with a maximum occurring slightly later than that of solar radiation. Air water vapour pressure is constant and vapour pressure deficit (*VPD*) increases with leaf temperature exponentially.

13.6.2 The Influence of Boundary Layer Conductance

Boundary layer conductance over the leaf surface is determined mainly by the thickness of the boundary layer. For a given species over a short time frame (daily), leaf shape and size can be considered constant and g_b is determined mostly by wind speed. A change in g_b may cause a change in concentrations of water vapour and CO_2 immediately over

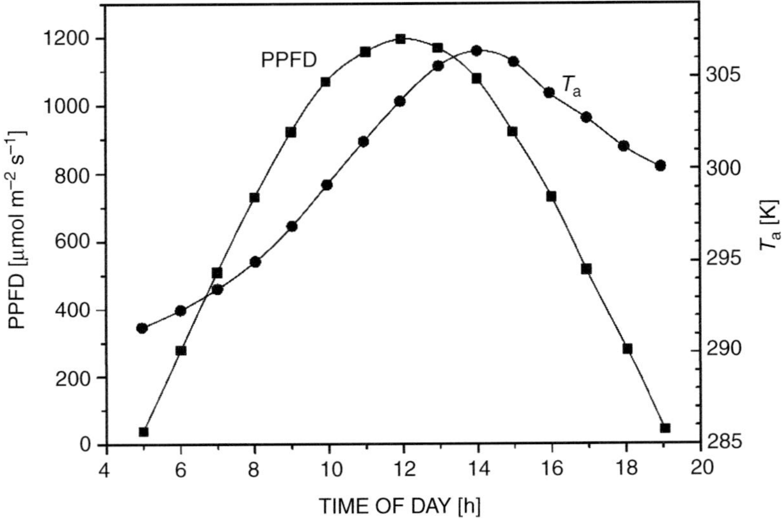

Figure 13.8. Diurnal variations of the photon flux density and air temperature used in subsequent simulations in Figures 13.9 to 13.12. Redrawn from Yu *et al.* (1998).

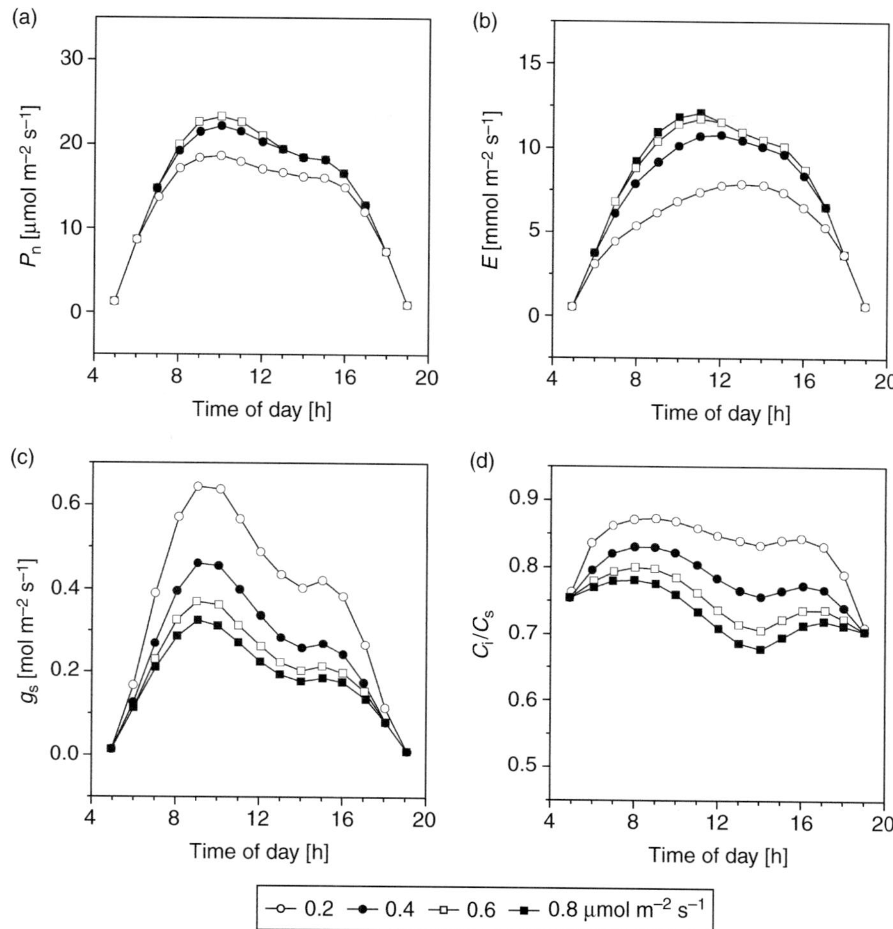

Figure 13.9. The influence of boundary layer conductance on diurnal changes in photosynthesis, transpiration and stomatal conductance (daily maximum temperature is 35°C and solar radiation is 1300 μmol m^{-2} s^{-1}). Different symbols refer to different values of g_b. Adapted from Yu *et al.* (1998).

the leaf surface and this leads to a series of physiological response. Figure 13.9 shows the physiological response to different values of g_b. It is seen that midday depressions occur for all three variables of P_n, E, and g_s, but the time of occurrence and the amplitudes of the depression differ. The decrease in g_s begins first and ends last and has the largest amplitude. The depression of E has the smallest extent and lasts for the shortest duration. P_n and E increase with increases in g_b, while g_s decreases because of its sensitivity to CO_2 concentration and both c_s and c_i and are raised by increase in g_b. This is particularly evident when g_b is small, which is lower enough to limit gaseous exchanges.

13.6.3 Vapour Pressure Deficit (VPD)

In this simulation, atmospheric water vapour pressure (e_a) does not change during the day. With different values given for e_a there will be different trajectories for *VPD*.

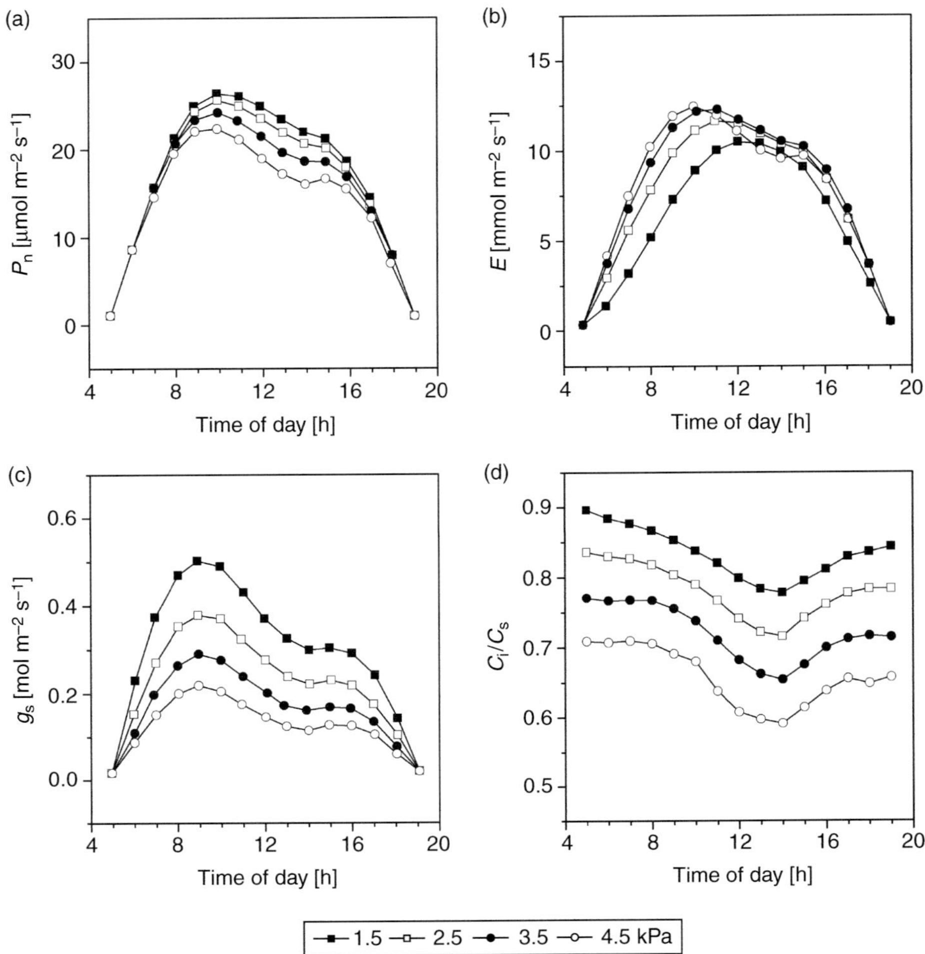

Figure 13.10. The influence of vapour pressure deficit (*VPD*) on diurnal variations of photosynthesis, transpiration and stomatal conductance (daily maximum temperature 35°C and solar radiation is 1200 μmol m^{-2} s^{-1}). Different symbols correspond to different values of *VPD*. Adapted from Yu *et al.* (1998).

Vapour pressure deficit is identical to *LAVPD* when leaf and air temperatures are taken as the same. Figure 13.10 shows the changes in key physiological variables for the plant in a day under different conditions of *VPD*.

When *VPD* is large, photosynthesis tends to be reduced and the midday depression is larger than when *VPD* is small. Photosynthesis does not vary much under different *VPD* in the morning when irradiance is low to moderate. Stomatal conductance decreases substantially with increased *VPD* and transpiration increases with increased *VPD* for the low to moderate range of *VPD* but when *VPD* exceeds a certain limit, further increases in *VPD* cause E to decline because g_s declines substantially at larger values of *VPD* (Chapter 2). This three-phase response of g_s to increasing *VPD* is discussed in Chapter 2.

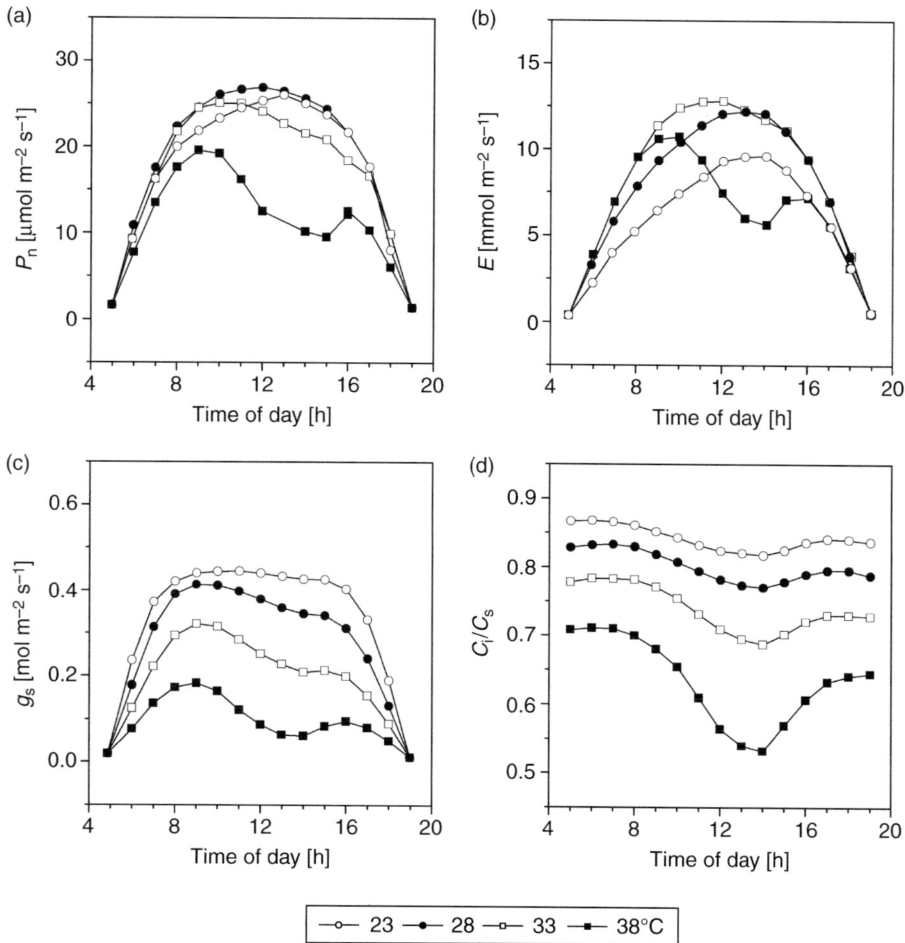

Figure 13.11. The influence of air temperature on diurnal variations of photosynthesis, transpiration and stomatal conductance (e_a = 1000Pa and daily I_{max} = 1200 μmol m^{-2} s^{-1}). Adapted from Yu *et al.* (1998).

13.6.4 The Influence of Air Temperature

Temperature affects photosynthesis in two ways: first, through its influence on rates of biochemical processes of photosynthesis; and, second, through its impact on *LAVPD* through its effect on the saturated intercellular vapour pressure. As previously discussed, photosynthesis decreases with increased *LAVPD* because of a reduction in stomatal aperture.

The ratio C_i/C_a is determined mainly by the changes in g_{sc} caused by changes in *LAVPD;* consequently an increase in temperature causes a monotonic decrease in C_i. When atmospheric water vapour pressure is held constant and air temperature is moderate (in Fig. 13.11a, T_{max} = 23°C and 28°C), P_n does not decrease at midday, nor does it increase significantly with an increase in irradiance. This is because light is already saturating for P_n. Under high temperature conditions, (e.g. $T_{a,max}$ = 33°C

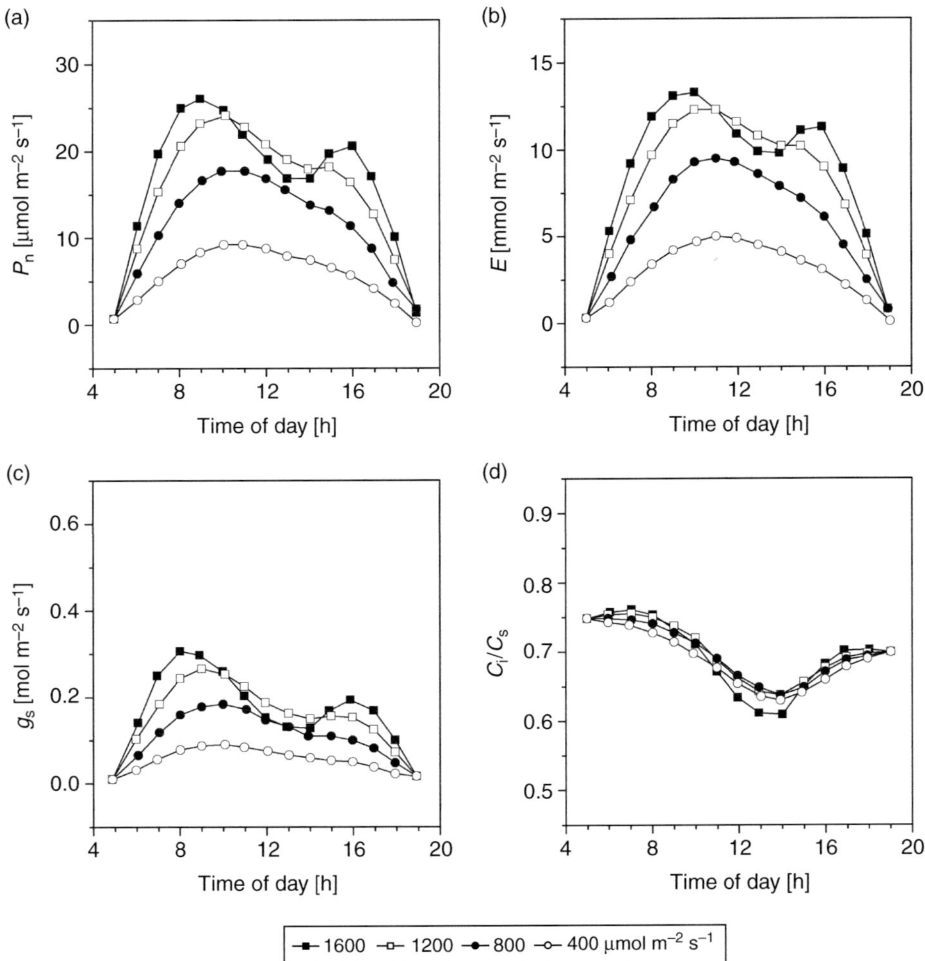

Figure 13.12. The influence of solar radiation on the diurnal variation of photosynthesis, transpiration and stomatal conductance (e_a is 1 kPa and daily maximum temperature is 35°C). Adapted from Yu *et al.* (1998).

in Fig. 13.11), midday depression appears and when temperature is very high, (e.g.
$T_{a,max}$ = 38°C), midday depression becomes very pronounced. The forms of diurnal
changes in g_s are similar to those in P_n (Fig. 13.10). The decrease in transpiration at
midday is slight, because of the increase in *LAVPD* (Fig. 13.11c). When temperature
is low ($T_{a,max}$ = 23°C and 28°C), transpiration changes with temperature and E occurs
at about the same as that of temperature and *LAVPD*. At higher temperature, with
$T_{a,max}$ = 33°C, maximal E occurs around noon and is low when *LAVPD* is the maximum in afternoon (14:00 h). When air temperatures are much higher (e.g. $T_{a,max}$ =
38°C), transpiration decreases and its amplitude is smaller than that of g_s because
LAVPD increases with an increase in leaf temperature. The ratio C_i/C_a decreases
with stomatal closure.

13.6.5 The Influence of Solar Radiation

Solar radiation received by the leaf is the energy source of photosynthesis and also the major determinant of a leaf's energy balance. When air temperature is far lower than optimal for growth and photosynthesis, increases in solar radiation result in increased leaf temperatures and increased rates of photosynthesis, assuming no other factor is limiting photosynthesis (for example frozen or very dry soil [Chapters 14 and 15]). Around noon on clear days solar radiation supply is non-limiting to photosynthesis and any further increase will not affect P_n significantly. If air temperature achieves or surpasses the temperature optimum P_n can decline with increasing radiation input. In Figure 13.12, under higher solar radiation P_n is reduced because of this effect. The pattern of diurnal variance of g_s is similar to that of P_n except that its midday depression lasts longer and falls more steeply. The change in E with solar radiation is mainly determined by g_s. Under high light intensities E is high in the morning and late afternoon when g_s is high; in early afternoon E decreases because g_s decreases with increases in irradiance, although this is partly compensated by an increase in *VPD*.

13.7 References

Aphalo PJ and PG Jarvis, (1993). The boundary-layer and the apparent responses of stomatal conductance to wind-speed and to the mole fractions of CO_2 and water-vapour in the air. *Plant Cell and Environment* 16(7), 771–783.

Ball JT, IE Woodrow and JA Berry, (1987). A model predicting stomatal conductance and its contribution to the control of photosynthesis under different environmental conditions. In: J Biggins (Editor), Progress in Photosynthesis Research. Martinus Nijhoff Publishers, Dordrecht–Boston–Lancaster, pp. 221–224.

Barton CVM, RA Duursma, BE Medlyn, DS Ellsworth, D Eamus, DT Tissue, MA Adams, J Conroy, KY Crous, M Liberloo, M Low, S Linder and RE McMurtrie, (2012). Effects of elevated atmospheric [CO_2] on istantaneous transpiration efficiency at leaf and canopy scales in Eucalyptus saligna. *Global Change Biology* 18(2), 585–595.

Collatz GJ, Ball JT, Grivet C and Berry JA, (1991). Physiological and environmental-regulation of stomatal conductance, photosynthesis and transpiration–a model that includes a laminar boundary-layer. *Agricultural and Forest Meteorology* 54(2–4), 107–136.

Collatz GJ, Ribas-Carbo M and Berry JA, (1992). Coupled photosynthesis-stomatal conductance model for leaves of C4 plants. *Australian Journal of Plant Physiology* 19(5), 519–538.

Comstock J and JE Ehleringer, (1993). Stomatal response to humidity in common bean (*Phaseolus vulgaris*): implications for maximum transpiration rate, water-use efficiency and productivity. *Australian Journal of Plant Physiology* 20, 669–691.

Dubbe DR, GD Farquhar and K Raschke, (1978). Effect of abscisic-acid on gain of feedback loop Involving carbon-dioxide and stomata. *Plant Physiology* 62(3), 413–417.

Eamus D, (1991). The Interaction of rising CO_2 and temperatures with water-use efficiency. *Plant Cell and Environment* 14(8), 843–852.

Eamus D and PG Jarvis, (1989). The direct effects of increase in the global atmospheric CO_2 concentration on natural and commercial temperate trees and forests. *Advances in Ecological Research* 19, 1–55.

Farquhar GD, SV Caemmerer and JA Berry, (1980). A biochemical-model of photosynthetic CO_2 assimilation in leaves of C-3 species. *Planta* 149(1), 78–90.

Harley PC, RB Thomas, JF Reynolds and BR Strain, (1992). Modeling photosynthesis of cotton grown in elevated CO_2. *Plant Cell and Environment* 15(3), 271–282.

Jarvis PG, 1980 Stomatal response to water stress in conifers. In: N.C. Turner and P.J. Kramer (Editors), *Adaptation of plants to water and high temperature stress*. Wiley-Interscience, New York, pp. 105–122.

Leuning R, (1990). Modeling stomatal behavior and photosynthesis of *Eucalyptus grandis*. *Australian Journal of Plant Physiology* 17(2), 159–175.

Leuning R, (1995). A critical-appraisal of a combined stomatal-photosynthesis model for C-3 plants. *Plant Cell and Environment* 18(4), 339–355.

Morison JIL, (1987). Climatology–plant-growth and CO_2 history. *Nature* 327(6123), 560–560.

Morison JIL and RM Gifford, (1983). Stomatal sensitivity to carbon-dioxide and humidity–a comparison of 2 C-3 and 2 C-4 grass species. *Plant Physiology* 71(4), 789–796.

Morison JIL and RM Gifford, (1984). Plant-growth and water-use with limited water-supply in high CO_2 concentrations. 2. Plant dry-weight, partitioning and water-use efficiency. *Australian Journal of Plant Physiology* 11(5), 375–384.

Mott KA, (1990). Sensing of atmospheric CO_2 by plants. *Plant Cell and Environment* 13(7), 731–737.

Paw U KT, (1987). Mathematical analysis of the operative temperature and energy budget. *Journal of Thermal Biology* 12 (3) 227–233.

Raschke K, (1976). How stomata resolve dilemma of opposing priorities. *Philosophical Transactions of the Royal Society of London Series B-Biological Sciences* 273(927), 551–560.

Schuepp PH, (1993). Tansley review No. 59 Leaf boundary-layers. *New Phytologist* 125(3), 477–507.

Schulze ED, RH Robichaux, J Grace, PW Rundel and JR Ehleringer, (1987). Plant water-balance. *Bioscience* 37(1), 30–37.

Sharkey TD, (1987). Stomatal responses to light. In: E. Zeiger, G.D. Farquhar and I.R. Cowan (Editors), *Stomatal function*. Stanford University Press, Stanford.

Sharkey TD and K Raschke K, (1981). Separation and measurement of direct and indirect effects of light on stomata. *Plant Physiology* 68(1), 33–40.

Shi JZ and TD Wang, (1994). Experimental study and mathematical sinmulation of water use efficiency in wheat leaves affected by some environmental factors. *Acta Botanica Sinca* 36, 940–946 (in chinese).

Tracy CR, FH Vanberkum, JS Tsuji, RD Stevenson, JA Nelson, BM Barnes and RB Huey, (1984). Errors resulting from linear-approximations in energy-balance equations. *Journal of Thermal Biology* 9(4), 261–264.

Vogel CA, DD Baldocchi, AK Luhar and KS Rao, (1995). A comparison of a hierarchy of models for determining energy-balance components over vegetation canopies. *Journal of Applied Meteorology* 34(10), 2182–2196.

Wong SC, IR Cowan and GD Farquhar, (1979). Stomatal conductance correlates with photosynthetic capacity. *Nature* 282(5737), 424–426.

Wong SC, IR Cowan and GD Farquhar, (1985a). Leaf conductance in relation to rate of CO_2 assimilation .1. Influence of nitrogen nutrition, phosphorus-nutrition, photon flux-density and ambient partial-pressure of CO_2 during ontogeny. *Plant Physiology* 78(4), 821–825.

Wong SC, IR Cowan and GD Farquhar, (1985b). Leaf conductance in relation to rate of CO_2 assimilation. 2: Effects of short-term exposures to different photon flux densities. *Plant Physiology* 78(4), 826–829.

Wong SC, IR Cowan and GD Farquhar, (1985c). Leaf conductance in relation to rate of CO_2 assimilation. 3: Influences of water-stress and photoinhibition. *Plant Physiology* 78(4), 830–834.

Yu Q, YF Liu, JD Liu and TD Wang, (2002). Simulation of leaf photosynthesis of winter wheat on Tibetan Plateau and in North China Plain. *Ecological Modelling* 155(2–3), 205–216.

Yu Q, BH Ren, TD Wang, SF Sun, (1998). A simulation of diurnal variations of photosynthesis of C3 plant leaves. *Chinese Journal of Atmospheric Sciences* 22(3): 256–268.

Yu Q and TD Wang, (1998). Simulation of the physiological responses of C-3 plant leaves to environmental factors by a model which combines stomatal conductance, photosynthesis and transpiration. *Acta Botanica Sinica* 40(8), 740–754.

Section Four

Case Studies

14

Boreal Forests

14.1 Introduction

Boreal (meaning northern) forests occupy a circumpolar belt (Fig. 14.1) in the northern hemisphere. They are mostly dominated by evergreen needle-leaved species, including pines (*Pinus*), spruce (*Picea*), and fir (*Abies*), but also the deciduous conifer larch (*Larix*), and cover about 17 percent of the earth's land surface. They are a dominant land cover in Russia, Canada and northern Europe. In Russia boreal forests are called Taiga.

An example of maximum, minimum, and mean temperatures, day length, and total precipitation (rainfall plus snowfall) for a Canadian boreal forest is given in Figure 14.2. Because boreal forests are found in high latitudes, winter temperatures are very low, summer temperatures are low-to-moderate, snowfall occurs every year, and day length varies substantially between seasons (Fig. 14.2). The growing season is confined to late spring, summer, and early autumn (100–200 days long). Where high elevation mountain ridges extend south from high latitude regions, boreal forest can also extend south–for example in New Jersey and the Appalachians in the United States. Boreal forests store approximately double the amount of carbon found in rainforests.

14.2 Coping with Freezing Winters: Photosynthetic C Gain and Transpiration

Consideration of the fluxes of C and water in boreal forests requires an understanding of how evergreen trees tolerate the stresses associated with very low winter temperatures. There are three major stresses that evergreen trees in boreal forests need to tolerate. The first is freezing temperatures and this stress induces two additional stresses: water stress and photo-inhibition.

When air and soil temperatures are sub-zero for prolonged periods, the potential for the formation of ice within cells is high. When water freezes it expands and this ruptures living cells and xylem. When the ice melts, the ruptured cell is dead (and the xylem is dysfunctional). Therefore intra-cellular freezing must be avoided in boreal

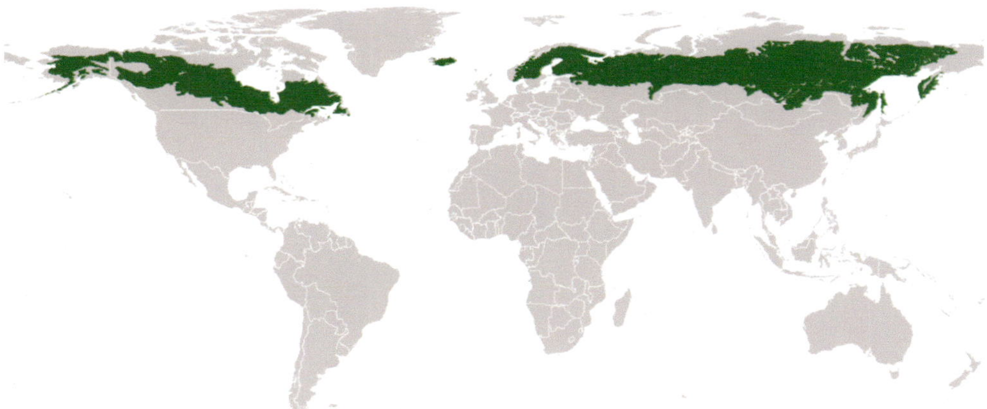

Figure 14.1. The global distribution of boreal and taiga forest (in green). From: http://upload.wikimedia.org/wikipedia/commons/5/5d/Taiga_ecoregion.png. Created by Mark Baldwin-Smith.

trees. As autumnal temperatures decline and day length shortens, evergreen trees undergo a process of hardening whereby the temperature that can induce freezing damage progressively declines (that is, frost hardiness increases). Accumulation of solutes within cells depresses the freezing point of intra-cellular water by several degrees but this is insufficient protection; more important are the poorly understood biochemical changes that prevent intra-cellular ice nucleation. By preventing ice nucleation cells can tolerate temperatures as low as −38°C, which is the spontaneous nucleation temperature of water. However, in some boreal forest regions winter temperatures much lower than −38°C occur. To tolerate these very low temperatures, living cells accumulate solutes in their apoplast and water is drawn out of the symplast into the apoplast. Only a thin, molecular-scale, layer of water plus associated cryo-protecting solutes is held tightly to the surface of proteins and membranes in the cell, and this layer doesn't freeze. In contrast water in the apoplast does freeze but induces no cellular damage. Thus the biochemical machinery within the cell is protected from freezing and these tissues can tolerate temperatures of −100°C or lower.

Freezing conditions induce a significant water stress in the canopy because of the relocation of intra-cellular water to the apoplast of needles. Frozen stems, roots, and the upper soil profile do not allow for the transport of liquid water and this further exacerbates the water stress experienced by the tree. However, the presence of the thin layer of liquid water and cryo-protectants around proteins and membranes prevents fatal dehydration. It is important to note that for every 1°C decline in ice temperature of the apoplast, the water potential of water vapour within needle air spaces decline by 1.2 MPa, and this imposes a significant threat of dehydration to any unfrozen cells nearby.

Water in the xylem, if frozen, will expand and rupture xylem cells. Consequently sapwood of boreal trees must be protected from freezing to the same degree as living cells in leaves and needles. At sub-zero temperatures water in xylem is metastable in

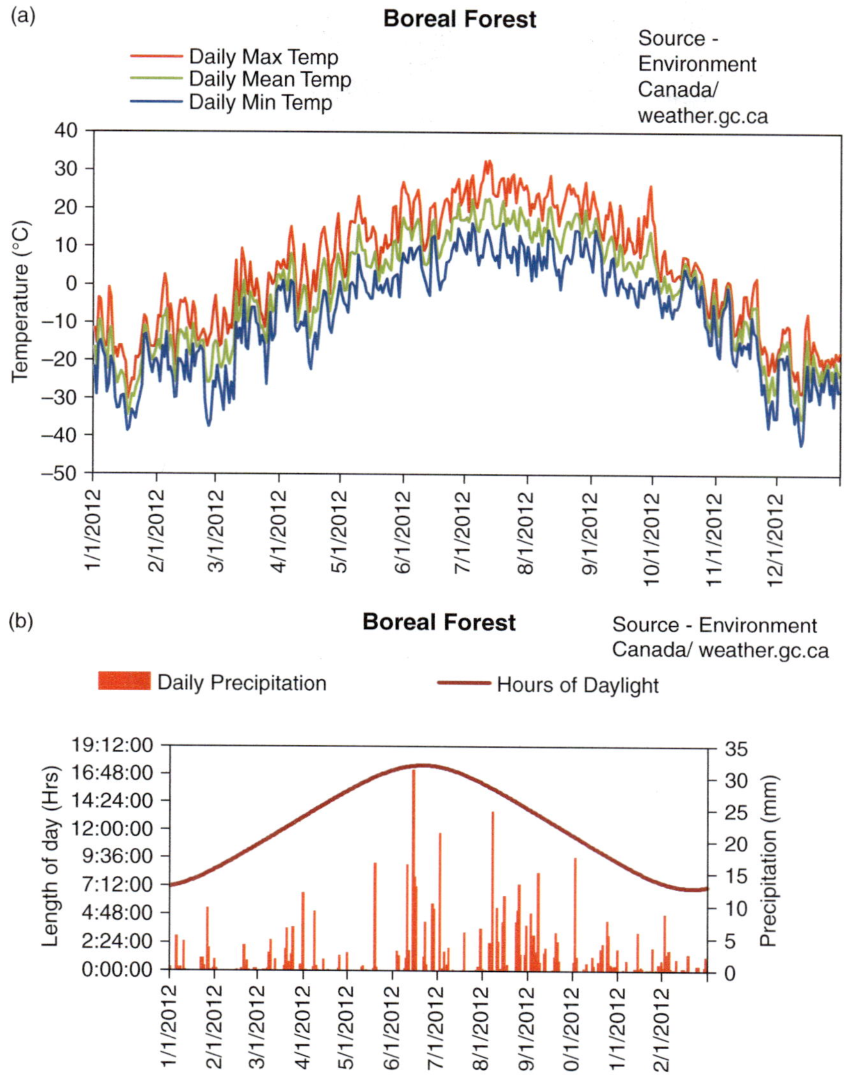

Figure 14.2. a) Maximum, minimum and mean temperatures and (b) day length and rainfall for a southern site in the boreal forest of Canada.

two directions: it is supercooled (and hence its temperature is below the ice nucleation temperature) but also it is under tension (and hence the pressure in the liquid water column is lower than the saturation vapour pressure). This metastable condition is tenable only if nucleation sites (such as gas bubbles) are absent.

It has recently been shown (Lintunen *et al.* 2013) that the temperature at which ice nucleation occurs in xylem increases as the conduit (vessel or tracheid; Chapter 2) radius increases (Fig. 14.3). Importantly, the narrowest conduit radii are found in conifers, conferring increased frost resistance and thereby allowing their dominance in boreal forests.

Plate 14.1. A Boreal forest of northern Europe. Note the presence of both evergreen and deciduous species.

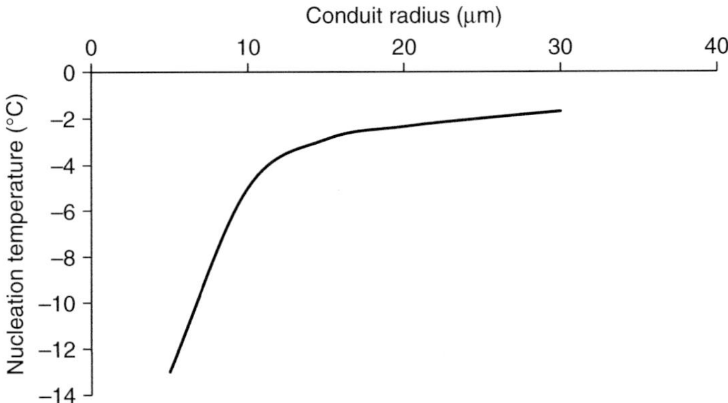

Figure 14.3. As conduit radius increases nucleation temperature of xylem water increases. From Lintunen *et al.* (2013).

How can the correlation between conduit radius and nucleation temperature be explained? The maximum supercooling temperature attainable for water is inversely proportional to the size of the particle acting as a nucleus for ice formation. Similarly it has been mathematically and empirically shown that the inter-conduit pit pores are effective filters for any particles or gas bubbles and prevent anything larger than a few nanometers from moving between conduits. As particle radius increases (2–20 nm) the ice nucleation temperature increases curvilinearly, in the same way that nucleation temperature increases with conduit radius. As conduit radius increases, pit pore radius increases. Thus the curvilinear relationship between maximum supercooling temperature and conduit radius is a function of the radius of the pit pores in the conduits. Maintaining liquid water in xylem in winter has the important benefit of allowing for the potential of supply to the canopy on warm sunny days. This is discussed in Section 14.2.1.

A third stress to which boreal conifers are exposed is photo-inhibition. This occurs when high light levels and low temperatures are combined. This is discussed after we have discussed photosynthesis of boreal trees more generally.

14.2.1 Can Conifers Photosynthesise during Winter?

One of the benefits of not having frozen xylem is the ability to conduct to the canopy water that was lost during transpiration during brief, warm, sunlit periods in winter. Despite air temperatures declining to $< -15°C$ every night, Owston *et al.* (1972) demonstrated water flow in stems of *Pinus* and *Abies spp*, but such transpirational water flow is only beneficial if photosynthesis occurs. But can boreal forest trees photosynthesise in winter when average air temperatures are sub-zero? Down regulation of photosynthesis in winter is generally observed in boreal forests (see below) but up-regulation in response to brief warm periods can also occur within a couple of days. What does this mean for canopy C uptake in winter in boreal forests?

Sevanto *et al.* (2006) used eddy covariance (EC) and leaf-scale measurements of photosynthetic C uptake during winter for a Scots pine (*Pinus sylvestris*) boreal forest in Finland. Mean air temperatures were sub-zero from approximately Julian day 300 to Julian day 80 of the following year. Four warm periods (on days 299–304, 48–50, 70–88, and 82–89) in the winter of 2002/2003 were identified when a positive NEE flux at night was followed by a negative NEE flux in the day; that is, photosynthesis was occurring and detected at the canopy scale with the EC system. Confirmation of photosynthetic carbon uptake was provided by use of leaf-scale gas flux analysers. The authors found that photosynthesis generally occurred when air temperature exceeds 3°C, even when temperature the day prior was −13°C (and night-time temperature even lower). More surprisingly photosynthesis always continued until air temperature was $< -7°C$. Stem conductance of water always lagged behind photosynthesis and this presumably reflects: (a) the low rate of transpiration occurring (because of low temperatures and low light levels); and, (b) a likely contribution of stored stem water, especially in the first few days of photosynthesis in winter. Stem freezing and thawing during cold (sub-zero) and warmer periods in winter were apparent. Similarly, photosynthetic C gain in black spruce (*Picea* spp.) growing in Alaska can occur from March onwards, despite the ground being covered in snow and soil temperatures below freezing (Ueyama *et al.* 2006; see later for discussion of sites with permafrost).

In contrast to the results of Sevanto *et al.* (2006), Goodine *et al.* (2008) observed that photosynthesis of Balsam fir (*Abies balsamea*) in Canada never occurred if soil temperatures were < 0°C throughout the root zone even if air temperature was as high as 10°C.

Thus the answer to the question: 'Can boreal forest trees photosynthesise during winter?' is: it probably depends upon the species and location in question. The *Abies* study (Goodine *et al.* 2008) showed that liquid water is required in the soil profile and that if the root zone remains frozen, warm canopy temperatures do not induce photosynthesis, while for the Scots pine study, stem capacitance is able to supply sufficient water for short bursts of photosynthesis on warm winter days after a suitable period of up-regulation of photosynthesis in response to increased air temperature. But what is up-regulation of photosynthesis and why is it needed during winter in boreal forest species?

14.2.2 *Photo-Protection of Chloroplasts During Winter in Boreal Forests*

Cold canopy temperatures inhibit enzyme activity, including enzymes of the Calvin-Benson cycle and this inhibits photosynthesis. Similarly, frozen soils prevent water uptake and transport to the canopy. However, the dark green needles of conifers and the snow-shedding structure of conical canopies means that light interception by chloroplasts occurs during winter days. Any solar energy absorbed by chloroplasts that is not channelled into photosynthetic processes has the potential to damage chloroplasts. This is because unused solar energy can reach oxygen molecules to produce highly reactive oxygen molecules ($O_2^{\cdot}$) which can damage membranes. To prevent this photo-protective mechanisms have evolved, including the presence of many enzymes capable of neutralising reactive oxygen molecules. For example, superoxide dismutase is an enzyme that combines O_2 and reactive $O_2^{\cdot}$ to produce hydrogen peroxide which is then broken down to release water and oxygen (Adams *et al.* 2004).

An additional photo-protective mechanism commonly used by evergreen boreal forest species involves thermal dissipation to rapidly de-excite excited pigments (Chapter 2). It is likely that 90–100 percent of absorbed photosynthetically-active radiation is lost as thermal energy in boreal forests. Thermal dissipation involves the xanthophyll cycle where zeaxanthin, antheraxanthin, and violaxanthin are inter-converted according to the degree of light and temperature stress experienced. This is discussed in more detail in Chapter 2. This allows for a rapid initiation of photosynthesis once conditions are more favourable (warmer canopy temperatures and a thawing soil profile, for example). Because the growing season can be short in many boreal regions, waiting for regrowth of the canopy, as observed in deciduous species, is probably not an evolutionary winning option.

14.3 Climate and Vegetation Interactions in a Canadian Boreal Forest: ET and WUE

Bruemmer *et al.* (2012) examined the effects of climatic factors and vegetation type (coniferous boreal forest and deciduous boreal forest) on evapotranspiration (*ET*) and water-use-efficiency using eddy covariance analyses across five boreal forest sites in Canada. In particular, they addressed the question: 'Is transpiration more strongly determined by net radiation (R_n) or vapour pressure deficit (*VPD*)?'

Despite low mean annual temperatures (Table 14.1) the growing season (defined as the number of days when net ecosystem productivity was larger than zero for 24 h averages) of each coniferous forest was at least 4 months long, and at one site it was almost 7 months long. In contrast the *deciduous* boreal forest growing season was only 3–4 months. Both the coniferous and deciduous boreal forests maintained a large leaf area index (LAI), although for the deciduous trees this was evident only during the summer. Because of the similarity in temperature, rainfall, LAI and canopy conductance values across the coniferous and deciduous forests, annual rates of *ET* were broadly comparable (Table 14.1) despite the shorter growing season of the deciduous forest.

Table 14.1. Range of climate and canopy variables across four coniferous boreal forests and one deciduous boreal forest in Canada

	Range for four boreal coniferous forests across 4 y study	Deciduous boreal forest across 4 y study
30 y mean annual air temp (°C)	0.4 to −2.6	0.5
30 y annual rainfall (mm)	448 – 971	403
Net radiation (GJ m^{-2} y^{-1})	1.85 – 2.82	1.72–1.94
Annual ET (mm)	239 – 325	262 – 452
Canopy conductance (mm s^{-1})	3.23 – 4.93	2.96–4.94
Decoupling coefficient	0.13	0.1–0.16
LAI / canopy height (m)	3.4 – 5.6 / 9 – 14	3.8 / 21
Length of growing season (days)	125 – 200	80–125

Source: From Bruemmer *et al.* (2012).

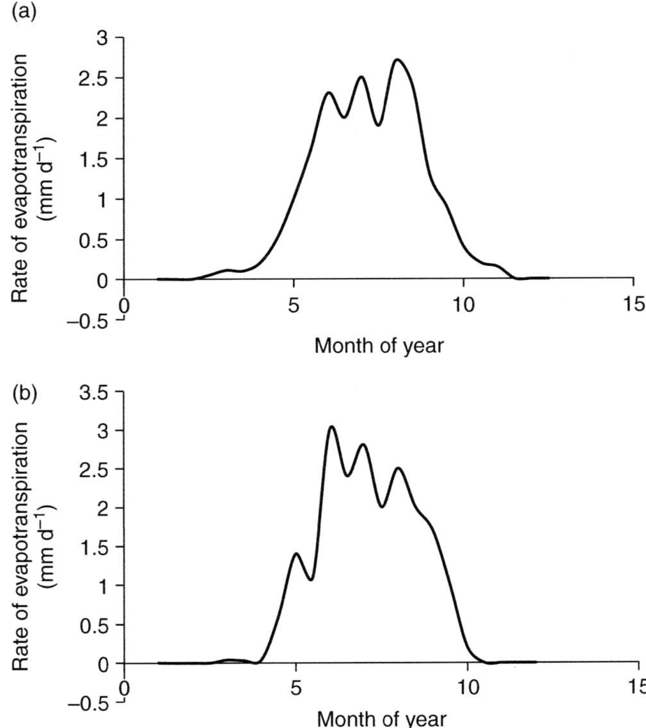

Figure 14.4. Seasonal patterns in rates of evapotranspiration for coniferous (upper panel) and deciduous (lower panel) boreal forests. Note the larger maximum rate for the deciduous forests but the shorter growing season. Redrawn from Bruemmer *et al.* (2012).

Rates of *ET* peaked in the summer for both coniferous boreal forest (CBF) and deciduous boreal forests (DBF) (Fig. 14.4). Typically the CBF exhibited maxima of about 2 mm d^{-1} although occasionally this increased to 3–4 mm d^{-1} for short periods. In contrast the DBF exhibited larger rates of *ET*, typically 3–4.5 mm d^{-1}, but for fewer days of each year (Fig. 14.4), consistent with the "live-fast, die-young" strategy of deciduous species. See Chapter 2 for further discussion of the cost-benefit analyses of deciduous *versus* evergreen phenologies.

Across all sites, rates of *ET* increased slightly with increasing rainfall (Fig. 14.5), with a slightly larger rate observed for the deciduous compared to the coniferous boreal forest. At each site, 80–90 percent of annual *ET* occurred during the growing season.

Monthly *ET* was significantly and linearly correlated with monthly mean *VPD* (Fig. 14.6) but the slope was larger for the DBF compared to the CBF. In contrast, monthly *ET* did not show a simple linear correlation with mean monthly R_n; a lag was observed between R_n and *ET* of nine to thirty-six days.

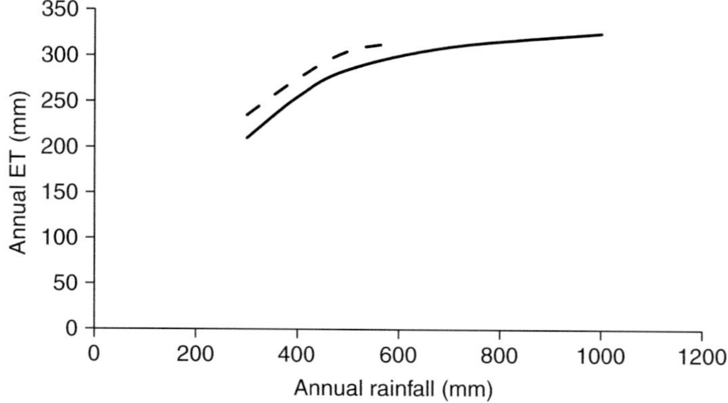

Figure 14.5. Annual ET increases with annual rainfall curvilinearly, with a slightly larger rate observed for the deciduous boreal forest (dashed line) than the coniferous boreal forest (solid line). Redrawn from Bruemmer *et al.* (2012).

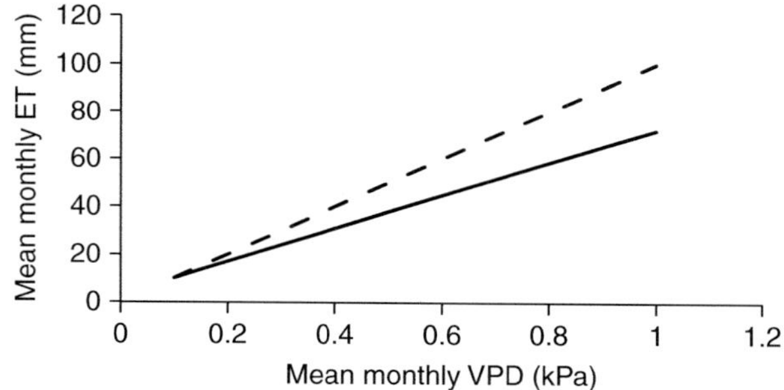

Figure 14.6. Mean monthly evapotranspiration is linearly correlated with mean monthly VPD for both coniferous (solid line) and deciduous (dashed line) boreal forests. Redrawn from Bruemmer *et al.* (2012).

This study demonstrated several key aspects of the patterns and regulation of *ET* by rainfall, length of growing season, and vegetation type, in boreal forests. Thus:

1. The seasonal pattern and maximum rates of *ET* differ between deciduous boreal and coniferous boreal forests, with larger maxima but a shorter growing season evident in deciduous boreal forests than coniferous boreal forests.
2. Annual total *ET* for both types of boreal forest was relatively insensitive to inter-annual variation in rainfall: two-fold inter-annual variation in rainfall resulted in annual *ET* variation of less than 50 percent. Across all five sites (four coniferous and one deciduous forest), annual *ET* ranged from 268–491 mm.
3. The high degree of coupling (low Ω) of both the coniferous and deciduous forests reflects the aerodynamically rough surface of boreal forests and the high linear correlation of *ET* with *VPD*. Thus variation in *VPD* is a major driver of variation in *ET* in these forests, and the radiative term in the Penman-Monteith equation was less important than the aerodynamic term in explaining *ET*.

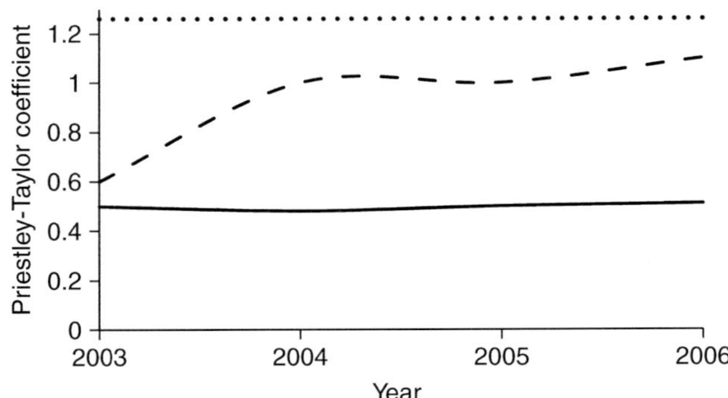

Figure 14.7. The Priestley-Taylor coefficient for coniferous (solid line) and broad-leaf (dashed lie) boreal forest. The dotted line at 1.26 is the theoretical maximum value obtained for well watered forests where radiation limits *ET*. Redrawn from Bruemmer *et al.* (2012).

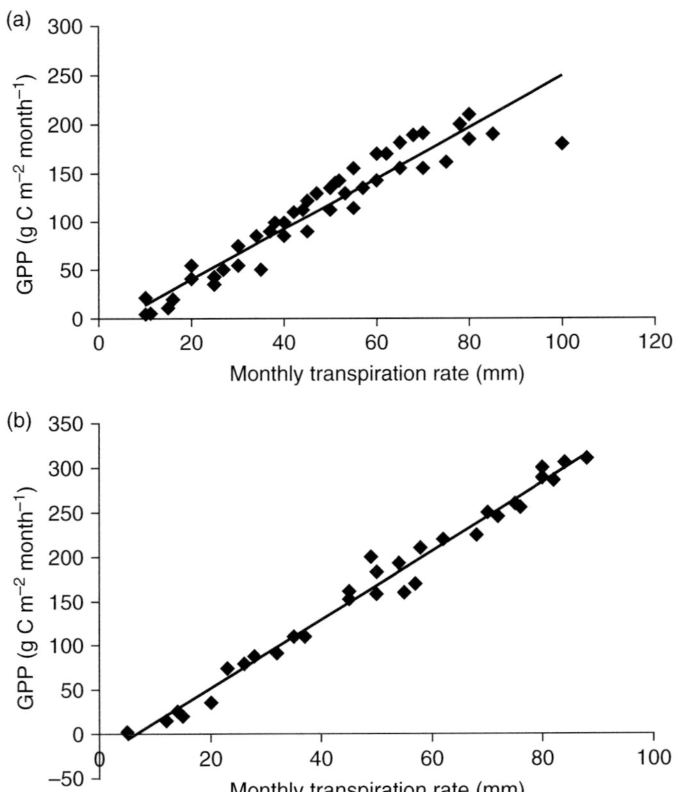

Figure 14.8. Strong linear correlations between GPP and transpiration for coniferous boreal forest (upper panel) and deciduous boreal forest (lower panel) show minimal intra- or inter-annual variation in WUE. Redrawn from Bruemmer *et al.* (2012).

4. Stomatal limitation of transpiration was common at the coniferous forest and the low Priestley-Taylor coefficient (α; Chapter 12) for these sites (Fig. 14.7) reflects the water-limited nature of *ET* in these forests. In contrast, the deciduous forests exhibited larger values of α, reflecting the reduced impact of water limitation. The increase in α after 2003 is the result of a breaking of the drought that occurred in 2003 at the deciduous forest site (Fig. 14.7).

5. The sensitivity of α to canopy conductance (g_c) at the deciduous forest was larger than that observed at the coniferous forest but no direct control of *ET*, α, or g_c by soil water content was evident.

6. Strong linear correlations between monthly GPP and monthly ET (Fig. 14.8) result in little variation in water-use-efficiency (WUE) across the year or between years. WUE was larger in the deciduous forest than the coniferous forest because deciduous leaves live by the "live-fast, die-young" philosophy (see cost-benefit analyses in Chapter 2) and have to photosynthesise at a fast rate for their short leaf lifespan.

14.4 Controls of *ET* and Carbon Flux in a Scots Pine (*Pinus sylverstris*) Forest

Using standard eddy covariance and micrometeorological methods, Zha *et al.* (2013) examined the rates and controls of C and water fluxes in the growing season of a Scots pine forest in Finland. Diurnal patterns in vapour pressure deficit (*VPD*), transpiration rate (*E*), photosynthetically active radiation (PAR), CO_2 uptake, canopy conductance (g_c), and the coupling factor (Ω), across three representative days in summer are shown in Figure 14.9. Variations in PAR observed around noon for two of the three days are reflected in corresponding changes in *E*, thereby highlighting the importance of energy supply in driving rates of transpiration. Similarly the shorter day length and lower maximum PAR apparent in September compared to May or July is reflected in the later onset of photosynthetic C uptake and the lower maximum rate of C uptake observed in September compared to May or July (Fig. 14.9). Rates of transpiration and CO_2 uptake (F_c) increased from the end of April through to the end of July (Fig. 14.10) as a result of increasing availability of PAR and increasing air and canopy temperatures. To support increasing rates of *E* and F_c in early summer, canopy conductance increased. The question: 'Which factors were most important in explaining observed seasonal patterns in *ET* and F_c?' is discussed in the next section.

14.4.1 Controls of Fluxes

Changes in *VPD* and canopy conductance accounted for 26 percent and 33 percent, respectively, of variation in *E*, while changes in PAR and air temperature accounted

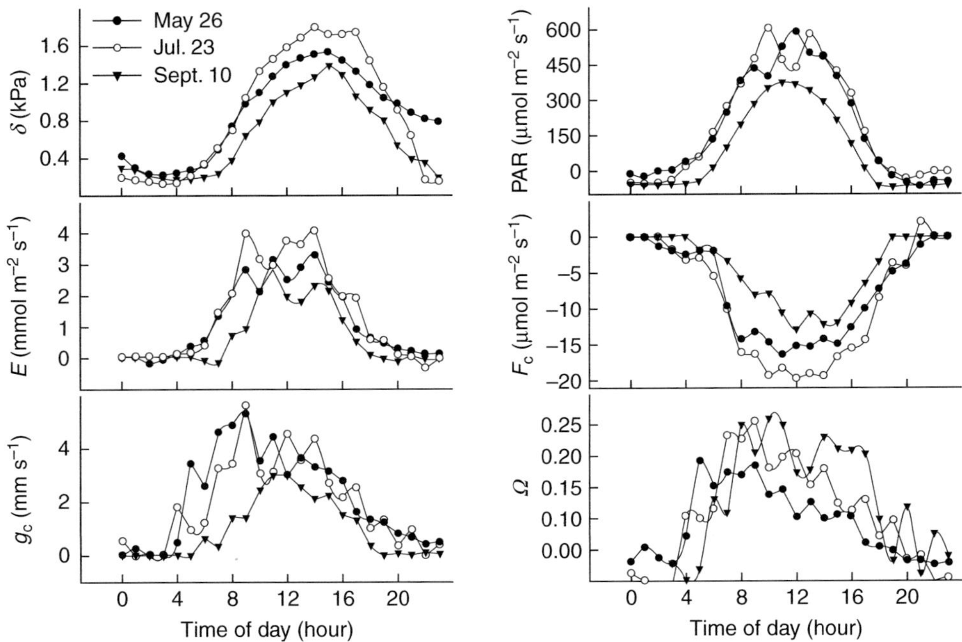

Figure 14.9. Diurnal changes in *VPD* (δ) transpiration rate (E), canopy conductance (g_c), photosynthetically active radiation (PAR), canopy CO_2 uptake (F_c) and coupling factor for a Scots pine forest in Finland. From Zha *et al.* (2013).

for 42 percent and 19 percent, respectively, of changes in E (Fig. 14.11). In total, 80 percent of variation in E was explained by variation in PAR, canopy conductance, *VPD*, and temperature, together. Similarly low coefficients of determination for the response of transpiration and *ET* to temperature, *VPD*, and net radiation, have been observed elsewhere for boreal forests (e.g. Ge *et al.* 2011). Very low dependence of *ET* on soil moisture content is commonly observed because soil water content is rarely (but see the discussion of permafrost below) limiting in these environments.

Canopy C uptake was regulated by canopy conductance and PAR, with each accounting for 25 percent and 36 percent, respectively (Fig. 14.12), of variation in CO_2 uptake. Canopy conductance was weakly dependent on *VPD* and this dependency increased with increasing PAR. The low values for the decoupling coefficient (Ω) show that the canopy is well coupled to the atmosphere (as is common for coniferous forests). Canopy conductance and the main abiotic factors of PAR, *VPD*, and temperature, together explained 80 percent of the variation in *ET*. This was a much larger fraction than that observed by environmental factors combined (48%) or separately (19–42%). Thus both biotic (g_c) and abiotic (PAR, *VPD*, temperature) factors control *ET* in this boreal forest. The importance of the role of g_c in regulating gas fluxes is underscored by the low values of Ω. However, Ω increased as the day progressed, indicating increasing stomatal limitations to gas flux throughout the day.

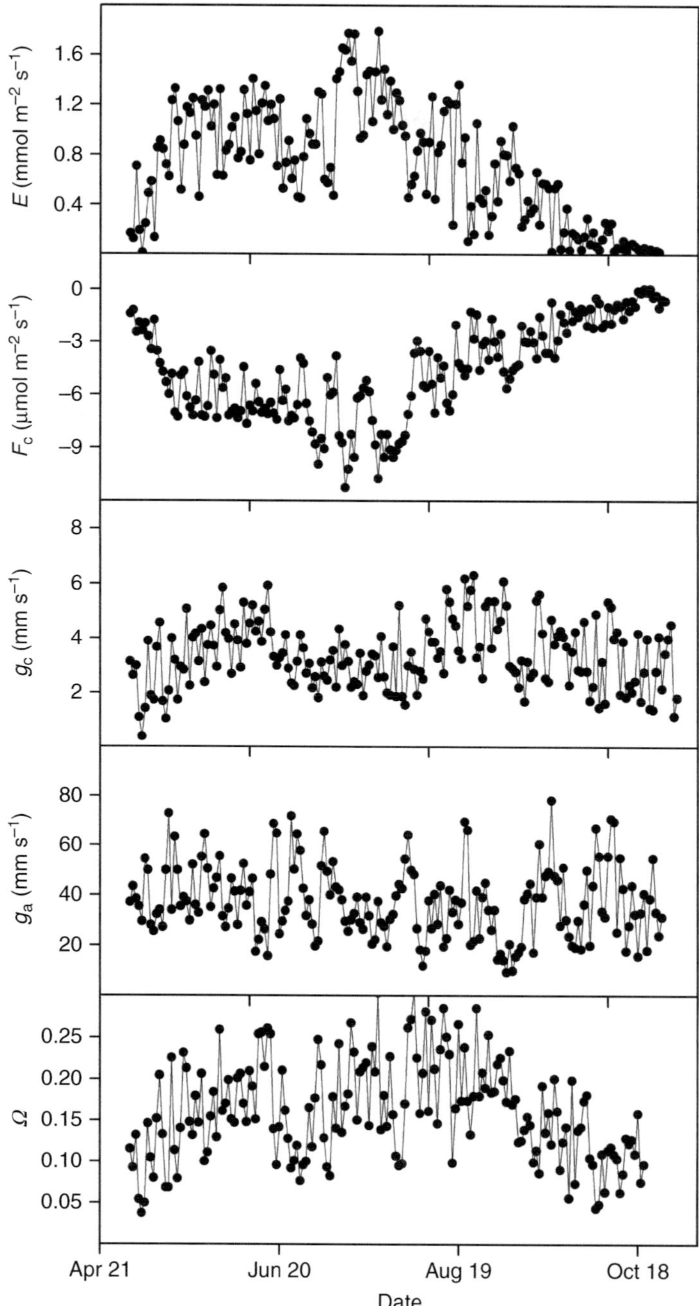

Figure 14.10. Seasonal trends in transpiration rate (E), CO_2 flux (F_c), canopy conductance (g_c), aerodynamic conductance (g_a) and the coupling coefficient (Ω) for a Finnish boreal forest. From Zha *et al.* (2013).

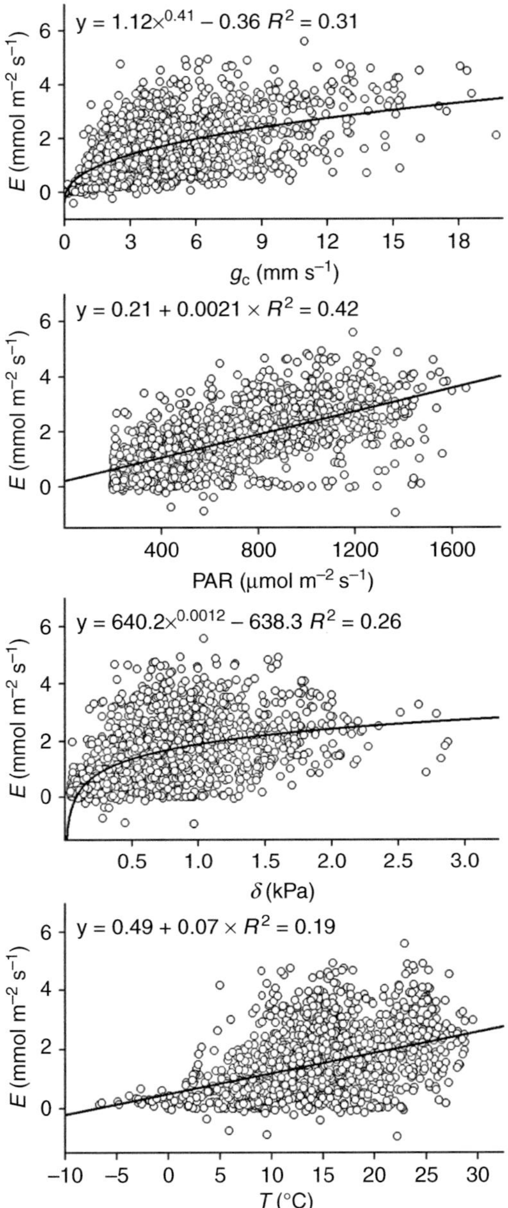

Figure 14.11. Control of evapotranspiration *(E)* by canopy conductance (g_c), PAR, vapour pressure deficit (δ), and temperature. From Zha *et al.* 2013).

14.5 Comparing Carbon Balances of Boreal Humid Evergreen Forests with Semi-Arid Boreal Forests

Boreal forests are characterised by freezing winter temperatures, low-to-moderate summer temperatures, a short day length in winter, and a long day length in summer (Fig. 14.2; Tables 14.1, 14.2). The relationship between gross primary productivity

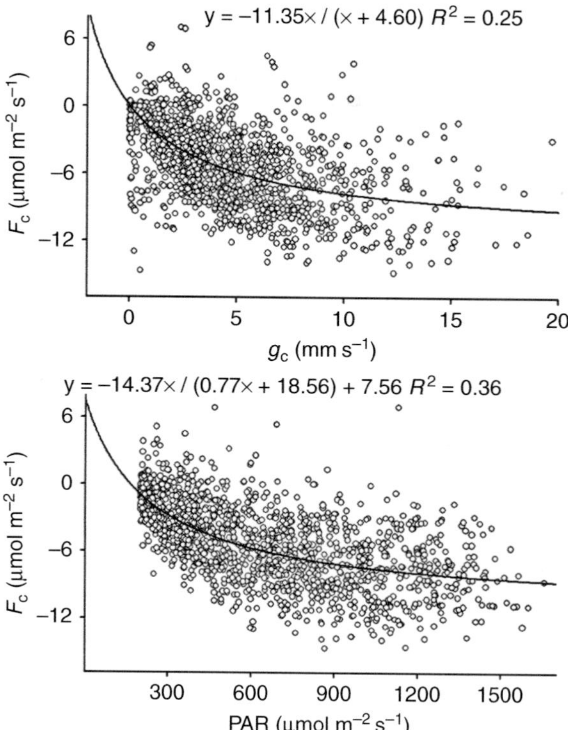

Figure 14.12. Control of CO_2 uptake (F_c) by canopy conductance and PAR. From Zha *et al.* (2013).

(GPP) and mean annual temperature for boreal forests is linear and moderately strong (Luyssaert *et al.* 2007) because temperature is a principle limiting factor to rates of enzymic reactions (such as photosynthesis) for the low-to-moderate temperatures experienced in boreal forests. In contrast the response of GPP to increased rainfall is modest because, in the absence of a sufficient number of warm sunny days, foliage has little opportunity to use the extra water to fix C.

Despite the colder winter and lower rainfall experienced in semi-arid deciduous boreal forests compared to humid evergreen boreal forest (Table 14.2), GPP and NPP for the former is significantly larger than for the latter (Table 14.3). This reflects the "live-fast, die-young" strategy of deciduous leaves (Eamus and Prichard 1998) compared to evergreen leaves (see Chapter 2 for a discussion of the relative costs/benefits of the evergreen *versus* deciduous strategies). This larger NPP is reflected in the larger standing biomass accumulated in boreal semi-arid deciduous forests compared to boreal humid evergreen or boreal semi-arid evergreen forests (Table 14.3). Despite the larger ecosystem respiration of the semi-arid deciduous forest (reflecting the larger GPP), the ratio of ecosystem respiration to GPP tends to be relatively consistent across all three types of boreal forests.

Table 14.2. Climate averages for humid boreal evergreen and semi-arid boreal evergreen and deciduous forests around the world

	Boreal humid evergreen	Boreal semi-arid evergreen	Boreal semi-arid deciduous
Mean winter temp (°C)	−9	−18	−20
Mean summer temp (°C)	13	13	13
Winter rainfall (mm)	295	52	47
Summer rainfall (mm)	144	183	156
Winter sum of net radiation (W m^{-2})	46	46	33
Summer sum of net radiation (W m^{-2})	216	359	348

Source: From Luyssaert *et al.* (2007).

Table 14.3. Mean C fluxes for GPP, NPP and R_e for humid and semi-arid boreal forest

	Boreal humid evergreen	Boreal semi-arid evergreen	Boreal semi-arid deciduous
LAI	4.1	3.4	3.5
Biomass (gC m^{-2})	7049	6370	8961
GPP	973	773	1201
NPP	271	334	539
R_e	824	734	1029
R_e/GPP	0.85	0.95	0.86

Source: From Luyssaert *et al.* (2007).

14.6 Permafrost, ET and NPP

Much of the boreal forest of the world is underlain by permafrost. Permafrost is soil and water that is frozen for more than two consecutive years and this tends to occur when average annual air temperature is less than −2°C. For the northern hemisphere about 24 percent of all ice-free land is influenced by permafrost. Sitting above permafrost is a layer of soil, typically 0.5–4 m thick, that thaws in the summer and this layer can support plant life, including trees. Plant roots cannot penetrate nor survive in frozen permafrost. Permafrosts store very large quantities of C (1400–1700 GtC; or twice the amount of C in the atmosphere) in the form of methane and peat and any large-scale thawing of permafrost therefore has the potential to release significant amounts of CO_2 and methane into the atmosphere.

Summer thawing of the upper soil profile above the permafrost releases water which cannot drain vertically and therefore saturation of the upper profile occurs, followed by run-off if slopes allow. However, the warmer temperatures and longer day lengths of summer, plus the pulsed release of melt water, is a resource than can be used by some tree species, for example Black Spruce (*Picea marinana*) in Alaska and larch forest in eastern Siberia (Sugimoto *et al.* 2002). Because of the limited depth of summer thaw and the propensity for saturation of soils with melt water, boreal forests on permafrost tend to be short in stature, often less than 4 m in height, but a significant understorey (both evergreen and deciduous) is often present which means that large seasonal increases in total LAI occur in the summer (e.g. Iwata *et al.* 2012).

Energy partitioning (between latent and sensible heat fluxes) is strongly influenced by the melting of snow and thawing of the upper soil profile in regions with permafrost. Incoming solar radiation is increasingly used to melt snow and the latent heat used for snowmelt is a significant fraction of the energy balance of these sites, accounting for up to 40 percent of incoming energy. Once snow melt has occurred much of the incoming radiation is partitioned into sensible heat flux, especially at the start of the growing season and the upper soil profile begins to melt. At this time soil available water content increases, root water uptake occurs, and *ET* begins so that the partitioning of energy in λE (that is, *ET*) increases. The Bowen ratio (the ratio of sensible to latent heat fluxes) decreases from early-to-late growing season as LAI increases, and then declines as LAI decreases in the late summer. Despite large variations in rainfall between years, *ET* tends to show only small inter-annual variation because *ET* is driven more by the water supplied as melt water during the summer, than rainfall. Towards the end of the summer soil water content tends to increase as melt water accumulates and this acts as a buffer for the supply of water in the following year (Iwata *et al.* 2012) after it thaws in the subsequent summer.

Although bulk canopy conductance declines with increased vapour pressure deficit, the magnitude of the decline is relatively small, reflecting the relatively small values of VPD experienced in these forests. While cumulative growing season *ET* tends not to be correlated with annual rainfall nor annual snowfall (which contributes to soil water content when it melts), cumulative *ET* is strongly correlated with net radiation in boreal forests underlain by permafrost (as it is in boreal forest that is not underlain by permafrost).

Three seasonal patterns in *ET* can be seen across the year for boreal coniferous forests underlain by permafrost (Fig. 14.13). Low, but positive rates of *ET* occur in winter, larger rates occur in the early growing season as leafing-out occurs, followed by the largest rates of ET in the warmest months when LAI in maximal (July–Sept). The slope of the relationships in Figure 14.13 increases through the year, presumably reflecting the influences of increasing temperature, *VPD*, and needle expansion from immature to mature stages.

Despite the presence of snow on the ground, frozen upper and lower soil profiles, average air temperatures being around $-10°C$, and minimal rainfall for the first two months of the year, photosynthesis in these permafrost boreal sites has been recorded from March onwards (Ueyama *et al.* 2006), with rates of C uptake increasing from early March through to about mid-July when LAI is maximal. Positive rates of C uptake are maintained through to the end of October (Fig. 14.14). These sites, despite being 120-year-old forests, maintained a positive annual C balance (they are a net C

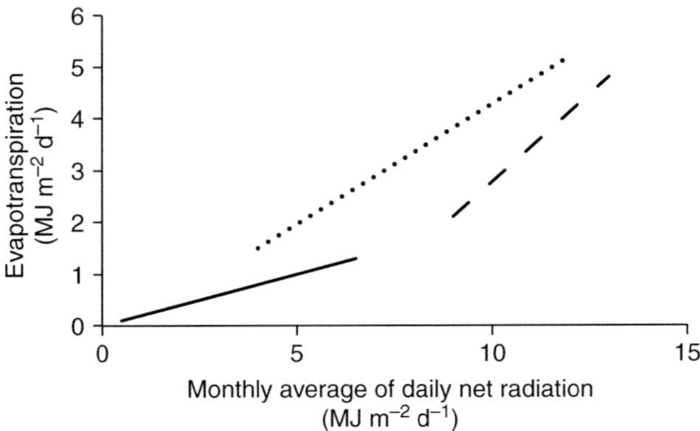

Figure 14.13. Changes in rates of evapotranspiration as a function of daily net radiation for three periods across the year: dormancy (Oct–April): solid black line; leafing out (May, June): dashed line; and summer (Jul–Sept): dotted line. Redrawn from Iwata *et al.* (2012).

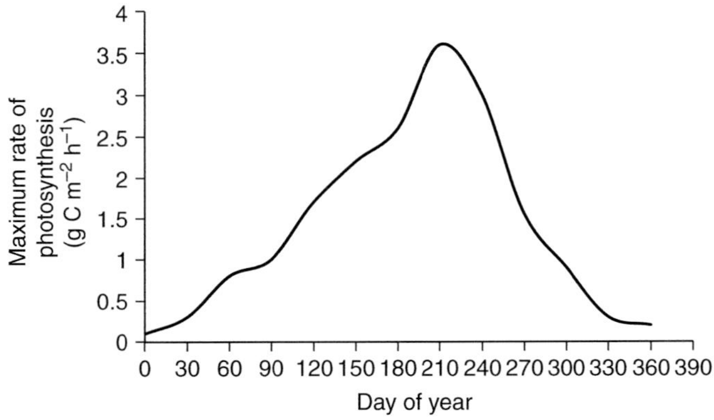

Figure 14.14. Changes in maximum rate of photosynthesis for a boreal forest on permafrost in Alaska. Redrawn from Ueyama *et al.* (2006).

sink). The observation of C uptake from March onwards supports the independent observation of positive *ET* for this period; that is, significant rates of transpiration occurring in March and April co-occurred with significant rates of C gain. A contribution of stored stem water on occasional warm early spring days is likely to support the low rates of *ET* and C gain observed from March onwards.

14.7 Modelling Controls of ET in a Scots Pine Boreal Forest

Several soil-vegetation-atmosphere-transfer models (SVATs) and soil-plant-atmosphere (SPA) models can be used to predict stand-scale water budgets of vegetated landscapes (Chapters 12 and 13), although few such models have been applied to boreal forests. However, Kellomaki and Wang (2000) and Wang *et al.* (2004) have parameterised a detailed process-based model for boreal Scots pine. More recently Ge *et al.*

(2011) have combined the Wang *et al.* (2004) hydrological model with a forest growth model to examine the relative contributions of leaf area index, rainfall, and temperature, to the control of rates of *ET* of a Scots pine boreal forest. The key features of the sub-routines within the SVAT and forest growth model are summarised thus:

1. Evapotranspiration was calculated as the sum of canopy transpiration, canopy evaporation, and evaporation from the ground surface at hourly or daily times-steps.
2. Net solar radiation for energy above the canopy was partitioned into net absorption by the canopy and the ground surface.
3. Tree transpiration was modelled using the multi-layer big-leaf model whereby the canopy was divided into sunlit and shaded fractions. Air temperatures and needle temperatures were assumed to be equal and the Penman–Monteith equation was used to calculate *ET* for sunlit and shaded fractions of the canopy. Needle conductance was similarly calculated for sunlit and shaded fractions.
4. The response of g_s to irradiance, *VPD*, soil moisture content, and temperature were modelled using a Jarvis-Stewart formulation (Chapter 2).
5. Canopy interception was modelled using the Penman-Monteith equation and an empirical relationship between LAI, rainfall, and interception rate.
6. Ground surface evaporation was calculated using a combination of empirical and physical approaches to solving the surface energy balance. The Penman-Monteith equation was used to calculate surface evaporation rates, given net radiation levels at the ground surface.
7. The forest growth model was a generic model of plant development requiring extensive site calibration and parameterisation. The biochemical model of Farquhar *et al.* (1980) was used to predict total C gain and the allocation of biomass amongst roots, stems, branches and leaves determined from allometric relationships established for Scots Pine at the study site.
8. Eddy covariance measurements were used to validate the model with a high degree of success.

The results of this modelling study demonstrated that:

1. Combining the hydrological process model with a forest growth model yielded significant improvements in understanding the temporal dynamics of the water balance of this boreal forest.
2. Close agreement was seen at monthly and annual time-scales for modelled and measured rates of *ET* (Fig. 14.15).
3. Partitioning the canopy into sunlit and shaded components was extremely important to improving the accuracy of the simulation.
4. The model was able to disaggregate total *ET* into component parts, something that eddy covariance tower measurements are unable to do.
5. The omission of understorey *ET* may have been a critical oversight to the model's structure.

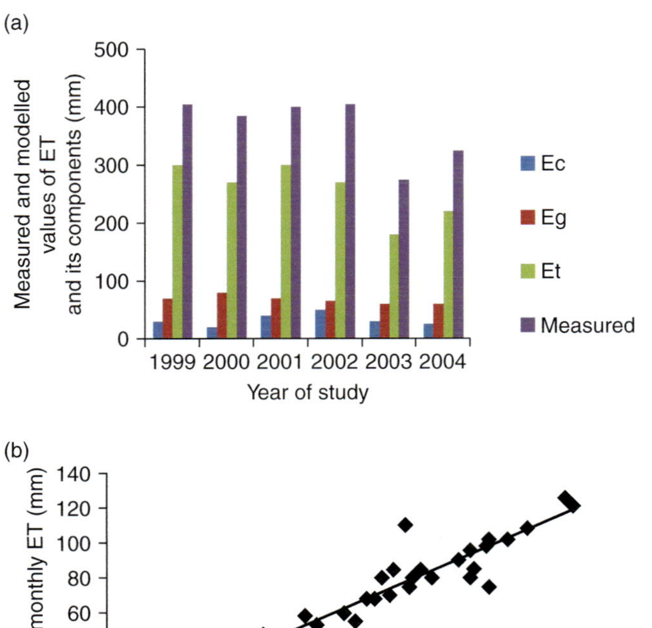

Figure 14.15. Measured and modelled rates of *ET* for a Scots Pine boreal forest in Finland shows good agreement at annual (upper panel) and monthly (lower panel) time-scales. The model was able to disaggregate total *ET* into component parts (c = canopy; g = ground: t = tree). Redrawn from Ge *et al.* (2011).

6. The fraction of total *ET* attributable to tree transpiration, canopy interception losses and ground surface evaporation was 67 percent, 11 percent, and 23 percent over the 10-year study period.

7. Air temperature and *VPD* were the principle drivers of *ET*, with little impact of soil moisture content except during drought years. Such results are in close agreement with field studies of boreal forests (Luyssaert *et al.* 2007, Bruemmer *et al.* 2012, Zha *et al.* 2013).

This chapter has discussed the influence of temperature, soil moisture content and solar radiation as controls of carbon and water fluxes in boreal forests. The importance of freezing stress and strategies used by trees to minimise the effect of freezing were emphasised.

14.8 References

Adams WW, CR Zarter, V Ebbert and B Denmig-Adams, (2004). Photoprotective strategies of overwintering evergreens. *BioScience* 54, 41–49.

Bruemmer C, TA Black and RS Jassal, (2012). How climate and vegetation type influence ET and WUE in Canadian forest, peatland and grassland ecosystems. *Agricultural and Forest Meteorology* 153, 14–30.

Eamus D and H Prichard, (1998). A cost benefit analysis of leaves of four Australian savanna species. *Tree Physiology* 18, 537–545.

Farquhar GD, ED Schulze, M Kuppers, (1980). Responses to humidity by stomata of *Nicotiana glauca* L. and *Corylus avellana* L. are consistent with the optimization of carbon dioxide uptake with respect to water loss. *Australian Journal of Plant Physiology* 7, 315–327.

Ge Z-M, X Zhou, S Kellomaki, H Peltola and K-Y Wang, (2011). Climate, canopy conductance and leaf area development controls on evapotranspiration in a boreal coniferous forest over a 10 year period: a united model assessment. *Ecological Modelling* 222, 1626–1638.

Goodine G, MB Lavigne and MJ Krasowski, (2008). Springtime resumption of photosynthesis in balsam fir (*Abies balsamea*). *Tree Physiology* 28, 1069–1076.

Iwata H, Y Harazono and M Ueyama, (2012). The role of permafrost in water exchange of a black spruce forest in interior Alaska. *Agricultural and Forest Meteorology* 161, 107–115.

Kellomäki S and KY Wang, (2000). Modelling and measuring transpiration from Scots pine with increased temperature and carbon dioxide enrichment. *Annals of Botany* 85, 263–278.

Lintunen A, T Holtta and M Kulmana, (2013). Anatomical regulation of ice nucleation and cavitation helps trees to survive freezing and drought stress. *Scientific Reports* 3, 2031, DOI 10.1038/srep02031.

Luyssaert S, I Inglima, M Jungs, AD Richardson and M Reichstein, (2007). CO_2 balance of boreal, temperate and tropical forests derived from a global dataset. *Global Change Biology* 13, 2509–2537.

Owston PW, JL Smith and HG Halverso, (1972). Seasonal water movement in tree stems. *Forest Science* 18, 266–272.

Sevanto S, Suni T, Pumpanen J, Gronholm T, Kolari P, Nikinmaa E, Hari P and Vesala T, (2006). Winter-time photosynthesis and water uptake in a boreal forest. *Tree Physiology* 26, 749–757.

Sugimoto A, N Yanagisawa, D Naito, N Fujita and TC Maximob, (2002). Importance of permafrost as a source of water for plants in east Siberian taiga. *Ecological Research* 17, 493–503.

Ueyama M, Y Harazono, E Ohtaki and A Miyata, (2006). Controlling factors on the inter-annual CO_2 budget at a subarctic black spruce forest in interior Alaska. *Tellus* 58B, 491–501.

Wang KY, S Kellomäki, TS Zha and H Peltola, (2004). Seasonal variation in energy and water fluxes in a pine forest: an analysis based on eddy covariance and an integrated model. *Ecological Modelling* 179, 259–279.

Zha T, C Li, S Kellomaki, H Peltola, K-Y Wang and Y Zhang, (2013). Controls of evapotranspiration and CO_2 fluxes from Scots Pine by surface conductance and abiotic factors. *PLoS One* 8, e69027, DOI 10.1371/journal.pone.0069027.

15

Arid and Semi-Arid Grasslands

15.1 Introduction

Grasslands are a major biome globally and are found on all continents except Antarctica. The dominant life form is grasses, a highly adaptable group of plants showing both annual and perennial life histories and using C3 and C4 photosynthetic pathways (Chapter 2). Because of this wide amplitude in life histories they can be found in cold (sub-zero annual average temperatures), temperate and tropical climates across a wide range of annual rainfall (300 mm to more than 1500 mm) (e.g. Fig. 15.1). Rainfall is often (but not always) highly seasonal and disturbance by fire a common feature. Grasslands can grade into savannas where rainfall is sufficient to support woody shrubs or trees, or they grade into deserts where rainfall is insufficient to support extensive plant life. Temperate grasslands experience larger seasonal amplitude in temperature than observed in tropical grasslands and savannas. Veldts in South Africa, steppes of Russia, The Mitchell Plains grasslands of northern Australia, Pampas grasslands of South America and the prairies of central North America are examples of grasslands: all support significant numbers of grazing animals, both farmed and unmanaged.

In this case study, we focus exclusively on fluxes of C (productivity) because these are the most important fluxes in terms of management of rangelands in arid and semi-arid landscapes, where water fluxes are very much linked to C fluxes.

15.2 Intra-Annual Patterns of Carbon Flux in Grasslands

Dormant periods when grass growth is absent is a common feature of arid and semi-arid grasslands, when lack of rain and/or low temperatures, inhibit growth. The onset of spring rains and increased temperatures are associated with increased grass growth resulting in increased LAI, light interception and NPP. Figure 15.2 shows how soil temperature increases through spring and summer (months 4–9) and, at this site, volumetric soil moisture content declined in the early spring because of rapid increases in LAI during the early spring growth period with little rainfall. Rainfall

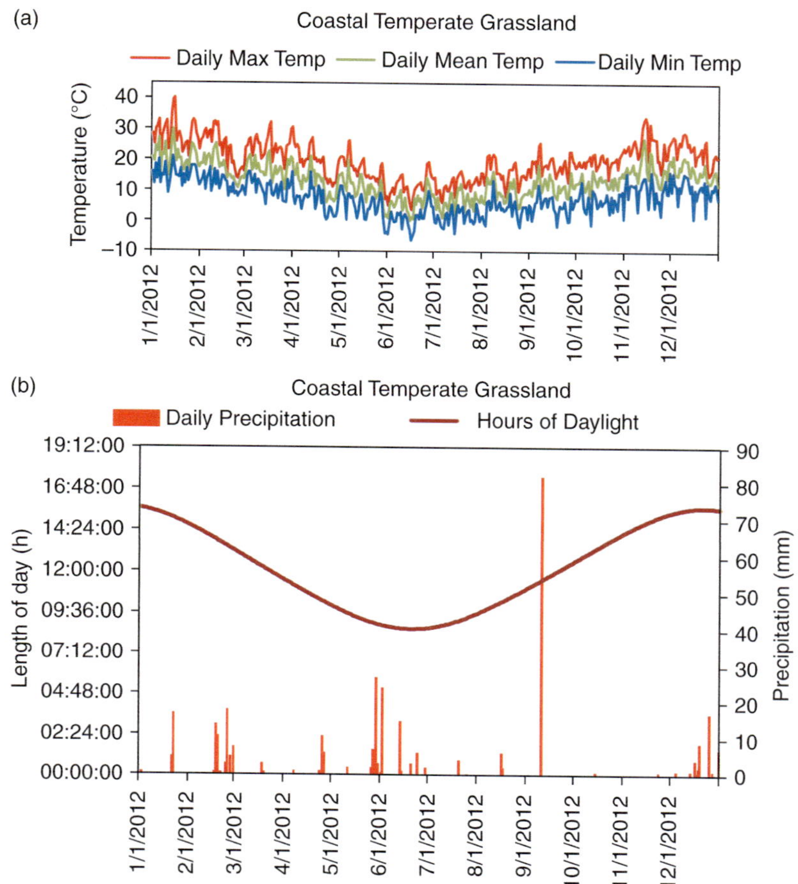

Figure 15.1. An example of the range of temperatures and day lengths experienced in a coastal temperate grassland in Argentina. *Source*: Servicio Meteorologico Nacional-Argentina

in August and September increased soil moisture content in the latter half of the summer.

Increased temperatures and a high soil moisture content in spring result in increasing grass growth and increasing rates of NEE (larger negative values indicate increased rates of C uptake) and this was associated with non-linear increased rates of soil respiration. A time-lag of increased NEE resulting in increased rates of autotrophic soil respiration of four to six days generally occurs, as discussed in Chapter 2. Long time-lags may reflect poorly-draining, high clay content soils of some grasslands, where gas diffusion is slow. Similar seasonality in LAI and NEE is a common feature of grasslands in arid and semi-arid regions (Fig. 15.3; Bremer and Ham 2010, Parton *et al.* 2012).

The positive impact of increased soil moisture content in the growing season is illustrated in Figure 15.4 (Parton *et al.* 2012) and this figure also illustrates the positive

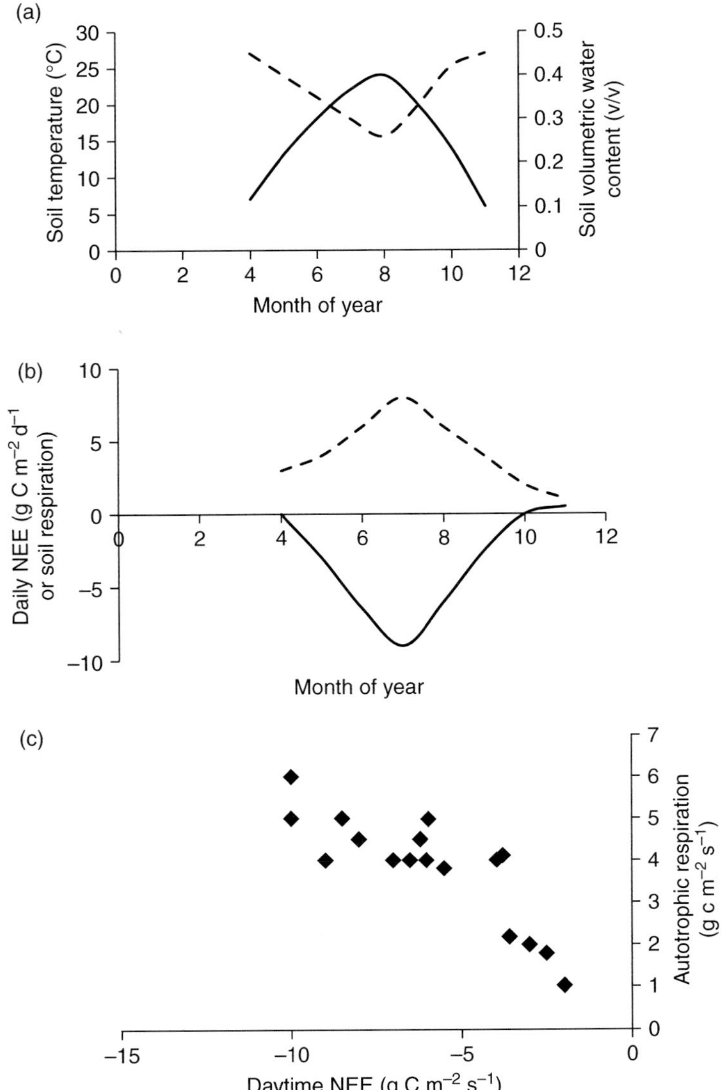

Figure 15.2. Upper panel: changes in soil temperature (solid line) and soil moisture content (dashed line) for a prairie grassland in the USA. Middle panel: Changes in daily net ecosystem exchange (NEE) (solid line) or soil respiration (dashed line). Lower panel, NEE and autotrophic respiration are correlated. Redrawn from Gomez-Casanovas *et al.* (2012).

impact of increased daytime NEE on soil respiration. Soil respiration is influenced by both temperature and soil moisture interactively such that low temperatures or low soil moisture content effectively inhibit the response of respiration to increased soil moisture content or temperature, respectively (Fig. 15.4).

The size of the rainfall event largely determines the relative responses of C gain and soil respiration. During the growing season there is a large respiratory loss of C

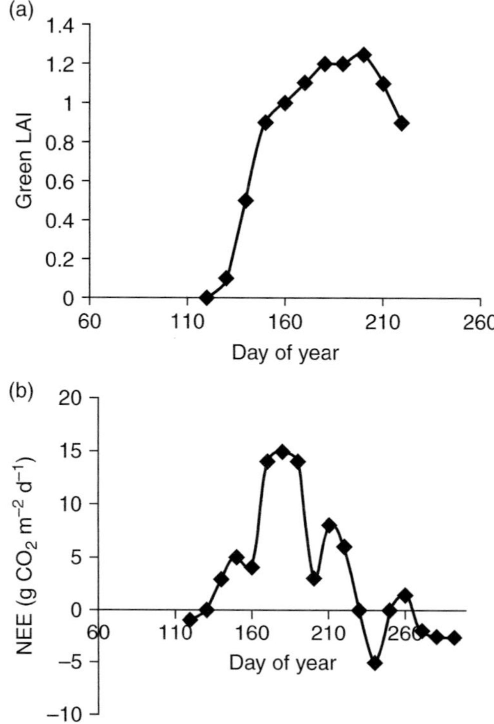

Figure 15.3. Changes in green leaf area index (LAI) and net ecosystem exchange (NEE) for a tallgrass Prairie in the United States. Redrawn from Bremer and Ham (2010).

from soil after rainfall. If the rainfall event is small, the period when C uptake exceeds respiration is short and when the rainfall event is large, the period when C uptake exceeds respiration is longer. These patterns in NEP are caused by: (a) rapid increases in heterotrophic respiration in response to rain; and, (b) increased vegetative C uptake which lags behind the response of soil respiration. As surface soil moisture declines, heterotrophic respiration declines, but root access to deeper soil moisture for several days after rain allows increased photosynthetic C gain (Parton *et al.* 2012).

15.3 Inter-Annual Patterns of Carbon Flux in Grasslands

Large inter-annual variation in primary productivity is a common feature of grasslands and much of this is attributed to variation in growing season rainfall (Weltzin *et al.* 2003). For large regions of the globe, including many grasslands in Canada, the USA, Australia and South America, variations in the El Niño Southern Oscillation (ENSO; Chapter 1) result in large variations in annual rainfall and temperature. Thus in the Canadian prairies El Niño years are associated with cooler wetter summers. The question is: 'Does this inter-annual variation in the ENSO influence productivity

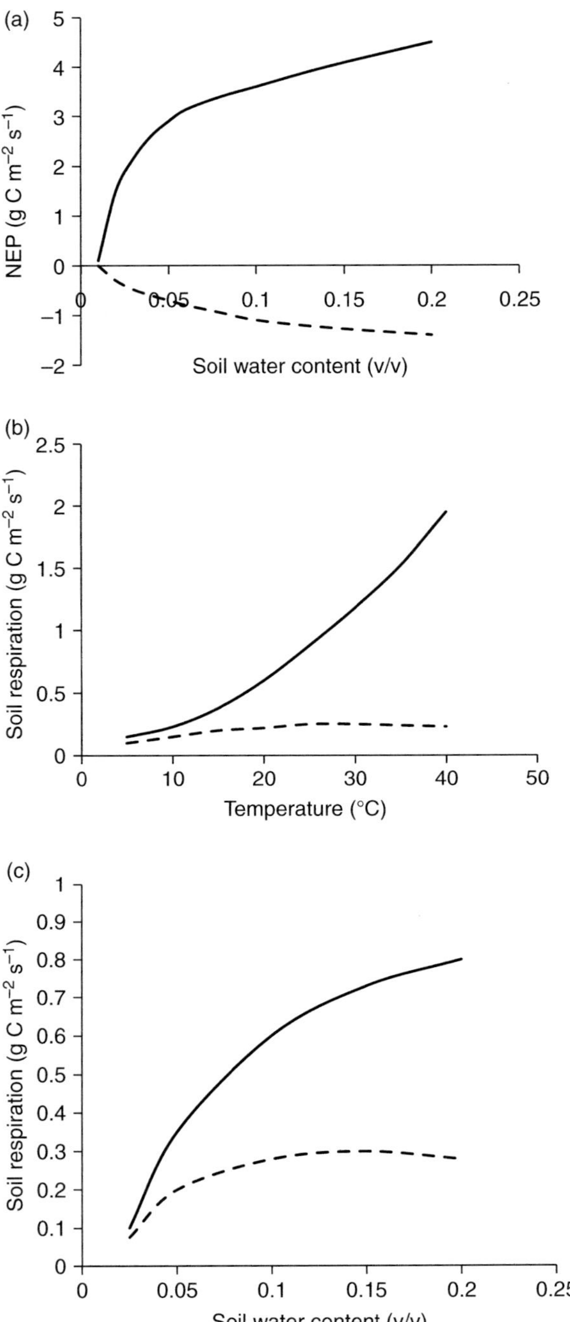

Figure 15.4. Upper panel: Increased soil moisture in a short-grass steppe in the United States increases both daytime net ecosystem exchange (NEE) (solid line) and night-time soil respiration (dashed line). Middle panel: soil respiration increases as a function temperature exponentially when soil moisture is high (solid line) but shows marginal increase when soil is dry (dashed line). Lower panel: Soil respiration increases significantly with increasing soil moisture content when temperatures are high (>15°C; solid line) but less so when soil temperature is low (<15°C; dashed line). Redrawn from Parton *et al.* (2012).

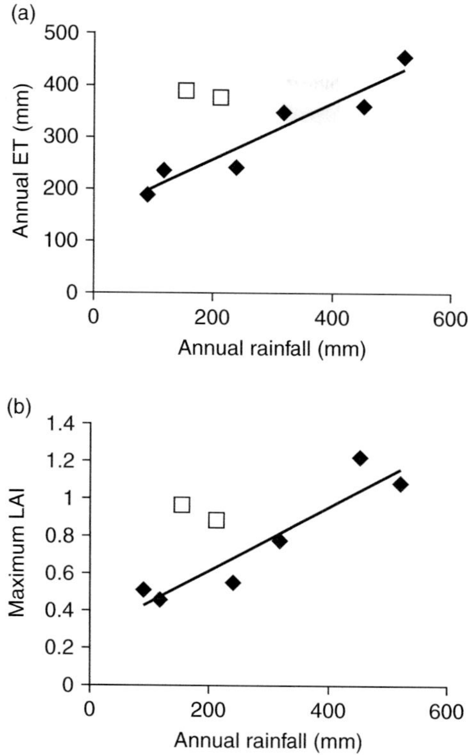

Figure 15.5. Inter-annual variation in rainfall for the period 1999–2006 is strongly and linearly correlated with growing season evapotranspiration (*ET*), when two years are excluded (2003 and 2006; open squares). Similarly maximum growing season leaf area index (LAI) is linearly correlated with annual rainfall, when the same two years are excluded. See text for why these two years don't fit the regression. Redrawn from Flanagan and Adkinson (2011).

of temperate arid grassland in Canadian prairies?' A related question arising from this is: 'To what extent does the "carry-over" of soil moisture from one year to the next influence the productivity of grasslands and can this dampen the effects of variations in ENSO?' These questions are discussed below.

Annual rainfall across grassland sites varies significantly across years, especially in arid and semi-arid regions. In arid central Australia, for example, rainfall can vary between less than 100 mm and more than 750 mm. Similarly in a Californian open grassland in the foothills of the Sierra Nevada, growth season rainfall varied between 20 mm and 255 mm between 2001 and 2006 (Ma *et al.* 2007) while in a short, mixed species, grassland in the Great Plains of Canada, rainfall for the period 1999–2006 varied between 90 and 520 mm (Flanagan and Adkinson 2011). This variation in the Canadian prairies is reflected in the total growing season (May–Oct) *ET* for the grassland (Fig. 15.5) when two of the years are excluded (2003 and 2006). The reason for the strong correlation of *ET* with annual rainfall is the strong response of maximum LAI to inter-annual variation in rainfall, with the same

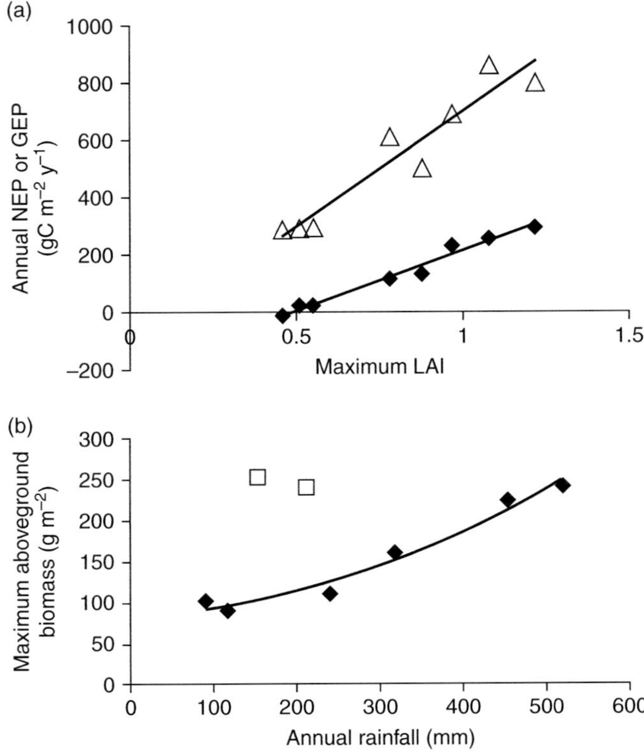

Figure 15.6. Annual net ecosystem exchange (NEP) (black diamonds) and gross ecosystem productivity (GEP; open triangles) are strongly correlated with maximum leaf area index (LAI) and biomass accumulation is strongly correlated with annual rainfall, when years 2003 and 2006 are excluded from the regression. Redrawn from Flanagan and Adkinson (2011).

two years not fitting the trend. The reason for the two outlying years (2003 and 2006) having a larger *ET* and LAI than expected given the rainfall in those years was the large carry-over of soil moisture from 2002 into 2003 and from 2005 into 2006. In 2002 and 2005, *ET* was significantly smaller than rainfall because rainfall was much larger than the long-term average. This carry-over meant that the total water available in 2003 and 2006 was larger than indicated from a consideration of rainfall alone. The water carried over was available for uptake and supported a larger *ET* and LAI than expected given the rainfall received in 2003 and 2006 (Flanagan and Adkinson 2011).

Because increased LAI increases the amount of solar radiation intercepted by the grass canopy, productivity (annual NEP or annual GEP) and biomass accumulation are strongly and linearly correlated with LAI and rainfall, respectively, (Fig. 15.6). Changes in foliar *N* and *P* contents did not differ across years in this study and could not explain the variation in productivity or biomass accumulation between years (Flanagan and Adkinson 2011).

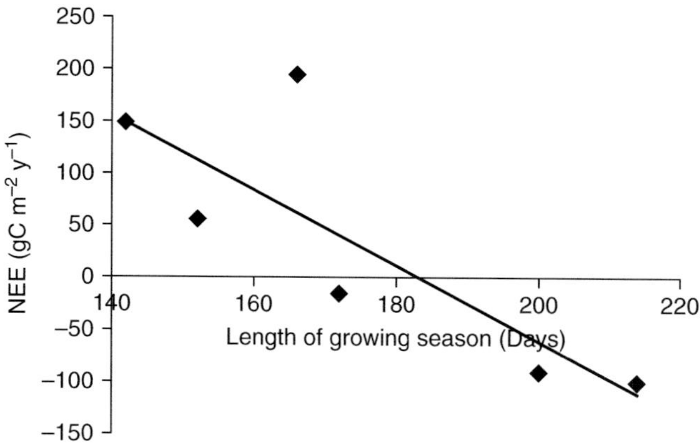

Figure 15.7. Increases in the length of the growing season result in increased C gain (net ecosystem exchange; NEE) by a Californian grassland. Redrawn from Ma *et al.* (2007).

Similarly increased growth season rainfall causes a linear increase in GPP and NEE (and respiration) in Californian open grassland (Ma *et al.* 2007) and other grasslands globally.

The length of the growing season is influenced by soil moisture and temperature in arid and semi-arid grasslands. In the Northern Territory of Australia in the southern, arid and semi-arid interior grass-dominated lands of Australia (south of latitude 17.5°S), the start, end-date and length of the growing season is strongly influenced by the amount of rainfall. This is in marked contrast to the northern, mesic, tree-dominated landscapes, where there is little variation in the length of the growing season (Ma *et al.* 2013). Although the start date of the monsoonal rains varies inter-annually, this does not cause large variation in the total amount of rainfall. A strong positive relationship between growing season length and NEE is evident in the Californian open grassland (Ma *et al.* 2007; Fig. 15.7) and reflects the increased time for light interception and hence C gain. For each day increase in growing season, an extra 4 g C m⁻² are sequestered. Although spring rainfall is a major determinant of the length of the growing season, temperature can also be a factor.

During wet years in eastern Australia, (El Niño years), average temperatures tend to be cooler than dry (La Niña) years because of increased cloud cover and the cooling effect of a larger partitioning of incoming radiation to latent heat flux rather than sensible heat flux. Both temperature and rainfall interact to control productivity and biomass accumulation in temperate climates (less so in tropical environments where temperatures are rarely limiting to photosynthesis). Garnett *et al.* (2006) has also shown strong correlations among summer rainfall, growing season temperatures and crop productivity as a function of El Niño and La Niña years (Chapter 1) across Canadian prairies. Similarly the impact of drought for the Northern Great Plains of the United States (covering approximately 72.8 million hectares) was recently analysed

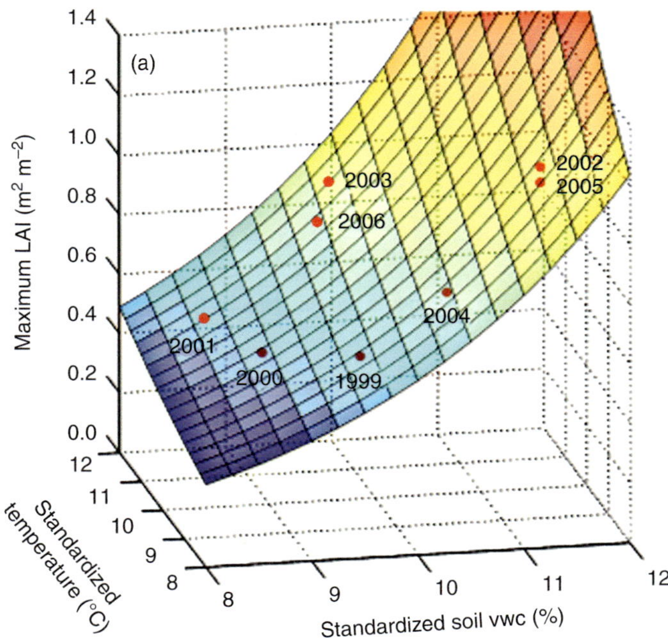

Figure 15.8. The interaction of soil moisture content and temperature in determining maximum leaf area index (LAI) in a Canadian prairie across an 8-year study period. Reprinted from Flanagan and Adkinson (2011) with permission.

(Zhang *et al.* 2010) using eddy covariance data and a regression model. Net ecosystem exchange was highly variable across years (ranging from –35 to 32 g C m^{-2} y^{-1}) and fluctuated from being a weak C source during drought years to a weak C sink during wet years (Zhang *et al.* 2010). Transitions between C sink and C source are also evident in other grasslands, including for example the Californian open grasslands of California (Ma *et al.* 2007) and European grasslands (Gilmanov *et al.* 2007).

Not only are the amount of rainfall and the temperature important variables explaining variation in LAI and productivity of grassland (Fig. 15.8), the timing of rainfall and temperature extremes is also important (Craine *et al.* 2012). Drought reduces productivity but its impact declines as the growing season progresses and the impact of drought tends to be largest when soils are wettest at the start of the growing season (Craine *et al.* 2012) and towards the end of the growing season drought has minimal impact on grass productivity. Similarly the negative impact of supra-optimal temperatures occurs predominantly in the middle of the growing season (July in the multi-decade study of Craine *et al.* (2012)) and this tends to be observed most during wet years rather than dry years. These insights suggest that accurate modelling of the impact of future climate change on grass (and hence crop) productivity requires consideration not only of the changes in average temperatures and annual rainfall, but consideration of the timing of changes in distribution of rainfall and temperatures as these subtle factors also impact grass growth and productivity (Craine *et al.* 2012, Ma *et al.* 2007).

15.3.1 Respiration, Temperature and Photosynthesis as Drivers of Inter-Annual Variation in GPP

Does respiration, or photosynthesis, or temperature, contribute the most to inter-annual variation in NEE arid and semi-arid regions? NEE is the difference between C uptake (photosynthesis) and C release (respiration). Valentini *et al.* (2000) show that NEE declines (less net C uptake) with increasing latitude (moving from the equator to the poles). In contrast, GPP was independent of latitude. The explanation for this observation lies in the relative importance of ecosystem respiration in relation to NEE. The ratio of NEE to ecosystem respiration increases with latitude and thus as latitude increases the relative importance of respiration in determining C fluxes increases (remember that NEE is expressed as a negative value so decreasing NEE means that the value of NEE tends to zero). Thus GPP tends to be constant with increasing latitude but respiration increases and therefore NEE declines with increasing latitude and the ratio NEE/respiration increases. While it would be expected that colder climates (increasing latitude) would have lower respiration rates because of the strong dependency of respiration on temperature, this isn't so across latitudinal gradients (but is seen within a single site) because boreal forests have very large soil stores of labile C and during the summer when water supply and temperatures increase, the rate of respiration increases substantially.

In arid and semi-arid zones, water supply is a major determinant of rates of soil respiration (Ma *et al.* 2007). During dry years water deficiency limits both GPP (through low LAI and low stomatal conductance) and respiration (through lack of substrate supply to soil microbes) and the response of respiration to temperature is small (because limited water supply limits the potential for microbial activity to respond to increased temperature). In contrast in wet years, respiration is increased and there is a more pronounced response of respiration to temperature. Thus in dry years both GPP and respiration are important determinants of NEE and in wet years temperature and respiration are important determinants of NEE.

15.4 Responses of Arid and Semi-Arid Zones to Pulses of Rainfall

Arid and semi-arid regions are characterised by large variability in rainfall across season and between years. Both the timing and the magnitude of rainfall events are highly variable in arid and semi-arid regions. Four conceptual models have been proposed to explain and predict ecosystem responses to such variation. These are the:

1. Two-layer hypothesis of Walter (Walter 1971)
2. Westoby-Bridges' pulse-reserve hypothesis (Noy-Meir 1973)
3. Threshold-delay model (Ogle and Reynolds 2004)
4. Bucket model (Knapp *et al.* 2008).

The two-layer model (Walter 1971) proposes that seasonality of rainfall in semi-arid savannas and the partitioning of water between shallow and deep soil profiles (with concomitant differential distribution of roots of grasses (relatively shallow) and trees (relatively deep)) minimises competition for water and enables co-existence of trees and grasses in these savannas. When wet-season rains are less than 200 mm grasses are expected to dominate because little rainfall penetrates to sufficient depths to allow trees to access sufficient water to support their growth. When wet-season rains exceed 200 mm, sufficient water percolates to depth and trees are supported. This model does not consider the impact of individual pulses of rain on ecosystem function. Consistent with this model, Kutiel *et al.* (2000) showed than annual herbs and grasses were more common at low rainfall sites and woody species are more dominant at the wetter sites, across an 1100 mm rainfall gradient in Israel. In contrast Ogle and Reynolds (2004) showed that grasslands are dominant in the western United States with higher (*ca* 300 mm) summer rainfall but shrubs are dominant in the western United States with lower (*ca* 150 mm) summer rainfall. No relation was observed with winter rainfall. It is likely that for arid and semi-arid regions, total annual rainfall is likely to be inadequate as a descriptor to explain the relative distributions of life forms (grasses *versus* shrubs *versus* trees). A major reason for this is because of the extreme plasticity observed in the distribution patterns (both laterally and vertically) of roots for shrubs and woody species and because total biomass distribution of roots may not be an adequate descriptor of root function (Ogle and Reynolds 2004). It is also likely that the timing for rainfall is as important as the amount of rainfall in determining species distribution and function.

The pulse-reserve model (Noy-Meir 1973) describes the response of different plant functional groups to pulses of rainfall. It relies on the idea that there are "biologically important" rainfall events and these are defined as events that trigger a growth/reproduction/germination response in vegetation. A significant fraction of the growth response is allocated to "reserves" within the plant, including seeds or storage of carbohydrates. However, this model also ignores plasticity in rooting depth and it is primarily applicable to annual rather than perennial species. Additionally it ignores the effect of antecedent conditions in determining both the timing and magnitude of vegetation responses to rainfall and the seasonality of rainfall, which are important drivers of ecosystem structure (composition; phenological patterns) and function (transpiration and C gain).

The threshold-delay model (Ogle and Reynolds 2004) is a relatively simple phenomenological model with six parameters. The model assumes that there are lower and upper thresholds in rainfall that elicit a response and that there is an upper limit to the magnitude of responses. Rainfall below the lower threshold does not cause a response and increased rainfall above the upper threshold does not cause any further increase in the response. A delay is built into the model such that it may take from one to several days before a response is observed. A conceptual diagram of this model is given below (Fig. 15.9). An application of this interpretation is given in Eamus *et al.* (2013) for the semi-arid region of central Australia.

It is clear that in many examinations of ecosystem responses to pulses of rain, the key diagnostics of a delay, an upper limit to a response and minimum and maximum

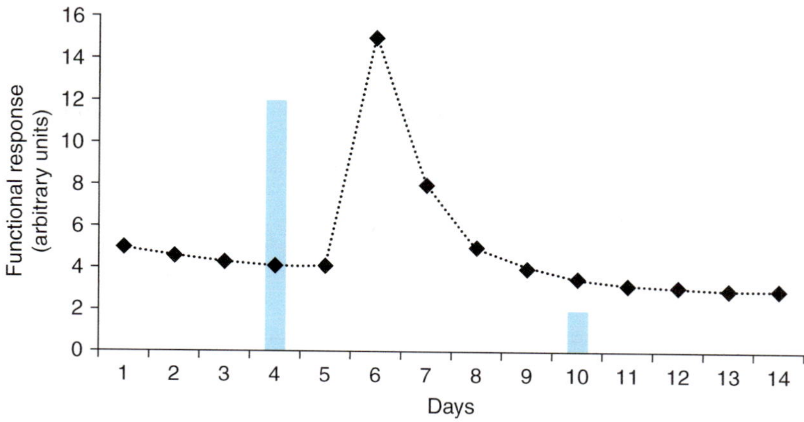

Figure 15.9. A representation of the threshold-delay model. The first rainfall event (light blue bar) exceeds the minimum threshold. The second rainfall event is below the minimum threshold. Redrawn from Eamus *et al.* (2013).

thresholds to rainfall, in terms of eliciting a response are discernible and that different plant functional types may differ in their characteristic responses, as shown by Burgess (2006) and Mitchell *et al.* (2008). See Chapter one for further discussion of this.

The Bucket model (Knapp *et al.* 2008) represents the soil profile that contains roots as a simple bucket that receives rainfall and which can store a certain volume of water that can be accessed by plants. Supra-optima soil water contents lead to anoxia. Sub-optimal water contents results in a decrease in some ecosystem processes (e.g. *ET*, C gain, respiration). Between these two extremes ecosystem function is highly dependent on changes in soil water content. Most importantly this model predicts that the response of xeric, mesic and hydric ecosystems differ in their response to, for example, less frequent but larger rainfall events (Fig. 15.10).

For xeric environments the larger rainfall events mean that more time is spent within the optimal range of water supply and NEE, for example, increases. For mesic environments a reductions in NEE is predicted because more time is spent below the lower threshold of water supply. Indeed the model can be applied to a large range of processes, from NEE, to run-off, to soil respiration and the combination of responses for a large range of processes differs between xeric, mesic and hydric environments (Knapp *et al.* 2008). In a recent test of this model, Thomey *et al.* (2011) varied both the amount of rainfall and the size distribution of rainfall events in a Chihuahuan desert grassland and found that when rainfall was increased, productivity increased but the increase in productivity was larger when the same amount of rainfall was received in large but infrequent events rather than in smaller but more frequent events. Pre-dawn leaf water potential was closer to zero in the treatment with infrequent but larger rainfall events than the treatment with frequent and smaller events; similarly leaf-scale photosynthesis was larger in the large rainfall event treatment than the small event treatment., Soil respiration was larger and total above ground net primary productivity was stimulated more in the large event treatment than in the small event

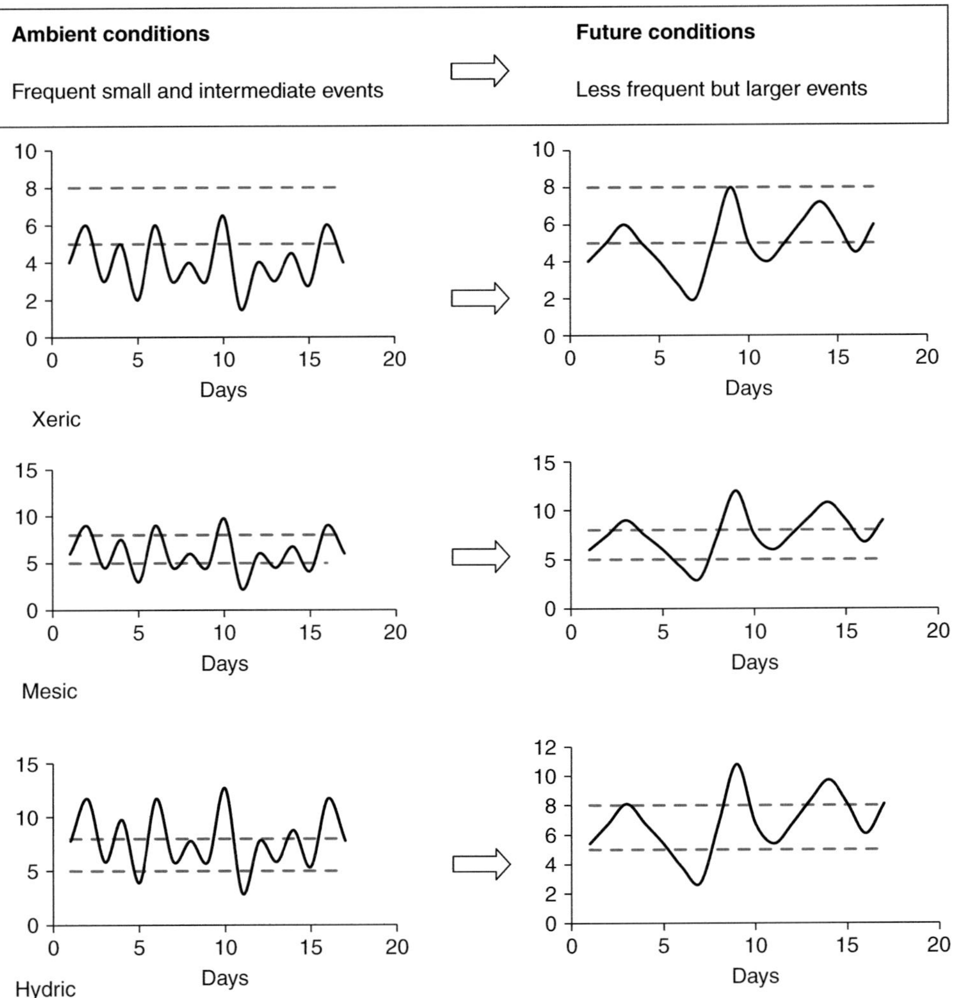

Figure 15.10. Conceptual representation of the Bucket model. The solid black line represents the diurnal course of an ecosystem process (e.g. net ecosystem exchange (NEE). The two horizontal lines represent the lower and upper bounds within which environmental conditions are optimal. In moving from a current to future pattern of rainfall the frequency at which the ecosystem process is within the optimum range increases for xeric environments. Redrawn from Knapp *et al.* (2008).

treatment. Both large and small rainfall event treatments increased these processes relative to the control which received only ambient rainfall (<200 mm annually).

Across a range of arid and semi-arid ecosystems, several generalities can be stated for flux responses to pulses of rain (Huxman *et al.* 2004a, 2004b; Parton *et al.* 2012; Craine *et al.* 2012). These are:

1. CO_2 in soil air spaces is physically displaced in a rapid but short-term "puff" of CO_2 release from the soil.

2. Carbon is released from labile soil stores and this stimulates microbial growth and activity and breakdown of soil C, resulting in a longer stimulation of soil respiration. This stimulation in soil respiration can be larger, across the year, than the stimulation of photosynthesis (see (4) below) and so the ecosystem can be a C source despite stimulation of photosynthesis.

3. Germination can be stimulated by sufficiently large rainfall events leading to increased LAI of the ecosystem and hence increased C uptake through photosynthesis.

4. Stomatal aperture can be increased, dormancy broken, photosynthesis up-regulated, leaf expansion stimulated and any combination of these can result from pulses of rain. All lead to increased rates of photosynthesis. Maximum ecosystem productivity occurs after soil respiration rates have peaked and thus the balance between soil respiration and leaf photosynthesis favours respiration early after the rainfall event, but photosynthesis dominates later.

5. The lag between the pulse and the response varies according to the response examined. Biological crusts can respond within an hour of rainfall; stomatal opening can be enhanced within 24 h; germination can occur within 48 hours; leaf-scale photosynthesis can be up-regulated within 72 h; and, increased LAI can occur within seven to fourteen days.

6. Approximately exponential declines in *ET* and NPP can be observed after a rainfall event and the half-time of this decline can range from two to twenty days.

7. Thresholds of rain amounts to induce a response are of the order of 5–20 mm per event.

8. In general, shallow rooted species tend to be more responsive (both faster and larger response) to rain pulses than deeply rooted species. Not all trees are deeply rooted.

9. Arid and semi-arid ecosystems can fluctuate between being net C sinks and net C sources on an intra-annual timeframe and an inter-annual timeframe.

15.5 References

Bremer DJ and JA Ham (2010). Net carbon fluxes over burned and unburned native tallgrass prairie. *Rangeland Ecology and Management* 63, 72–81.

Burgess SSO, (2006). Measuring transpiration responses to summer precipitation in a Mediterranean climate: a simple screening tool for identifying plant water-use strategies. *Physiologia Plantarum* 127, 404–412.

Craine JM, JB Nippert, AJ Elmore, AM Skibbe, SL Hutchinson and NA Brunsell, (2012). The timing of climate variability and grassland productivity. *Proceedings of the National Academy of Sciences* 109, 3401–3405.

Eamus D, J Cleverly, N Boulain, N Grant, R Faux and R Villalobos-Vega, (2013). Carbon and water fluxes in an arid-zone *Acacia* savanna woodland: An analyses of seasonal patterns and responses to rainfall events. *Agricultural and Forest Meteorology* 182–183, 225–238.

Flanagan LB and AC Adkinson, (2011). Interacting controls on productivity in a northern Great Plains grassland and implications for response to ENSO events. *Global Change Biology* 17, 3293–3311.

Garnett R, N Nirupuma, CE Haque and TS Murty, (2006). Correlates of Canadian Prairie summer rainfall: implications for crop yields. *Climate Research* 32, 25–33.

Gilmanov TG, JE Soussana and L Aires, (2007). Partitioning European grassland net ecosystem CO_2 exchange into gross primary productivity and ecosystem respiration using light response function analysis. *Agricultural Ecosystems and Environment* 121, 93–120.

Gomez-Casanovas N, R Matamala, DR Cook and MA Gonzalez-Meler, (2012). Net ecosystem exchange modifies the relationship between the autotrophic and heterotrophic components of soil respiration with abiotic factors in prairie grasslands. *Global Change Biology* 18, 253–545.

Huxman TE, JM Cable, DD Ignace, JA Eilts, NB English, J Weltzin and DG Williams, (2004a). Response of net ecosystem gas exchange to a simulated precipitation pulse in a semi-arid grassland: the role of native *versus* non-native grasses and soil texture. *Oecologia* 141, 295–305.

Huxman TE, KA Snyder, D Tissue, AJ Leffler, K Ogle, WT Pockman, DR Sandquist, DL Potts and S Schwinning, (2004b). Precipitation pulses and carbon fluxes in semi-arid and arid ecosystems. *Oecologia* 141, 254–268.

Knapp AK, C Beier, DD Briske, AT Classen, Y Luo and M Reichstein, (2008). Consequences of more extreme precipitation regimes for terrestrial ecosystems. *BioScience* 58, 811–821.

Kutiel, P, H Kutiel and H Lavee, (2000). Vegetation response to possible scenarios of rainfall variations along a Mediterranean-extreme arid climatic transect. *Journal of Arid Environments* 44, 277–290.

Ma S, DD Baldocchi, L Xu and T Hehn, (2007). Inter-annual variability in carbon dioxide exchange of an oak/grass savanna and open grassland in California. *Agricultural and Forest Meteorology* 147, 157–171.

Ma X, A Huete, Q Yu, N Restrepo-Coupe, K Davies, M Broich, P Ratana, J Beringer, LB Hutley, J Cleverly, N Boulain and D Eamus, (2013). Spatial patterns and temporal dynamics in savanna vegetation phenology across the North Australian Tropical Transect. *Remote Sensing of Environment* 139, 97–115.

Mitchell PJ, EJ Veneklaas, H Lambers and SSO Burgess, (2008). Using multiple trait associations to define hydraulic functional types in plant communities of south-western Australia. *Oecologia* 158, 385–397.

Noy-Meir I, (1973). Desert ecosystems: environment and producers. *Annual Review of Ecology and Systematics* 4, 25–52.

Ogle K and JF Reynolds, (2004). Plant responses to precipitation in desert ecosystems: integrating functional types, pulses, thresholds and delays. *Oecologia*, 141, 282–294.

Parton W, J Morgan, D Smith, S Gel Grosso, L Prihodko, D Lecain, R Kelly and S Lutz, (2012). Impact of precipitation dynamics on net ecosystem productivity. *Global Change Biology* 18, 915–927.

Thomey ML, SL Collins, R Vargas, JE Johnson, RF Brown, DO Natvig and MT Friggens, (2011). Effect of precipitation variability on net primary production and soil respiration in a Chihuahuan Desert grassland. *Global Change Biology* 17, 1505–1515.

Valentini R, G Matteucci, AJ Dolman, E-D Schulze, C Rebmann, (2000). Respiration as the main determinant of carbon balance in European forests. *Nature* 404, 861–865.

Walter H, (1971). *Ecology of Tropical and Subtropical Vegetation.* Oliver and Boyd, Edinburgh, UK.

Weltzin JF, ME Loik, S Schwinning and DG Williams, (2003). Assessing the response of terrestrial ecosystems to potential changes in precipitation. *BioScience* 10, 94–52.

Zhang L, BK Wylie, W Ji, TG Gilmanov and LL Tieszen, (2010). Climate-driven inter-annual variability in net ecosystem exchange in the Northern Great Plains grasslands rangeland. *Ecology and Management* 63, 40–50.

16

Savannas

16.1 Introduction

Savannas are biomes with a discontinuous tree layer and a grass understorey. They are generally tropical or sub-tropical with a distinctly seasonal distribution of rainfall. The two major plant functional types (trees and grasses) have contrasting strategies for surviving the dry season and this is reflected in the seasonality of productivity, leaf area index and patterns in water-use. Woody thickening, whereby the standing biomass of trees increases over time, is a globally observed phenomenon which has implications for carbon (C) storage and catchment water balances of savannas.

Key questions addressed in this case study include: How do tree and grass productivities vary at daily, seasonal and inter-annual time-scales? What are the major drivers of variability in C uptake and water-use at a single site and along rainfall gradients? What is causing woody thickening globally?

A range of field based studies at leaf, canopy and plot scales are discussed in this chapter to illustrate the relationships among meteorological variables at hourly, daily, seasonal and inter-annual time-scales. Remotely sensed data are used to examine: (a) the relative importance of various environmental drivers in explaining differences in productivity; and, (b) large spatial and temporal patterns in phenology along an aridity gradient in north Australia.

Variation in soil moisture content and vapour pressure deficit are vitally important as drivers of daily and seasonal patterns of water-use and C fluxes of savannas. Explanations for inter-annual variations in phenological patterns differ according to aridity and the relative dominance of monsoonal *versus* convective storm rainfall. Productivity of savannas is very strongly influenced by water availability and much of this is related to changes in LAI of the understorey.

16.2 What Are Savannas?

Savannas are characterised by: (i) a high degree of predictability in the occurrence of wet and dry seasons (although timing and amount of rainfall can vary and this can have important impacts on savanna function); (ii) moderate to high temperatures; (iii) the coexistence of a tree dominated overstorey and grass dominated understorey; (iv) the occurrence of fire as a frequent disturbance; and, (v) for tropical and sub-tropical savannas, only small variation in day length. This is contrast to Boreal forests Chapter 14), which are characterised by winters of extreme cold and short day length, and arid and semi-arid grasslands (Chapter 15), which are characterised by very low inputs of rain and extreme variability in the timing and amount of rainfall received each year.

Savannas (or miombo in Angola, cerrado and caatinga in Brazil and llanos in Venezuela and Colombia) are classically defined as tropical or sub-tropical ecosystems having a discontinuous tree canopy above a seasonally continuous grassy understorey and experiencing a distinct wet and dry season. Increasingly, the "tropical and sub-tropical" characteristic is being replaced by "warm", which can mean having an average annual temperature above 20°C (Eamus *et al.* 2001) or above 15°C (Ma *et al.* 2007). Dry seasons might be as short as 1 month or as long as 8 months but they occur each year. Annual rainfall in savannas typically range from less than 300 mm to more than 1700 mm. Savannas cover about 15×10^{12} m^2 of the Earth's land surface and are a major biome supporting human habitation.

There is a large degree of variation across global savannas as to which phenological guild is dominant in the tree strata. In north Australian savannas, evergreen eucalypts dominate in terms of the contribution to the LAI of the overstorey. Although deciduous, brevi-deciduous and semi-deciduous tree species occur in north Australian savannas, their contribution to total tree LAI is relatively small (Williams *et al.* 1997). In contrast African savannas are mainly deciduous, as are those of India. In the Llanos of South America, evergreen species are dominant, while in seasonally dry open woodlands of Costa Rica, deciduous species are dominant. In all savannas the dry season is characterised by loss of the grass understorey as the upper soil profile dries out.

16.3 Daily and Seasonal Patterns in C and Water Flux

Figure 16.1 is an example of the annual distribution of rainfall in the humid (northern) region of Australian savannas. Rainfall shows a bi-modal distribution with a summer wet season (Nov–Mar inclusive) and a winter dry season. Annual rainfall at this site is typically 1500–2000 mm per annum. Such bi-modal distribution of rainfall is common in savannas globally and results in parallel fluctuations in soil moisture content (Fig. 16.2d). For tropical savannas, day length shows little variation but variation increases as latitude increases. A similar trend is observed for temperatures also.

Seasonality in rainfall is reflected in seasonality of ecosystem C fluxes (NPP; GPP, measured with eddy covariance technologies; Baldocchi 2008). During wet seasons, peak fluxes are large and typically peak around solar noon (Fig. 16.2). Wet season fluctuations within a day reflect the influence of extensive, deep, but rapidly moving cloud cover. However, as the dry season progresses, maximum rates of C uptake

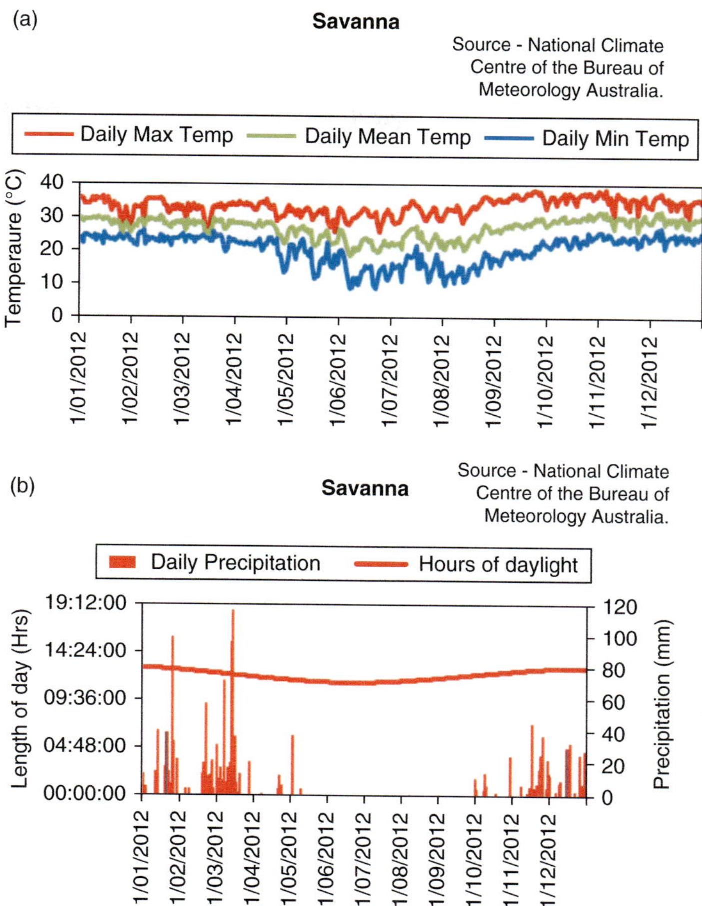

Figure 16.1. Average temperatures are high for tropical and sub-tropical savannas (upper panel), whilst rainfall is highly seasonal (lower panel) and can be uni-modal or bi-modal, as seen for Australian savannas (lower panel). These data are for the northern (humid) end of the distribution of Australian savannas.

decline and the period over which maximum rates are observed, declines as the dry season progresses (Fig. 16.2).

Similar seasonal patterns in canopy C flux are observed in the Orinoco lowland savannas (San Jose *et al.* 2008), oak/grass savanna in California (Ma *et al.* 2007) and Mopane savanna of Botswana (Arneth *et al.* 2006). Reasons for this seasonal decline are discussed in Section 16.3.1.

The tight coupling of net ecosystem exchange and rainfall is also clearly apparent in the Sudanian savanna study of Brummer *et al.* (2008), presented in Figure 16.3, where significant C uptake by the savanna only occurs after the start of the wet season. Similarly latent heat flux (i.e. ET) is largest in the wet season (June–October) when soil moisture is sufficient to support grass growth and transpiration by trees and grasses. The low level of latent heat flux in December and January to May reflects

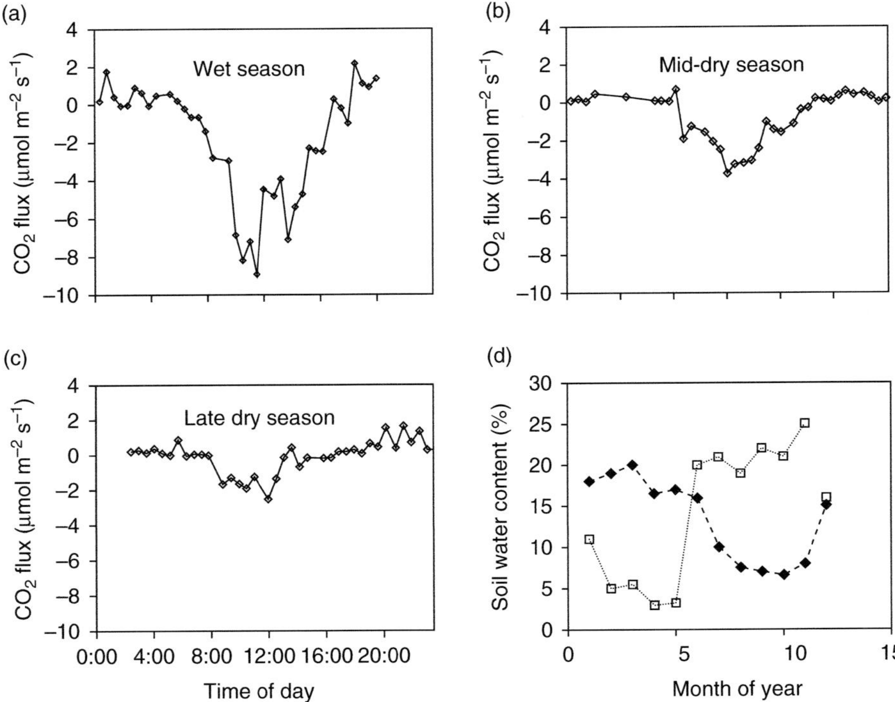

Figure 16.2. Changes in the diurnal pattern of net ecosystem exchange (CO_2 flux) across the wet, mid–and late dry seasons in a north Australian savanna (a–c). Also shown are changes in upper soil moisture content across a year (d) for a north Australian savanna (solid black line) and a Venezuelan savanna (dashed line; redrawn from San Jose *et al.* 2008). Note the reverse of the timing of the wet season for the southern hemisphere and northern hemisphere savannas.

the low level of transpiration supported by the deep rooted trees. When latent energy fluxes are low, sensible heat fluxes are larger and *vice versa*. Such complementarity of latent and sensible heat fluxes are commonly observed in savannas.

Not only do ecosystem-scale rates of net ecosystem exchange (NEE) vary across seasons, but leaf-scale rates of photosynthesis also display wet-season maxima and dry-season minima across a range of savanna sites and species (Prior *et al.* 1997, Arneth *et al.* 2006, Veenendaal *et al.* 2008). The causes of seasonal changes in leaf and canopy scale C fluxes are now discussed.

16.3.1 Drivers of Seasonal Patterns in C and Water Flux

Seasonal patterns of C and water fluxes are driven principally by four abiotic factors and two biotic factors and these operate across two time-scales: sub-daily and seasonally. These factors, many of which are inter-related, include:

1. Rainfall and soil moisture content
2. Vapour pressure deficit

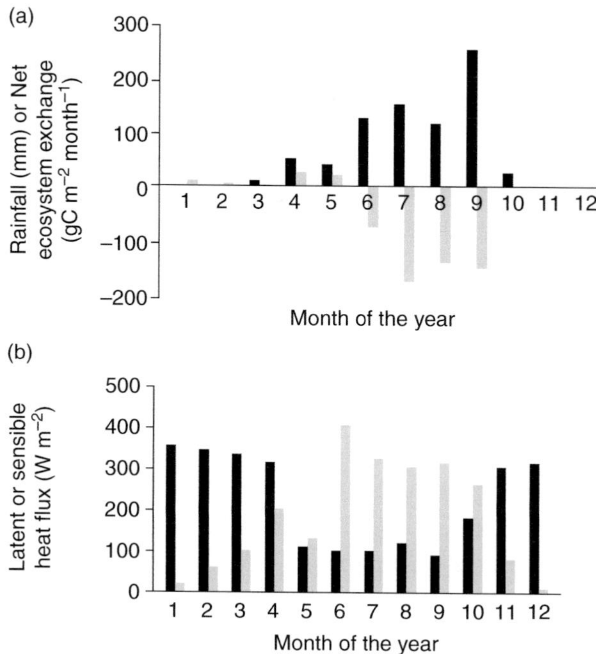

Figure 16.3. Top panel: Seasonal patterns in rainfall (positive values; black bars) and net ecosystem exchange (grey bars, generally negative). Net ecosystem exchange data for Oct–Dec are omitted because of the impact of fire at this site in these months. Bottom panel: Seasonal patterns in sensible heat flux (solid black bars) and latent heat flux (grey bars). Data are for a Sudanian savanna, redrawn from Brummer *et al.* (2008).

3. Solar radiation
4. Temperature
5. Stomatal conductance
6. Leaf area index.

Rainfall is highly seasonal in all savannas, with distinct wet and dry seasons (Figs. 16.1 and 16.3). Consequently soil moisture content shows distinct seasonality with periods of high and low soil water content evident (Fig. 16.2d). As soil moisture content declines, pre-dawn leaf water potential and leaf-scale stomatal conductance (g_s) decline (Chapter 2). Such responses are observed in Australian, American, Asian and African savannas. Declining soil water content increases the supply of abscisic acid from roots to the canopy, increases xylem cavitation (Chapter 3) and causes the minimum daily leaf water potential to decline (which increases abscisic acid synthesis in the leaf); all three factors reduce stomatal aperture. Consequently canopy scale conductance (G_c) declines in the dry season compared to wet seasons and this reduces CO_2 uptake and hence canopy productivity (NEE, GPP).

Reduced soil moisture content and reduced ET result in increased partitioning of incoming energy into sensible heat flux. This, along with reduced ET, results in increased vapour pressure deficits (VPD). Stomata and hence G_c, show a negative response to increased VPD (Chapter 2). Sensitivity of g_s and G_c to VPD increases as soil moisture content declines (Thomas and Eamus 1999, Kutsch *et al.* 2008, Eamus *et al.* 2013), an adaptive mechanism to reduce canopy water loss during periods of low soil water availability and large evaporative demands.

Large changes in leaf area index occur seasonally in savannas. At one extreme complete loss of grass understorey (grass LAI is zero) plus large changes in tree LAI in savannas dominated by deciduous trees occurs in the dry season, with concomitant large reductions in water and C fluxes in the dry season relative to the wet season. At the other extreme, complete loss of the grass understorey occurs but only small changes in tree LAI because of the dominance of evergreen species (Bucci *et al.* 2008, Williams *et al.* 1997). When the grass understorey is lost a very large fraction (up to 60% in Whitley *et al.* 2011) of total savanna C uptake is lost (Whitley *et al.* 2011) because the grasses are often C4 with larger LAI and larger rates of photosynthesis (per unit leaf area) than C3 trees. Similarly, rates of evapotranspiration decline with the loss of leaf area in the understorey.

In a geographically extensive analysis of MODIS enhanced vegetation index (eVI) data (Chapter 6), Ma *et al.* (2013) examined inter-annual variation and latitudinal gradients in patterns of seasonal changes in LAI along an 1100 km transect comprising a 1300 mm rainfall gradient in northern Australia. They observed several key features in temporal and spatial patterns of change in the LAI of savannas (Fig. 16.4):

1. LAI varies seasonally along the entire gradient and this is dominated by changes in the LAI of the understorey.
2. For the northern or top half of the transect (approximately 12°S to 17.7°S corresponding to a rainfall gradient from 1700 mm to 700 mm), the start, peak and end dates of seasonal greening-up periods (that is, when the understorey canopy is regrowing and understorey LAI is increasing rapidly) is progressively delayed from north-to-south along the gradient (that is, with increasing aridity and increasing distance from the equator). This means that the length of the "green-season" (corresponding essentially with the wet season) was constant for this part of the transect.
3. Inter-annual variability in the timing of the start, peak and end dates increased from north to south in the top half of the transect. This is because of the weakening of the influence of the inter-tropical convergence zone monsoonal weather pattern.
4. The major factor driving phenological patterns in the top half of the transect is the timing of the start and end of the wet season, but this shows only small inter-annual variation. Larger inter-annual variation in total wet season rainfall has minimal impact on phenological patterns (i.e. timing).
5. The southern (xeric) half of the transect (south of 15°S) shows distinctly different patterns in phenological timing compared to the northern half. Thus the timing of phenological patterns is strongly influenced by the amount of rainfall in each wet

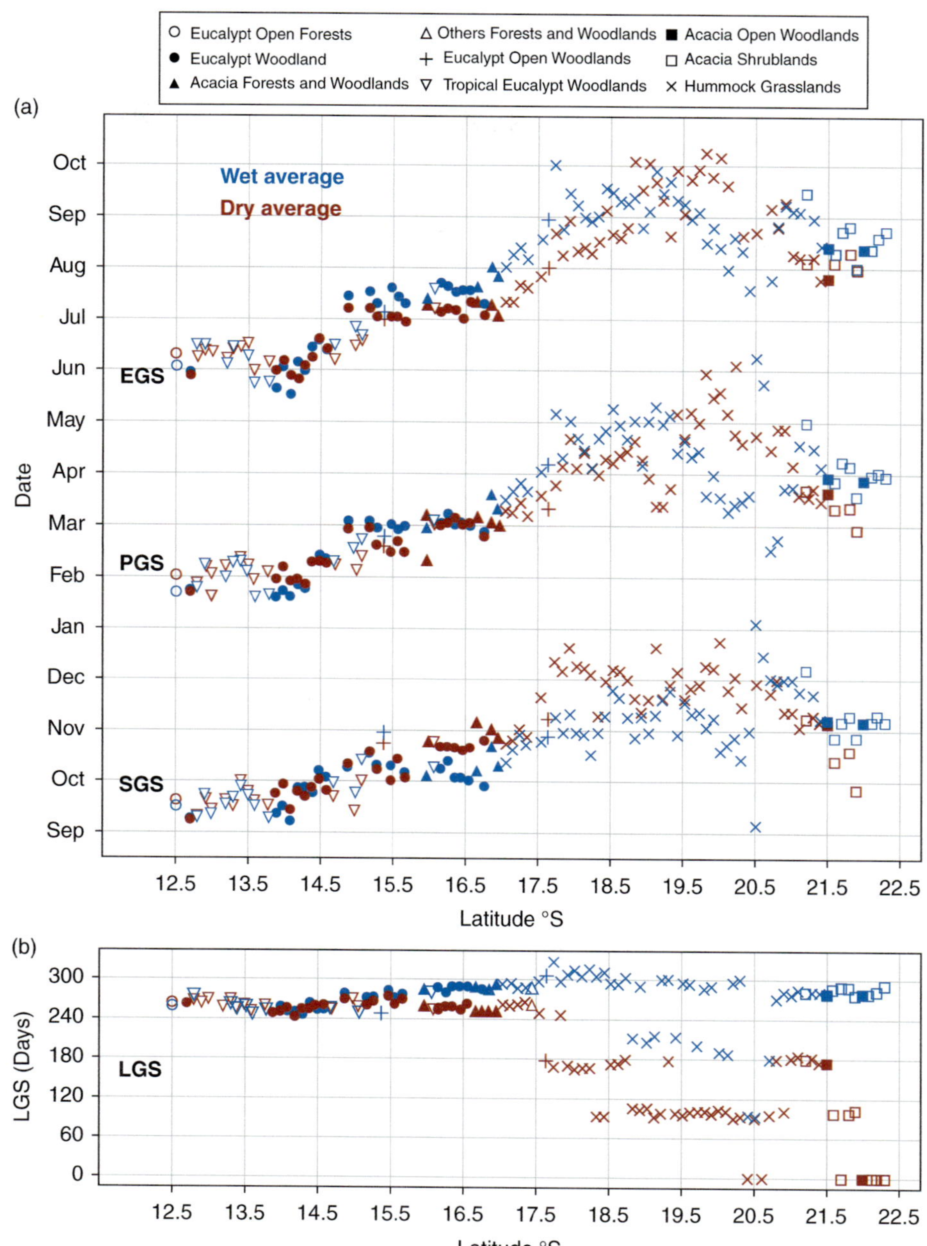

Figure 16.4. (a) Changes in the date of the start, end and peak growing season as a function of latitude in north Australian savannas for an average wet and average dry year. (b) The length of the growing season is approximately constant for the northern (mesic) half of the transect but shows a distinct step-wise decrease in the southern (xeric) half of the transect in drier-than-average years. Reproduced from Ma *et al.* (2013) with permission.

season. Inter-annual variability in both the timing of the start and end of wet seasons and variability in the total amount of rainfall is much higher in the southern half. This is because the monsoonal weather pattern interacts with large convective storms in the southern half and these can deliver either very low, moderate, or very large fractions of total rainfall in different years. This highlights the need to understand climate patterns across savannas when interpreting spatial and temporal patterns in savanna function.

16.3.2 Seasonal Patterns in Tree Water-Use Do Not Always Mirror Those of C Flux

At landscape/plot scales, declines in rates of photosynthesis, as measured by eddy covariance techniques, are observed in the dry season compared to the wet season. The major reason for this is the decline in LAI observed in the dry season. Most of this decline is in the grass understorey, although in seasonally dry forests dominated by deciduous trees loss of the tree canopy LAI is also apparent. Similarly at landscape/plot scales, rates of ET also decline, for the same reasons. However, for evergreen trees in some savannas (but not all) rates of transpiration from trees do not decline in the dry season. In a multi-site study of neotropical savannas in Brazil, Bucci *et al.* (2008) observed no decline in stand (tree) transpiration (Fig. 16.5) in the dry season compared to the wet season despite a pronounced dry season with zero rain for the months of June–Aug each year. In this study, seasonal changes in tree canopy LAI were minimal (Fig. 16.5) and changes in canopy conductance were not the result of changes in tree LAI but were the result of changes in stomatal conductance in response to declining soil moisture content and increased VPD in the dry season.

Similarly, in a three site study along a rainfall gradient (1700 mm to 500 mm) in north Australia, Eamus *et al.* (2000) showed that dry season rates of tree water-use did not decline compared to wet season rates and in many instances rates of water-use in the dry season were larger than those in the wet season (Fig. 16.6), despite the observed decline in predawn leaf water potential in the dry season at all three sites (Fig. 16.6). Three features are apparent in this north Australian study (Fig. 16.6):

1. The large rainfall gradient across sites did not result in very large differences in daily rates of transpiration (expressed per m² sapwood).
2. Significant declines in predawn water potential occur in the dry season compared to the wet season at all three sites and the driest site (Newcastle Waters) had a lower dry season predawn water potential than the wettest site (Darwin).
3. In four of six comparisons, rates of transpiration were larger in the dry season than the wet season.

The absence of a decline in transpiration rate in the dry season compared with the wet season is contrary to previous findings based on leaf-scale (Pitman 1996, Eamus

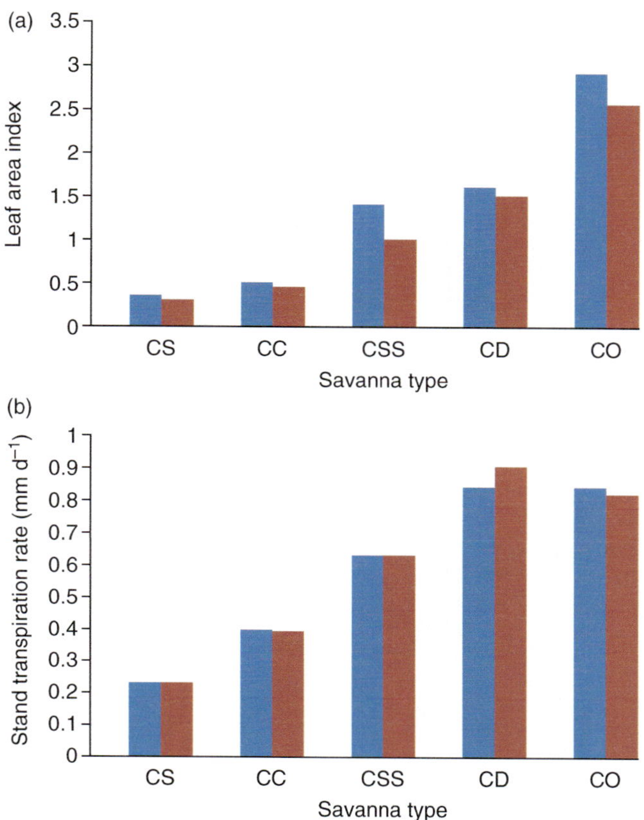

Figure 16.5. Comparisons of seasonal changes in (a) tree leaf area index (LAI) and (b) tree stand transpiration rate for 5 savanna types in Brazil. Wet seasons are blue and dry seasons are red. CS = Campo sujo; CC = Campo cerrado; CSS = Cerrado *sensu stricto*; CD = Cerrado denso; CO = Cerradao. Redrawn from Bucci *et al.* 2008.

and Cole 1997, Franco 1998), tree-scale (Dye 1996) and canopy-scale (Miranda *et al.* 1997) measurements elsewhere, where significant declines in leaf and tree water-use in the dry season were observed. How do we explain this seemingly anomalous result?

It is clear that soil moisture content in the dry season is less than in the wet season since (a) rainfall is highly seasonal and (b) predawn water potentials decline in the dry season. Access to deep stores of water are insufficient to prevent large declines in predawn water potentials and therefore are unlikely to explain the increase in transpiration in the dry season, although Leuning *et al.* (2005) and Bucci *et al.* (2008) have suggested that deep stores of water can sustain transpiration in the dry season at seasonally arid sites.

Eamus *et al.* (2000) proposed that water availability in the dry season determines rates of transpiration in both wet and dry seasons, resulting in aseasonal patterns of water-use at all sites along the rainfall gradient. Several studies suggest a hydraulic limit to water-use. First, leaf area/sapwood area ratio, sapwood area, leaf area and

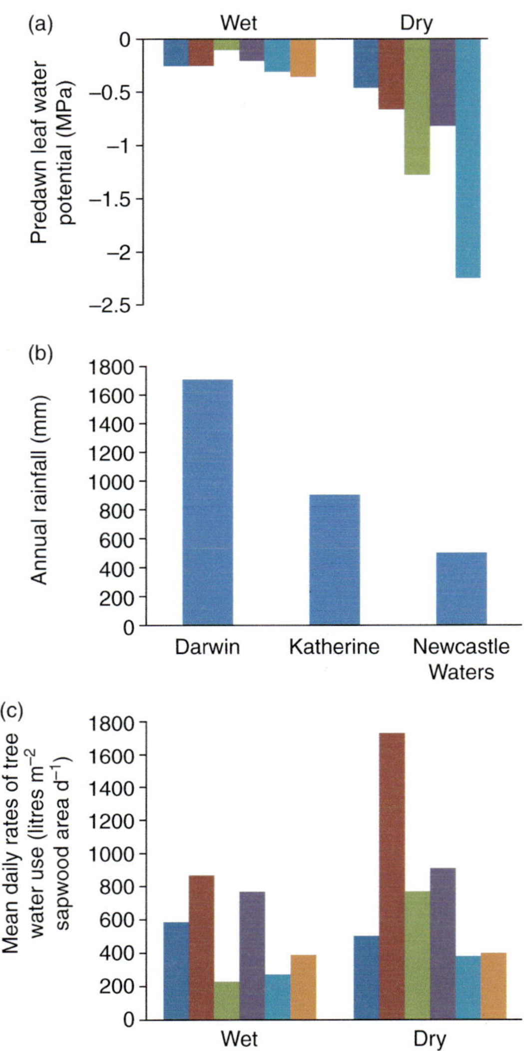

Figure 16.6. (a) Pre-dawn leaf water potential is lower in the dry season than the wet season across three sites in north Australia; (b) these three sites differ in annual rainfall; (c) however, this is not reflected in seasonal patterns of transpiration. Redrawn from Eamus *et al.* (2001). For Darwin: Blue bars are *Eucalyptus tetrodonta*, red bars are *E. miniata*; for Katherine: green bars are *E. tetrodonta*, purple bars are *E. latifolia*; for Newcastle Waters, turquoise bars are *E. capricornia*, orange bars are *E. terminalis*.

hydraulic conductivity are all influenced by differences in vapour pressure deficit across sites (Chapter 3). Second, the hydraulic architecture of trees is structured such that runaway emboli are minimised (Tyree and Ewers 1996). Third, the hydraulic conductance of stems can limit stomatal conductance in some species (for example *Pinus ponderosa*), and finally, stomatal responses to soil and atmospheric water content can be explained by hydraulic architecture in some species (Hubbard *et al.* 1999).

In the dry season, in the study of Eamus *et al.* (2000), evergreen trees transpired at rates equal to (for the mesic Darwin site), slightly higher than (for the semi-arid Katherine site) or the same as (for the arid Newcastle Water site) the corresponding rates in the wet season. To explain the lack of a decline in transpiration in the dry season, despite the absence of rain, Eamus *et al.* (2000) suggest that the hydraulic architecture of these evergreen eucalypt trees is constructed such as to prevent runaway embolism under dry season conditions and that as site aridity increases, whole-tree hydraulic conductance decreases. Vulnerability to embolism is primarily determined by xylem properties (Chapter 3; Tyree and Ewers 1996). Wide xylem conduits and pit pores have a low resistance to water flow but are very sensitive to embolism. As annual rainfall declines along a rainfall gradient and the dry season progresses, soil water availability declines and therefore plant water potential must decline to continue to extract water. Xylem and pit-pore diameter must be small enough to prevent excessive embolism at the water potentials experienced during the dry season. Consequently resistance to flow will be larger at drier sites and this sets an upper limit on the rate of water-use per tree, assuming a sustainable leaf water potential is maintained. Myers *et al.* (1997) noted that the difference between predawn water potential and midday water potential is the same in both wet and dry seasons (Fig. 3.10). Thus, within a species, the gradient in water potential up a tree did not vary seasonally and transpiration must remain constant if hydraulic conductance does not vary seasonally. Conversely, because transpiration rates do not vary seasonally and the water potential gradient between root and leaf appears to be relatively fixed, hydraulic conductance of the conducting system must also be aseasonal. Seasonal changes in percentage embolism are very small in *E. tetrodonta* in these north Australia savannas (Prior and Eamus 2000), indicating that xylem tissues formed at different times of year behave similarly. Tognetti *et al.* (1997) have shown that whole-tree hydraulic conductance is lowest in trees derived from seed taken from the most arid sites, as predicted from this mechanism and suggesting a genetic component to xylem characteristics. Recently, Do *et al.* (2008) observed minimal variation in rates of transpiration of *Acacia tortilis* in Africa and attributed this to (i) deep roots and (ii) constraints on transpiration by the conductance of xylem, thus supporting the hypothesis that dry season conditions may limit wet season rates of water-use (Eamus *et al.* 2000).

16.4 Modelling Seasonal Changes in Canopy C Uptake: Application of Optimality Theory to Savannas

Models of photosynthetic gas exchange have been extensively developed and applied over the past twenty years, especially to homogenous crops where variation among leaves and canopy height is relatively small. Application to natural vegetation is more difficult because of the range of species, canopy heights, leaf attributes and light climates that occur through a natural canopy such as a savanna, where the upper layer can contain evergreen and deciduous species and the understorey can contain a mixture of C3 and C4 species.

A recent development has been the application of optimisation theory to model native canopies (Schmanski *et al.* 2007, Medlyn *et al.* 2011). Optimisation theory assumes that stomatal conductance is optimised at short time-scales (sub-daily or daily) to maximise C gain per unit water transpired (Cowan and Farquhar 1977). A major hurdle to the application of optimisation theory to native canopies has been the inability to define correctly the objective function, that is, to define the correct function that vegetation maximises by adjusting, in this case, stomatal conductance. Schmanski *et al.* (2007) assume that optimal adaptation of vegetation occurs when net C profit is maximised. Net C profit is the difference between gross C uptake and investment of C in structural tissues (for example cell walls, roots and xylem) *and* respiratory losses associated with building and maintaining these structures. To make their solution tractable, Schmanski *et al.* (2007) focused on canopy properties and assumed that observed transpiration rates were a reflection of an optimisation of investments in roots and stems that had already occurred, given the long-term conditions of soil and climate at a given site. Two key developments were, therefore, the treatment of the canopy as both a single, sunlit layer of leaves when considering photosynthetic processes, but also as a series of sunlit and shaded layers, when considering light penetration and LAI and the use of C profit rather than net primary production. The aims of the optimisation model were to: (a) accurately represent daily and seasonal patterns of ecosystem C flux for a mesic savanna; and, (b) accurately represent the LAI of a seasonally wet-dry savanna. Key features of this approach were:

1. Calculation of the light level through the canopy and the sunlit and shaded fraction of each canopy layer.
2. Calculation of the rate of electron transport for each canopy layer.
3. Modelling of canopy rates of photosynthesis using the standard Farquhar and von Caemmerer model; C4 photosynthesis was omitted; only electron limited rates of photosynthesis were considered.
4. Respiration was set as 7 percent of the maximum rate of photosynthesis (Givnish 1988).
5. Transpiration was modelled as a diffusive process based on stomatal conductance and the gradient in water vapour pressure between sub-stomatal cavity and air.
6. Foliage turnover costs were derived from the Global Plant Trait Network (Wright *et al.* 2004).

This approach was remarkably successful. Key findings were:

1. The optimal distribution of photosynthetic capacity follows the vertical distribution of sunlit leaf area; a result in agreement with field studies.

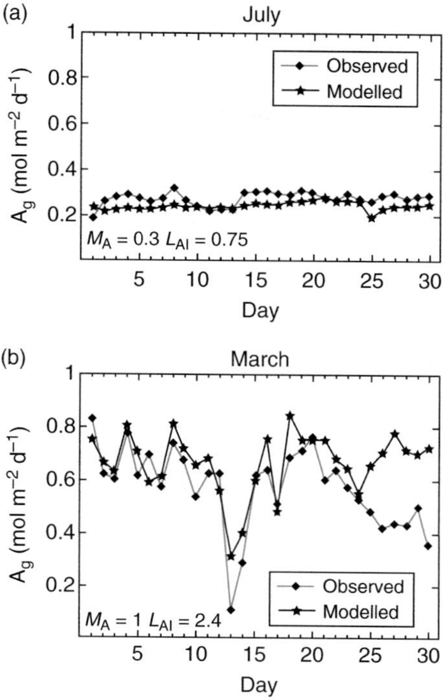

Figure 16.7. Modelled and observed daytime canopy C uptake (A_g) in the dry season (July) and wet season (March) for a mesic savanna in north Australia. Reproduced with permission from Schymanski *et al.* (2007).

2. Very good agreement between observed and modelled rates of net canopy C uptake were observed in the wet season but dry season agreement was only observed when the vegetation fraction of the land surface was prescribed using *in-situ* estimates. Examples of the correspondence of daytime canopy C uptake in the dry season (July) and wet season (March) are given in Figure 16.7.

3. Maximisation of net C profit (as used by Schymanski *et al.* 2007) differs from the usual optimisation model of maximising net primary production. This might represent a new principle for self-organisation of plant communities (Schymanski *et al.* 2007).

4. The costs associated with transport of water and nutrients in the wet season appear to be small, but are likely to be larger in the dry season (because of the increased depth to available soil water in the dry season).

Recent work (Medlyn *et al.* 2011) has similarly applied the optimality approach to the development of a new, unified model of stomatal conductance. This unified model of stomatal conductance provides a theoretical interpretation of model parameters which, when tested against field data, were highly successful. This is further discussed in the stomatal modelling chapter (Chapter 11).

16.5 Productivity Along Rainfall Gradients

Savannas frequently occur along large-scale gradients of rainfall, from mesic to semi-arid regions. This is associated with gradients in standing biomass, GPP, basal area and leaf area index along the rainfall gradient. Figure 16.8 shows the relationships among annual rainfall, annual potential evapotranspiration, tree height and tree density (Schulze *et al.* 1998) for north Australian savannas. As rainfall declines and temperatures remain high across the entire gradient, aridity increases as indicated by the increasing ratio of potential evapotranspiration to rainfall across the transect. Because of declining supplies of water, tree density, tree growth and tree height decline because of the coupling of fluxes of water and CO_2 through stomata.

Figure 16.8 shows the relationship between annual rainfall and GPP for a large number of sites from mesic through to semi-arid regions of north Australia. Results from a dynamic global vegetation model (ORCHIDEE) of the vegetation of Africa reveal a similar relationship for the extensive savannas of Africa (Ciais *et al.* 2009), while Merbold *et al.* (2009) similarly found a positive correlation between maximum rate of canopy photosynthesis and rainfall across nine savanna/grassland/bushland sites in Africa (Fig. 16.9).

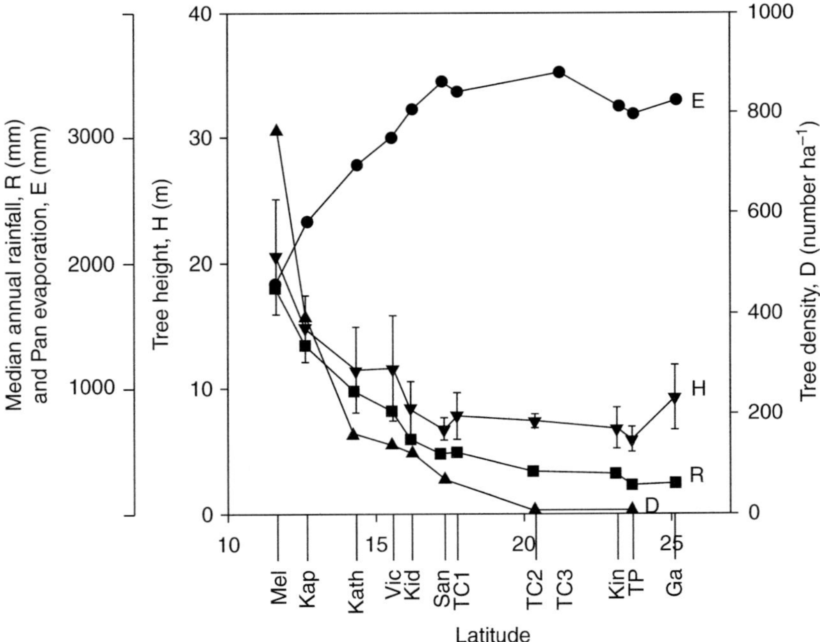

Figure 16.8. Changes in annual rainfall, (R), potential evapotranspiration (E), tree height (H) and tree density (D) along a north Australian tropical transect. Reproduced from Schulze *et al.* (1998) with permission.

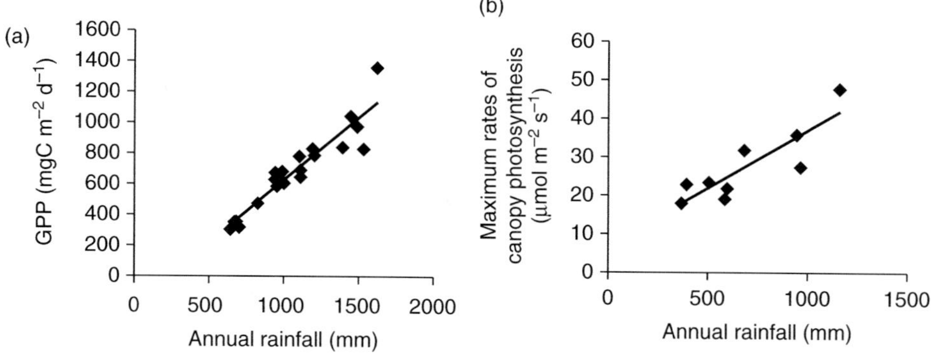

Figure 16.9. (a) The relationship between annual rainfall and gross primary productivity (GPP) for a range of savanna sites in northern Australia and (b) a range of sites in Africa. Redrawn from Kanniah *et al.* (2011) and Merbold *et al.* (2009).

16.5.1 Drivers of Changes in Productivity Along Rainfall Gradients

Both NPP and ET decline as aridity increases (i.e. as rainfall declines). There are five principle causes of these declines in NPP and ET, namely:

1. Soil moisture availability declines
2. Vapour pressure deficit increases
3. LAI of trees and grasses decline
4. Species composition changes
5. Change in the balance between GPP and respiration (see Section 16.6).

Canopy conductance declines with declining soil moisture content and increasing VPD (Chapter 2) and this reduction in conductance reduces both C and water fluxes. The increase in stomatal limitation to C gain is evident in the leaf C isotope data of Miller *et al.* (2001), where a strong increase in ^{13}C discrimination occurred as rainfall declined along the north Australian rainfall gradient (Fig. 16.10). Because total annual C gain is reduced with reduced total soil water availability along the rainfall gradient, stem growth and leaf growth are reduced and total LAI declines with declines in annual rainfall (Fig. 16.10). Reduced LAI results in reduced light interception and hence reduced GPP. Reduced LAI can be the result of reduced leaf area per tree as trees are shorter with increasing aridity and also because of reduced stand density (tree stems per hectare).

As rainfall declines from mesic to semi-arid zones, species composition changes significantly. Species adapted to mesic (high rainfall zones) have a lower water-use-efficiency, typically shallower roots and a larger shoot:root ratio. In contrast species adapted to xeric (low rainfall) zones have a higher water-use-efficiency and have either deeper roots in some woody species, or shallower, but having a wider lateral development, root system in other woody species and grasses. In addition they

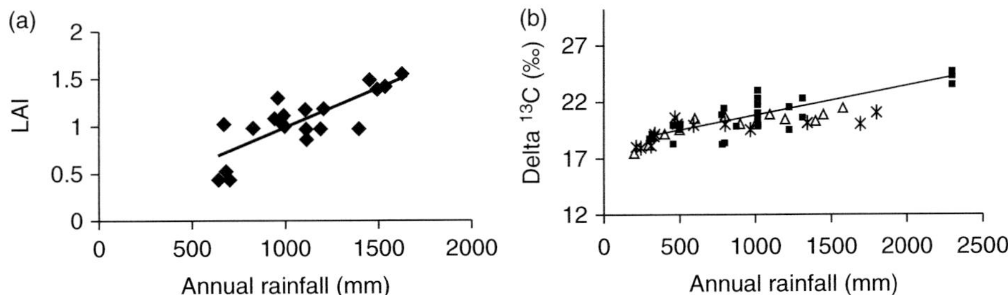

Figure 16.10. (a) leaf area index (LAI) declines with decreasing rainfall along a north Australian rainfall gradient. Redrawn from Kanniah *et al.* (2011). (b) discrimination against ^{13}C increases with increasing aridity, reflecting increasing stomatal limitation on photosynthesis with declining rainfall. Redrawn from Miller *et al.* 2001. Different symbols reflect different rainfall gradients in Australia.

have a smaller shoot:root ratio. They also display adaptations in leaf morphology (e.g. increased reflectance, needle-like leaves (e.g. *Acacia*), xylem anatomy (increased wood density and narrower xylem vessels and pit pores; Chapter 3) and other traits (for example, osmoregulation)). These traits tend to limit maximum rates of transpiration and hence maximum rates of C gain (GPP), further contributing to reduced rates of productivity of xeric sites.

16.5.2 Interactions Amongst Factors Driving Productivity: RS and Modelling Assessments

Campo-Bescos *et al.* (2013) used NDVI data across a 1000 mm rainfall gradient across three catchments in Southern Africa and combined these data with monthly values of several key environmental variables, including rainfall, temperature, soil moisture content and potential evapotranspiration rate. By using dynamic factor analyses it was possible to establish the relative importance of each factor in explaining seasonal and inter-annual variation in NDVI and hence productivity across the rainfall gradient. The key findings can be summarised thus:

1. Fire is an important determinant of productivity. High rates of fire frequency reduce landscape productivity; however the importance of fire as a determinant of annual productivity is largest in the high rainfall region and smallest in the low rainfall region. This is because the accumulation of biomass is larger in higher rainfall regions and this supports more destructive fires on long-lived and massive (relative to grasses) trees. Such a result is commonly observed in savannas globally. Low rates of fire frequency can increase productivity (Buis *et al.* 2009).

2. Soil moisture content is a major determinant of productivity. In contrast to the response observed for fire, the importance of soil moisture as a determinant of productivity decreased with increasing rainfall. Such a result is commonly observed globally in savannas and reflects the increasing importance of factors

other than water supply in high rainfall regions in determining savanna productivity, including grazing, fire and temperature.

3. Annual rainfall was a strong determinant of productivity along the rainfall gradient. Rainfall and soil moisture are physically linked, but surprisingly poorly correlated at small-scales because of the influence of small-scale variation in three factors (soil depth, soil texture and temperature) in determining soil moisture content independently of rainfall.

4. Annual average temperature was important in determining productivity but unlike the trends observed for fire, rainfall and soil moisture content, there was no north-south gradient mirroring the rainfall gradient. The importance of temperature is likely to be most sensitively felt by C3 species (trees and some grasses) and less sensitively by C4 grasses.

5. Potential ET was negatively correlated with productivity and its importance was largest in the region having intermediate rainfall (*ca* 900 mm). Potential ET is a measure of atmospheric demand for water and therefore includes a contribution from temperature and vapour pressure deficit.

6. Perhaps the key insight derived from this study was the fact that the importance of each variable (fire, rainfall, soil moisture content, temperature, potential ET) varied along the rainfall gradient in predictable ways. Where rainfall was less than 750 mm (where grasses dominate), soil moisture and precipitation were the principle drivers of variation in productivity. Fire and temperature and potential ET had little impact on productivity in these regions. A saturating response for productivity with increasing rainfall is observed globally across a range of biomes. For regions where rainfall > 900 mm, that is regions in which trees dominate ecosystem biomass, fire and mean temperature were the main drivers of variability in productivity. Potential ET was more important than rainfall in these regions. In the transitional zone between 900 mm > P > 750 mm, dominance by one or two factors in driving variability in productivity was not seen. These results are summarised graphically in Figure 16.11.

The two ends of the rainfall gradient can be characterised as either being soil moisture and low rainfall dominated regions, where grasses dominate, or temperature dominated high rainfall regions where trees are the dominant life form. This strongly mirrors the analyses by Yi *et al.* (2010) who showed that 125 sites of varying biomes can be classified into one of three contrasting groups, based upon their response of NEE to temperature, aridity or a combination of both (Fig. 16.12). Group 1 is those biomes where variation in temperature is the primary determinant of variation in NEE. In these biomes variation in temperature can account for 84 percent of spatial variation in NEE; Group 2 is those biomes where aridity (the ratio of radiation to rainfall) is the primary driver of variation in NEE. Variation in aridity can account for 81 percent of variation in NEE in this group; and Group 3 where both temperature and aridity co-limit productivity. Savannas occur across large gradients of rainfall and

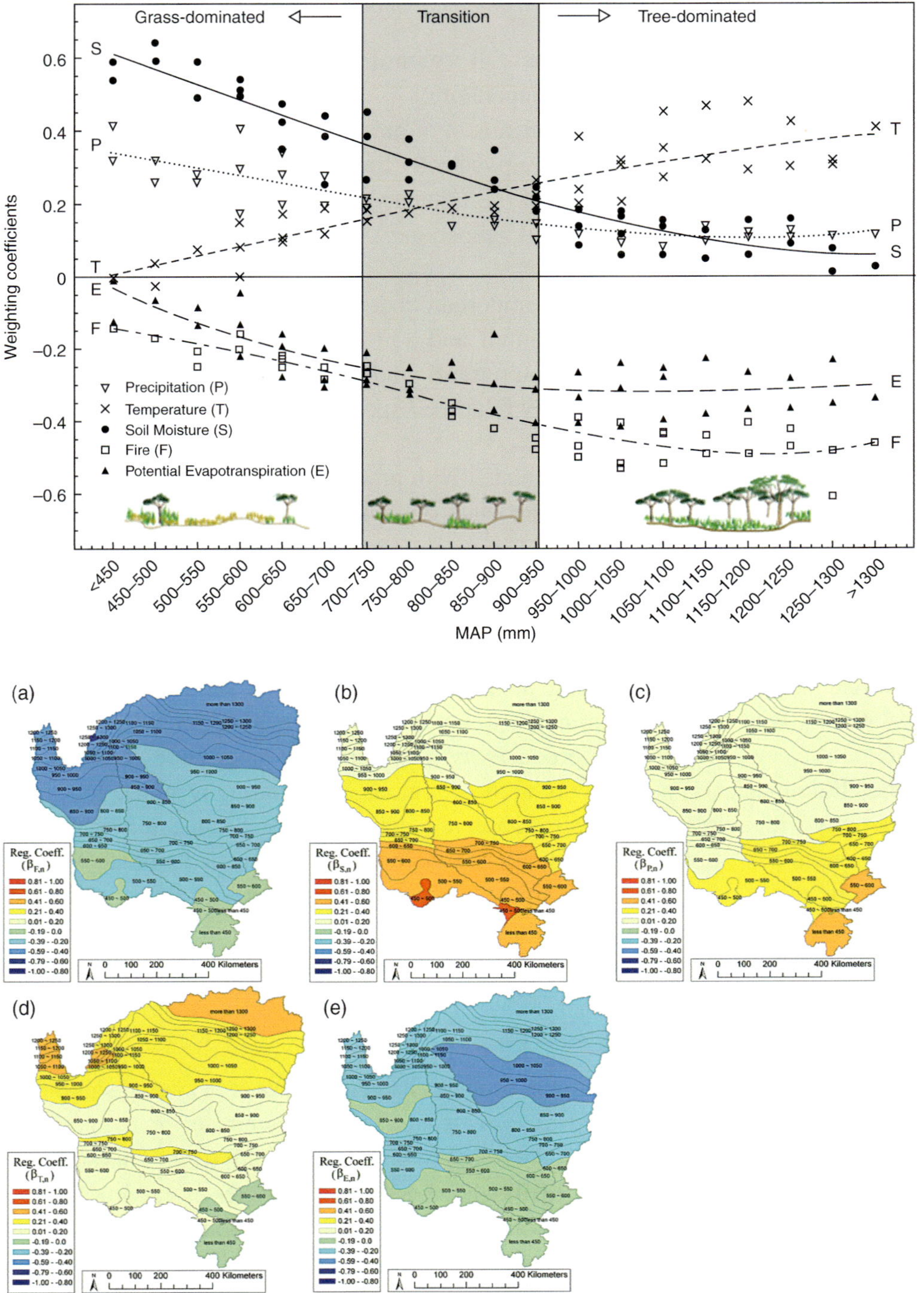

Figure 16.11. Upper panel: The relative importance of rainfall and soil moisture in explaining variation in normalized difference vegetation index (NDVI) and hence productivity, of African savannas along a 1000 mm rainfall gradient is large (large values of the weighting coefficient) in the southern, low rainfall regions where grasses dominate. In contrast the importance of temperature and fire is larger in the more mesic regions. Lower panel: Importance of the five variables is represented by the distribution of the $\beta_{k,n}$ regression coefficients (values −1 to 1) for each variable: (a) fire, F; (b) soil moisture, S; (c) precipitation, P; (d) mean temperature, T; and (e) potential evapotranspiration, E. Reproduced from Campo-Bescoc *et al.* (2013) with permission.

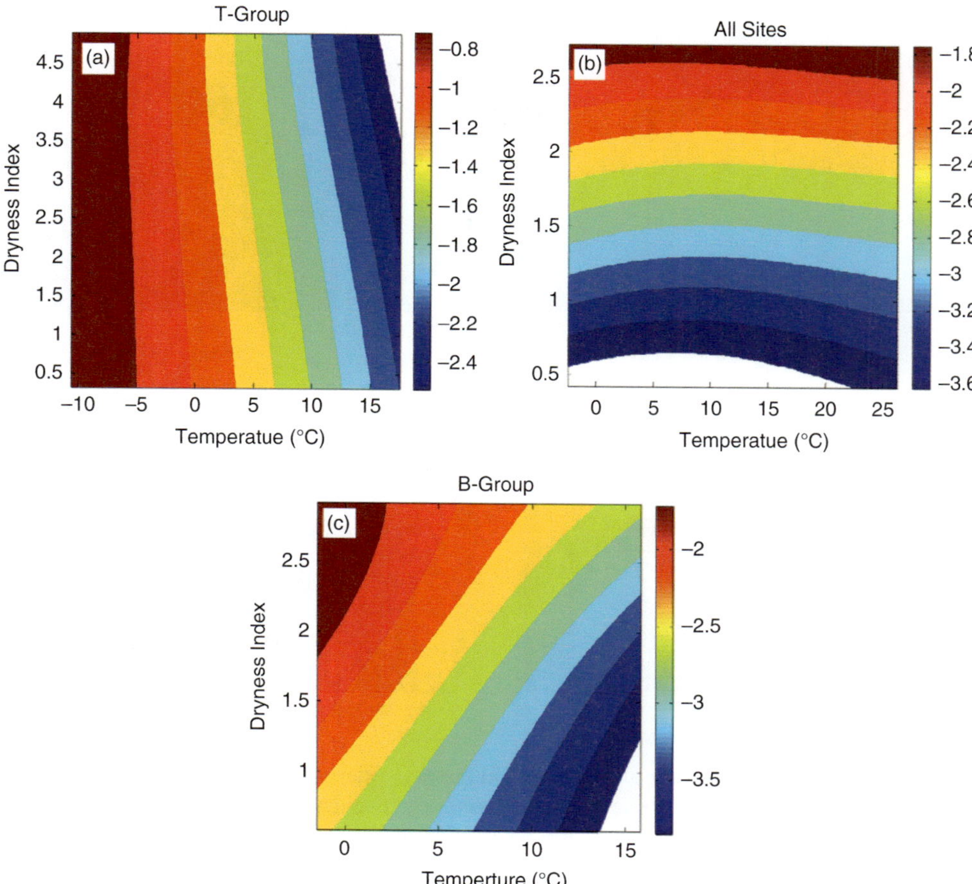

Figure 16.12. Contour plots of net ecosystem productivity (NEE) (gC ha^{-1} y^{-1}) for a range of biomes where (a) temperature is the primary driver of variation in productivity; (b) where aridity is the primary driver; and, (c) where both temperature and aridity co-limit productivity. Boreal forests are in group (a), tropical arid biomes are in group (b) and, mid-latitude forests are in group (c). Reproduced from Yi *et al.* (2010) with permission.

temperature and therefore the principal drivers of variation in productivity can vary across sites, as observed by Campo-Bescos *et al.* (2013) and others (Ma *et al.* 2007, Kanniah *et al.* 2011).

The response of NEE to temperature and aridity, singly, for temperature- and aridity-limited biomes respectively and the temperature and aridity responses for the mid-latitude biomes are shown in Figure 16.13.

16.5.3 Do Changes in Photosynthetic Capacity Along a Rainfall Gradient Explain Declining Productivity with Declining Rainfall?

Do changes in foliar photosynthetic biochemistry (for example electron transport capacity) account for the decline in canopy C uptake along rainfall gradients? While

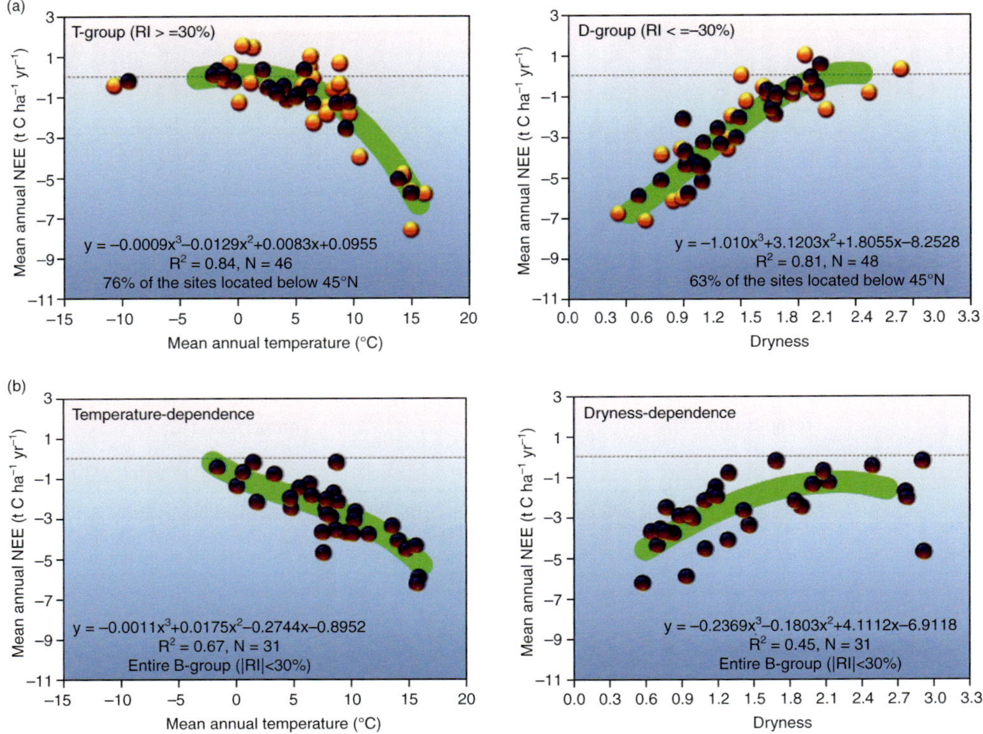

Figure 16.13. The response of mean annual net ecosystem exchange (NEE) to mean annual temperature and a dryness index for temperature-dependent, dryness dependent and co-dependent ecosystems. Reproduced with permission from Yi *et al.* 2010.

there is minimal variation in the Δ^{13}C of leaf material along the north Australian rainfall gradient (Miller *et al.* 2001, Schulze *et al.* 1998) and this might be interpreted to reflect a lack of variation in photosynthetic capacity at the leaf-scale, (because in C3 plants the Δ^{13}C of leaf material provides information about the c_i/c_a ratio), this conclusion may be incorrect. The c_i/c_a ratio is determined by changes in rates of photosynthesis (A) *and* stomatal conductance (g_s). Minimal change in Δ^{13}C suggests little variation in the c_i/c_a ratio but it is unclear as to whether this is because A and g_s remain unchanged, or because both change in proportion. Finally, the Δ^{13}C could reflect the c_i/c_a ratio that occurred during periods of favourable conditions (the wet season) rather than an annual integrated value reflecting changes between wet and dry seasons.

When expressed on a leaf area basis or a leaf dry weight basis (Prior *et al.* 2005), rates of net photosynthesis measured under light-saturating conditions along the north Australian rainfall gradient do not differ significantly (Cernusak *et al.* 2011, Prior *et al.* 2005). Typically rates of net photosynthesis ranged from about 8–16 μmol m^{-2} s^{-1}. When measured under CO_2 and light-saturating conditions there was again little difference along the rainfall gradient, with rates ranging from about 25 to 40 μmol m^{-2} s^{-1}. Similarly, apparent quantum yield did not differ along the gradient.

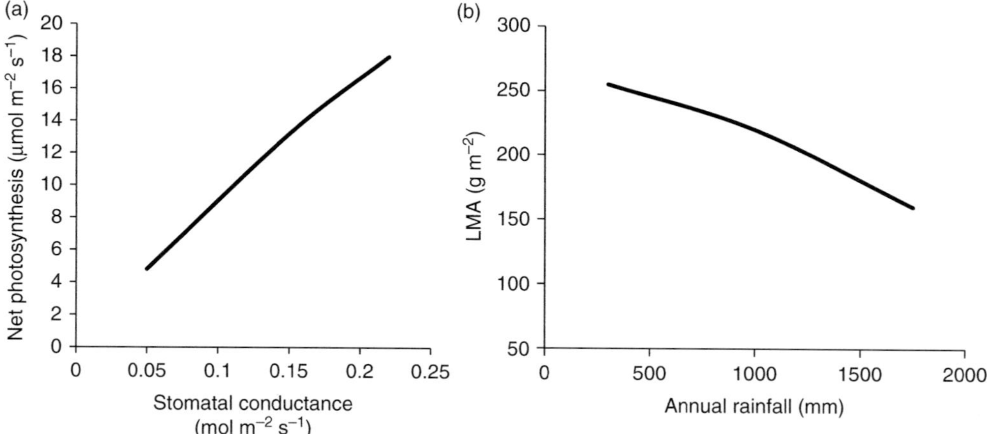

Figure 16.14. Changes in: (a) net photosynthesis with stomatal conductance; and (b) leaf mass area ratio (LMA) with rainfall along a rainfall gradient in northern Australia. Redrawn from Cernusak *et al.* (2011).

In contrast there was a strong decrease in leaf mass ratio (the ratio of leaf dry mass to leaf area) with increasing rainfall (Fig. 16.14). As a consequence, rates of photosynthesis per unit dry weight of leaf declined with a decrease in rainfall. However, this was compensated for to some extent by an increase in V_{cmax} with increasing LMA (i.e. V_{cmax} was larger in low rainfall sites than high rainfall sites; Fig. 16.14b). Very strong and linear increases in net photosynthesis with increasing g_s occur, as is generally observed across a very wide range of species globally.

The conclusion from these studies is that productivity gradients along aridity gradients in savannas (see later) is not caused by variation in leaf photosynthetic capacity but is caused by changes in leaf area index with aridity. This is further discussed later.

16.6 GPP, NEE and Respiration Differ in Their Response to Temperature and Aridity

Determining what regulates NEE (and hence net productivity) is difficult because NEE is the difference between two fluxes having very different biochemistries: photosynthetic carbon fixation (GPP) and ecosystem respiration (R_{eco}). GPP is influenced by canopy conductance, photosynthetic active radiation and LAI and is plant driven; in contrast R_{eco} is the sum of autotrophic (vegetative) and heterotrophic (especially microbial) respiration, much of which occurs within soils. Autotrophic respiration shows an approximately exponential increase with increasing temperature, with no optimal value identifiable. In addition, heterotrophic respiration of soils also shows an exponential increase with temperature, but soil moisture content is a very important determinant of soil respiration. As soil moisture content increases, respiration rate increases across a large range of temperatures (Fig. 16.15; Kutsch *et al.* 2008). In contrast to the exponential response of respiration to temperature, GPP shows an

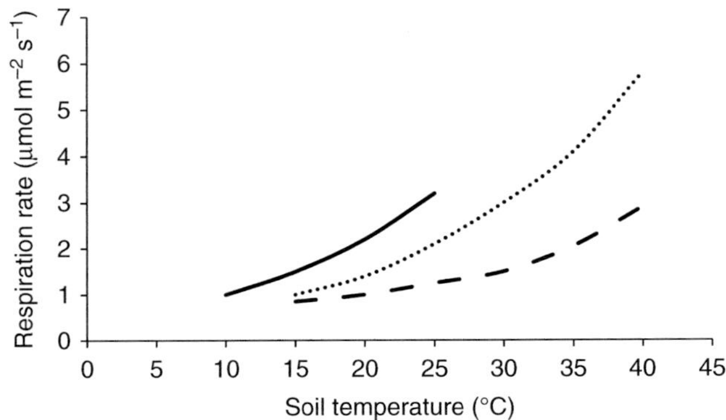

Figure 16.15. Rates of microbial respiration are larger when soil moisture content is large (solid line) and declines as soil moisture content declines to medium (dotted line) and small values (dashed line) across a wide range of soil temperatures. Redrawn from Kutsch *et al.* (2008).

optimal temperature. Consequently at temperatures above or below this optimum, rates of C fixation decline.

Because of the interaction of rainfall, temperature and photosynthesis and respiration, LAI and light interception, understanding patterns of change in C fluxes above savannas is relatively complex. Figure 16.16 summarises the principle interactions across sites and across years.

16.7 Inter-Annual Variations in Rainfall and Productivity: Comparisons Within a Single Site

For many locations, within a single site, there is a strong positive relationship between growth season rainfall or annual rainfall and LAI (Fig. 16.17) or productivity for savannas (and other biomes). Figure 16.18 shows data for an oak savanna/grassland in California (Ma *et al.* 2007) where increased growing season rainfall results in increased GPP. Much of this increase in GPP is the result of an increase in the length of the growing season. Similarly, Leuning *et al.* (2005) found a strong relationship between the amount of rainfall and growing season length in a semi-arid savanna in north eastern Australia. This was principally because of the response of the LAI of the C4 grasses to rainfall. Indeed, their study site was a C source for one year when the wet season delivered very low rainfall, but was a C sink when rainfall was average or better-than-average.

16.7.1 Drivers of Inter-Annual Variation in Savanna Productivity

Four principle causes of inter-annual variation in savanna productivity can be identified. All are linked to rainfall and the amount of rainfall is generally closely linked to

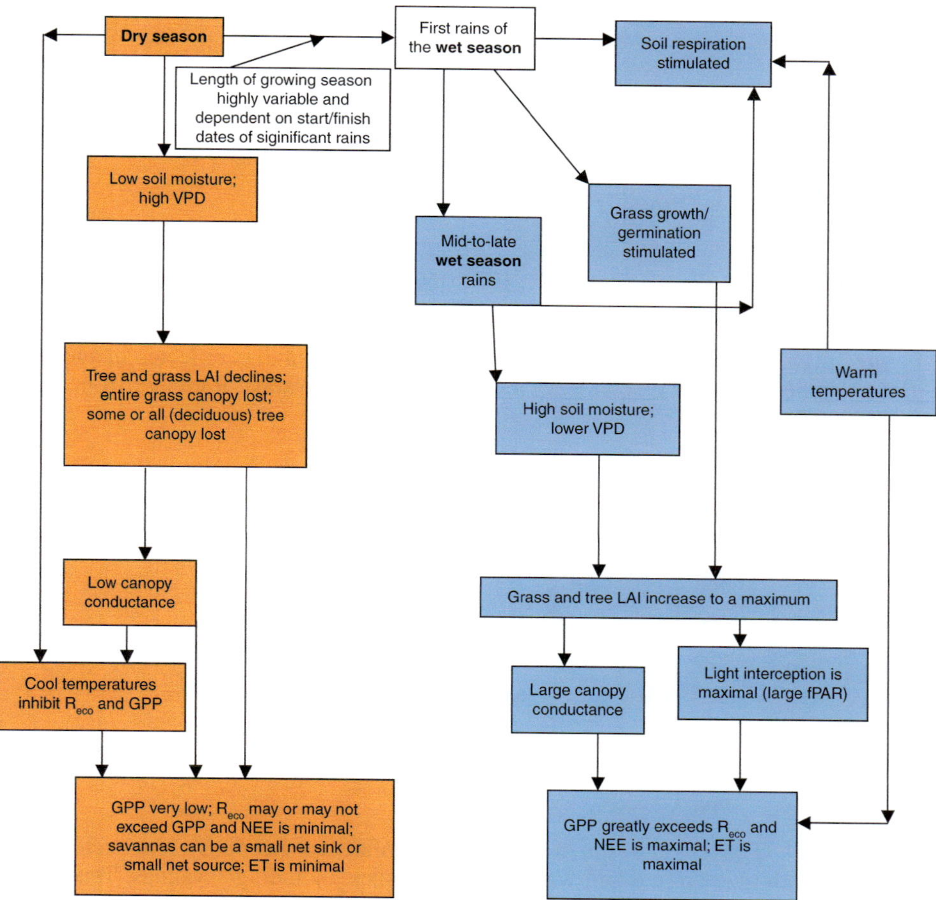

Figure 16.16. A summary of some of the interactions amongst abiotic and biotic drivers of C and water fluxes of savannas.

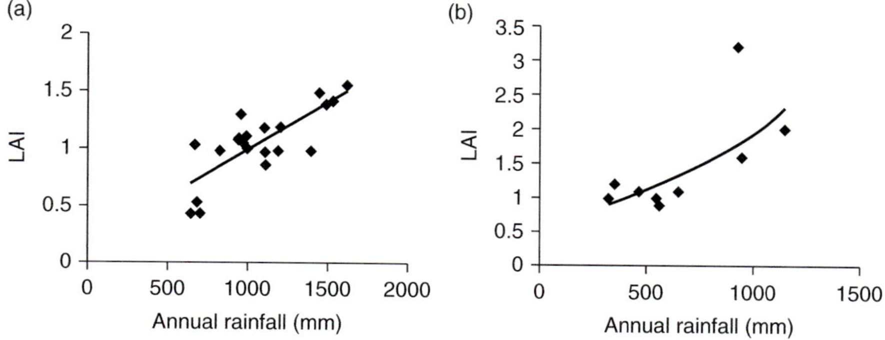

Figure 16.17. (a) The relationship between annual rainfall and leaf area index (LAI) for a range of savanna sites in north Australian and (b) African sites. Redrawn from Kanniah *et al.* (2011) and Merbold *et al.* (2009).

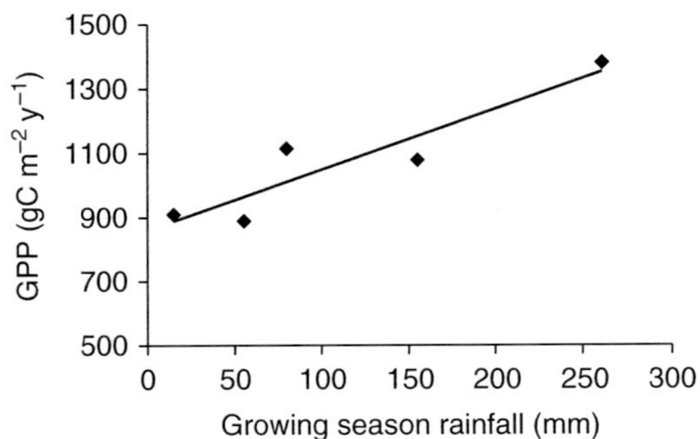

Figure 16.18. Inter-annual variation in growing season rainfall is reflected in annual gross primary productivity (GPP) in a Californian oak savanna. Redrawn from Ma *et al.* (2007).

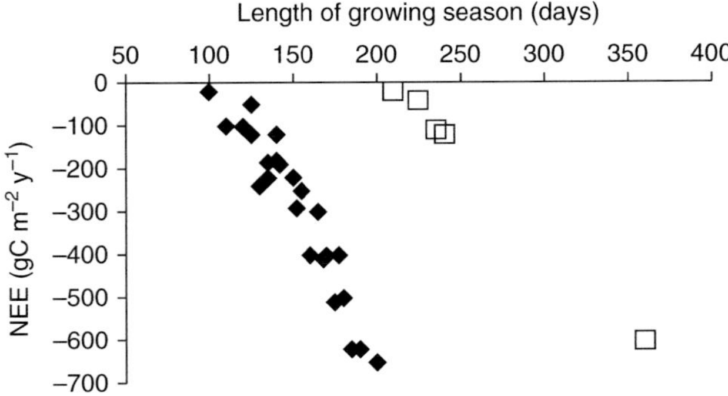

Figure 16.19. The length of the growing season influences net ecosystem exchange (NEE) of both temperature and water limited biomes, including savannas. Temperate and boreal ecosystems (black diamonds) and deciduous and evergreen savannas (open squares). Redrawn from Baldocchi (2008).

large-scale weather patterns, such as the El Niño and La Niña Southern Oscillation Index (Chapter 1). During El Niño, the west coast of South America receives more rainfall than average and savanna productivity in this region increases. Similarly in northern east Africa, rainfall is increased and productivity increased. In contrast, during El Niño, southern and central eastern Africa and eastern Australia experience warmer, drier conditions and productivity declines. From their analyses, Ciais *et al.* (2009) showed that inter-annual variation in rainfall, rather than variation in temperature, was the driver of variation in GPP in African savannas.

The four main causes of inter-annual variation in savanna productivity (excluding the effect of anthropogenic land-use change, for example, converting savanna to cropland or grazing land after clearing), are:

1. Length of the growing season (Fig. 16.19)
2. Changes in LAI of the understorey
3. Fire
4. Atmospheric CO_2 concentrations: see wood thickening discussed below.

The length of the growing season is principally determined by the timing of: (a) the "greening-up" and "browning-down" phases of the grass understorey; (b) the "leafing-out" (or bud-burst) and leaf-fall of deciduous and semi-deciduous shrubs and trees; and, (c) the recovery of full photosynthetic potential of the evergreen canopy as soil moisture reserves are replenished at the onset of the wet season. It has been recognised for many years that leaf flushing in many tree species occurs prior to wet season rains (Williams *et al.* 1997). The stimulus for this has been variously postulated to include day length, temperature and vapour pressure deficit as environmental cues. The influence of the timing of the start and end of the wet season on grass growth and productivity (Fig. 16.19) is discussed more fully in the arid and semi-arid grassland case study (Chapter 15). Suffice to state here that the timing of greening up is determined by the start of the wet season, which varies at each savanna site inter-annually. Recovery of photosynthetic potential of evergreen trees occurs through the gradual replacement of old (previous wet season) leaves with young (current wet season) leaves which contain more nitrogen, are less sclerophyllous and have far less injury from fungal, insect and herbivore damage than older leaves. The length of the growing season is responsive to the start and end of the wet season, which is the principle determinant of the LAI of the understorey that can be developed.

Fire has four principle impacts on savanna productivity. First is the loss of tree canopy foliage if the fire is of sufficient intensity. Second is the death of trees and saplings if the fire is of sufficient intensity. Third is the release of nutrients into the soil which can stimulate subsequent regrowth of vegetation; and, fourth, fire can stimulate seed fall and seed germination in many savanna species, thereby facilitating regrowth (resilience) in the vegetation. Fire, in most savannas, varies in intensity, frequency and spatial distribution across years. Few sites burn every year and this inter-annual variability produces significant inter-annual variability in productivity at any given site. Fire can also stimulate NPP of a site relative to unburnt sites, in both North American and South African savannas (Buis *et al.* 2009), but only when managed grazing is absent and when soil depths are not too shallow so that soil moisture stores are available to support rapid growth after the passing of fire.

The impact of long-term changes in atmospheric CO_2 concentration is discussed in detail in Section 16.8.

A fifth variable driving inter-annual variation in productivity is the ratio of annual direct beam to diffuse beam radiation (Kanniah *et al.* 2012). Illumination of leaves within canopies increases when diffuse beam radiation is the principle source of radiation rather than direct beam radiation because shelf-shading is reduced compared to solely direct beam radiation inputs. When Mount Pinatubo erupted and large amounts of aerosols were in the upper atmosphere, for example, increased rates of NEE were measured at Harvard Forest (Gu *et al.* 2003). More commonly it is changes in cloud cover that cause changes in the diffuse and direct beam environment of savannas.

16.8 Woody Thickening and Atmospheric CO$_2$ Concentrations

Woody thickening (also called vegetation thickening) is the increase in woody standing biomass in a landscape that already contains woody biomass. Woody encroachment is the movement of trees into grasslands. Both phenomena have been increasingly identified in the past twenty years (Macinnis-Ng *et al.* 2011, Eldridge *et al.* 2011) and are commonly observed in arid and semi-arid regions in Australia, in tropical rainforests of Central and South America and in grasslands of western North America. In southern Africa 13 million hectares of land are experiencing woody encroachment. Scholes and Hall (1996) estimate that the conversion of grasslands to woodlands could increase the terrestrial C sink by 94 PgC. Figure 16.20 shows an example of the change in tree stem density in South Africa for the period 1950–2010, where the mesic savanna has shown significant increases, whilst the xeric savanna has not. The xeric savanna does not receive sufficient rainfall to support an increased tree density but the mesic savann does.

Multiple causes for woody thickening and woody encroachment have been identified. These include: (1) overgrazing and recovery from anthropogenic disturbances from earlier land management practices; (2) reduced fire frequency; and, (3) long-term climate change.

Changes in land use (converting native savanna to grazed grassland or cropland, for example) are associated with the anthropogenic loss of tree cover globally. However, when sites are abandoned from human intervention these sites will generally return

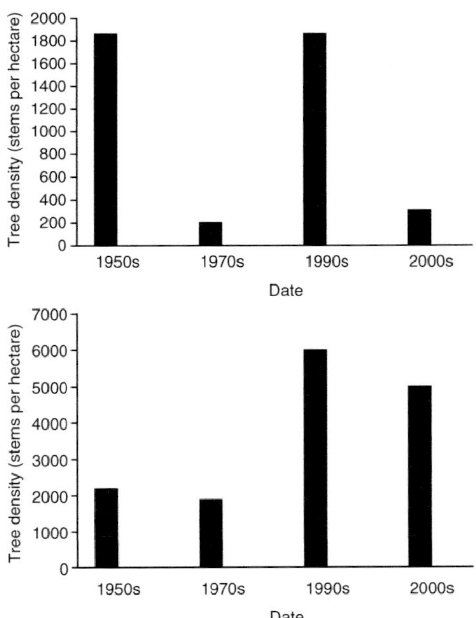

Figure 16.20. Changes in tree stem density for a xeric (upper panel) and a mesic (lower) savanna in South Africa. Note the difference in scale for the two graphs.

to a more woody composition and this is recorded as an increase in tree stem density. Similarly changes in land management practices can influence the frequency and severity and extent of fire and this is usually undertaken to reduce fire frequency. This is also associated with increased tree stem density. However, we are not concerned with these aspects of woody thickening in this book. Rather, we focus on the potential influence of climate change and in particular, increased atmospheric CO_2 concentrations as a potential cause of woody thickening.

Eamus and Palmer (2007) proposed that climate change and in particular, increased atmospheric CO_2 concentration and decreased evaporative demand (the Pan Evaporation Paradox; Gifford 2005) may explain the global phenomenon of woody thickening. They proposed a conceptual model based around six well-established observations:

1. The concentration of atmospheric CO_2 has increased since the industrial revolution (Keeling *et al.* 1995).
2. This increase accelerates the rate of photosynthesis (Eamus *et al.* 1995) and this increase is larger in C3 plants (i.e. trees) than C4 grasses.
3. Stomatal conductance declines in response to elevated concentrations of CO_2 in the atmosphere (Eamus and Jarvis 1989).
4. The growth rate of young trees and shrubs is enhanced by CO_2 enrichment (Berryman *et al.* 1993) and the proportional stimulation of growth is larger in xeric than mesic sites (Eamus and Ceulemans 2001).
5. Pan evaporation rates have declined globally (Roderick and Farquhar 2002, 2004).
6. Because of (3) and (5), there is increased water availability in the short-to-medium term which can support a larger standing biomass of woody species, that is, woody thickening.

This conceptual model was strongly supported in a recent modelling analysis of woodland function (Macinnis-Ng *et al.* 2011). These researchers modelled GPP, soil water content, stomatal conductance and leaf-scale rates of photosynthesis of a temperate woodland under four scenarios: (a) control, (the current climate); (b) a pre-industrialised climate with reduced atmospheric CO_2 concentration and increased vapour pressure deficit; (c) a future climate with an enriched atmospheric concentration of CO_2 and a reduced vapour pressure deficit; and, (d) a future climate with a 30 percent increase in LAI.

They showed that a future climate with CO_2 enrichment and a small decrease in evaporative demand (Table 16.1) results in:

1. Decreased canopy conductance and decreased rates of tree water-use (sapflow)
2. Increased soil moisture content for much of the year
3. Increased GPP and increased water-use-efficiency
4. An increase in LAI of 30% could be sustained with an unaltered pattern of rainfall and this supports an increased tree stand density.

Table 16.1. Total annual sapflow, GPP and mean daily WUE for four scenarios modelled using the SPA model

	Annual sapflow (mm y^{-1})	Annual GPP (gC m^{-2} y^{-1})	Average daily WUE ($\times$ 10^3)
Current ambient	400	1485	4.7
Past, pre-industrialised climate (275 ppm CO_2 and 125% of current average D)	427	1210	3.5
Future climate (561 ppm CO_2, 75% of current D	357	1727	6.8
Future climate (as above) but with LAI increased 30%	398	1811	6.3

Source: From Macinnis-Ng *et al.* (2011).

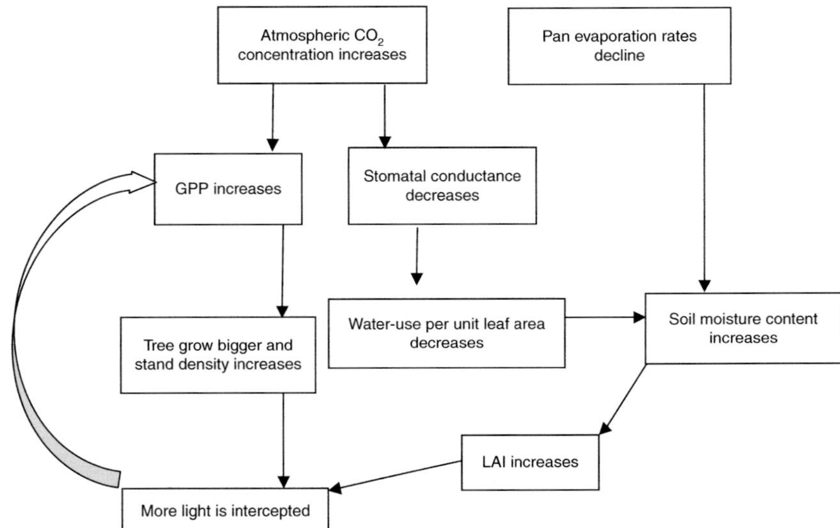

Figure 16.21. A conceptual model for the interactions of CO_2 increase, decreased vapour pressure deficit (VPD), soil moisture content, increased gross primary productivity (GPP) and increased tree stand density. Redrawn from Macinnis-Ng *et al.* (2011).

A conceptual model for the mechanisms linking climate change (increased CO_2 concentration and decreased VPD) and woody thickening is shown in Figure 16.21. It was observed in this modelling study that an increase of 30 percent in LAI could be supported through the increase in soil moisture content arising from decreased rates of tree water-use and soil evaporation arising from reduced stomatal conductance and pan evaporation rates. Support for this mechanism can be found in the analyses of Berry and Roderick (2002) who have demonstrated an increase in the tree density of Australia over the past two hundred years, a result they ascribe to increased atmospheric CO_2 concentrations. Similarly, annual water savings in response to CO_2 enrichment have been observed under a range of field-based CO_2 enrichment studies (Wullschleger *et al.* 2002). Finally, continental-scale run-off may be increasing (Labat *et al.* 2004, Gedney *et al.* 2006), a result ascribed to reduced rates of vegetation water-use because of reduced stomatal conductance.

This chapter has examined the patterns and controls of water and C flux at daily, seasonal and inter-annual time-scales. The importance of seasonality of water supply was highlighted, but interactions with temperature and fire were also discussed. The global phenomenon of woody thickening was discussed, with a potential mechanism presented. The savanna biome contrasts significantly from Boreal and tropical montane forests, which are the topics of other case study chapters.

16.9 References

Arneth A, EM Veenendaal, C Best, W Timmermans, O Kolle and L Montagnani, (2006). Water-use strategies and ecosystem-atmosphere exchange of CO_2 in two highly seasonal environments. *Biogeosciences Discussion* 3, 421–437.

Baldocchi DD, (2008). Breathing of the terrestrial biosphere: lessons learned from a global network of carbon dioxide flux measurement systems. *Australian Journal of Botany* 56, 1–26.

Baldocchi DD and Y Ryu, (2011). Synthesis of Forest Evaporation Fluxes–from Days to Years–as Measured with Eddy Covariance. In *Forest Hydrology and Biogeochemistry: Synthesis of Past Research and Future Directions*, DF Levia and D Carlyle-Moses (Eds.), *Ecological Studies*, 216, Springer Science Business Media. DOI 10.1007/978-94-007-1363-5.

Berry Sl and ML Roderick, (2002). CO_2 and land-use effects on Australian vegetation over the last two centuries. *Australian Journal of Botany* 50, 511–531.

Berryman CA, D Eamus and GA Duff, (1993). The influence of carbon dioxide enrichment on growth, nutrient content and biomass allocation of *Maranthes corymbosa*. *Australian Journal of Botany* 41, 195–209.

Bruemmer C, U Falk, H Papen, J Szarzynski and R Wassmann, (2008). Diurnal, seasonal and interannual variation in carbon dioxide and energy exchange in shrub savanna in Burkina Faso (West Africa). *Journal of Geophysical Research-Biogeosciences* 113, GO2030.

Bucci SJ, FG Scholz, G Goldstein, WA Hoffmann, FC Meinzer, AC Franco, T Giambelluca and F Miralles-Wilhelm, (2008). Controls on stand transpiration and soil water utilization along a tree density gradient in a Neotropical savanna. *Agricultural and Forest Meteorology* 148, 839–849.

Buis GM, JM Blair, DE Burkepile, CE Burns, (2009) Controls of above-ground net primary production in mesic savanna grassland: an inter-hemispheric comparison. *Ecosystems* 12, 982–995.

Campo-Bescos MA, R Munoz-Carpena, DA Kaplan, J Southworth, L Zhu, PR Waylen, (2013). Beyond precipitation: physiographic gradients dictate the relative importance of environmental drivers on savanna vegetation. *PLOS One* 8, e72348. DOI: 10.1371/journal.pone.0072348

Cernusak LA, LB Hutley, J Beringer, JAM Holtum, BL Turner *et al.*, (2011). Photosynthetic physiology of eucalypts along a sub-continental rainfall gradient in northern Australia *Agricultural and Forest Meteorology* 151, 1462–1470.

Ciais P, S-L Piao, P Cadule, P Friedlingstein and A Chedin, (2009). Variability and recent trends in the African terrestrial carbon balance. *Biogeosciences* 6, 1935–1948.

Cowan IR and GD Farquhar, (1977). Stomatal function in relation to leaf metabolism and environment. *Symposium of the Society for Experimental Biology* 31, 471–505.

Do FC, A Rocheteau, AL Diagne, V Goudiaby, AL Granier and JP Homme, (2008). Stable annual pattern of water-use by *Acacia tortilis* in Sahelian Africa. *Tree Physiology* 28, 95–104.

Dye PJ, (1996). Climate, forests and streamflow relationships in South African afforested catchments. *Commonwealth Forestry Review* 75, 31–38.

Eamus D and R Ceulemans, (2001). Effects of greenhouse gases on the gas exchange of forest trees. In *The Impact of CO2 and Other Greenhouse Gases on Forest Ecosystems*, Karnosky D and R Ceulemans (Eds.), CABI Publishing, UK, pp. 17–56.

Eamus D, J Cleverly, N Boulain, N Grant, R Faux and R Villalobos-Vega, (2013). Carbon and water fluxes in an arid-zone *Acacia* savanna woodland: An analyses of seasonal patterns and responses to rainfall events. *Agricultural and Forest Meteorology* 182–183, 225–238.

Eamus D and SC Cole, (1997). Diurnal and seasonal comparisons of assimilation, phyllode conductance and water potential of three *Acacia* and one eucalypt species in the wet–dry tropics of Australia. *Australian Journal of Botany* 45, 275–290.

Eamus D, G Duff and CA Berryman, (1995). The effects of increased atmospheric CO_2 concentration on the response of assimilation to temperature, LAVPD, light flux density and CO_2 concentration of *Eucalyptus tetrodonta*. *Environmental Pollution* 68, 133–140.

Eamus D, LB Hutley and A O'Grady, (2001). Daily and seasonal patterns of carbon and water fluxes above a north Australian savanna. *Tree Physiology* 21, 977–988.

Eamus D and PG Jarvis, (1989). The direct effects of increase in the global atmospheric CO_2 concentration on natural and commercial temperate trees and forests. *Advances in Ecological Research* 19, 1–55.

Eamus D, A O'Grady and L Hutley, (2000). Dry season conditions determine wet season water-use in the wet-dry tropical savannas of northern Australia. *Tree Physiology* 20, 1219–1226.

Eamus D and AR Palmer, (2007). Is climate change a possible explanation for woody thickening in arid and semi-arid regions? *Research Letters in Ecology* DOI: 10.1155/2007/37364.

Eldridge DJ, MA Bowker, FT Maestre, E Roger, JF Reynolds and WG Whitford, (2011). Impacts of shrub encroachment on ecosystem structure and functioning: towards a global synthesis. *Ecology Letters* 14, 709–722.

Franco AC, (1998). Seasonal patterns of gas exchange, water relations and growth of *Roupala montana*, an evergreen savanna species. *Plant Ecology* 136, 69–76.

Gedney N, PM Cox, RA Betts, O Boucher, C Huntingford and PA Stott, (2006). Detection of a direct carbon dioxide effect in continental river runoff records. *Nature* 439, 835–838.

Gifford RM, (2005). (Ed) Pan evaporation: An example of the detection and attribution of trends in climate variables. Proceedings of a workshop held at the Shine Dome, Australian Academy of Science, Canberra 22–23 November 2004.

Givnish J, (1988). Adaptation to sun and shade–a whole plant perspective. *Australian Journal of Plant Physiology* 15, 63–92.

Gu LH, DD Baldocchi, SC Wofsy, JW Munger, JJ Michalsky, SP Urbanski and TA Boden, (2003). Response of a deciduous forest to the Mount Pinatubo eruption: enhanced photosynthesis. *Science* 299, 2035–2038.

Hubbard RM, BJ Bond, MG Ryan, (1999) Evidence that hydraulic conductance limits photosynthesis in old *Ponderosa* trees. *Tree Physiology* 19, 165–172.

Kanniah KD, J Beringer and LB Hutley, (2011). Environmental controls on the spatial variability of savanna productivity in the Northern Territory. *Agricultural and Forest Meteorology* 151, 1429–1439.

Kanniah KD, J Beringer, P North and L Hutley, (2012). Control of atmospheric particles on diffuse radiation and terrestrial plant productivity: A review. *Progress in Physical Geography* 36, 209–237.

Keeling C, TP Whort, M Wahlen and J van der Plicht, (1995). Interannual extremes in the rate of rise of atmospheric carbon dioxide since 1980. *Nature* 375, 666–670.

Kutsch WL, N Hanan, B Schoiles, I McHugh, W Kubheka, H Eckhardt and C Williams, (2008). Response of carbon fluxes to water relations in a savanna ecosystem in South Africa. *Biogeosciences* 5, 1797–1808.

Labat D, Y Godderis, JL Probst and JL Guyot. (2004). Evidence for global runoff increase related to climate warming. *Advances in Water Resources* 27, 631–642.

Leuning R, HA Cleugh, SJ Zegelin and D Hughes, (2005). Carbon and water fluxes over a temperate *Eucalyptus* forest and a tropical wet/dry savanna in Australia: measurements and comparison with MODIS remote sensing estimates. *Agricultural and Forest Meteorology* 129, 151–173.

Ma S, DD Baldocchi, L Xu and T Hehn, (2007). Inter-annual variability in carbon dioxide exchange of an oak/grass savanna and open grassland in California. *Agricultural and Forest Meteorology* 147, 157–171.

Ma X, A Huete, Q Yu, N Restrepo Coupe, K Davies, M Broich, P Ratana, J Beringer, LB Hutley, J Cleverly, N Boulain and D Eamus, (2013). Spatial patterns and temporal dynamics in savanna vegetation phenology across the North Australian Tropical Transect. *Remote Sensing of Environment* 139, 97–115.

Macinnis-Ng C, M Zeppel, M Williams and D Eamus, (2011). Applying a SPA model to examine the impact of climate change on GPP of open woodlands and the potential for woody thickening. *Ecohydrology* 4, 379–393.

Medlyn BE, RA Duursma, D Eamus, DS Ellsworth, CI Prentice, CVM Barton, KY Crous, P de Angelis, M Freeman and L Wingate, (2011). Reconciling the optimal and empirical approaches to modelling stomatal conductance. *Global Change Biology* 17, 2134–2144.

Merbold, L, J Ardo, A Arneth, RJ Scholes, Y Nouvellon, A de Grandcourt, S Archibald, JM Bonnefond, N Boulain, N Brueggemann, C Bruemmer, B Cappelaere, E Ceschia, HAM El-Khidir, BA El-Tahir, U Falk, J Lloyd, L Kergoat, V Le Dantec, E Mougin, M Muchinda, MM Mukelabai, D Ramier, O Roupsard, F Timouk, EM Veenendaal and KL Kutsch, (2009). Precipitation as driver of carbon fluxes in 11 African ecosystems. *Biogeosciences* 6, 1027–1041.

Miller JM, RJ Williams and GD Farquhar, (2001). Carbon isotope discrimination by a sequence of *Eucalyptus* species along a sub-continental rainfall gradient in Australia. *Functional Ecology* 15, 222–232.

Miranda AC, HS Miranda, J Lloyd, J Grace, RJ Francey, JA McIntyre, P Meir, P Riggan, R Lockwood and J Brass, (1997). Fluxes of carbon, water and energy over Brazilian cerrado: an analysis using eddy covariance and stable isotopes. *Plant, Cell and Environment* 20, 315–328.

Myers BA, GA Duff, D Eamus, IR Fordyce, A O'Grady and RJ Williams, (1997). Seasonal variation in water relations of trees of differing leaf phenology in a wet–dry tropical savanna near Darwin, northern Australia. *Australian Journal of Botany* 45, 225–240.

Pitman JE, (1996). Ecophysiology of tropical dry evergreen forest, Thailand: measured and modelled stomatal conductance of *Hopea ferrea*, a dominant canopy emergent. *Journal of Applied Ecology* 33, 1366–378.

Prior L and D Eamus, (2000). Seasonal changes in hydraulic conductance, xylem embolism and leaf area in *Eucalyptus tetrodonta* and *E. miniata* saplings in a north Australian savanna. *Plant, Cell and Environment* 23, 800–810.

Prior LD, DMJS Bowman and D Eamus, (2005). Intra-specific variation in leaf attributes of four savanna tree species across a rainfall gradient. *Australian Journal of Botany* 53, 323–335.

Prior LD, D Eamus and GA Duff, (1997). Seasonal trends in carbon assimilation, stomatal conductance, pre-dawn leaf water potential and growth in *Terminalia ferdinandiana*, a deciduous tree of northern Australian savannas. *Australian Journal of Botany* 45, 53–69.

Roderick ML and GD Farquhar, (2002). The cause of decreased pan evaporation over the past 50 years. *Science* 298, 1410–1411.

Roderick ML and GD Farquhar, (2004). Changes in Australian pan evaporation from 1970 to 2002. *International Journal of Climatology* 24, 1077–1090.

San Jose J, R Montes, J Grace and N Nikonova, (2008). Land-use changes alter CO_2 flux patterns of a tall-grass Andropogon field and a savanna-woodland continuum in the Orinoco lowlands. *Tree Physiology* 28, 437–450.

Scholes RJ and DO Hall, (1996). The carbon budget of tropical savannas, woodlands and grasslands. In *Global Change: Effects on Coniferous Forests and Grasslands*, SCOPE, Vol. 56, Breymeyer AI, DO Hall, JM Melillo and GI Agren (eds). Wiley: Chichester; pp. 69–100.

Schulze ED, RJ William, GD Farquhar, ED Schulz, J Langridge, JM Miller, BH Walker. (1998). Carbon and nitrogen isotope discrimination and nitrogen nutrition of trees along a rainfall gradient in northern Australia. *Australian Journal of Plant Physiology* 25, 413–425.

Schymanski SJ, ML Roderick, M Sivapalan, LB Hutley and J Beringer, (2007). A test of the optimality approach to modelling canopy properties and CO_2 uptake by natural vegetation. *Plant, Cell and Environment* 30, 1586–1598.

Thomas DS, D Eamus and S Shanahan, (1999). Influence of season, drought and xylem ABA on stomatal responses to leaf-to-air vapour pressure difference of trees of the Australian wet-dry tropics. *Australian Journal of Botany* 48, 143–151.

Tognetti R, M Michelozzi and A Giovannelli, (1997). Geographical variation in water relations, hydraulic architecture and terpene composition of Aleppo pine seedlings from Italian provenances. *Tree Physiology* 17, 241–250.

Tyree MT and FW Ewers, (1996). Hydraulic architecture of woody tropical plants. In *Tropical Forest Plant Ecophysiology*. SS Mulkey, RL Chazdon and AP Smith (Eds). Chapman and Hall, New York, pp 217–243.

Veenendaal EM, KB Mantlana, NW Pammenter, P Weber, P Huntsman-Mapoila and J Lloyd, (2008). Growth form and seasonal variation in leaf gas exchange of *Colophospermum mopane* savanna trees in northwest Botswana. *Tree Physiology* 28, 417–424.

Whitley RJ, CMO Macinnis-Ng, LB Hutley, J Beringer, M Zeppel, M Williams, D Taylor and D Eamus, (2011). Is productivity of mesic savannas light limited or water limited? Results of a simulation study. *Global Change Biology* 17, 3130–3149.

Williams RJ, BA Myers, MJ Muller, GA Duff and D Eamus, (1997). Leaf phenology of woody species in a north Australian tropical savanna. *Ecology* 78, 2542–2558.

Wright IJ, PB Reich, M Westoby and DD Ackerly, *et al.*, (2004). The worldwide leaf economics spectrum. *Nature* 428, 821–827.

Wullschleger SD, CA Gunderson, PJ Hanson, KB Wilson and RJ Norby, (2002a). Sensitivity of stomatal and canopy conductance to elevated CO_2 concentration–interacting variables and perspectives of scale. *New Phytologist* 153, 485–496.

Wullschleger SD, TJ Tschaplinski and RJ Norby, (2002b). Plant water relations at elevated CO_2 implications for water-limited environments. *Plant, Cell and Environment* 25, 319–331.

Yi C, D Ricciuto, R Li, J Wolbeck, X Xu, M Nilsson and L Aires, (2010). Climate control of terrestrial C exchange across biomes and continents. *Environmental Research Letters* 5, 034007.

17

Seasonal Behaviour of Vegetation of the Amazon Basin

17.1 Introduction

The Amazon basin represents a major component of regional and global carbon and water cycles. Understanding seasonal and spatial variations of tropical forest structure and functioning in Amazonia is important for understanding and predicting the fate of Amazon forests under climate change.

Seasonality in productivity of tropical forests is more subtle than that observed in temperate zones that are dominated by an active growing and a dormant season. Field-based studies of tree phenology in the humid tropics show flowering and flushing of new leaves along with increases in leaf area during the dry season when solar radiation availability is larger relative to the more cloudy wet season.

Eddy covariance flux tower measurements show enhanced photosynthesis and larger rates of evapotranspiration in the dry season. Tower measurements are crucial for understanding the mechanisms of functional phenology in tropical forests.

Satellite-based remote sensing data also show forest greening at the basin-scale during the dry season, thereby greatly extending site-based studies. Earth system models have generally represented Amazon forests as water-limited or with year-round constant leaf area and thus predict dry season declines in productivity. However, some models show enhanced productivity in the dry season.

There are many challenges in field, satellite and model assessments of tropical forest phenology, with field methods time consuming and limited to narrow spatial and temporal scales, satellite data subject to cloud contamination and optical artefacts and models subject to uncertainties.

The paradigms of light-limitation or water limitation in tropical forests remain controversial, yet increasingly there is a consensus that basin-wide predictions of modelled carbon and water fluxes can be constrained by incorporation of remote sensing data and local flux measurements.

17.2 Biogeography of the Amazon Basin

Knowledge of spatial and seasonal variations in Amazon tropical forests resulting from meteorological drivers, soils, topography, and disturbance (fire, logging, deforestation) is important for understanding their ecological function and future fate with climate change. The Amazon Basin encompasses an area of 7.5×10^6 km^2 and is covered mostly of dense tropical evergreen broadleaf rainforests from wet equatorial forests to tropical dry forests. These primary forests are mostly closed canopies with high LAI of 5–7, and are approximately 40 m tall (Goulden *et al.* 2004).

Topographic elevation variations are from zero to approximately 400 m and the basin is drained by the Amazon River of 6,800 kilometres length (Fig. 17.1). The western Amazon has the highest elevations, while the central region is at the lowest elevation with seasonally inundated forests and upland (*terra firme*) forests present (Fig. 17.2). The eastern Amazon Basin consists of mostly upland forests dissected by broad, slow-flowing river valleys. There is a soil fertility and productivity gradient from highly productive western margins of the basin to the nutrient-poor clay Oxisols and sandy Ultisols in central and eastern forests (Malhi *et al.* 2009; Quesada

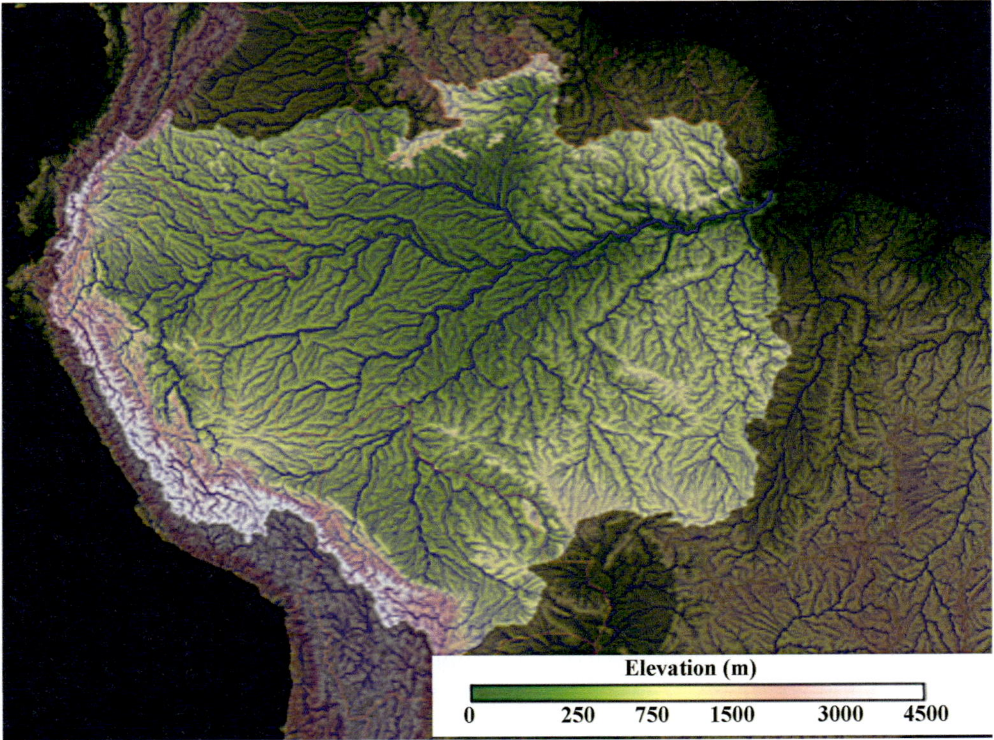

Figure 17.1. The Amazon Basin and the watersheds feeding it. The rim of higher terrain along the western edge of the Amazon is where tributaries drain off mountain peaks in the Andes. The Amazon is 6,800 kilometres long. Image from Earth Observatory (http://earthobservatory.nasa.gov/Features/LBA/).

et al. 2010). To the south are transitional or ecotone forests, and dry tropical forests (Fig. 17.3).

Amazon forests are also subject to land cover changes resulting from increasing land use pressures, forest disturbance and conversion. High levels of deforestation have occurred at the eastern and southern margins of the basin forming a surrounding arc of deforestation (Fig. 17.3). As a result, a more complex regional mosaic of forest

(a)

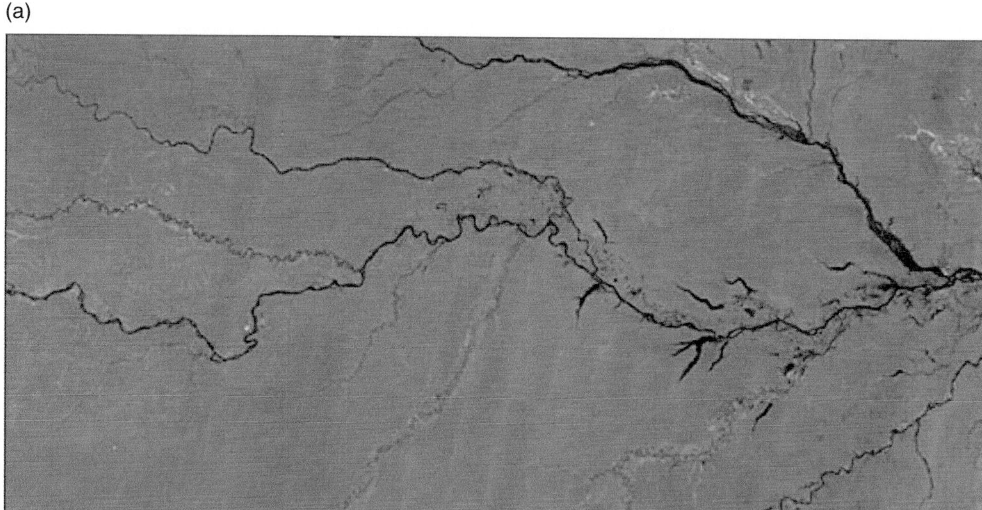

Dry Season (September–November 1995)

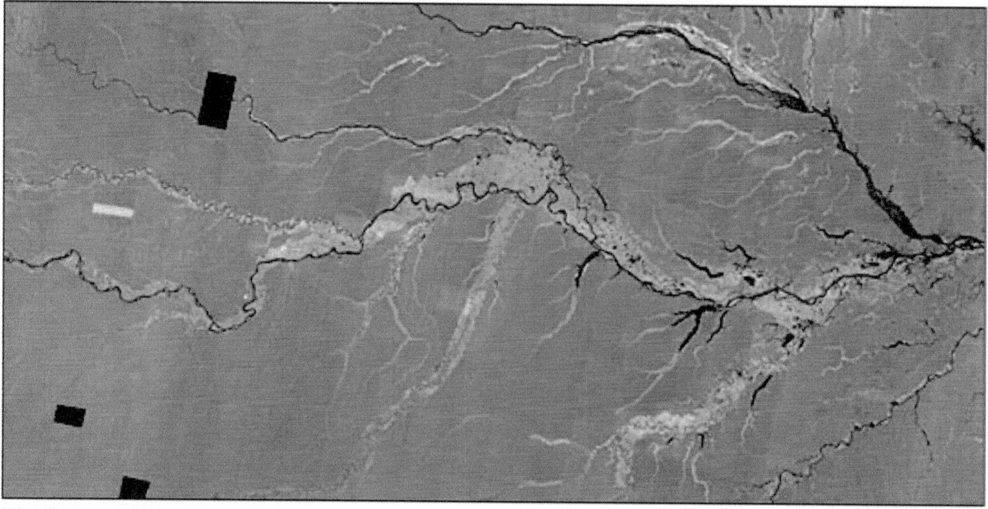

Wet Season (May–June 1996)

Figure 17.2. (a) Radar maps of the Amazon Basin reveal the seasonally flooded forest. In the pair of images, black represents permanent waterways, dark grey represents forest, and light grey represents flooded areas; (b) Ground photographs of seasonal flooding. (Image from Earth Observatory http://earthobservatory.nasa.gov and with data provided by the Global Rainforest Mapping Project).

(b)

Dry Season (June–November)

Wet Season (December–May)

Figure 17.2. (*cont.*)

types and vegetation structures can be expected from land use activities and forest encroachment (Trancoso *et al.* 2010; Asner *et al.* 2004).

Meteorological conditions vary extensively across the Amazon basin with diverse rainfall regimes and mean annual precipitation ranging from 1900 to over 3000 mm. The Inter-tropical Convergence Zone (ITCZ) is centred over the Amazon Basin from December to February, and the shift of the ITCZ to the northwest (Horel *et al.* 1989) results in a dominant seasonal rhythm in rainfall with a distinct dry season (defined as precipitation inputs less than 100 mm month^{-1}) varying in length from zero months

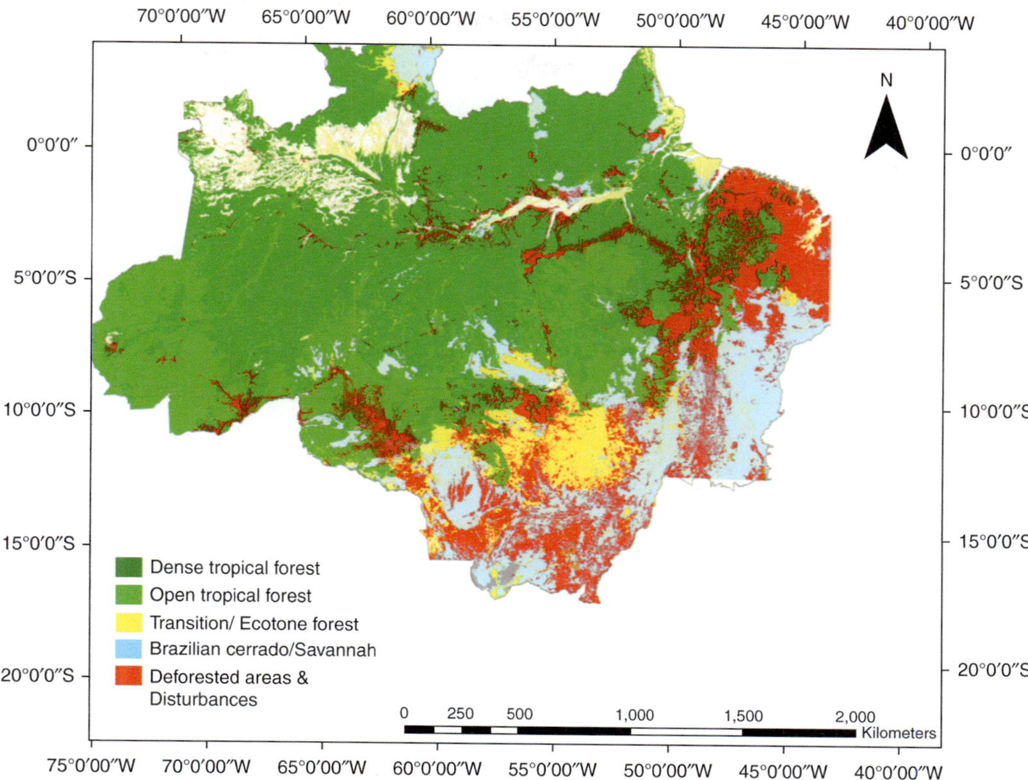

Figure 17.3. Distribution of the dominant forest types in the Amazon basin. *Source*: RADAMBRASIL and the deforestation from PRODES project 2005.

(no dry season in the ever-wet northwestern Amazon) to 5 months in central eastern equatorial Amazonia and the southwestern edge (Sombroek 2001) (Fig. 17.4). The eastern Amazon Basin has an annual rainfall of about 2000 mm with a distinct four- to five-month dry season period from July to November. The central equatorial forests have a similar annual rainfall but a shorter dry season (<2 months). The western Amazon lacks a dry season, although there remains appreciable seasonality in available radiation and rainfall.

The dry season brings a decrease in cumulus cloud cover and in general increases in solar radiation at the canopy surface. Smoke from biomass burning can become prevalent as the dry period progresses (Koren *et al.* 2004), shifting the proportions of direct and diffuse radiation onto the canopy surface.

17.3 Seasonality in Tropical Forest Function

The functioning of Amazon rainforests plays an essential role in global carbon, water and energy fluxes and the Earth's climate. Globally significant variations in these cycles are vulnerable to climate change (Betts *et al.* 2004). Our ability to project

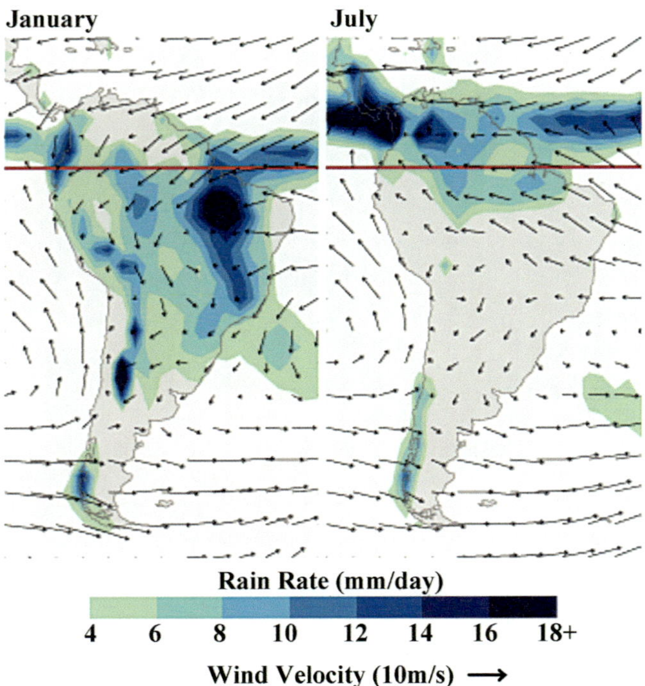

Figure 17.4. The prevailing winds over the equatorial Amazon flip from blowing north to south in January to blowing south to north in July. The reversal of the winds brings moisture from the Atlantic and the Caribbean into different parts of the Amazon, shifting the rainy season back and forth between areas north of the equator and areas to the south. (Image from Earth Observatory http://earthobservatory.nasa.gov/Features/LBA/) adapted from images provided by Wenhong Li).

the role of forest functional feedbacks on future climate critically depends upon our understanding of these tropical forest ecosystems, their tolerance to climate extremes and tipping points of ecosystem collapse. Climate change scenarios projected for the twenty-first century show tropical forests in the Amazon becoming drier due to stronger El Niño conditions and predictions for the future of Amazon forests under climate change vary widely with some models showing tropical forest die-back (Betts *et al.* 2004), while other models imply forest persistence (Friedlingstein *et al.* 2006).

Further, land use and land cover change alter species composition and thereby influence ecosystem functioning, carbon and water cycling, and regional climate (Lambin *et al.* 2003; Nobre *et al.* 1991). There is increasing concern about Amazon forest responses to global warming and land use pressures with climate-carbon cycle feedbacks due to deforestation and rising temperatures having potentially catastrophic consequences to regional and global climate (Betts *et al.* 2004; Cox *et al.* 2004). Fires and forest disturbance result in large amounts of carbon released to the atmosphere while regenerating secondary forests can rapidly sequester carbon through rapid regrowth.

Diverse evidence from ecological field studies, eddy-flux towers, and satellites suggests that many tropical forests may be more light-limited, rather than water-limited, and hence "green-up" during the higher sunlight dry season (Restrepo Coupe *et al.* 2013). Short-term flushes of greening have also been observed during inter-annual drought events (Saleska *et al.* 2007). This suggests that tropical forests, in responding positively to increased light availability in dry periods, are more complex, and more robust than most earth system models suggest, with implications for long-term vulnerability to climate change. Further, the vulnerability of tropical forest systems to climate change depends, not just on the physical climate system acting upon the forest but also on the biological response of forests to initial climate changes.

17.4 Field Phenology Studies

Local field studies of tropical tree species have generally reported synchronized flushing and exchange of new leaves, with periods of increased foliage density, senescence, and litterfall. Wright and Schaik (1994) found leaf and flower production at sites across eight different rainforests, including the Ducke Reserve near Manaus in the central Amazon, to coincide with seasonal peaks in solar irradiance in the dry season. This avoids herbivory by insects in the wet season, if leaves emerge in the dry season (Fig. 17.5). In the seasonally dry, central and eastern Amazon rainforests, the envelopment of mature leaves by epiphylls (moss, algae) in the late wet season inhibits photosynthesis (Roberts *et al.* 1998). This is followed by an onset of leaf senescence and litterfall in the early dry season succeeded by a period of rapid leaf turnover with significant increases in leaf area and photosynthesis, coincident with dry season increases in radiation.

Trumbore (1995) reported seasonally-dry evergreen forests of eastern Amazonia to remain active through an extended dry season as their biologic available water is supplemented by facultative use of deeper soil moisture layers. Solar radiation intercepted by clouds may be the limiting factor to forest productivity in the wetter areas with shorter dry seasons (longer duration of clouds; see Tropical Montane Forest case study). Species-specific shoot and leaf growth with little evidence of stress occurs possibly because deep tree roots allow continuous access to deep soil moisture layers, even during the dry season (Nepstad *et al.* 1994).

Borchert *et al.* (2005a) noted that from equatorial to temperate latitudes, the development of perennial plants is characterized by distinct periods of dormancy, shoot growth (flushing), and synchronous flowering, a prerequisite for efficient cross-pollination of tree species. The timing of these phenological periods tends to be well adapted to the local seasonality of the environment, however, such signals are nearly absent along the equator, despite observed synchronous flowering and leaf flushing periods in these regions.

Borchert *et al.* (2005b) investigated over one hundred tree species in equatorial South America and Africa and reported synchronous tree species flushing or flowering that corresponded to periods of increasing solar insolation, a property of sun angle and mean day length (see Chapter 2). They noted that photoperiodic controls of plant

Figure 17.5. Leaf flushing, epiphylls, and insect herbivory in the dry season in the Amazon, July 2009. (Photos A Huete).

phenology along the Equator were implausible, since day length is 12 h year-round and further reported bimodal greening patterns that were coincident with the two annual periods of increasing or declining insolation (two equinox periods of March 21 and September 21). They concluded that at all latitudes seasonal changes in daily insolation, rather than day length, induce synchronous plant development that starts at the same time each year.

By contrast, tropical forest phenologies can shift with variations in seasonal climate and with water limitations becoming more dominant in the drier tropical areas. Herrerias-Diego *et al.* (2006) found more synchronous flowering and fruiting events in the early wet season in tropical dry forests of Mexico noting that these forests may directly brown-down and have lower photosynthetic capacity due to reduced water availability. Other factors that may alter seasonal functioning are inundation periods in lowland tropical forests (Fig. 17.2).

Tropical rainforests are light (that is, energy) limited as long as there is access to sufficient deep soil water reserves. Each functional layer in a canopy has its own hydrologic environment. In contrast to tropical evergreen forests, the photosynthetic capacity and productivity of more shallow-rooted converted areas, such as pastures and agriculture ecosystems, are more directly driven by precipitation inputs, and thus are more vulnerable to dry periods.

17.5 Flux Tower Measurements in the Amazon

Tower-based estimates of photosynthetic flux derived from observations of net ecosystem exchange of carbon dioxide are the most reliable and direct measure of ecosystem-scale photosynthesis to test models and "ground truth" remote sensing observations. A network of eddy flux measurements, begun as part of the Large-scale Biosphere-Atmosphere Experiment in Amazonia (LBA) (Keller *et al.* 2004), has shown that gross ecosystem production (GEP) and evapotranspiration (ET) at central Amazon sites are not water limited and seasonal variations in GEP and ET are largely driven by the availability of net radiation (da Rocha *et al.* 2009; Fisher *et al.* 2009).

Individual tower studies have shown a range of results, from significant increases in dry season forest photosynthesis in the central eastern Amazon (Hutyra *et al.* 2007), to dry season decreases in photosynthesis in the southwest Amazon (von Randow *et al.* 2004), and no detectable seasonality in photosynthesis in the far eastern Amazon. Malhi *et al.* (1998) reported depressed photosynthetic activity in the dry season due to moisture stress, in central Amazon rainforests near Manaus. However, their study coincided with an El Niño year when rainfall was lower than average. Other tower flux measurements have shown increased photosynthesis activity in the dry season at both central eastern and eastern sites (Saleska *et al.* 2003).

These flux studies have pointed to a range of possible biophysical drivers for photosynthetic carbon uptake, including length of dry season, cloud cover, access to deep soil water and disturbance history, among others. To move beyond these site-specific results requires a consistent, integrated analysis of ecosystem-scale photosynthesis across sites and biomes in order to achieve a coherent picture of spatial variability and drivers across the Amazon basin.

Restrepo-Coupe *et al.* (2013) investigated the seasonal patterns of Amazonian forest photosynthetic activity and the effects thereon of variations in climate and land-use, by integrating data from the network of LBA eddy flux towers in Brazil. They found that the degree of water limitation, as indicated by seasonality in the ratio of sensible to latent heat flux (Bowen ratio) plays a dominant role in, and is predictive of, seasonal patterns of photosynthesis.

In equatorial Amazonian forests, water limitation is absent and photosynthetic fluxes (or gross ecosystem productivity, GEP) exhibit high or increasing levels of photosynthetic activity as the dry season progresses, likely a consequence of allocation to growth of new leaves. In contrast, forests along the southern flank of the Amazon, pastures converted from forest, and mixed forest-grass savanna, exhibit dry-season declines in GEP, consistent with increasing degrees of water limitation.

They found GEP fluxes largely followed the phenology of canopy photosynthetic capacity (P_c), with only smaller deviations from this primary pattern driven by variations in PAR.

Restrepo-Coupe *et al.* (2013) found significant seasonal variation in standing canopy P_c within the tropical evergreen forests. Thus, canopy phenology, and not environmental variability, was the dominant control on the seasonality of tropical forest photosynthesis. GEP deviations from the primary pattern set by P_c are explained by variations in availability of sunlight, which effectively regulates the fraction of photosynthetic capacity utilized (GEP/P_c). This suggests that if canopy photosynthetic capacity were constant (as many ecosystem models assume for tropical evergreen forests), then the seasonality of GEP at non water-limited sites would indeed be well-explained by seasonality of sunlight.

Estimates of leaf flush at three non-water-limited equatorial forest sites demonstrate a peak of activity in the dry season. This may be correlated with high dry season light levels arising from the lack of cloud cover. The larger P_c that follows persists into the wet season, driving high GEP that is out of phase with sunlight, explaining the asynchronous relationship with sunlight. Overall, these patterns suggest that at sites where water is not limiting, light interacts with adaptive mechanisms to determine P_c indirectly through seasonality in leaf flush and litterfall.

17.6 Satellite-Based Studies of Landscape Seasonality

Satellite remote sensing is integral to assessments of basin-wide Amazon forest seasonality and functioning, given the size and environmental complexity of the basin and the difficulties associated with *in situ* sampling and ground measurements. When properly calibrated, satellite products can be used to extend results at the field and tower scale to more regional scales.

Moderate and fine scale optical remote sensing data (e.g. Landsat) have been used extensively to map tropical forest classes and disturbance, including deforestation, regrowth, and fire (Kalacska *et al.* 2005; Fig. 17.6) as well as to spectrally map and distinguish individual crowns of tropical tree species Clark *et al.* (2004).

Early successional forest canopies that follow forest conversion consist of fast-growing pioneer trees with an understory of small shrubs and herbaceous vegetation which appear bright in the NIR. Older regenerating forests are characterized by increasing species richness, a multi-layered tree canopy and an understory vegetation increasingly limited by low light levels. Mature, primary tropical forests are generally darkest in the NIR due to their complex stand structure, with different tree strata at different heights and high levels of intra–and inter-species shading. Hence, in progressing from younger to older successional forests and mature forests, many structural traits vary, including number of canopy layers, height and continuity of the upper canopy and the presence and number of emergent trees. Spectrally, successional

(a)

(b)

Figure 17.6. Deforestation in Rondonia. Images from Earth Observatory http://earthobservatory.nasa.gov/Features/WorldOfChange/deforestation.php

forests are a transitional class between pasture and primary tropical forest, with the initial stages of forest regeneration spectrally similar to pasture, while the final stages of forest regeneration are more spectrally related to primary tropical forest.

Much of what is known about seasonal vegetation dynamics in the tropics comes from coarse resolution satellite measurements, given the difficulties in obtaining sufficient cloud-free images to characterize forest phenology from the less frequent availability of finer resolution satellite imagery. Coarse resolution monitoring satellites provide daily images that can be composited for large-scale vegetation monitoring and vegetation -climate studies (Running *et al.* 1995). Satellite observations with the NOAA Advanced Very High Resolution Radiometer (AVHRR) have characterized the seasonality of moist tropical evergreen broadleaf forests as "flat" and seasonally invariant (Justice *et al.* 1985). Such results may have been partly due to saturation of the normalized difference vegetation index (NDVI) signal resulting from a low optical depth of penetration through densely vegetated forest canopies. This renders more subtle phenology characteristics difficult to measure. In a study of tropical successional forests, Sader *et al.* (1989) found the NDVI to be a poor predictor of biomass in these rapidly growing regenerating forests, which quickly maximize the absorption of red radiation and saturate the NDVI signal.

Some of the more recent, advanced coarse resolution satellite sensors, such as SPOT VEGETATION (VGT) and the Moderate Resolution Imaging Spectroradiometer (MODIS) have much better sensor capabilities for seasonal assessments of tropical forest productivity. The MODIS sensor offers improved calibration, atmosphere correction, narrower spectral bands without water vapour influences, and finer resolution (250 m–1 km) observations that facilitate cloud-filtering and noise removal. In contrast to the AVHRR-NDVI data, strong local–and region-wide patterns in phenology were found in Amazon forests with MODIS and SPOT-VGT enhanced vegetation index (EVI) data, an index of canopy photosynthetic capacity (Huete *et al.* 2006; Brando *et al.* 2010). The EVI is designed to improve upon the NDVI signal by not saturating so readily at high LAI. The EVI shows a strong correlation with gross primary productivity (GPP) across a range of ecosystems and is therefore suited to answer the question: 'How does GPP vary seasonally and inter-annually in Amazonian forests?'

Amazon studies utilising MODIS remote sensing products have detected positive 'greening' forest responses to seasonal drying (Huete *et al.* 2006; Myneni *et al.* 2007) (Fig. 17.7). Huete *et al.* (2006) found Amazon primary forests to green-up from the onset to end of the dry season (July–October) with continued greening into the early wet season (December), with dry season defined as <100 mm/month precipitation (Sombroek 2001). The larger MODIS EVI vegetation response during the dry season compared with the wet season was indicative of radiation-limited forest ecosystems. This was opposite that found in adjacent pasture sites where greenness values were closely coupled to rainfall and decreased by as much as 25 percent in the dry season in response to soil moisture stress.

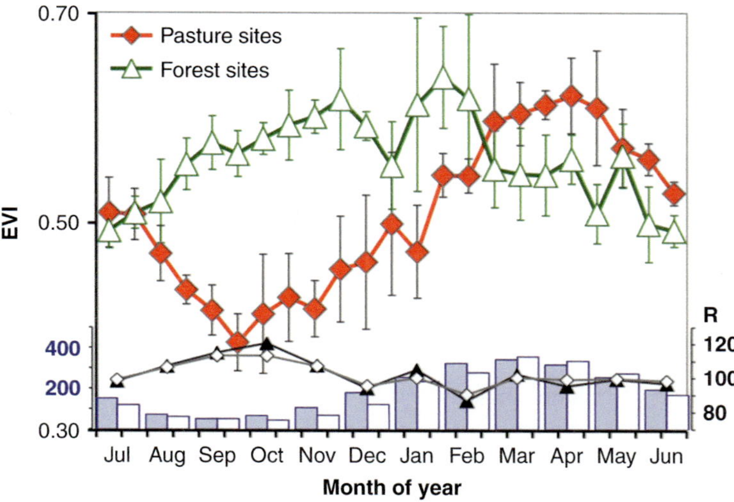

Figure 17.7. Five-year mean seasonal changes in EVI (green triangles and red diamonds), rainfall (bars), and PAR (black triangles and open diamonds), for pasture and forested sites in the Amazon. Reproduced with permission from Huete *et al.* (2006).

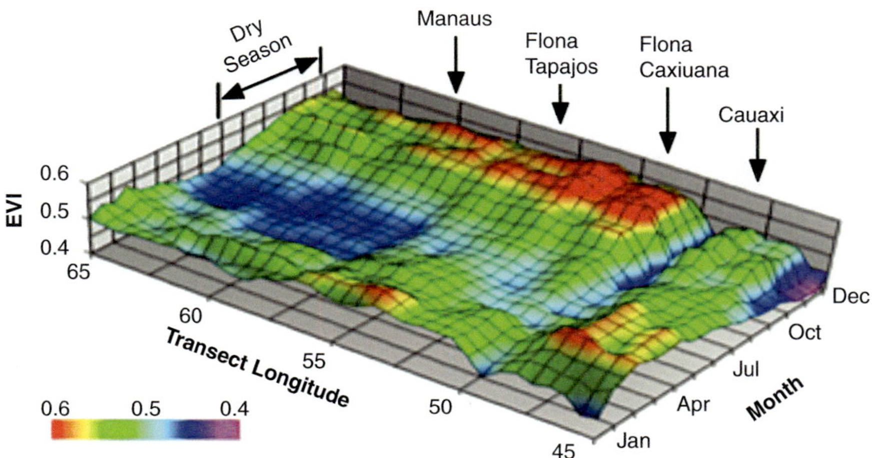

Figure 17.8. Phenology profiles along an Amazon equatorial transect with dry seasons ranging from zero months in the west to four to five months to the east, and forest converted areas to the extreme east at 45° to 50° Longitude. Used with permission from Huete *et al.* (2006).

Huete *et al.* (2006) found widespread greening in the dry season along a 2000 km equatorial transect that spanned seasonally-dry central and eastern Amazonian rainforests to the ever-wet tropical forests of the western Amazon (Fig. 17.8).

The largest increase in EVI (25% increase) is observed in the more seasonally dry eastern Amazon, where the dry season is typically four to five months. In contrast, seasonal changes in EVI are weaker in central Amazon forests (65° to 60° Long.) with shorter dry season periods (1–3 months). Seasonal inundation in the lower elevation central forests (65° to 58° Long.) have the smallest EVI values. A complete reversal in phenology ('brown-down') can be seen in the easternmost portion of the transect where extensive forest conversion has resulted in more shallow-rooted vegetation with reduced access to deep soil water stores that persist through the dry season. A basin-wide analysis showed that almost 50 percent of the basin green-up in the dry season, coincident with increased PAR availability due to reduced cloud cover.

Myneni *et al.* (2007) used the physical-based MODIS leaf area index (LAI) product and showed net leaf flushing during the early to mid-part of the light-rich dry season, followed by net leaf abscission during the cloudy wet season. Remote sensing of regional-scale chlorophyll solar-induced fluorescence (SIF) in the Amazon (by the new Japanese Greenhouse gases Observing Satellite, GOSAT), which may give a more direct index of photosynthesis than existing sensor technologies (Frankenberg *et al.* 2011), suggests patterns broadly similar to those observed via EVI (SIF *versus* EVI, $R^2 = 0.52$) (Lee *et al.* 2013).

Deep tree roots (Nepstad *et al.* 1994) combined with hydraulic redistribution by roots (Oliveira *et al.* 2005) and increased availability of sunlight have been suggested as explanations for increasing greenness throughout the dry season, while in the forest converted areas the removal of deep-rooted forest trees reduces access to deep soil water, and results in decreasing greenness through the dry season.

17.6.1 Dry Tropical Forests and Water Limitations

In contrast to humid rainforests that are more light-controlled, forest phenology and photosynthetic activity are more closely associated with moisture limitations in the transitional and drier tropical forests of the southern Amazon that have lower biomass and leaf area index (LAI). Von Randow *et al.* (2004) found seasonality in photosynthesis to be small at the Jaru Biological Reserve tower flux site in Rondonia, and Vourlitis *et al.* (2001) reported strong positive correlations between photosynthesis and water availability in a transitional forest flux tower site near Sinop, Mato Grosso. Ratana *et al.* (2006) examined tropical forest phenology across the humid-dry forest transition zone in the southern Amazon and found mixed light, moisture and anthropogenic controls on seasonal patterns of MODIS EVI. These studies suggest tropical forests are highly dynamic, with strong phenology responses to light and moisture as well as land use and land cover change.

17.6.2 Satellite and Flux Tower Coherence

Satellites (with broad spatial coverage) and networks of eddy flux towers and forest plots (with limited spatial coverage but critical information about biological processes and mechanisms) provide potentially powerful observations of whole ecosystem patterns that can test mechanisms of forest-climate interactions.

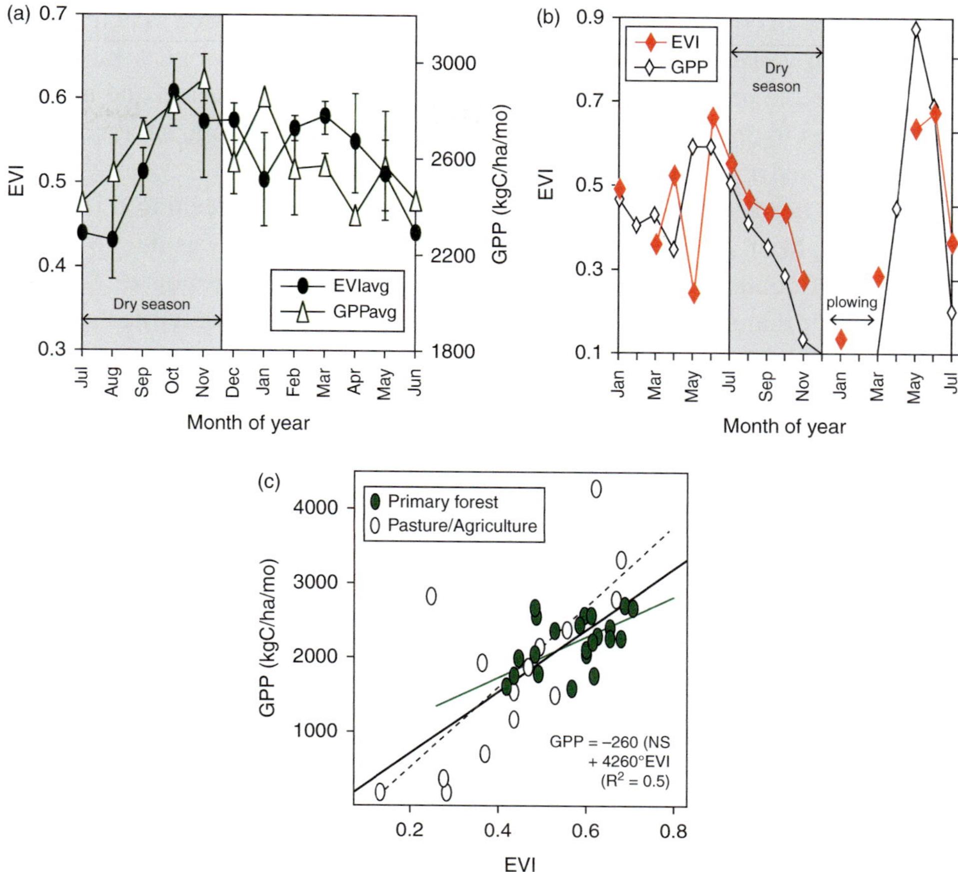

Figure 17.9. Seasonal phenological patterns and GPP for: (a) a primary forest in the Amazon, (b) pasture in the Amazon, and: (c) regression of eddy covariance estimates of GPP *versus* EVI, for the two sites. Used with permission from Huete *et al.* (2006).

Huete *et al.* (2006) confirmed satellite EVI seasonality to be synchronous with local-scale tower fluxes of gross primary production (GPP) in both a primary forest near Santarem and a nearby converted pasture site. A strong correlation between EVI and GPP was observed with both satellite and tower measurements showing forest greening and simultaneous pasture browning in the dry season, confirming the interpretation of satellite data (Fig. 17.9). The proximity of the forest and pasture sites (~15 km apart), along with close agreement between satellite-EVI and tower-GPP results, were strong evidence against artefacts (residual clouds, sensor geometry) in the satellite-observed phenology.

17.6.3 Satellite Data Controversies

Remote sensing observations are crucial to monitor vast and inaccessible tropical forests, however, controversies persist in the interpretation and understanding of remote sensing data over tropical forest areas that are prone to cloud contamination in the wet season and aerosol contamination (biomass burning) in the dry season. Satellite studies have shown variable and inconsistent seasonal patterns over tropical rainforests with some satellite products showing canopy drying in the dry season and negative forest responses to drought, while other products show leaf flushing and greening in the dry season and a positive response to drought.

There have been suggestions that satellite tropical forest greening during the Amazon drought of 2005, as reported in Saleska *et al.* (2007) was due to atmospheric aerosol contamination, rather than a true vegetation response (Samanta *et al.* 2010), or that spectral vegetation measures observed by satellites are simply insufficient to detect vegetation responses (Atkinson *et al.* 2011). In order to interpret the satellite data correctly, it is essential to ensure that observations are not a result of artefacts resulting from aerosol or sub-pixel cloud contamination. The main limitations to resolve such debates have been the lack of available field sites and high-quality *in situ* data sets to validate and calibrate remote sensing results.

Previously, coarse-scale AVHRR satellite data, at 8 km pixel resolution, was particularly constrained by cloud contamination, cloud shadows, and difficulties in atmosphere correction of seasonally varying atmosphere water vapour and aerosols. Kobayashi and Dye (2005) showed there were strong seasonal signals from clouds and aerosols in the AVHRR–NDVI data sets over the Amazon, which dominated the relatively weak forest biologic signal from the tropical forests themselves. Such external artefacts along with a dense forest saturated NDVI response will alter the perceived phenology of Amazon forests.

More recently, Morton *et al.* (2014) reframed the debate over climate controls on Amazon forest productivity by presenting an alternative hypothesis that explains observed seasonality in greenness as optical artefacts due to shifting sun-sensor view geometries. They demonstrated that the convergence of satellite observations upon the principal plane from the June solstice to the September equinox, results in increasing proportions of sunlit leaves (due to higher sun elevation angles) yielding higher NIR reflectances, EVI greenness and LAI estimates. The three-month sun-sensor geometry shift from the solstice (June 21) to the equinox (September 21) was coincident with the start and end of the dry season, thereby generating ambiguity in the causes of greening.

Using their own correction model to adjust MODIS satellite data to a fixed sun angle geometry, they reported that seasonal changes in surface reflectances and spectral greenness measures, were eliminated, and concluded that Amazon forests are aseasonal and without light nor water limitations. They further confirmed their results with canopy model simulations of forest structure and satellite LiDAR retrievals of relative forest heights. These results remain controversial since no ground truth verification (which generally show enhanced greening at local scales) was presented and previous satellite-based studies showed dry season greening continuing to increase beyond the September equinox and peaking in December, close to the solstice

(Fig. 17.7). If the satellite EVI were purely artefact, then the profile should be aligned with sun angle geometries both before and after the equinox, which does not appear to be the case.

Both tower flux and satellite EVI seasonality are subject to seasonal sun angle geometries, given that day length will increase from the solstice to the equinox thereby potentially enabling greater values of daily integrated GPP fluxes. A consistent correlation between tower GPP and corrected satellite EVI in both forest and converted land suggests that basin-wide carbon models can be constrained by integrating remote sensing and local flux measurements.

17.7 Model Results

A fundamental question for earth system models (ESMs; see Chapter 2) to address is the vulnerability of tropical forests to climate change. Many climate change scenarios projected for the 21st century show the Amazon becoming drier due to stronger El Niño conditions and an enhanced North-South Atlantic sea surface temperature gradient (Cox *et al.* 2008). Widespread Amazon forest collapse due to global warming-induced drought is predicted by some coupled carbon/climate models (Betts *et al.* 2004; Malhi *et al.* 2009), while other models imply forest persistence (Friedlingstein *et al.* 2006).

Understanding how tropical forests will be impacted by changes in future climate is difficult to observe or measure until climate change happens and hence needs to be modelled. However, key modelled mechanisms should be testable by observing forest responses to seasonal and inter-annual climate variability (e.g. ENSO events). Thus, understanding seasonal and spatial variation of forest metabolism is an important basis for understanding ecological responses to climate generally. These mechanisms are poorly represented in ecosystem models and represent an important challenge to efforts to predict tropical forest responses to climatic variations at both seasonal and interannual time-scales.

Current earth system models have been unable to correctly depict seasonal forest functioning nor predict the impacts of anthropogenic disturbance (Fig. 17.10). A fundamental understanding of the functioning of these tropical systems across seasonal and inter-annual remain poorly understood and is the subject of continuing controversy in the remote sensing and modelling literature (Werth and Avissar 2002; Malhi *et al.* 2009).

At the landscape-scale many climate and growth models characterize tropical rainforests as having no seasonal variation in biophysical plant properties such as greenness, leaf area index (LAI), fraction of absorbed photosynthetically active radiation (FAPAR), and albedo. Many vegetation modelling studies have represented Amazon forests as water-limited and so predict dry season declines in transpiration and/or photosynthesis (Lee *et al.* 2005). Early empirical studies near Manaus both supported (Malhi *et al.* 2009, 2014) and opposed the water-limitation view (Shuttleworth 1988).

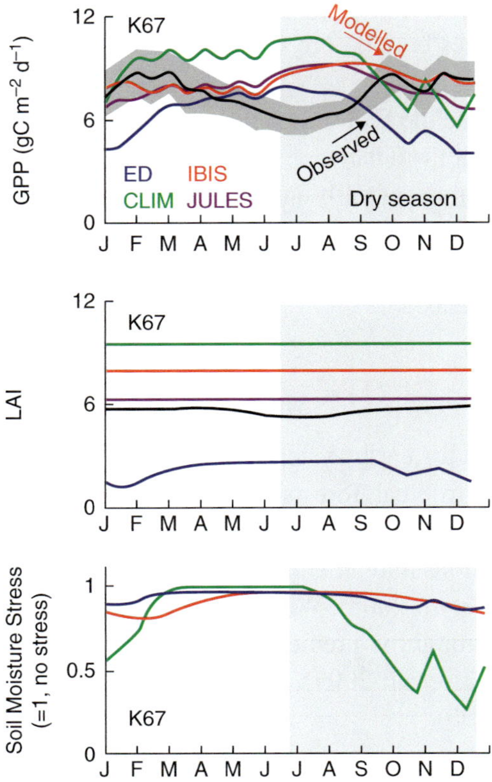

Figure 17.10. (a) Modelled and observed seasonal cycles of photosynthetic metabolism in the Amazon (Santarem, K67) from eddy flux tower and satellite EVI show strong observed GPP and greenness increases during the dry season, whereas models show declining GPP; (b) model representations of LAI; and (c) soil water stress resulting from meteorological driven forest canopy models (Courtesy of Restrepo-Coupe, unpublished).

Other models get different functional seasonalities (or same seasonalities for different reasons), that range from forest productivity declines in the dry season to enhanced productivity in the dry season.

Differing modelled fates of the forest are due to model differences in representation of forest function, along with differences in representations of climate (Sitch *et al.* 2008). Thus, divergence between models and observations become apparent when modelled photosynthetic seasonality is driven by environmental variability, in which case soil-water stress suppresses dry-season GPP, or when LAI is assumed constant or with dry season declines (Fig. 17.10). Observed photosynthetic seasonality, by contrast, appears driven by changes in canopy phenology (biologic drivers). Current knowledge is insufficient to determine which forest representations are most consistent with real forest ecosystems.

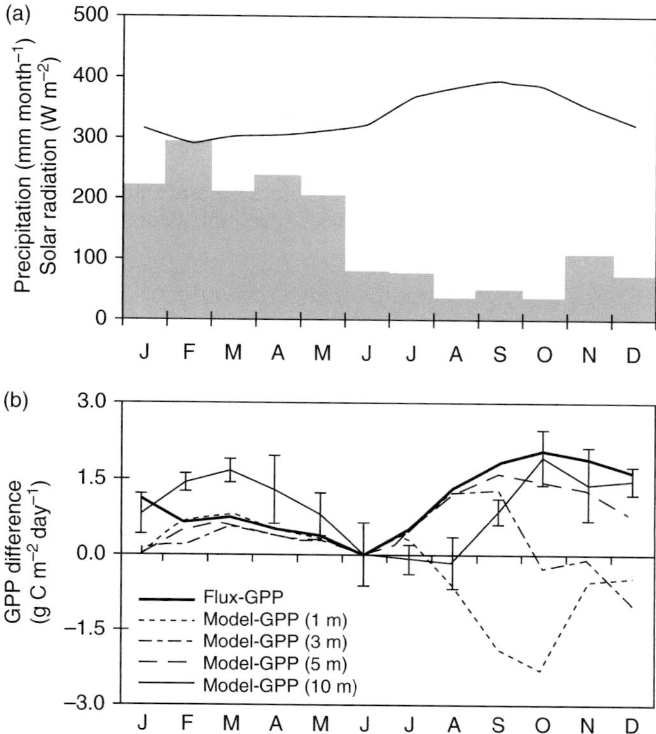

Figure 17.11. Precipitation and solar radiation vary across an annual timeframe at an evergreen broadleaf forest in the Amazon Basin. By varying the assumed root depth of the vegetation within the Biome-BGC model, it was possible to obtain a match between the observed eddy covariance tower GPP data and the model output. Reproduced with permission from Ichii *et al.* (2007).

Ichii *et al.* (2007) presented an excellent example of how models, combined with remote sensing, can provide information of rooting depth through sensitivity analyses. They applied a simple root length model to show that by adjusting the rooting depth one can get the models and satellite data to agree (Fig. 17.11). Similarly, Fisher *et al.* (2007) used a combination of field observation and a soil-plant-atmosphere model to estimate rooting depth of eastern Amazonian forest and concluded that deep roots contributed significantly to a reduced sensitivity of such forests to drought.

It is worth noting that although remote sensing cannot directly monitor soil and plant root properties, it can help parameterize models, and with extrapolated climatic data, make it possible to estimate whole ecosystem carbon exchange under conditions not yet encountered. If such models predict large changes in EVI and the seasonal display of leaves, then remote sensing vegetation indices can and should play a role in validating ecosystem responses to climate change.

17.8 Modelling, Remote Sensing, Ecophysiology and Drought in the Amazon

The 2009/2010 drought in the Amazon was one of the most severe in recent history, with more than 40 percent of vegetation experiencing low rainfall, with concomitant declines in river flows across the Basin. Most global climate models predict an increased frequency of El Niño-like conditions and reduced rainfall for the eastern Amazon. Given the importance of the Amazon to regional climate and global biodiversity, it is vital that we understand the response of the Amazon to drought. Changes in the enhanced vegetation index (EVI) from the MODIS satellite data have been used in the Carnegie Ames Stanford Approach (CASA) model to model Amazonian carbon fluxes for the Amazon Basin for the years 2008 (pre-drought), 2009 (drought) and 2010 (drought) (Potter *et al.* 2011). This model estimates monthly C fluxes and biomass increment and has been validated with measurements of net ecosystem production (NEP) obtained from eddy covariance towers located across the Basin. From this modelling analyses, net primary production (NPP) across a number of Amazonian forest classes declined by 7 percent in 2010 compared to 2008, or a decline in uptake of C of 0.5 Pg C y^{-1} in 2010 compared to 2008. Total Amazonian NPP was about 7.6 Pg C y^{-1} in 2008. Across the Basin NEP changed from being neither a net sink nor net source of carbon (so zero net flux of carbon) in 2008, to being a strong source in 2009 and 2010 (0.29 PgC released in 2009 and 0.42 PgC released in 2010; Potter *et al.* 2011).

An alternative modelling approach to understanding the impacts of drought on Amazonian forests is to use a soil-plant-atmosphere model that includes a detailed mechanistic understanding of the processes governing net ecosystem exchange (NEE) and GPP of forests. Fisher *et al.* (2007) used the Edinburgh SPA model to examine the impacts of drought of Amazonian rainforests. They used two years of tree-scale sapflow data and three years of soil moisture data collected at the *Caxiuana* throughfall exclusion experiment (CTFE) site to parameterise their detailed SPA model. The SPA model was able to closely represent seasonal changes in soil moisture content (reductions in the dry season; June–November) and also the impact of throughfall exclusion (reductions of about 250–300 mm of water across 0–3 m soil depth). The reduction in soil moisture content also caused an almost three-fold reduction in the soil-to-root hydraulic conductance. Because of this reduced ability for soil to supply water to roots, leaf water potential declined to a minimum that approached the critical value associated with increased xylem embolism. Consequently stomatal conductance declined to minimise the development of embolism and this resulted in reduced rates of transpiration (Fisher *et al.* 2007); and annual total water use in the control plots (no throughfall exclusion) was relatively constant across the three year study (1223–1316 mm y^{-1}). In contrast total annual tree water use declined in each year (1258, 953, and 805 mm, for 2001, 2002 and 2003 respectively). Because of reduced stomatal conductance, g_s, GPP declined by 50–60 percent in the dry season for the

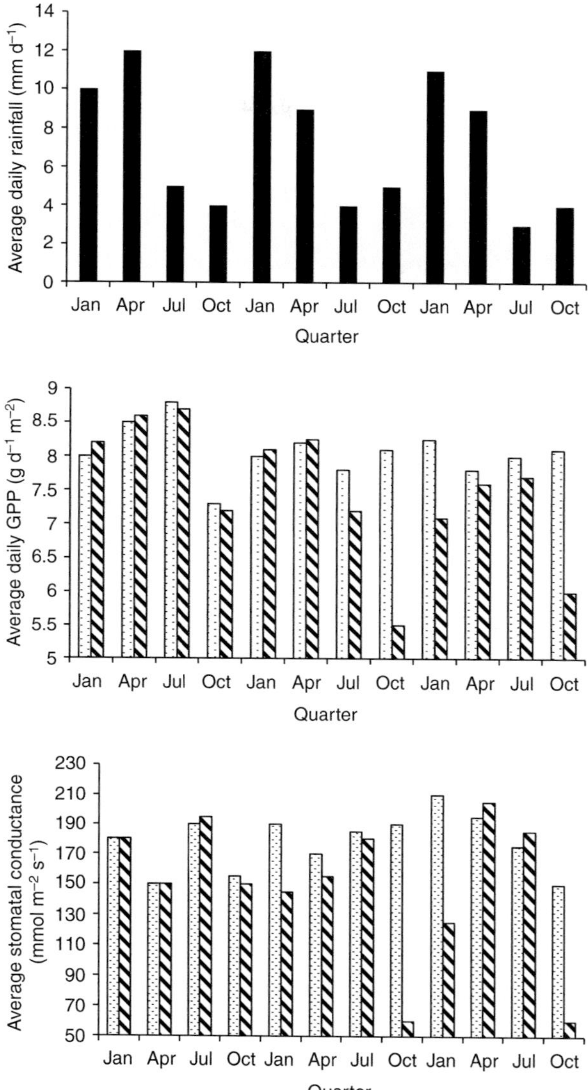

Figure 17.12. Daily rainfall, stomatal conductance of the control (horizontal dash) and throughfall exclusion site (thick diagonal stripes) and modelled GPP of the control and throughfall exclusion site in the Amazon. Reproduced with permission from Fisher *et al.* (2007).

CTFE plots compared to dry season GPP of control plots (Fig. 17.12), or a 14 percent reduction on average over the two-year experiment.

Rates of water flux, both modelled and measured (with sapflow sensors) were larger in the dry season than the wet season (Fisher *et al.* 2007). This is a surprising result since soil water content declines in the dry season. However, such results are consistent with satellite observations and eddy covariance measurements of water and

carbon flux. Most importantly, satellite data, eddy covariance data, and SPA modelling data, are consistent with each other but contradict the majority of land surface
models of terrestrial function. Such land-surface models (e.g. Tian *et al.* 1998, Zeng
et al. 2005) predict significant reductions in water and carbon flux in the dry season.
This issue is now discussed.

17.8.1 Greening of the Amazon during Drought

Saleska *et al.* (2007) reported short-term greening flushes in Amazon forests during
the early stages of the 2005 drought. This short-lived response was suggested to be
indicative of light-limitations in forests, given the increase in solar radiation availability to the forest and the full soil moisture reserves available. This was highly controversial as converted areas quickly dried up and became vulnerable to fires and 2005
became one of the strongest years for aerosol-induced haze from biomass burning.
However, given the different functional classes of vegetation, it is not that surprising
that fires and forest greening can occur simultaneously.

Mechanisms controlling tropical rainforest phenology and productivity are poorly
understood. Many land-surface models of vegetation function (water and carbon
fluxes) predict reduced LAI and dry-season reductions in gas exchange because of
the assumption that a drying soil must cause reduced stomatal conductance and hence
reduced water and carbon fluxes (e.g. Zeng *et al.* 2005).

Potter *et al.* (2009) used the NASA-CASA model to predict terrestrial ecosystem
C fluxes across the Amazon Basin, using MODIS EVI data at 8 km resolution. The
model was validated using eddy covariance data in the Amazon Basin. NPP is modelled as a function of solar radiation, EVI, temperature, soil moisture content, and a
constant light-use-efficiency term. Time-varying stress scalar terms are included for
the effects of temperature and soil moisture stress (Potter *et al.* 2009).

The NASA-CASA model was able to capture seasonal variability in NEE for the
Tapajos National Forest site in east-central Brazilian Amazon. During the wet season,
NEE was positive (net carbon source) but during the dry season, the site became a
carbon sink (negative NEE). This pattern was consistent across the three-year study
period. This result was attributed to the tight coupling of ecosystem respiration to soil
moisture content, such that in the wet season, soil respiration dominated ecosystem
carbon fluxes and the site was a net carbon source. In the dry season soil respiration
declined significantly and this, together with the green-up of vegetation that occurs in
the dry season, resulted in the forest becoming a net carbon sink.

While large seasonal changes in NPP occurred for the Brazilian Amazon (Potter
et al. 2009), the annual average NPP showed minimal variation across a five-year
study period (2000–2004 inclusive). Similarly, NEP showed large variation seasonally, being strongly positive in the dry season and strongly negative in the wet season.
The Amazon rainforest was found to be net sinks for carbon in wet La Niña years and
net carbon sources in dry El Niño years. Some regions of the Amazon (western parts

of the states of Acre and Rondonia) show consistently high C sink fluxes and some regions (the state of Maranhao and the southern parts of Amazonas) show consistently high carbon source fluxes (Potter *et al.* 2009).

17.9 Conclusions

The question of photosynthetic seasonality in tropical evergreen systems is of fundamental importance, both in terms of basic ecology (understanding plant strategies for resource acquisition when resources, such as water or sunlight, are limiting), as well as the need to understand the practical problem of vegetation response and feedbacks to climate change.

The seasonality question is an important threshold test for our ability to predict tropical forest response to climate change, for if we do not know the answer to such a basic question as, what is the ecosystem-scale response of these forests to regular seasonal variation in climate (which is observable every year), there is less hope we can confidently predict their response to as yet unobserved future climates.

Three levels of information are needed to resolve questions about seasonality of tropical forests:

(1) detailed information about *environmental drivers* (especially light, including its diffuse and direct components and impacts of aerosols and clouds, and precipitation);
(2) detailed process information at the individual tree, branch, and leaf scale to quantify canopy phenological patterns and water relations that are inadequately represented in models;
(3) information at the canopy-to-landscape scale that capture emergent patterns driven by the mechanisms outlined above, with a goal to better use satellite products in tropical forests to understand what changes in *greenness* signify and thereby bridge the "scaling gap" between individual trees and remote sensing measurements.

Forest greening patterns involve complex interactions among heterogeneous forest types, dry season duration and intensity and the timing of major drought events. Continuing observations from satellites and from the network of eddy flux towers provide tools that can rigorously test mechanisms of forest-climate interactions and hence provide critical insight into the potential future of tropical forests.

17.10 References

Atkinson PM, J Dash and C Jeganathan, (2011). Amazon vegetation greenness as measured by satellite sensors over the last decade. *Geophysical Research Letters*, 38(19), L19105.
Asner G, M Keller and J Silva, (2004). Spatial and temporal dynamics of forest canopy gaps following selective logging in the eastern Amazon. *Global Change Biology*, 10 (5), 765–783.

Betts R, P Cox and M Collins, (2004). The role of ecosystem-atmosphere interactions in simulated Amazonian precipitation decrease and forest dieback under global climate warming. *Theoretical and Applied Climatology*, 78 (1), 157–175.

Borchert R, K Robertson and MD Schwartz, (2005a). Phenology of temperate trees in tropical climates. *International Journal of Biometeorology* 50, 57–65.

Borchert R, SS Renner, Z Calle, D Navarrete, A Tyne, L Gautier, R Spichiger and P von Hildebrand (2005b). Photoperiodic induction of synchronous flowering near the Equator. *Nature*, 433(7026), 627–629.

Brando PM, SJ Goetz, A Baccini, DC Nepstad, PS Beck and MC Christman, (2010). Seasonal and inter-annual variability of climate and vegetation indices across the Amazon. *Proceedings of the National Academy of Sciences of the United States of America*, 107 (33), 14,685–90.

Clark DA and DB Clark, (2011). Assessing tropical forests' climatic sensitivities with long-term data. *Biotropica*, 43(1), 31–40.

Cox P, R Betts and M Collins, (2004). Amazonian forest dieback under climate-carbon cycle projections for the 21st century. *Theoretical and Applied Climatology*, 78 (1), 137–156.

Cox PM PP Harris, C Huntingford, RA Betts, M Collins, CD Jones, TE Jupp, JA Marengo and CA Nobre, (2008). Increasing risk of Amazonian drought due to decreasing aerosol pollution. *Nature*, 453(7192), 212–215.

Fisher RA, M Williams, AL da Costa, Y Malhi Y, RF da Costa, S Almeida, PW Meir, (2007). The response of an Eastern Amazonian rainforest to drought stress: Results and modelling analyses from a through-fall exclusion experiment. *Global Change Biology* 13, 2361–2378.

Frankenberg C, A Butz and GC Toon, (2011). Disentangling chlorophyll fluorescence from atmospheric scattering effects in O_2 A-band spectra of reflected sun-light. *Geophysical Research Letters*, 38(3), L03801.

Friedlingstein P, P Cox, R Betts, L Bopp, W von Bloh, V Brovkin, P Cadule, S Doney, M Eby, I Fung, G Bala, J John, C Jones, F Joos, T Kato, M Kawamiya, W Knorr, K Lindsay, HD Matthews, T Raddatz, P Rayner, C Reick, E Roeckner, K-G. Schnitzler, R Schnur, K Strassmann, AJ Weaver, C Yoshikawa and N Zeng, (2006). Climate-carbon cycle feedback analysis, Results from the C4MIP model inter-comparison. *Journal of Climate*, 19(14), 3337–3353.

Goulden ML, SB Miller, HR da Rocha, MC Menton. HC de Freitas, AMS Figueira and CAD de Sousa, (2004). Diel and seasonal patterns of tropical forest CO_2 exchange. *Ecological Applications*, 14(sp4), 42–54.

Herrerias-Diego Y, M Quesada, KE Stoner and JA Lobo, (2006). Effects of forest fragmentation on phenological patterns and reproductive success of the tropical dry forest tree *Ceiba aesculifolia*. *Conservation Biology*, 20(4): 1111–1120.

Horel J, A Hahmann and J Geisler, (1989). An investigation of the annual cycle of convective activity over the tropical Americas. *Journal of Climate*, 2 (11), 1388–1403.

Huete A, K Didan, YE Shimabukuro, P Ratana, SR Saleska, LR Hutyra, W Yang, RR Nemani and R Myneni, (2006). Amazon rainforests greenup with sunlight in dry season. *Geophysical Research Letters* 33, L06405, doi:200610.1029/2005GL025583.

Hutyra LR, JW Munger, SR Saleska, E Gottlieb, BC Daube, AL Dunn, DF Amaral, PB de Camargo, SC Wofsy, (2007). Seasonal controls on the exchange of carbon and water in an Amazonian rain forest. *Journal of Geophysical Research Biogeosciences* 112, doi:10.1029/2006JG000365

Ichii K, H Hashimoto, MA White, CS Potter, LR Hutyra, AR Huete, RB Myneni, RR Nemani, (2007). Constraining rooting depths in tropical rainforests using satellite data and ecosystem modeling for accurate simulation of GPP seasonality. *Global Change Biology*. 13, 67–77.

Justice CO, JRG Townshend, BN Holben and CJ Tucker, (1985). Analysis of the phenology of global vegetation using meteorological satellite data. *International Journal of Remote Sensing*, 6(8), 1271–1318.

Kalacska MER, GA Sánchez-Azofeifa, JC Calvo-Alvarado, B Rivard and M Quesada, (2005). Effects of season and successional stage on leaf area index and spectral vegetation indices in three meso-american tropical dry forests. *Biotropica*, 37(4), 486–496.

Keller M, A Alencar, GP Asner, B Braswell, M Bustamante, E Davidson, T Feldpausch, E Fernandes. M Goulden, P Kabat. B Kruit, (2004). Ecological research in the large-scale biosphere-atmosphere experiment in Amazonia, Early results. *Ecological Applications*, 14(sp4), 3–16.

Kobayashi H and DG Dye, (2005). Atmospheric conditions for monitoring the long-term vegetation dynamics in the Amazon using normalized difference vegetation index. *Remote Sensing of Environment*, 97(4), 519–525.

Koren I, YJ Kaufman, LA Remer and JV Martins, (2004). Measurement of the effect of Amazon smoke on inhibition of cloud formation. *Science,* 303(5662), 1342–1345.

Lambin EF, HJ Geist and E Lepers, (2003). Dynamics of land-use and land-cover change in tropical regions. *Annual Review of Environment and Resources*, 28(1), 205–241.

Lee J-E, C Frankenberg, C van der Tol, JA Berry, L Guanter, CK Boyce, JB Fisher, E Morrow, JR Worden. S Adefi, G Badgley and S Saatchi, (2013). Forest productivity and water stress in Amazonia, observations from GOSAT chlorophyll fluorescence. *Proceedings of the Royal Society B, Biological Sciences*, 280(1761), 20130171.

Lee J-E, RS Oliveira, TE Dawson and I Fung, (2005). Root functioning modifies seasonal climate. *Proceedings of the National Academy of Sciences of the United States of America*, 102(49), 17576–17581.

Malhi Y, AD Nobre, J Grace, B Kruijt, MGP Pereira, A Culf and S Scott, (1998). Carbon dioxide transfer over a Central Amazonian rain forest. *Journal of Geophysical Research* D24, 31593–31612.

Malhi Y, LEOC Aragão, DB Metcalfe, R Paiva, C a. Quesada, S Almeida, L Anderson, P Brando, JQ Chambers, ACL da Costa, LR Hutyra, P Oliveira, S Patinõ, EH Pyle, AL Robertson and LM Teixeira, (2009). Comprehensive assessment of carbon productivity, allocation and storage in three Amazonian forests. *Global Change Biology*, 15 (5), 1255–1274.

Malhi Y, F Amézquita, CE Doughty, JE Silva-Espejo, CA Girardin, DB Metcalfe, LE Aragão, LP Huaraca-Quispe, I Alzamora-Taype, L Eguiluz–Mora, TR Marthews, K Halladay, CA Quesada, AL Robertson, JB Fisher, J Zaragoza-Castells, CM Rojas-Villagra, Y Pelaez-Tapia, N Salinas, P. Meir and OL Phillips (2014). The productivity, metabolism and carbon cycle of two lowland tropical forest plots in south-western Amazonia, Peru. *Plant Ecology and Diversity*, 7 (1–2), 85–105.

Morton DC, J Nagol, CC Carabajal, J Rosette, M Palace, BD Cook, EF Vermote, DJ Harding and PRJ North, (2014). Amazon forests maintain consistent canopy structure and greenness during the dry season. *Nature*, 506 (7487), 221–4.

Myneni RB, W Yang, RR Nemani, AR Huete, RE Dickinson, Y Knyazikhin, K Didan, R Fu, RI Negr´on Ju´arez, SS Saatchi, H Hashimoto, K. Ichii, NV Shabanov, B Tan, P Ratana, JL Privette, JT Morisette, EF Vermote, DP Roy, RE Wolfe, MA Friedl, SW Running, P Votava, N El-Saleous, S Devadiga, Y Su and VV Salomonson, (2007). Large seasonal swings in leaf area of Amazon rainforests. *Proceedings of the National Academy of Sciences of the United States of America*, 104 (12), 4820–3.

Nepstad DC, CR de Carvalho, EA Davidson, PH Jipp, PA Lefebvre, GH Negreiros, ES da Silva, TA Stone, SE TE Trumbonre and S Vieira (1994). The role of deep roots in the hydrological and carbon cycles of Amazonian forests and pastures. *Nature*, 372, 666–669.

Nobre CA, PJ Sellers and J Shukla, (1991). Amazonian deforestation and regional climate change. *Journal of Climate*, 4(10), 957–988.

Oliveira R, TE Dawson, SO Burgess and D Nepstad, (2005). Hydraulic redistribution in three Amazonian trees. *Oecologia*, 145(3), 354–363.

Potter C, S Klooster, A Huete, V Genovese, M Bustamante, LG Ferrerira, RS de Oliveira and R Zepp, (2009). Terrestrial carbon sinks in the Brazilian Amazon and Cerrado region predicted from MODIS satellite data and ecosystem modeling. *Biogeosciences*, 6(6), 937–945.

Potter C, S Klooster, C Hiatt, P Gross, V Brooks-Genovese and JC Castilla-Rubio, (2011). Changes in the carbon cycle of Amazon ecosystems during the 2010 drought. *Environmental Research Letters* 6, doi:10.1088/1748-9326/6/3/03402.

Quesada CA, J Lloyd, M Schwarz, S Patiñõ, TR Baker, C Czimczik, NM Fyllas, L Martinelli, GB Nardoto, J Schmerler, AJB Santos, MG Hodnett, R Herrera, FJ Luizaõ, A Arneth, G Lloyd, N Dezzeo, I Hilke, I Kuhlmann, M Raessler, WA Brand, H Geilmann, JO Moraes Filho, FP Carvalho, RN Araujo Filho, JE Chaves, OF Cruz Junior, TP Pimentel and R Paiva, (2010). Variations in chemical and physical properties of Amazon forest soils in relation to their genesis. *Biogeosciences*, 7 (5), 1515–1541.

Ratana P, AR Huete and K Didan, (2006). MODIS EVI based Variability in Amazon Phenology across the Rainforest-Cerrado Ecotone, *IEEE Geoscience and Remote Sensing Symposium*, 2006, 1942–1944.

Ray DK, VS Manoharan and RM Welch, (2011). Cloud cover conditions and stability of the Western Ghats montane wet forests. *Journal of Geophysical Research-Atmospheres* 116, D12104.

Restrepo-Coupe N, HR da Rocha, LR Hutyra, AC da Araujo, LS Borma, B Christoffersen, OM Cabral, PB de Camargo, FL Cardoso, ACL da Costa, DR Fitzjarrald, ML Goulden, B Kruijt, JM Maia, YS Malhi, AO Manzi, SD Miller, AD Nobre, C von Randow, Sa´ LDA, RK Sakai, J Tota, SC Wofsy, FB Zanchi and SR Saleska (2013). What drives the seasonality of photosynthesis across the Amazon basin? A cross-site analysis of eddy flux tower measurements from the Brazil flux network. *Agricultural and Forest Meteorology*, (182–183), 128–144.

Roberts DA, BW Nelson, JB Adams and F Palmer, (1998). Spectral changes with leaf aging in Amazon caatinga. *Trees*, 12(6), 315–325.

da Rocha HR, AO Manzi, OM Cabral, SD Miller, ML Goulden, SR Saleska, N Restro-Coupe, SC Wofsy, LS Borma, L Artaxo, G Vourlitis, JS Nogueira, FL Cardosa, AD Nobre, B Kruijt, HC Freitas HC, C von Randow. RG Aguiar and JD Maia, (2009). Patterns of water and heat flux across a biome gradient from tropical forest to savanna in Brazil. *Journal of Geophysical Research* 114, G00B12.

Running SW, TR Loveland, LL Pierce, RR Nemani and ER Hunt, (1995). A remote sensing based vegetation classification logic for global land cover analysis. *Remote Sensing of Environment*, 51(1), 39–48.

Sader SA, RB Waide, WT Lawrence and AT Joyce, (1989). Tropical forest biomass and successional age class relationships to a vegetation index derived from Landsat TM data. *Remote Sensing of Environment*, 28(0), 143–198.

Saleska SR, K Didan, AR Huete and HR Da Rocha, (2007). Amazon forests green-up during 2005 drought. *Science*, 318(5850), 612.

Saleska SR, SD Miller, DM Matross, ML Goulden, SC Wofsy, HR da Rocha, PB de Camargo, P Crill, BC Daube, HC de Freitas, L Hutyra, M Keller, B Kirchoff, M Menton, JW Munger, EH Pyle, (2003). Carbon in Amazon forests, unexpected seasonal fluxes and disturbance-induced losses. *Science*, 302(5650), 1554–1557.

Samanta A, S Ganguly, H Hashimoto, S Devadiga, E Vermote, Y Knyazikhin, RE Nemani and RB Myneni, (2010). Amazon forests did not green-up during the 2005 drought. *Geophysical Research Letters*, 37(5).

Shuttleworth WJ, (1988). Evaporation from Amazonian rainforest. *Proceedings of the Royal Society of London B: Biological Sciences* 233(1272), 321–346.

Sitch S, C Huntingford, N Gedney, PE Levy, M Lomas, SL Piao, R Betts. P Ciais, P Cox, P Froedlingstein. CD Jones, IC Prentice and FI Woodward, (2008). Evaluation of the terrestrial carbon cycle, future plant geography and climate-carbon cycle feedbacks using five Dynamic Global Vegetation Models (DGVMs). *Global Change Biology*, 14(9), 2015–2039.

Sombroek, W, (2001), Spatial and temporal patterns of Amazon rainfall. AMBIO: *A Journal of the Human Environment*, 30 (7), 388–396.

Tian H, JM Melillo, DW Kicklighter, AD McGuire, JVK. Helfrich, B Moore and CJ Vörösmarty, (1998). Effect of inter-annual climate variability on carbon storage in Amazonian ecosystems. *Nature*, 396(6712), 664–667.

Trancoso RA. F Carneiro, J Tomasella, J Schietti, BR Forsberg and RP Miller, (2010). Deforestation and conservation in major watersheds of the Brazilian Amazon. *Environmental Conservation*, 36 (04), 277–288.

Trumbore SE, EA Davidson, P Barbosa de Camargo, DC Nepstad and LA Martinelli, (1995). Belowground cycling of carbon in forests and pastures of eastern Amazonia. *Global Biogeochemical Cycles*, 9(4), 515–528.

von Randow C, AO Manzi, B Kruijt, PJ de Oliveira, FB Zanchi, RL Silva, MG. Hodnett, JHC Gash, JA Elbers, MJ Waterloo, FL Cardoso and P Kabatvon, (2004). Comparative measurements and seasonal variations in energy and carbon exchange over forest and pasture in South West Amazonia. *Theoretical and Applied Climatology*, 78(1–3), 5–26.

Vourlitis GL, NP Filho, MMS Hayashi, J De S Nogueira, FT Caseiro and JH Campelo, (2001). Seasonal variations in the net ecosystem CO_2 exchange of a mature Amazonian transitional tropical forest. *Functional Ecology*, 15(3), 388–395.

Werth D, and R Avissar, (2002). The local and global effects of Amazon deforestation. *Journal of Geophysical Research*, 107 (D20), 1–8.

Wright SJ, and CP Schaik, (1994). Light and phenology of tropical trees. *American Naturalist* 143, 192–199.

Zeng XD, XB Zeng, SSP Shen, RE Dickinson and QC Zeng, (2005). Vegetation–soil water interaction within a dynamical ecosystem model of grassland in semi-arid areas. *Tellus B*, 57(3), 189–202.

18

Tropical Montane Cloud and Rainforests

18.1 Introduction

Tropical montane cloud and rainforests are forests that occur at high elevation in tropical regions. They differ from tropical lowland forests because they experience low temperatures and because extensive mist and cloud cover alter the water and light environments compared to lowland forests. This chapter discusses the structure and behaviour of tropical montane forests (Fig. 18.1) and briefly compares these to lowland tropical forest.

Due to their terrain and relative isolation, tropical montane forests have not received the same level of intense ecophysiological, modelling, or remote sensing, study that lowland forests have received, although this is slowly changing. Consequently the majority of studies have been leaf- and tree-scale and field-based. Eddy covariance methods have not been routinely applied to-date but developments in using remote sensing techniques with corrections for the influence of altitude, aspect and slope are occurring.

Light, nutrient and temperature effects have major impacts on the ecophysiology, productivity and structure of tropical montane forests. Foliar uptake of water makes a significant contribution to the water status of the canopy.

The questions addressed in this chapter include the following: Where do tropical montane forests occur? What leaf- and canopy-scale adaptations occur that allow montane forests to grow at altitude? Does photosynthetic capacity vary with altitude and what are the patterns in ET and productivity in these forests? Finally, this chapter addresses the question: do mist and clouds supply water directly to the canopy through foliar uptake?

18.2 Types of Tropical Montane Cloud and Rainforests

The case study of Amazonian rainforest (Chapter 17) focuses on seasonality of rainfall, phenological patterns and the impacts of drought on Amazonian rainforest productivity. That chapter examined the environmental controls of productivity in lowland

Table 18.1. *Some attributes of lowland evergreen rainforest compared to three montane forest types*

	Lowland evergreen rainforest	Lower montane rain/ cloud forest	Upper montane cloud forest	Sub-alpine cloud forest
Canopy height (m)	25–45	15–33	1.5–18	1.5–9
Tree buttresses	Common and large	Uncommon, small	Rare	Absent
Compound leaves	Abundant	Occasional	Rare	Absent
Average leaf size	Large-to-very large	Medium	Small	Very small
Vascular epiphytes	Frequent	Abundant	Frequent	Rare
Non-vascular epiphytes (mosses, liverworts)	Occasional	< 50%	70– 80%	> 80%

Source: From Bruijnzeel *et al.* (2011).

tropical rainforest. In contrast, this case study examines *tropical montane cloud* and *rainforest* where the controls of productivity are generally temperature, light availability and sometimes, soil nutrient supply.

Tropical montane cloud and rainforests occur at elevation in the tropics and are characterised as being shrouded in fog/cloud for a large fraction of the year. They have very large levels of endemism, very high levels of species richness, and are the source of large volumes of high quality drinking water for lower elevation sites. Cloud forests may increase the water yield of high altitude catchments through canopy interception and capture of cloud/fog water and through the suppression of evapotranspiration that occurs because of the presence of cloud/fog.

Three classes of tropical montane cloud forests are recognised by Scatena *et al.* (2010), but four classes are recognised by Bruijnzeel *et al.* (2011). Scatena and co-workers identify: (i) lower montane cloud forest; (ii) upper montane cloud forest; and, (iii) sub-alpine cloud forest (Scatena *et al.* 2010). These differ from lowland rainforests in several ways, including maximum tree height, leaf type, the presence/ absence of non-vascular epiphytes and the absence of persistent cloud within the tree canopy in rainforests (see climate later). These classes are compared in Table 18.1. Bruijnzeel and co-workers identify: (i) lower montane rainforest beneath the cloud belt; (ii) tall lower montane cloud forest; (iii) upper montane cloud forest of intermediate height; and, (iv) a combination of sub-alpine cloud forest and elfin cloud forest with low stature.

The transition from lower to upper montane forest (or types (i) and (ii) in Bruijnzeel's classification) coincides with the altitude at which cloud formation becomes most consistent, usually at about 2000–3000 m for equatorial inland regions. Sub-alpine montane forest occurs where the average maximum temperature is less than 10°C; on large equatorial mountains this occurs at about 3000 m elevation. Tree height declines substantially, epiphytes are absent, and leaf size is much reduced in sub-alpine montane forests compared to lower and upper montane forests. The altitude at which cloud condensation consistently occur increases as distance from the coast increases. This is because atmospheric water content declines as distance from the coast increases (Chapter 1) and therefore the dew point temperature falls.

18.3 The Climate of Tropical Montane Cloud Forests

The climate of tropical montane cloud forests (TMCF) is similar to but different from that of lowland tropical rainforests. Lowland tropical rainforests (also called tropical moist broadleaf forests) are consistently warm (mean monthly temperature exceeds 18°C) and receive abundant rainfall (more than 1700 mm *pa*), which may, or may not, have some degree of seasonality. Lowland tropical rainforests occur at low elevations (<800 m). In contrast, TMCFs occur at altitude of between 1200 m and 3500 m (Jarvis and Mulligan 2011) and are therefore cooler (mean minimum temperature <18°C), may or may not have seasonality in rainfall, and are characterised by being covered in cloud/fog at ground level for a large fraction (up to 100%) of the year. Cleveland *et al.* (2011) determined that temperature was able to explain a significant fraction of variation in net primary productivity (NPP) for TMCFs but for lowland tropical rainforests, temperature was not an important variable. Soil nutrient status, and especially P supply, was more important in explaining variation in NPP for these lowland forests.

As with all mountains, air and soil temperatures decrease with increasing altitude (Chapter 1). If the temperature at sea-level is 28°C in the tropics, temperatures decline to about 20°C at 1000 m altitude. If the air is not saturated with water vapour, the rate of cooling is approximately 1°C per 100 m increase in elevation (the dry adiabatic lapse rate; Chapter 1). When the air becomes saturated with water vapour (because as air rises and cools, the water vapour pressure increases towards saturation), water condenses and forms fog or cloud. Condensation of water releases energy and warms the air and consequently the moist adiabatic lapse rate is reduced to about 0.5°C per 100 m. The range 1000 m to 3000 m is the altitude over which high rainfall and extensive cloud cover is common. Wetter slopes of mountains are cooler than drier slopes (for example the Pacific side of Costa Rica is drier and hence 2°C warmer than the wetter and cooler Atlantic side of Costa Rica).

The *Massenerhebung* effect ("mountain mass elevation") is the name given to the phenomenon whereby the altitude at which the tree line occurs varies with the size of the mountain and whether other mountains are close by. Thus smaller mountains have a steeper temperature decline with altitude than larger mountains, and isolated mountains tend to have tree lines at a lower altitude than occurs in mountain ranges such as the Andes. This is because of the beneficial effects of heat retention and wind shadows that can occur in mountain ranges but which don't occur in isolated mountains. As an example, in Borneo, Gunung Palung, located on the coast, has moss covered cloud forest at 900 m but this doesn't occur on Gunung Mulu until 1200 m (inland) and at 1800 m on Mount Kinabalu.

In temperate zones, rainfall usually increases with altitude but on tropical mountains, rainfall increases to a maximum at an intermediate altitude and then declines with further increases in altitude. As the annual total of rainfall declines, the altitude of the maximum rainfall rises. For wet eastern slopes of the Columbian Andes maximum rainfall occurs at 1500 m but for the drier Northern Rift in Ethiopia, maximum rainfall occurs at 2500 m. These differences in altitude for maximum rainfalls are correlated with different altitudes of cloud belts and hence explain the different altitudes for cloud forests in different locations.

Extensive and almost continuous cloud cover significantly reduces the level of solar radiation available to vegetation, by up to 55 percent in "elfin" cloud forest (upper tropical montane cloud forest) of Serrania de Macuira and up to 74 percent in the lower montane cloud forest of Colombia (Letts *et al.* 2010). This reduction in light availability has a significant impact on forest productivity.

Interception of cloud by vegetation results in significant canopy drip and stem flows of water and this source of water can be up to approximately 90 percent of the

Figure 18.1. A tall lower montane forest in Borneo. Photo D Eamus.

total annual water supply (the other 10% being rainfall) (Ray *et al.* 2011). Hutley *et al.* (1997) examined the water balance of a tropical lower montane forest in NE Australia and found that interception of fog/cloud could increase water availability by more than 33 percent above rainfall (896 mm rainfall was increased by 343 mm of fog/cloud interception).

Canopy interception of mist and fog can also contribute to water uptake by trees through foliar absorption. In a recent detailed study of a TMCF in Central Veracruz in Mexico, Gotsch *et al.* (2013) found that foliar uptake of water occurred for about 35 percent of all dry season hours and the amount of water absorbed increased with increasing duration of the fog event across all eight tree species examined. On average, foliar absorbed water contributed almost 10 percent of total daily transpiration. By absorbing water directly into leaves, tree canopies can become temporarily decoupled from soil water stores and this may contribute to rehydration of the canopy after short periods without rain (Gotsch *et al.* 2013).

18.4 Leaf Structure Varies with Altitude

Leaf size declines with increasing altitude on tropical mountains. This may result from declining temperatures with increasing altitude because cell expansion is sensitive to temperature, especially night temperature and exposure to frosts. Low nutrient supply and low stem hydraulic conductance have also been suggested as causing small leaf size in montane forests (Cavelier 1996).

Specific leaf area (m^2 g^{-1}) decreases with altitude. In a comparative study of forty tree species growing across a three-site altitudinal gradient on the eastern slopes of the southern Ecuadorian Andes, Wittich *et al.* (2012) found average specific leaf area (SLA) declined from 0.0091 m^2 g^{-1} at 100 m elevation to 0.0079 m^2 g^{-1} at 2000 m and declined further to 0.0046 m^2 g^{-1} at 3000 m elevation. This apparent increase in leaf thickness with increasing altitude may increase protection from ultraviolet (UV) damage (UV levels increase with increasing altitude and UV damages proteins, membranes and DNA).

Many species of montane plant possess a hypodermis, a layer of cells beneath the epidermis which may function as a "fast response" temporary water supply to the mesophyll to support rapid but short-term large increases in evaporative demand that occur during brief periods when fog dissipates and light levels suddenly increase. The hypodermis may also increase protection against harmful UV light.

18.5 Does Photosynthetic Capacity Vary with Altitude in Tropical Montane Trees?

Aboveground net primary productivity (ANPP) of tropical montane cloud forests is generally smaller than lowland tropical rainforests because of the following:

a) Lower mean temperatures; and

b) Reduced levels of PAR because of extensive and frequent cloud cover.

While these conditions are almost always observed in TMCFs, some sites also exhibit one or more of the following:

c) Low nutrient status of soils

d) High soil acidity

e) Persistent water logged soils and hence low nutrient uptake rates.

Are the inherent photosynthetic capacities of leaves of TMCFs lower than those of lowland rainforest species and could this contribute to the low ANPP of TMCFs? There are many reasons for thinking that this is not the cause of reduced ANPP in TMCF. First, Wittich *et al.* (2012) observed that the stand average rate of light saturated photosynthesis was larger at the mid-elevation (2000 m) montane site compared to the lower-elevation (1000 m) or higher-elevation (3000 m) site in Ecuador. This was the result of a larger availability of N and P at the mid-elevation site. Second, comparisons of leaf-scale attributes of tree species of TMCFs often reveal the same relationships as observed in lowland rainforests. Leaves of TMCFs are generally long-lived and sclerophyllous compared to leaves of lowland rainforests, traits which are globally associated with stressful environments and low rates of photosynthesis (see Chapter 1 for a discussion of cost-benefit analyses of deciduous *versus* evergreen strategies). A long leaf lifespan for TMCFs is an adaptation to ensure sufficient carbon return on the costs of construction of leaves where light levels are low (because of fog) and temperatures are low, compared to lowland tropical rainforests. However, sclerophylly is not only observed in TMCFs as it is common in arid and semi-arid regions globally (see savanna and semi-arid grassland case studies). Soil N and P contents are as variable in TMCF as lowland rainforests and are frequently not significantly lower than lowland rainforests. Indeed, in several comparative studies, soil N and P contents were larger in TMCFs compared to nearby lowland rainforest (Cavelier 1996; Malhi *et al.* 2009; Letts *et al.* 2010), although declining foliar N with altitude has been observed in some TMCFs (Letts and Mulligan 2005; Moser *et al.* 2010). Most importantly, in a log-log plot of foliar N content *versus* maximum photosynthetic rate, the regression of the global data set available in GLOPNET (Wright *et al.* 2004) was the same as the regression for lowland tropical rainforest and montane rainforest (Letts *et al.* 2010). This indicates that photosynthetic capacity (rather than the observed rate of photosynthesis under ambient conditions) may not always be inherently lower in TMCF compared to lowland rainforests (Fig. 18.2). Similarly, mean V_{cmax} and J_{max}, (Chapters 2 and 10) expressed on a leaf area basis for a range of tree species in Peruvian TMCFs were found to be at the high end of the range measured for lowland rainforest (van de Weg *et al.* 2012). The regression of V_{cmax} *versus* J_{max} for a global data set and for tropical rainforest saplings showed very little difference from that of the regression for the TMCF sites of van de Weg *et al.*

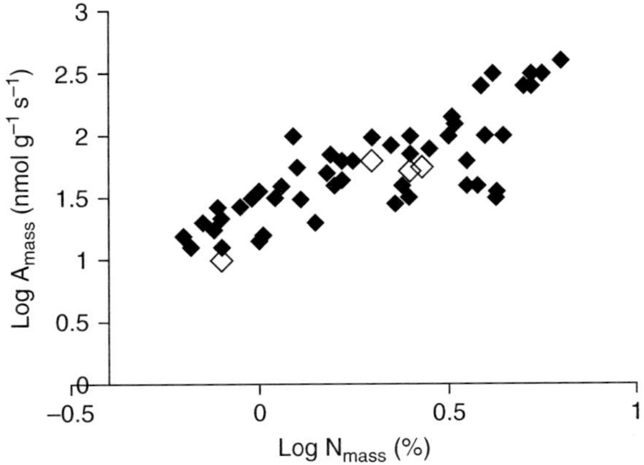

Figure 18.2. The log of rate of photosynthesis expressed on a leaf mass basis (A_{mass}) is linearly correlated with the log of foliar N content (N_{mass}) for a globally derived data set (solid symbols). The same relationship is observed for lower montane cloud forest species (open symbols; 4 tree species) in Colombia and Ecuador. Redrawn from Letts *et al.* (2010).

(2012), further supporting the conclusion that the photosynthetic capacity of leaves of TMCFs are not inherently lower than that of lowland rainforests. How, then, do we explain the lower NPP of TMCFs compared to lowland rainforests?

While the photosynthetic capacity of leaves of TMCFs can be as high as those in lowland rainforests, the observed rates of carbon (C) fixation are generally lower because of low levels of PAR (reduced by up to 80%) in cloud forests and lower temperatures in TMCF compared to lowland rainforests. In a pan-tropical literature review, Wittich *et al.* (2012) examined data for about 170 species and showed that rates of light saturated photosynthesis decline with elevation by about 1.3 µmol m^{-2} s^{-1} per 1000 m increase in altitude, or about 0.2 µmol m^{-2} s^{-1} per 1°C cooler mean annual temperature. Furthermore, respiration rates can be larger in TMCFs than lowland rainforests (van de Weg *et al.* 2012) which reduces net C gain to the vegetation. Finally, increased allocation of C to roots with increasing elevation (see Section 18.6) reduces the capacity for the canopy to intercept PAR and hence NPP is reduced relative to lowland tropical rainforest.

18.6 NPP and C Allocation Patterns in Tropical Montane Cloud Forests

Aboveground standing biomass declines with elevation in TMCFs in Borneo, Ecuador, Puerto Rico, Venezuela, Panama, Peru, and Hawaii (Leuschner *et al.* 2007, Girardin *et al.* 2010). In contrast, fine root biomass increases with elevation (Leuschner *et al.* 2007, Girardin *et al.* 2010). Table 18.2 shows an example of these changes for an altitudinal gradient in Peru. Here the trend of increasing fine root biomass with increasing altitude is observed, as is the decline in stem biomass with increasing altitude. However, because course root biomass declines with altitude and this fraction

Table 18.2. Standing carbon in biomass for course and fine roots and stems along an altitudinal gradient in Peru

Altitude (m)	Forest type	Fine root biomass (Mg C ha^{-1})	Course root biomass (Mg C ha^{-1})	Stem biomass (Mg C ha^{-1})	Soil organic C content (Mg C ha^{-1})
194	Lowland rainforest	1.5	31.5	118.5	3.55
210	Lowland rainforest	2.3	31.5	123.5	n/a
1000	Pre-montane	2.6	16.7	79.5	8.5
1500	Lower montane	n/a	21.6	102.8	27.6
1855	Lower montane	6.5	11.7	55.6	58.3
2020	Lower montane	6.8	8.1	38.6	52.1
2720	Upper montane	4.1	13.8	65.9	63.6
3020	Upper montane	5	9.9	47	68.8

Note: Stem biomass declines with increasing altitude and fine root biomass increases with altitude. However, course root biomass also declined with increasing altitude. The two lowest elevation sites are lowland rainforests.
Source: From Girardin *et al.* (2010).

weighs more than fine root biomass, the overall trend is for total belowground biomass to decline with increasing altitude.

The cause of the decline in standing aboveground biomass with increasing altitude is the decline in tree height, rather than a decline in tree stem density (that is, the number of trees per hectare). Because soil organic C content increased with increasing altitude, the total C stock of the forests (C content of soil plus belowground plus aboveground) did not vary with altitude; what does vary with altitude is the allocation pattern of C with an increased allocation to belowground C stocks with increasing altitude.

Total net primary productivity declines with altitude in TMCFs. Thus Aragao *et al.* (2009) estimate an average NPP of 12.8 Mg C ha^{-1} y^{-1} across ten lowland Amazonian forests while Girardin *et al.* (2010) found an average of 5.68 Mg C ha^{-1} y^{-1} across their TMCF sites. All components of NPP (canopy, stem, fine roots, course roots) were lower in the TMCF sites compared to the lowland sites. The reduction in total NPP in TMCFs compared to lowland forests is observed across a large number of altitudinal gradients globally. The principal cause of the decline in total NPP with increased altitude is declining average temperatures, with a secondary effect of declining PAR supply because of the presence of cloud at canopy heights for much of the year. Declines in temperature have two effects on TMCFs. The first is on plant physiology *per se* – when temperatures are lower than the optimum for photosynthesis, C gain is reduced. The second is on rates of soil mineralisation and litter decay, which are slower at colder temperatures, and this can reduce nutrient supplies to roots leading to reduced growth.

18.7 Transpiration, Evapotranspiration and Stomatal Conductance

18.7.1 Tree and Canopy-Scale Studies

The importance of the availability of sunlight in driving daily whole tree transpiration rates was highlighted in the study of transpiration of tropical montane rainforests in the southern Ecuadorian Andes (Motzer *et al.* 2005). Daily totals of tree water use were linearly correlated with daily sums of photosynthetically active radiation for three dominant upper story tree species (Fig. 18.3) and up to 70 percent of the variance in rates of tree water use was explained by variance in daily PPFD. At shorter time-scales (hourly) sapflow was also highly responsive to vapour pressure deficit. By combining vapour pressure deficit (*VPD*) and light supply (photosynthetic photon flux density; *PPFD*) into a single climatic index (*VPD/PPFD*), Motzer *et al.* (2005) were able to significantly reduce the variability in the regression of stomatal conductance against these variables (Fig. 18.4). Stomata of the overstory species were tightly coupled to the atmosphere, with low values for the decoupling factor (Chapter 2) and stomata therefore exerted a very tight regulation on transpiration rates at the sub-daily time-scale (Motzer *et al.* 2005).

Annual rainfall interception is largest where cloud extent is most frequent and LAI is largest. Thus, sub-alpine and elfin forests, which can receive large annual rainfalls (Table 18.3), have a low LAI and hence low rate of interception of rainfall (and fog). Rates of stand transpiration for these forests are also reduced, principally because of the low LAI (relative to lowland rainforest and lower and upper montane cloud forest; Fig. 18.5 and Table 18.3) but also because of their lower temperatures. A strong relationship of both transpiration and *ET* with LAI is commonly observed in montane

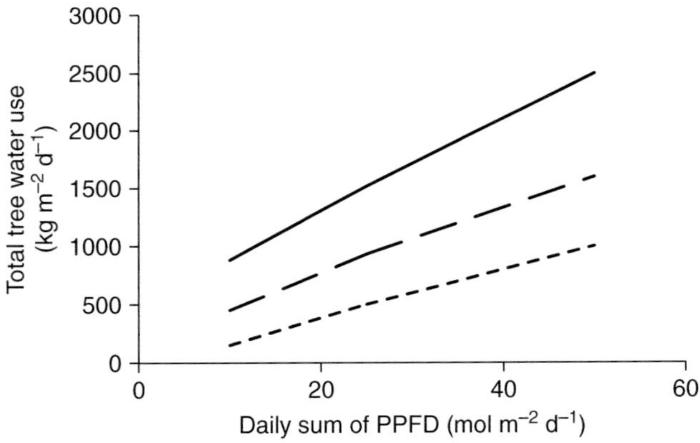

Figure 18.3. Total tree water-use for three species in an Andean tropical montane rainforest increase linearly with increased supply of solar radiation (PPFD = photosynthetic photon flux density). The solid line is the relationship for *Trichilia guianensis*; the dashed line is for *Ruagea pubescens* and the dotted line is for *Aniba muca*. Redrawn from Motzer *et al.* (2005).

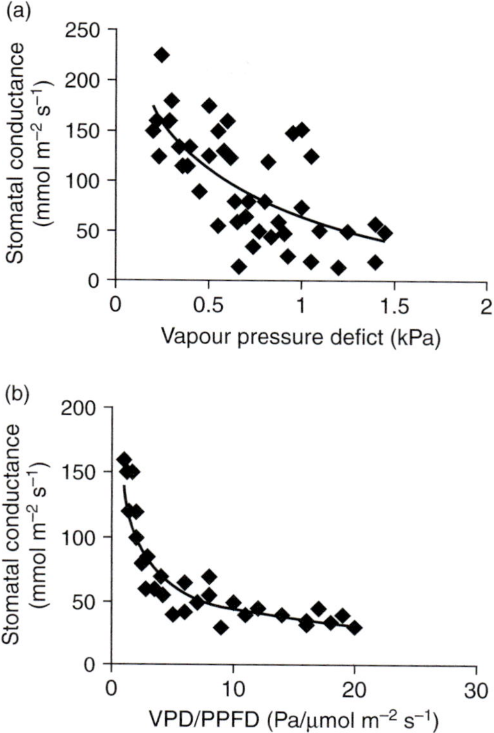

Figure 18.4. Stomatal conductance declines with increasing vapour pressure deficit (VPD) across a range of species in the Andean tropical montane rainforest. However, the range of variation is much reduced when the index *VPD/PPFD* is used to describe the response of conductance to changing environmental conditions. Redrawn from Motzer *et al.* (2005).

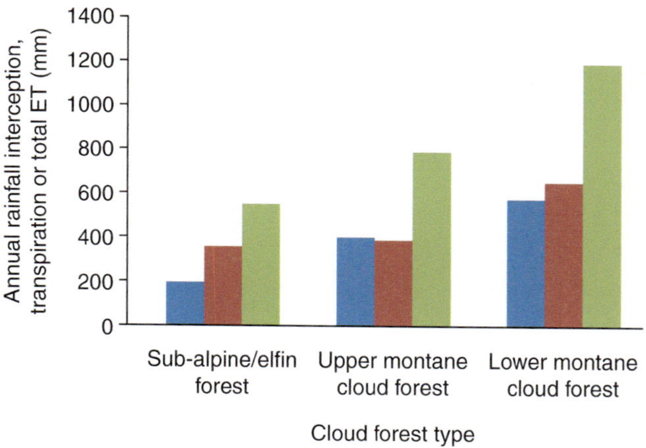

Figure 18.5. Annual rainfall interception (blue bars), transpiration (red bars) and total ET (green) for three classes of tropical montane cloud forest. Redrawn from meta-analyses of global data in Bruijnzeel *et al.* (2011).

Table 18.3. Indicative values for elevation, rainfall, LAI, annual transpiration, interception, and ET for a range of tropical montane rain and cloud forests

Forest type	Elevation (m)	Mean annual rainfall (mm)	LAI	Annual transpiration (mm)	Annual interception (mm)	Annual ET (mm)	ET/R_{net}
Lowland evergreen rainforest	50–100	1800–2750	6–7	900–1200	200–360	1300–1400	0.61
Lower montane rainforest	1000–2500	1900–2300	4–7	600–800	400–600	1100–1350	
Lower montane cloud forest	1000–2500	1300–3300	4–6	500–850	400–900	800–1500	0.49
Upper montane cloud forest	1000–3000	3000–7500	3–4.5	300–500	200–550	600–900	0.37
Elfin cloud forest	1000–1300	1000–4500	2–4	300–400	100–250	500–600	0.27

Source: Adapted from Bruijnzeel *et al.* (2011).

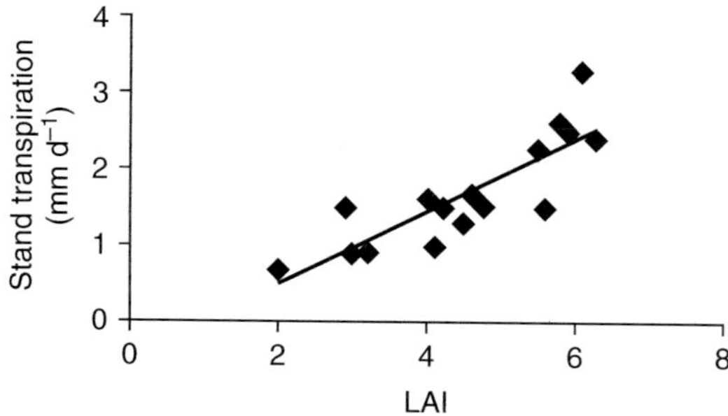

Figure 18.6. The relationship of daily stand transpiration rate with LAI for a range of lowland rainforest sites (LAI >4), and lower, upper, and elfin cloud forests. Upper and elfin cloud forests have LAI <4.

cloud forests (Fig. 18.6) because generally the supply of water is not limiting to *ET*. It is the supply of energy that limits *ET* (Fig. 18.3) and increased LAI increases the amount of radiation intercepted. Interestingly, the ratio of *ET* to net radiation (ET/R_n) declines with elevation: it is 0.61 for lowland evergreen rainforest, 0.49 for lower montane cloud forest and 0.37 and 0.265 for upper montane cloud forest and elfin cloud forest respectively (Table 18.3; Bruijnzeel *et al.* 2011). Therefore it is often not only a lack of radiation that can cause a decline in total *ET* along an altitudinal gradient. Additional factors that can reduce *ET* include: (a) root anoxia associated with excessive soil wetness; and (b) increased frequency of cloud/fog formation reducing evaporative demand of the atmosphere (Motzer *et al.* 2005).

Nocturnal transpiration has been documented in multiple ecosystems and is strongly affected by vapour pressure deficit (Chapter 3). Generally nocturnal transpiration occurs most frequently when (a) soils are wet; (b) vapour pressure deficits are small; and (c) ecosystems are not usually exposed to large soil or atmospheric water deficits (Dawson *et al.* 2007). All three conditions are commonly observed in TMCFs and nocturnal transpiration should therefore be expected. This expectation has been supported by studies in a TMCF in Mexico (Gotsch *et al.* 2013) and Hawai'i (Dawson *et al.* 2007). Gotsch *et al.* (2013) found that "high" night-time values of vapour pressure deficit (0.5–1.5 kPa) were associated with nocturnal transpiration and that almost 60 percent of all dry season night-time hours exhibited nocturnal transpiration. Consequently approximately 17 percent of total dry season transpiration occurred at night. The total volume transpired at night was approximately twice the volume of water absorbed through foliage (Gotsch *et al.* 2013).

18.8 Remote Sensing of ET

While point measurements of evapotranspiration (*ET*) or transpiration (*T*) are possible with sapflow sensors for tree-scale measurements and gas analysers for leaf-scale measurements, the rugged terrain and steep slopes commonly observed in TMCFs make application of eddy covariance techniques highly problematic. Estimating *ET* at catchment or regional scales in TMCFs is made additionally difficult because of the high diversity of vegetation. Application of remotely sensed data to estimate *ET* in TMCFs is also difficult because of the complex terrain. However, significant progress in combining remotely sensed data and digital elevation models is being made.

One way to estimate *ET* from remotely sensed data is to use the surface energy balance approach, whereby *ET* calculated from the following equation (Eq. 18.1):

$$ET = 3600 \ (\lambda E/\lambda) \tag{18.1}$$

where *ET* is expressed in mm h^{-1}, λE is latent heat flux, and λ is the latent heat of vaporization.

Wang *et al.* (2010) have developed a method that uses remotely sensed satellite images (Thematic Mapper, Advanced Spaceborne Thermal Emission and Reflection Radiometer data), surface meteorological data (rainfall, temperature, *VPD*) and topographic data, within digital elevation models to estimate *ET* of tropical montane forests in Thailand.

Comparisons of modelled surface energy fluxes with *in situ* measurements showed that inclusion of terrain corrections greatly improved the accuracy of the estimates of *ET*. Figure 18.7 summarises the sequence of data processing required to generate estimates of *ET*. The key terrain corrections account for the fact that slope and hence

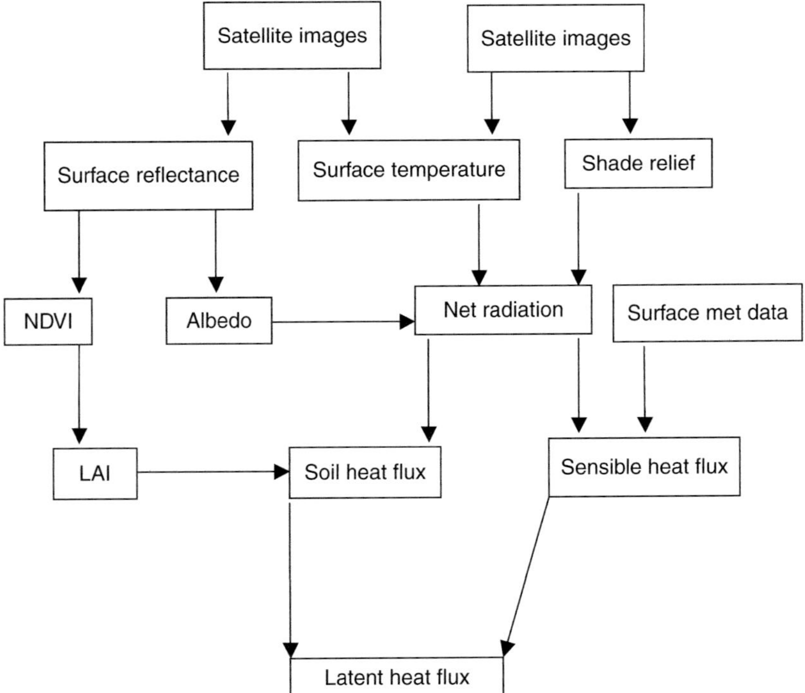

Figure 18.7. The flowchart for calculating latent heat flux using a surface energy and remote sensing approach. Redrawn from Wang *et al.* (2010).

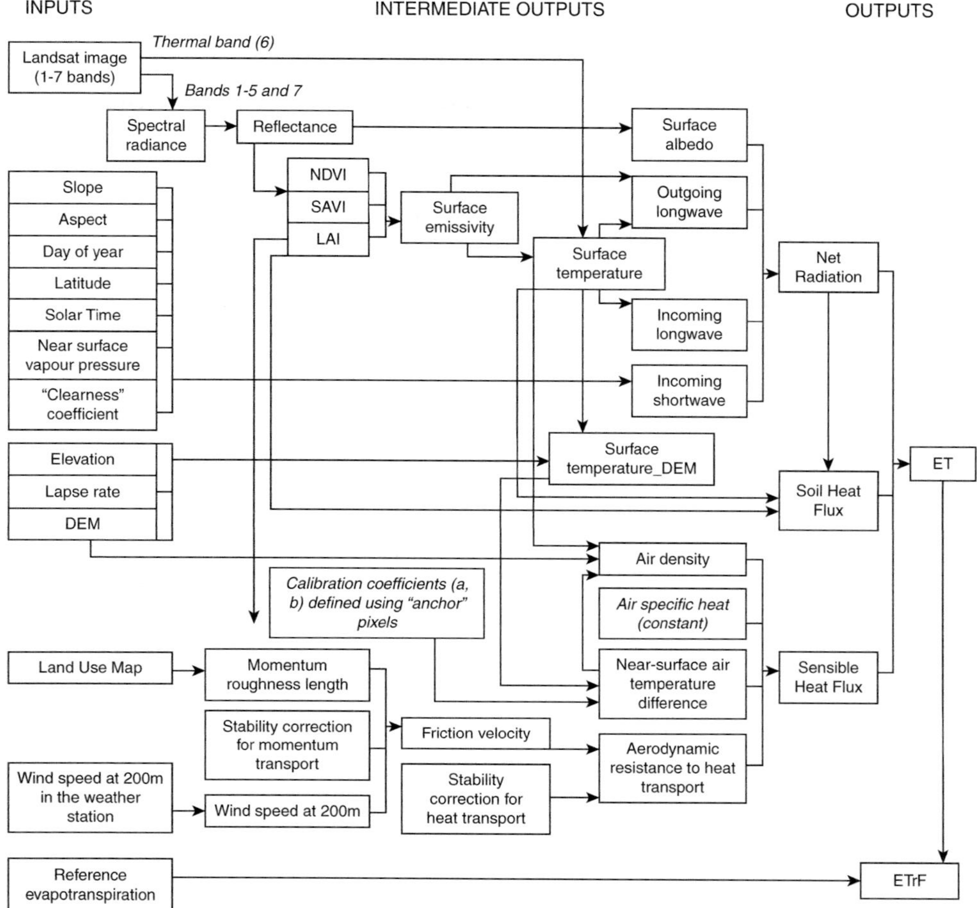

Figure 18.8. A summary of the data inputs, processing and outputs of the methodology of Pocas *et al.* (2013) for estimating *ET* in mountainous regions.

azimuth angle and aspect (Chapter 1) vary pixel by pixel according to local terrain, atmospheric transmissivity varies with altitude, and surface temperatures vary with altitude.

A conceptually similar approach for estimating *ET* in mountainous regions can be found in Pocas *et al.* (2013). Figure 18.8 provides a summary of the data inputs, processing, and outputs of the methodology of Pocas *et al.* (2013). A serious drawback of the surface energy balance approach is the need for very wet and very dry regions within the image of the region of interest and these are not always easy to find.

18.9 Climatological Links Between Tropical Lowland and Montane Forests

Tropical biodiversity hotspots are frequently located in TMCFs. While they may only cover about 0.4 percent of the world's land surface, TMCFs are home to about

20 percent of the world's plant species and 16 percent of the world's vertebrates (Myers *et al.* 2000). However, these regions are also increasingly threatened by human activities, including deforestation and conversion to agriculture. Furthermore it is now understood that changes in land cover in lowland forests, which are also increasingly subject to land clearance for agricultural and mining purposes, can affect montane forests and *vice versa*. The loss of TMCFs has been associated with reduced stream flow to lower elevations, for example in Sri Lanka where tea crops have replaced TMCFs (Dounege *et al.* 1995). But do changes in lowland forests have the capacity to influence upland forests? It is through the exchange of heat and moisture between land surface and atmosphere that changes in land cover affect the boundary layer air within which orographic clouds are formed.

Lawton *et al.* (2001) and Nair *et al.* (2003) have demonstrated, using remotely sensed images, that the formation of cumulus clouds in Costa Rica has been reduced as a result of deforestation of lowland forests. Numerical modelling confirmed these observational data. Not only was the extent and frequency of cloud cover reduced by lowland forest clearing, but the altitude of the cloud base (an important trait affecting the distribution of cloud forests) increased and cloud liquid water content was also altered, further exacerbating the lift in cloud base.

Ray *et al.* (2006) extended the work of Lawton *et al.* (2001) and Nair *et al.* (2003). They used remote sensing techniques (MODIS LAI and land classifications and satellite images of cloud cover) coupled with field observations and numerical atmospheric modelling to test the hypothesis that lowland deforestation produces warmer and drier air masses and that this results in higher altitude orographic cloud bases and less frequent orographic cloud formation in the dry season. They made the following key conclusions:

1. The numerical atmospheric model achieved moderate success in simulating cloud formation, when compared with remotely sensed measures of cloud presence.
2. Deforestation increase sensible heat fluxes and decrease latent heat fluxes in lowland regions. These have significant effects on boundary layer conditions and these have a "flow-on" effect to TMCFs (see 3–5).
3. Deforestation of lowlands lifted the cloud base significantly, often to altitudes that were larger than the altitude of the mountains, thereby removing this source of water entirely from the TMCF.
4. Deforestation of the lowlands reduced the total daily period of cloud formation, with an approximately 25 percent reduction in cloud duration.
5. Deforestation increased the air temperature and reduced the dew point temperature (that is, reduced the water content of the atmosphere).

The Monteverde cloud forests in Costa Rica (the study sites of Ray *et al.* 2006, Lawton *et al.* 2001 and Nair *et al.* 2003) are relatively well protected but declines (and extinctions) in frog and toad populations and the migration of birds to higher elevations have been documented. It is highly likely that these changes are the result of reduced cloud formation, higher elevation cloud bases, reduced cloud areal extent, and reduced cloud water content, all of which can be attributed to deforestation of lowland regions.

A similar combination of numerical atmospheric modelling, remote sensing of vegetation classes, LAI, and cloud cover (and cloud attributes) plus field assessments, has been applied to biodiversity hotspots in the Western Ghats of southwest India (Ray *et al.* 2011). In contrast to the Costa Rican study, Ray *et al.* (2011) demonstrated that lowland deforestation could increase precipitation over the TMCFs. This was attributed to deforestation of lowland regions resulting in stronger convective rainfall events. A major limitation of this study, however, was that the numerical simulations were confined to a single dry season month. However, in support of their conclusion Murugan *et al.* (2009) found decreasing seasonal average and annual total rainfall for three of four TMCF sites in the Western Ghats. Similarly, increased cloud cover and rainfall has been observed in deforested regions of the Amazon during the dry season Costa *et al.* (2007).

It is clear from these studies that not only can deforestation of TMCFs influence streamflow downslope, but deforestation of lowland tropical rainforests can impact the climate of upslope TMCFs and that this represents a significant threat to the biodiversity hotspots commonly associated with these forests.

18.10 References

Aragão LEOC, Y Malhi, DB Metcalfe, JE Silva-Espejo, E Jiménez, D Navarrete, S Almeida, ACL Costa, N Salinas, OL Phillips, LO Anderson, TR Baker, PH Goncalvez, J Huamán-Ovalle, M Mamani-Solórzano, P Meir, A Monteagudo, MC Peñuela, A Prieto, CA Quesada, Rozas-A Dávila, A Rudas, JA Silva Junior and Q Vásquez, (2009). Above- and below-ground net primary productivity across ten Amazonian forests on contrasting soils. *Biogeosciences* 6, 2441–2488.

Bruijnzeel LA, M Mulligan and FN Scatena, (2011). Hydrometeorology of tropical montane cloud forests: emerging patterns. *Hydrological Processes* 25, 465–498.

Cavelier J, (1996). Environmental factors and ecophysiological processes along altitudinal gradients in wet tropical mountains. In *Tropical Forest Plant Ecophysiology*, SS Mulkey, RL Chazdon and AP Smith (Eds). pps. 399–439. Chapman and Hall.

Cleveland CC, AR Townsend, P Taylor, S Alvarez-Clare, MMC Bustamante, G Chuyong, SZ Dobrowski, P Grierson, KE Harms, BZ Houlton, A Marklein, W Parton, S Porder, SC Reed, CA Sierra, WL Silver, EVJ Tanner, WR Wieder and R William, (2011). Relationships among net primary productivity, nutrients and climate in tropical rain forest: a pan-tropical analysis. *Ecology Letters* 14, 939–947.

Costa, MH, SNM Yanagi, PJOP Souza, A Ribeiro and EJP Rocha, (2007). Climate change in Amazonia caused by soybean cropland expansion, as compared to caused by pastureland expansion. *Geophysical Research Letters* 34, L07706, doi:10.1029/2007GL029271.

Dawson TE, SSO Burgess, KP Tu, RS Oliverira, LS Santiago JB Fisher, KA Simonin and AR Ambrose, (2007). Nighttime transpiration in woody plants from contrasting ecosystems. *Tree Physiology* 27, 561–575.

Doumenge C, D Gilmour, MR Perez and J Blochus, (1995). Tropical montane cloud forests: Conservation status and management issues. In *Tropical Montane Cloud Forests*, LS Hamilton, JO Juvik and FN Scatena (Eds), pp. 24–37, Springer, New York.

Girardin CAJ, Y Malhi, LEOC Aragao, M Mamani, WH Huasco, L Durand, KJ Feeley, J Rapp, JE Silva-Espejo, M Silman, N Salinas and RJ Whittaker, (2010). Net primary productivity allocation and cycling of carbon along a tropical forest elevational transect in the Peruvian Andes. *Global Change Biology* 16, 3176–3192.

Gotsch SG, H Asbhjornses, F Holwerda, GR Goldsmith, AE Weintraub and TE Dawson, (2013). Foggy days and dry nights determine crown-level water balance in a seasonal tropical montane cloud forest. *Plant, Cell and Environment* 37, 261–272.

Hutley LB, D Doley, DJ Yates and A Boonsaner, (1997). Water balance of an Australian subtropical rainforest at altitude: the ecological and physiological significance of intercepted cloud and fog. *Australian Journal of Botany* 45, 311–329.

Jarvis A and M Mulligan, (2011). The climate of cloud forests. *Hydrological Processes* 25, 327–343.

Lawton, RO, US Nair, RA Pielke and RM Welch, (2001). Climatic impact of tropical lowland deforestation on nearby montane cloud forests. *Science*, 294, 584–587.

Letts MG and M Mulligan, (2005). The impact of light quality and leaf wetness on photosynthesis in north-west Andean tropical montane cloud forest. *Journal of Tropical Ecology* 21, 549–57.

Letts MG, M Mulligan, ME Rincon-Romero and LA Bruijnzeel, (2010). Environmental controls on photosynthetic rates of lower montane cloud forest vegetation in SW Colombia. In *Tropical Montane Cloud Forests: Science for Conservation and Management*. LA Bruijnzeel, FN Scatena and LS Hamilton (Eds). Cambridge University Press, UK.

Leuschner C, G Moser C Bertsch, M Roderstein and D Hertel, (2007). Large altitudinal increase in tree root/shoot ratio in tropical mountain forests of Ecuador. *Basic and Applied Ecology* 8, 219–230.

Malhi Y, LEOC Arag, DB Metcalfe, R Paiva, C Quesada, S Almeida, L Anderson, P Brando, JQ Chambers, ACL da Costa, LR Hutyra, P Oliveira, S Patiño, EH Pyle, AL Robertson and LM Teixeira, (2009). Comprehensive assessment of carbon productivity, allocation and storage in three Amazonian forests. *Global Change Biology*, 15(5), 1255–1274.

Marthews TR, Y Malhi, CAJ Girardin, JES Espejo, LEOC Aragao, DB Metcalfe, JM Rapp, LAM Mercado, RA Fisher, DR Galbrait, JB Fisher, N Salinas-Revilla, AD Friend, N Restrepo-Coupe and RJ Williams, (2012). Simulating forest productivity along a neotropical elevational transect: temperature variation and carbon use efficiency. *Global Change Biology* 18, 2882–2898.

Motzer T, N Munz, M Kuppers, D Schmitt and D Anhuf, (2005). Stomatal conductance, transpiration and sap flow of tropical montane rain forest trees in the southern Ecuadorian Andes. *Tree Physiology* 25, 1283–1293.

Murugan M, PK Shetty, A Anandhi, R Ravi, S Alappan, M Vasudevan and S Gopalan, (2009). Rainfall changes over tropical montane cloud forests of southern Western Ghats, India. *Current Science* 97, 1755–1760.

Myers N, RA Mittermier, CG Mittermier, GA B da Fonesca and J Kent, (2000). Biodiversity hotspots for conservation priorities, *Nature* 403, 853–858.

Moser G, C Leuschner, M Roderstein, S Graefe, N Soethe and D Hertel, (2010). Biomass and productivity of fine and coarse roots in five tropical mountain forest stands along an altitudinal transect in southern Ecuador. *Plant Ecology and Diversity* 3, 151–64.

Nair, US, RO Lawton, RM Welch and RA Pielke, (2003). Impact of land use on Costa Rican tropical montane cloud forests: Sensitivity of cumulus cloud field characteristics to lowland deforestation. *Journal of Geophysical Research* 108(D7), 4206, doi:10.1029/2001JD001135.

Pocas I, M Cunha, LS Pereira and RG Allen, (2013). Using remote sensing energy balance and evapotranspiration to characterise montane landscape vegetation with focus on grass and pasture lands. *International Journal of Applied Earth Observation and Geoinformation* 21, 159–172.

Ray DK, US Nair, RO Lawton, RM Welch and RA Pielke, (2006). Impact of land use on Costa Rican tropical montane cloud forests: Sensitivity of orographic cloud formation to deforestation in the plains. *Journal of Geophysical Research* 111, D02108, doi:10.1029/2005JD006096.

Scatena SN, LA Bruijnzeel, P Bubb and S Das, (2010). Setting the stage. In *Tropical Montane Cloud Forests: Science for Conservation and Management*. LA Bruijnzeel, FN Scatena and LS Hamilton (Eds). Cambridge University Press, UK.

van de Weg MJ, P Meir, J Grace and GD Ramos, (2012). Photosynthetic parameters, dark respiration and leaf traits in the canopy of a Peruvian tropical montane cloud forest. *Oecologia* 168, 23–34.

Wang Y-C, TY Chang TY and Y-A Liou, (2010). Terrain correction for increasing the evapotranspiration estimation accuracy in a mountainous watershed. *IEE Geoscience and Remote Sensing Letters* 7, DOI: 10.1109/LGRS.2009.2035138

Wittich B, V Horna, J Homeier and C Leuschner, (2012). Altitudinal change in the photosynthetic capacity of tropical trees: A case study from Ecuador and a pantropical literature analysis. *Ecosystems* 15, 958–973.

Wright IJ, PB Reich, M Westoby and DD Ackerly, *et al.*, (2004). The world-wide leaf economics spectrum. *Nature* 428, 821–827.

19

Groundwater Dependent Ecosystems

19.1 Introduction

Much of the global land surface is arid or semi-arid, where water supplies limit plant growth, food supplies, and human population densities. Groundwater is a major resource to both humans and the ecology in such regions, and groundwater extraction is increasing exponentially in many regions. In the past three decades research has increasingly shown that in many places, terrestrial vegetation can also be reliant on the availability of groundwater. In order to allocate water to sustain these groundwater dependent ecosystems, which have commercial, ecological, and amenity value, knowledge of their location, their water requirements, their structure, and an understanding of how they respond to changes in water supply are required. Ecophysiological and remote sensing studies are now being combined to increase our understanding of these issues.

Field assessments of vegetation water-use using eddy covariance, sapflow sensors, and stable isotopes are combined with remotely sensed data and models to determine rates of vegetation water-use, rates of groundwater-use, and the location of groundwater dependent ecosystems across landscapes.

The location of groundwater dependent ecosystems (GDEs) can be difficult to identify, especially as they are often located in arid and semi-arid regions with low population densities and sparse infrastructure. Quantifying their rate of water-use has been equally problematic in the past but the application of remotely sensed data has provided a spatially extensive method to overcome these problems.

The three big questions addressed in this case study are: (1) how can we find GDEs in a landscape; (2) what is the rate of water-use for GDEs; and, (3) can remote sensing and ecophysiology be combined to study GDEs?

19.2 Groundwater and Groundwater Dependent Ecosystems

The water table represents the surface (interface) between the water-saturated, oxygen-depleted aquifer below and the aerated soil above. Roots are mostly confined to the aerated soil and the capillary fringe above the aquifer. The capillary fringe is

Table 19.1. Groundwater-use for the top five users globally

Country	Annual groundwater extraction (km³)	Total renewable groundwater resources (km³)	Extraction as a % of renewable resources	Extraction as a % of global extraction
India	190	419	45.3	28.9
USA	110	1,300	8.5	16.7
Pakistan	60	55	109.1	9.1
China	53	828	6.4	8.1
Iran	53	49	108.2	8.1
Global total	**658**	**11,282**	**5.8**	**100**

Source: From Giordano (2009).

the depth of soil that receives water moving vertically from the aquifer through capillary rise. In course-textured soils (sand) the capillary fringe is narrow (measured in centimetres) while in fine-textured soils (silt) it may be up to 2 m in depth.

Groundwater (water stored underground in an aquifer) is the largest store of liquid (i.e. not frozen) freshwater in the world, accounting for about 96 percent of global liquid freshwater (Shiklomanov 2008). For millennia, groundwater has been accessed by humans at naturally occurring springs/oases and as base flow discharge (the lateral flow of groundwater into a river). However, it has only been during the past one hundred years, with the application of electro-mechanical pumps, that groundwater-use has become of concern (Gleick and Palaniappan 2010). Table 19.1 summarises the rate of groundwater-use for the five largest users in the world. Only three of the top ten users are members of the Organisation for Economic Co-operation and Development, perhaps indicating the reliance of less developed nations on groundwater resources. This probably reflects the fact that nations in arid and semi-arid climates rely most on groundwater and these tend to be less well-developed economically than nations with a larger extent of mesic climates.

In a recent review Fan *et al.* (2013) summarised global patterns of groundwater depth from more than 1.6 million observations, in combination with a simple groundwater depth model. While extensive gaps in the data-set exist, several interesting features were apparent in the data:

1. The distribution of groundwater depth peaks between 2 and 7 m. This reflects the fact that human interest in groundwater depth declines with increasing depth because of increasing inaccessibility, and therefore observations tend to be more frequent in areas with shallow rather than deep groundwater.
2. The depth to the water table is more commonly shallow in humid regions, as opposed to arid regions, reflecting an influence of climate on groundwater recharge and discharge. However, shallow groundwater can be found in valleys in arid zones, reflecting the influence of topography on groundwater depth.

3. Groundwater depth tends to be shallow in coastal regions, especially where flat coastal plains meet the sea. This distribution is mirrored in the global distribution of large coastal wetlands.
4. About 15 percent of the land surface is covered by lakes, rivers and inundated wetlands that consistently receive groundwater inflows; about 2 percent of the land surface is covered by wetlands that receive less consistent supplies of groundwater; 5–15 percent of the land surface has groundwater that is likely to be within the rooting depth of terrestrial upland plants.

Over extraction of groundwater can result in four immediate problems. These are:

1. Loss of discharge of groundwater to wetlands, springs and streams/rivers, which results in loss of ecosystem structure and function, with associated loss of ecosystem services (Eamus *et al.* 2005, Murray *et al.* 2006, Evans 2007).
2. Increased depth of groundwater, thereby reducing its availability within the root zone of groundwater dependent vegetation.
3. Reduced availability of groundwater for direct human consumption; and
4. Reduced availability of groundwater for commercial use, including irrigation, stock watering and other industrial applications.

Almost half of the world's land surface is arid or semi-arid and almost half of the world's population lives there. Because rainfall is minimal and often highly unpredictable (both in timing and amount), reliance on groundwater is largest in arid and semi-arid regions. Thus about 40 percent of global groundwater abstraction but only 2 percent of groundwater recharge occurs in these regions (Wada *et al.* 2010).

Water is a geopolitical and strategic resource, with disputes between neighbouring nations occurring as population and industrial demands for groundwater increase. Declines in groundwater availability have major negative impacts on food supplies, food prices and concomitant social unrest, particularly in semi-arid and arid regions. For example, between 10 percent and 25 percent of the food produced in China and India (home to almost 30% of the global population) is at risk because of groundwater depletion (Brown 2007, Seckler *et al.* 1999).

In this chapter we focus on the links between ecohydrology and groundwater availability. In particular, reference is made to field studies, remote sensing and modelling to address the following four questions:

1. What types of groundwater dependent ecosystems (GDEs) occur across the globe?
2. How do we know where a groundwater dependent ecosystem is in the landscape? If their location is not known we cannot manage them and allocate groundwater resources appropriately.
3. How much groundwater is used by a GDE? An allocation of water by managers requires knowledge of rates of water-use.
4. What is the ecophysiological response of vegetation to GW extraction?

19.3 Classes of GDEs

A simple classification system for GDEs was proposed in 2006 by Eamus and co-workers (Eamus *et al.* 2006):

I. Aquifer and cave ecosystems where stygofauna (including crustacea, worms, bacteria and fish), reside. This class also includes the hyporheic zones of rivers and floodplains.

II. Ecosystems reliant on surface expression of groundwater. This includes base-flow rivers, streams and wetlands, springs, and estuarine seagrasses.

III. Ecosystems reliant on sub-surface presence of groundwater within the rooting depth of a terrestrial ecosystem (via the capillary fringe), including riparian forests and other woodlands (Fig. 19.1).

This classification scheme was recently adopted in the Australian National Atlas of Groundwater Dependent Ecosystems and its use can assist managers in identifying the appropriate techniques for assessing GDE structure, function and management regime (Eamus *et al.* 2006).

Figure 19.1. A tall dense eucalypt woodland growing above a 2 m deep aquifer in south-eastern Australia. The diameter and height of the trees and their growth rates are significantly enhanced during dry periods compared to nearby woodlands where groundwater depth is >30 m. Photograph Dr S Zolfaghar.

19.4 Identifying Groundwater Dependent Vegetation

Identifying the location of GDEs is the vital first step to managing them. However, identifying their location across a landscape is often difficult, time-consuming, and hence expensive. In this section, a range of field and remote sensing techniques that can be used to assist in this are discussed.

19.4.1 Inferential Assessments to Determine the Location of a GDE

Early assessments of groundwater dependency frequently relied on inference (Clifton and Evans 2001, Eamus *et al.* 2006). Answers in the affirmative to one or more of the following questions can be taken as supporting the hypothesis that at least some species in an ecosystem are using groundwater:

1) Do observations or modelling outputs indicate that groundwater or the capillary fringe above the water table is located within the rooting depth of any of the vegetation?

2) In locations with surface discharge of groundwater, is vegetation different (in terms of species composition, phenological pattern, leaf area index, or structure) compared to vegetation close-by but which is not accessing groundwater?

3) Is the annual rate of transpiration observed significantly larger than annual rainfall at the site?

4) During a prolonged dry period, are plant water relations (e.g. predawn and midday water potentials) indicative of less water stress (water potentials closer to zero; transpiration rate larger) than vegetation located nearby but not accessing the groundwater discharged at the surface?

5) Does the water balance of a site indicate that the sum of water-use plus interception loss plus run-off plus deep drainage is significantly larger that annual rainfall plus run-on?

6) Do key developmental stages of the vegetation (such as flowering, germination, seedling establishment) occur when groundwater discharge to the surface occurs?

7) Does a proportion of the vegetation remain green and physiologically active (principally, transpiring and fixing carbon, although stem diameter growth or leaf growth are also good indicators) during extended dry periods of the year? See Section 19.4.3.

8) Despite long periods of low or zero rainfall (and thus zero surface flows), does a stream or river flow all year?

9) Do salinity levels within an estuary fall below that of seawater in the absence of surface water inputs?

10) In the absence of inflow from a tributary or surface flows, does the total flux in a river increase downstream in the absence of inflow from a tributary or surface flow?

11) During extended dry periods are water levels in a wetland maintained at a high level?

12) Is groundwater discharged to the land surface for significant periods of time each year? If so, this groundwater is likely to be used by vegetation.

13) Within a small region (and thus an area having the same annual rainfall, temperature and vapour pressure deficit) and in an area not having access to run-on or stream or river water, do some ecosystems show large seasonal changes in leaf area index whilst others do not?

14) Are seasonal changes in groundwater depth larger than can be accounted for by the sum of lateral flows and percolation to depth (that is, is vegetation a significant discharge path for groundwater (Cook *et al.* 1998))? Clearly, if the error terms in the estimation of lateral flow and percolation to depth are of similar magnitude or greater than the rate of vegetation water, this method may not be appropriate.

Although affirmative answers to one or more of these questions leads to the inference that the system is a GDE, this does not provide information about the nature of the dependency (obligate or facultative) nor about the groundwater regime (e.g. timing of groundwater availability, volume utilised, location of surface expression, the pressure of the groundwater aquifer required to support the surface discharge of groundwater) needed to support the ecosystem.

For base-flow systems (that is, rivers and streams showing significant flows during periods of zero surface or lateral flows), measurements of the chlorofluorocarbon, magnesium or radon concentrations of river and groundwater supply can identify and quantify the amount and timing of groundwater inflows into the river (Cook *et al.* 2003).

When tracers are added to groundwater (for example deuterium or lithium), subsequent uptake by vegetation is proof of GW use by vegetation. Although the presence of a tracer in a shallow rooted species can also occur if neighbouring deep rooted species exhibit hydraulic lift (also called hydraulic redistribution) and the shallow rooted plants then "harvest" this water (Caldwell *et al.* 1998), this can still be taken as groundwater-use by the shallow-rooted species.

19.4.2 Field Observations of Groundwater-Use: Stable Isotope Analyses

Direct evidence that vegetation is using groundwater can be obtained by comparing the stable isotope composition of groundwater, soil water, surface water (where relevant), and vegetation xylem water (Kray *et al.* 2012, Lamontagne *et al.* 2005, O'Grady *et al.* 2006, Zencich *et al.* 2002). Where sufficient variation in isotopic composition among these sources occurs then it is possible to identify the single or the most dominant source of water being used by different species at different times of year (Zencich *et al.* 2002). An example of the use of ^{18}O isotope analyses of xylem water, soil water and groundwater is shown in Figure 19.2.

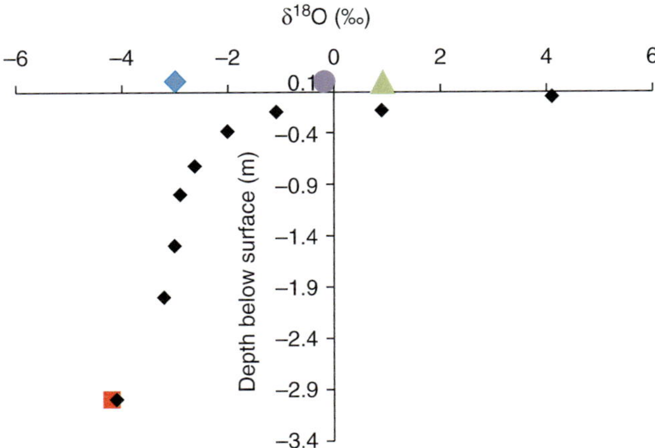

Figure 19.2. Comparative analyses of ^{18}O of xylem water, soil water and groundwater in a multi-species study in northern Yucatan (Mexico). The ^{18}O content of soil declines with depth (black diamonds) and eventually groundwater is reached (at 3 m; large red square). The xylem ^{18}O content of three species (*Ficus* spp. green triangle; *Spondias* spp. purple circle; and *Talisia* spp. blue diamond) reveals that *Ficus* was the least reliant on groundwater and *Talisia* was the most reliant. Redrawn from Querejeta *et al.* (2007).

Estimation of the relative contribution of multiple sources of water to the water absorbed by roots can be estimated from mixed-member models (Kolb *et al.* 1997, Phillips and Gregg 2003). Thus analyses of stable isotopes of soil water, xylem water, and groundwater can provide insight to spatial and temporal variations in groundwater dependency and also can be used to calculate rates of groundwater-use of co-occurring species within an ecosystem. For example, deep rooted *Banksia attenuata* growing on a dune crest of Western Australia sources between 20 and 95 percent of its water from groundwater and the proportion of groundwater-use increases significantly in the hot dry summer and declines in the cooler wetter winter (Fig. 19.3a; Zencich *et al.* 2002). Furthermore when this species was growing in the low-lying dampland, groundwater-use increased after soil water stores were depleted during the summer. In contrast the shallow-rooted *Hibbertia hypericoides* growing in the dampland used groundwater during the hot dry summer when rainfall was absent, but used rainfall in the wet winter. When this species was growing on the dune crest, it could not access groundwater at any time of the year because of its shallow roots (Fig. 19.3b).

19.4.3 *Application of Vegetation Indices Derived from Remote Sensing to Identify the Location of GDEs*

A common conceptual model in the application of remote sensing (RS) to identify GDEs across a landscape is that of "green islands". To apply this philosophy the

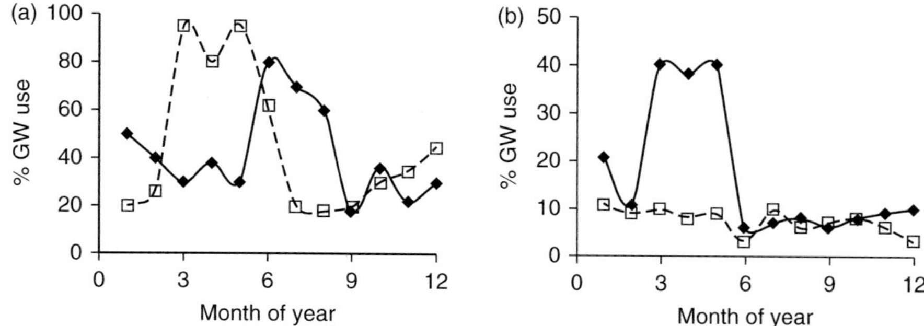

Figure 19.3. Changes in percentage groundwater-use for: (a) *Banksia attenuata* growing in a low lying dampland (solid line closed symbol) and on the dune crest (open symbol dashed line; and (b) *Hibbertia hypericoides* growing in a low lying dampland (solid symbol solid line) or dune crest (open symbol dashed line). Note that month "1" is October, month "4" is January. Months 3–5 (Dec–Feb) are the hot dry summer season in this Mediterranean climate. Redrawn from Zencich *et al.* (2002).

structure or function of one pixel in a RS image is compared to that of an adjacent pixel. If a GDE covers the area of one pixel but not the other it is assumed that during prolonged dry periods the structure/ function of the two vegetation types will diverge because of the different availabilities of water to the two ecosystems. For the vegetation accessing groundwater, dry periods do not induce soil dryness to the same extent (if at all) as experienced by vegetation that is not accessing groundwater. Assessments of vegetation *structure* or *function* are determined for the site of interest and compared to adjacent "control" sites, either at a single time, or, when possible, across several contrasting times (comparisons across "wet" and "dry" periods usually).

Münch and Conrad (2007) examined three sites in the northern Sandveld of South Africa. Landsat imagery was used to identify the presence/absence of wetlands and this imagery was combined with GIS terrain modelling to determine whether GDEs could be identified using a landscape "wetness potential". It is important to note that this application focused on Class II GDEs that is, those reliant on a surface expression of groundwater. The researchers applied the "green island" philosophy and compared the attributes of potential GDEs with the attributes of surrounding land cover three times: in July, after rains started at the end of a dry year; the winter (August) of a wet year; and, at the end of a dry summer. They showed that RS data could be used to classify landscapes and when this process was combined with a spatial GIS-based model using landscape characteristics, a regional-scale map of the distributions of GDEs could be produced.

In arid and semi-arid regions, plant density is generally correlated with water availability and plant density tends to be large when groundwater is available compared to adjacent areas where groundwater is unavailable. Lv *et al.* (2012) used remotely sensed images of a vegetation index (the Normalised Difference Vegetation Index; NDVI; Chapter 7) to assess changes in NDVI as a function of depth-to-groundwater

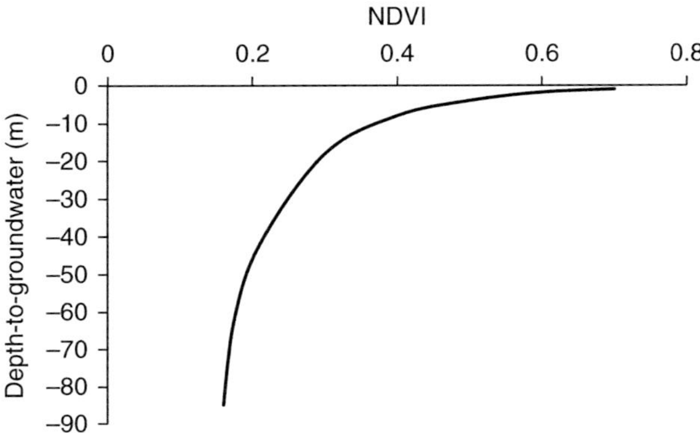

Figure 19.4. The relationship between the normalized difference vegetation index (NDVI) and depth-to-groundwater for the Hailiutu River catchment in northern China. Redrawn from Lv *et al.* (2012).

in northern China. Approximately 29,000 pixels of 300 m resolution of NDVI data were used across a 2600 km^2 catchment and they found a strong relationship between NDVI and groundwater depth (Fig. 19.4; Lv *et al.* 2012).

Clearly the largest NDVI values, which reflect higher vegetation cover, occurred at the shallowest depths of groundwater and cover declines with increasing depth-to-groundwater. A cut-off of approximately 10 m depth-to-groundwater was apparent; when the water table was lower than 10 m vegetation cover was relatively insensitive to further increase in groundwater depth. A similar method was used by Jin *et al.* (2011) for sites within the Gobi desert where oases support agriculture and native vegetation. Maximum NDVI were not observed at the shallowest groundwater sites but at intermediate (2.5–3.5 m) depths. A cut-off of 4.4 m depth-to-groundwater was apparent and vegetation was absent where groundwater depth exceeded 5.5 m.

Most recently Barron *et al.* (2014) mapped the distribution of GDEs by examining the response of two multispectral indices, the NDVI and the normalised difference wetness index (NDWI). The NDVI is a reliable indicator of the density and "health" of the vegetation in a pixel while the NDWI is an index of the moisture content of plant canopies in a pixel and is sensitive to changes in canopy water content. By monitoring and comparing the NDVI and the NDWI during a wet season and a prolonged dry season, Barron *et al.* (2014) were able to identify the location of four land-cover classes: (1) permanent water; (2) vegetation that did not dry during the dry season; (3) vegetation that dried only slowly during the dry season; and, (4) vegetation that dried rapidly during the dry season.

Class 1 (permanent water) may represent very large, rain-fed lakes and reservoirs, and are thus not GDEs, or may represent points of discharge of groundwater to the land surface and therefore the ecology of these ecosystems are groundwater dependent. The method applied by Barron *et al.* (2014) cannot differentiate between these two possibilities. Class 2 ecosystems (vegetation that did not dry out during the dry season) were deemed, through ground-truthing of the Landsat images, to be those with permanent access to groundwater, while Classes 3 and 4 were deemed to be ecosystems with slowly declining access to groundwater (because of increasing depth to the water table during the dry season) or ecosystems with no access to groundwater, respectively.

19.5 Ecophysiology of Terrestrial GDEs Subject to Groundwater Abstraction

Reductions in water availability impact rapidly on vegetation. In the short term reduced growth rates, reduced stomatal conductance and reduced rates of carbon gain, are apparent. In the longer term reductions in leaf area index occur, and in even longer time frames changes in species composition can occur. These responses are summarised in Figure 19.5.

The effects of groundwater abstraction on woodlands has been documented for the Gnangara Mound, a shallow unconfined aquifer of the Swan Coastal Plain in Western Australia (Canham *et al.* 2009, 2012, Stock *et al.* 2012). Over the past several decades increased depth-to-groundwater has been extensively documented and this is the result of a long-term decline in annual rainfall, increased rates of abstraction, and increased discharge (reduced recharge) arising from the development of a plantation industry in the region.

In 1985 widespread mortality (up to 80% mortality close to the abstraction bores) of the native *Banksia* woodland were recorded during the hot dry summer. To determine longer-term floristic changes arising from groundwater abstraction, a series of transect studies were initiated in 1988. Two years after the start of groundwater pumping from the bore field a 2.2 m increase in depth-to-groundwater, coupled to higher-than-normal summer temperatures resulted in a 20–80 percent adult mortality of overstorey species and up to 64 percent mortality in understorey species. Most importantly, control sites, not impacted by groundwater pumping, did not display increased mortality.

Using xylem embolism vulnerability curves (Chapter 3) as an indicator of sensitivity to water stress, Froend and Drake (2006) compared three *Bankisa* and one *Melaleuca* species. They showed that xylem vulnerability reflected the broad eco-hydrological distribution of each species across the topographic gradient present at the site. Froend and Drake were also able to identify a threshold leaf water potential below which increased mortality was likely.

Canham *et al.* (2009) examined Huber values (the ratio of sapwood to leaf area), leaf-specific hydraulic conductivity (k_l), and xylem vulnerability (Chapter 3), of two obligate phreatophytes and two facultative phreatophytes at the same sites used by Froend and Drake (2006). Where water availability was high (groundwater was readily available to the roots) there were no interspecific differences in vulnerability to water stress. However, at sites where groundwater depth was large the two facultative

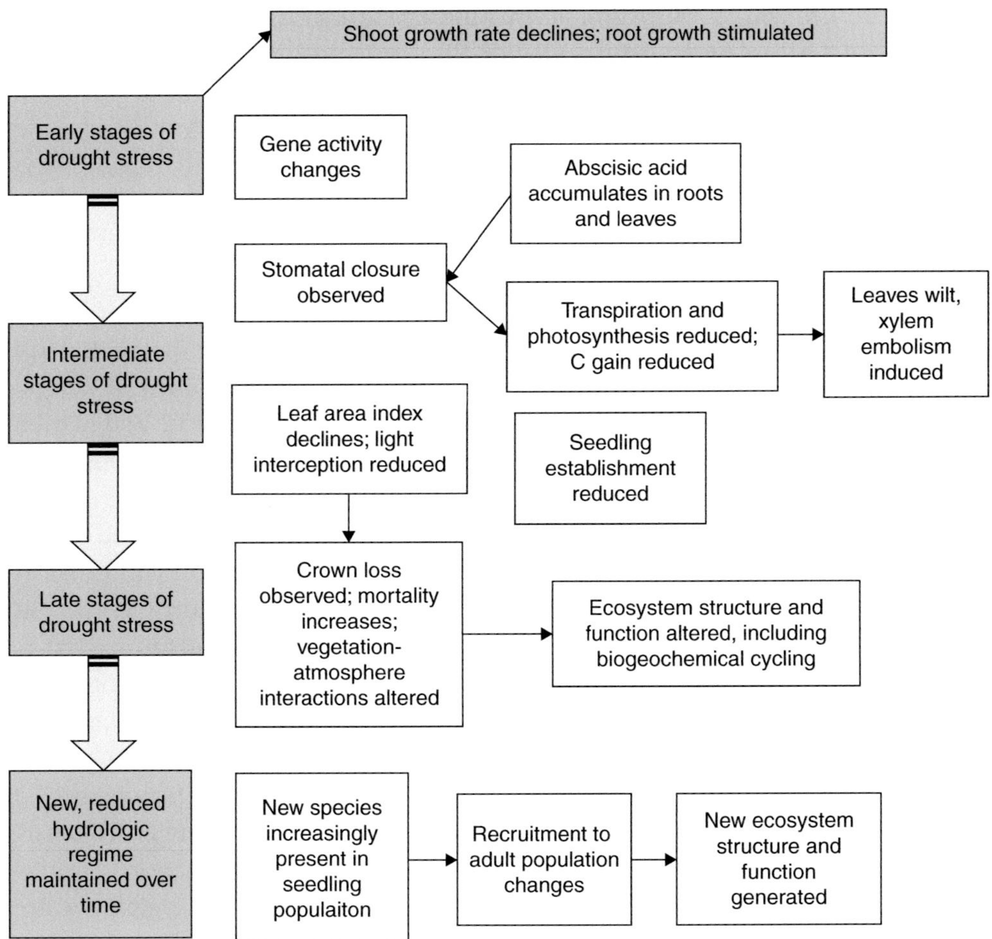

Figure 19.5. Schematic outline of some of the changes in plant physiology, eco-physiology, and ecology, associated with short-, medium–and long-term changes in water availability.

phreatophytes (but not the obligate phreatophytes) were more resistant to xylem embolism at the upper slope than the lower slope (Canham *et al.* 2009).

Changes in the depth-to groundwater impact GDEs directly and rapidly. The first plant organ to experience the change is, of course, the root system, and differences in the responses of roots to changes in depth-to-groundwater contribute to the relative responses of different species to changes in depth-to-groundwater. Where groundwater occurs at shallow depths, root growth can be in synchrony with above-ground growth patterns (Canham *et al.* 2012), which are determined by seasonal patterns in climate. In contrast root growth may occur all year and can be independent of aerial climate at sites where groundwater is too deep for access. Canham *et al.* (2012) observed that as depth-to-groundwater increased during the summer at a particular site (because the site receive only winter rainfall), roots grew increasingly deeper, accessing the capillary fringe. As groundwater recharge occurred in the winter and depth-to-groundwater declined, anoxia resulted in root death at depth. The ability to rapidly increase root depth during the summer is a critical attribute of phreatophytes occupying sites with seasonally dynamic depth-to-groundwater.

Long-term (>2 years) studies of the influence of changes in depth-to-groundwater are required to develop ecosystem response functions for the impact of groundwater abstraction. Froend and Sommer (2010) used results from a forty-year vegetation survey of the Gnangarra Mound in Western Australia. Although the long-term (1976–2008) average rainfall at this site is 850 mm, this has been declining for the past forty years and currently the annual average is about 730 mm. This decline has occurred simultaneously with significantly increased rates of groundwater abstraction. Consequently depth-to-groundwater has increased over the past fifty years by about 1 m. Seasonally, depth-to-groundwater fluctuates about 0.5–3 m, with a maximum depth occurring at the end of the dry summer. Three vegetation communities were identified along two transects: a control transect where gradual increases in depth-to-groundwater occurred (9 cm y^{-1}) and an impacted transect where large rates of change in depth-to-groundwater have occurred. The three sites along each transect correspond to lower-slope, mid-slope, and upper-slope, positions. These locations correspond to shallow, intermediate, and deep, depth-to-groundwater, respectively. Species known to have a high dependency on consistent water supplies (mesic species) were dominant at the down-slope site while xeric species dominated the upper-slope sites.

On the control transect, the hypothesis that groundwater abstraction would result in a replacement of the mesic by the xeric species was not supported. Most of the compositional and structural attributes of the three communities were unchanged. The principle community-scale response was a change in the abundance of mesic and xeric species rather than a complete replacement of one species for another. In contrast to the results of Shafroth *et al.* (2000), mesic species growing on sites with shallow groundwater were not more sensitive to increases in depth-to-groundwater than xeric species.

On the "experimental" transect where the increase in depth-to-groundwater was much faster, changes in composition were far more pronounced and mass mortality observed across all classes (mesic to xeric) of species. This result emphasises the importance of the rate of increase in depth-to-groundwater in determining the response of species and communities.

Small changes in groundwater depth can have large negative impacts on tree health (crown cover), growth (stem increment) and survival. Scott *et al.* (1999) examined the response of riparian cottonwoods (*Populus deltoides*) to a 1–1.5 m increase in depth-to groundwater (Table 19.2) and found rapid (within three years) negative impacts.

Table 19.2. The impact of an increase in depth-to-watertable on riparian cottonwood forest

	Control site	Sites where GW depth increased by 1–1.5 m
Crown volume (% of control)	100	40–80
Stem increment	35–65 cm^2/tree	10–20 cm^2/tree
% mortality	0–10	90–95%

Source: From Scott *et al.* (1999).

19.6 Estimating Rates of Water-Use of GDEs

Estimating groundwater-use by GDEs is a key step to sustainable management of GDEs and groundwater resources. However, it poses many methodological difficulties, including:

1. Up-scaling from tree-scale measurements to stand-scale estimates of tree water-use
2. Partitioning vegetation water-use into rain and groundwater sources
3. Quantifying variations in seasonal and life-cycle rates of groundwater-use
4. Quantifying climatic influences on rates of tree water-use.

Several approaches have been developed to estimate groundwater-use by Class III GDEs including a spreadsheet tool, measuring sub-daily fluctuation in groundwater depth, and, using remote sensing and stable isotopes. These are now discussed.

19.6.1 Field Measurements: Diurnal Fluctuations in GW Depth

Diurnal patterns of vegetation water-use can be detected in shallow unconfined aquifers where roots of vegetation are directly accessing the water table via the capillary zone (Gribovszki *et al.* 2010). The volume of water transpired can be calculated from the change in volume of water in the aquifer that would account for the observed changes in the depth of the water table on an hourly or daily basis. This calculation requires that the specific yield of the aquifer is known. An idealised representation of the deil pattern of groundwater depth in a shallow unconfined aquifer is shown in Figure 19.6.

In Figure 19.6 the solid continuous curve represents the cycle of groundwater drawdown arising from vegetation water-use during the day followed by the "rebound" of the water table when transpiration returns to zero (assuming no nocturnal transpiration) at night. The dashed straight line (with slope = r) is used to estimate the amount of water transpired by vegetation in twenty-four hours (0:00 h to 0:00 twenty-four hours later; indicated by the horizontal dotted arrow). The volume of water transpired in twenty-four hours is represented by the vertical arrow which is the difference between the groundwater depth that would have occurred in the absence of vegetation water-use and the observed groundwater depth. A decline of 0.1 m of the

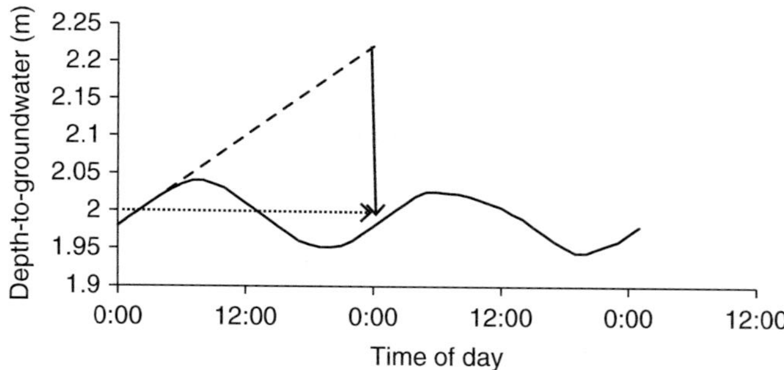

Figure 19.6. A schematic representation of changes in depth-to-groundwater over a 48 h period. The interpretation of this graph is explained in the text.

water table is not equivalent to 100 mm of water being transpired so the specific yield (S_y; the volume of water released from storage by an unconfined aquifer per unit surface area of the aquifer per unit fall of the water table) and the change in the depth of the water table (ΔD) over the 24 h period are used in the calculation of vegetation water-use (E; Eq. 19.1):

$$E = S_y(24r + \Delta D) \tag{19.1}$$

This approach has been successfully applied in many catchments, including the Okavango Delta in Botswana (Bauer *et al.* 2004), an upland grassland catchment in central Argentina (Engel *et al.* 2005), the Sopron Hills of western Hungary (Gribovszki *et al.* 2008), and various sites in the United States (Butler *et al.* 2007, Martinet *et al.* 2009).

Lautz (2008) provides a valuable analysis of the application of the White method to estimate groundwater-use. She shows that spatial differences in groundwater-use can be explained by differences in vegetation type (riparian wetland *versus* grassland) and specific yield of the aquifer. As expected, the ratio of groundwater-to-soil water extraction increased as soil moisture content declined.

19.6.2 Using Stable Isotopes to Estimate Rates of Groundwater-Use

Stable isotopes (^{18}O and deuterium in water) can be used to calculate the proportion of total vegetation water-use that is derived from groundwater (Kray *et al.* 2012, Máguas *et al.* 2011, McLendon *et al.* 2008). To do this, an independent estimate of rates of total vegetation water-use are needed and this can be obtained using eddy covariance (Eamus *et al.* 2013), sapflow sensor (Zeppel *et al.* 2008), and remote sensing technologies (Nagler *et al.* 2009). When only a single isotope is analysed (^{2}H or ^{18}O), a linear mixing model can distinguish between only two potential sources of water (groundwater and soil water). If both isotopes are used spatial resolution is increased and it is possible to distinguish between three sources of water, but only if the two isotopic compositions are independent of each other, which is often not the case.

The results of stable isotope studies of GDEs reveal some valuable insights. First, comparisons across species at a site confirm niche separation in patterns of water uptake. Thus co-occurring species display contrasting patterns (spatially or temporally) in water uptake and this minimises competition for water (Kray *et al.* 2012, Querejeta *et al.* 2007). Second, as depth-to-groundwater increases the proportion of total vegetation water-use that is derived from groundwater often (but not always) diminishes (O'Grady *et al.* 2006, McLendon *et al.* 2008). Finally, the proportion of groundwater used by vegetation usually (McLendon *et al.* 2008) but not always (Kray *et al.* 2012) increases as the upper soil profile dries out. Consequently seasonality of groundwater-use may occur when rainfall is highly seasonal and groundwater availability is maintained throughout the dry season (O'Grady *et al.* 2006; Fig. 19.3).

19.6.3 Remote Sensing Methods to Determine Rates of Vegetation Water-Use

Remote sensing (RS) allows rapid and spatially extensive assessment of vegetation structure (e.g. leaf area index, basal area) and vegetation function (e.g. canopy temperature, rates of evapotranspiration and "greenness"). It also allows determination of relationships among climate variables, vegetation function and structure.

The energy balance of a land surface can be written thus (Eq. 19.2):

$$LE + H = R_n - G \qquad (19.2)$$

where LE is latent energy flux (= ET), H = sensible heat flux, R_n is net radiation, and G is soil heat flux. By measuring differences in temperature between boundary air and canopy, rates of sensible heat flux can be calculated. If it is assumed that over a 24 hour cycle $G = 0$ and R_n is either measured or calculated from remote sensing data, then LE can be calculated by difference. Li and Lyons (1999) used three models based on surface temperatures to estimate ET. The first model only used differences in surface and air temperature; the second required NDVI and surface temperature data. This latter model requires the four extreme values of surface temperature and NDVI to be present within the area of study (i.e. patches of dry bare soils, wet bare soil, wet fully vegetated patches, and dry (water-stressed), fully-vegetated surfaces). The need for these four values makes application of the model problematic. The third model simply used the Priestley-Taylor equation (see Li and Lyons 1999) to estimate potential ET.

Two key functional attributes of terrestrial ecosystems are rates of water-use and rates of carbon fixation. The fluxes of these two gases are coupled through the action of stomata (Chapters 2 and 11) and it is because of this tight coupling that vegetation indices such as NDVI or EVI, which are good proxies of productivity and hence carbon flux, can be also applied in looking for GDEs because it is an increase in water supply that drives their structural and functional differences (compared to adjacent non-GDEs).

Remote sensing can also be used to estimate rates of water flux from landscapes and if comparisons are made across ecosystems within a single climate envelope, it is possible to identify locations with larger-than-expected (based upon consideration of local climate and rainfall) rates of water-use. These sites might, therefore, be GDEs. The key issue here is how to estimate vegetation water-use at a sufficiently small-scale to allow comparison of adjacent vegetation units, thereby allowing identification of those locations where water-use is larger than expected. This is now briefly discussed.

The water balance of a catchment can be written as:

$$P = E + Q + R + \Delta S \tag{19.3}$$

where P is precipitation, E is evapotranspiration, Q is overland flow, R is groundwater recharge, and ΔS is change in soil water storage. Re-arranging equation 19.3 for E highlights the fact that it assumes no contribution of groundwater to evapotranspiration. This is, of course, a failure of many water budget studies. For an excellent review of methods available to estimate land surface evapotranspiration using RS data, see Kalma *et al.* (2008). For a very brief overview of some methods for measuring vegetation water-use, including RS methods, see Whitley and Eamus (2009).

Spatially extensive estimates of vegetation water-use are needed by water resource managers but these cannot be measured using eddy covariance or by sapflow techniques. Remote sensing, in contrast, can provide such estimates. Nagler *et al.* (2005) used MODIS data (eVI and MODIS derived maximum daily air temperatures) to derive an empirical, scaled estimate of riparian *ET* and compared this estimate with point-scale measurements from eddy covariance (EC) data on the Middle Rio Grande of the United States. Their equation for *ET* is (Eq. 19.4):

$$ET = a(1 - \exp^{-bEVI})(c/ (1+ \exp^{-(T_a - d/e)}) + f \tag{19.4}$$

where *ET* is daily evapotranspiration; a, b, c, d, e, f are regression constants from the regression analyses; T_a is air temperature; and, *EVI* is the MODIS-derived enhanced vegetation index. Strong correlations between *EVI*, T_a and *ET* were observed and used to provide scaled estimates for larger areas of vegetation (Fig. 19.7).

A similar method and regression model was developed by Scott *et al.* (2008) that combined eddy covariance measurements with MODIS-derived values of *EVI* and land surface temperature to determine riparian *ET* and rates of groundwater-use along the San Pedro River in the United States. Their model is given in equation 19.5:

$$ET = a(1 - \exp^{-bEVI}) + c^{edT_s} + e \tag{19.5}$$

where *ET* is daily evapotranspiration derived from the eddy covariance measurements; *a, b, c, d* and *e* are regression parameters; and, T_s is night time surface temperature derived from MODIS. The R^2 of the regression of *ET* calculated from the MODIS RS data and the observed *ET* from the EC data was larger than 0.93 for all three years of the study, across all three sites (grassland, shrubland, and woodland),

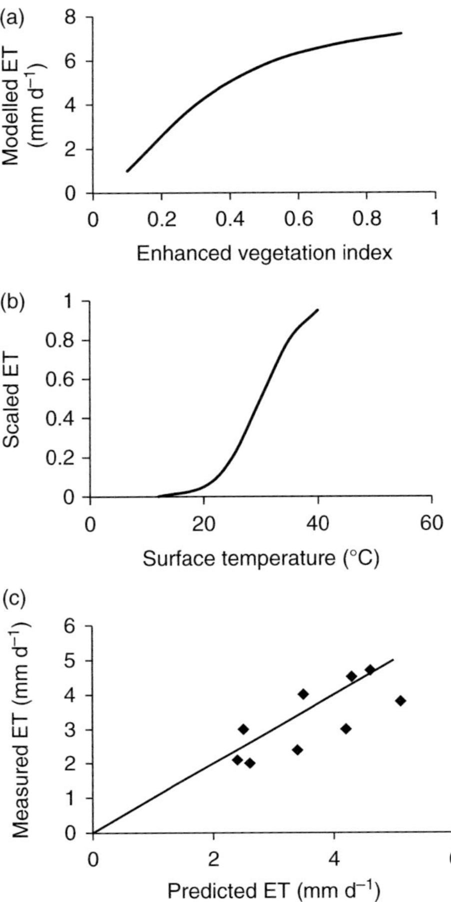

Figure 19.7. Relationships between (a) the enhanced vegetation index (EVI) and modelled evapotranspiration (*ET*); (b) surface temperature and scaled *ET*; and (c) modelled and measured daily *ET* for a riparian forest. Redrawn from Nagler *et al.* (2005). The line in (c) is the 1:1 relationship.

indicating the applicability of this method. The value of this approach is that it is has the ability to: (1) scale from single EC towers to the entire reach of the river; and, (2) estimate the difference between annual rainfall and annual *ET* and thereby estimate groundwater-use by the vegetation. Annual rates of groundwater-use across the three sites across all years ranged from 330 mm to 517 mm, while annual rainfall ranged from 227 to 312 mm. It is unlikely, however, that this empirical model will be applicable to other sites globally.

19.6.4 *Using Remote Sensing to Estimate Groundwater-Use*

It is important to quantify the water balance of arid and semi-arid groundwater basins to define safe yields for those resources. However, it is difficult, time

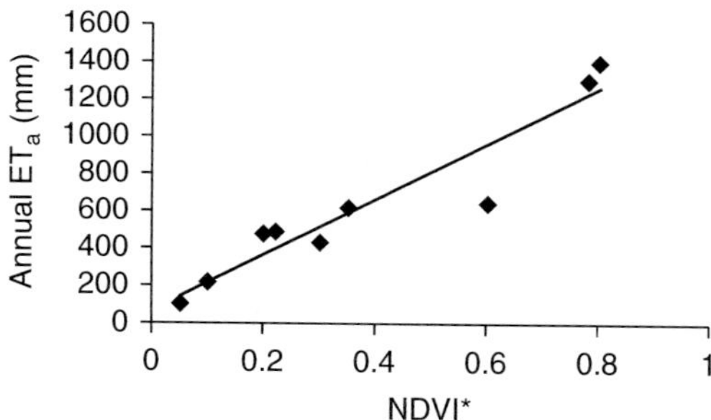

Figure 19.8. Mid-summer adjust normalized difference vegetation index (*NDVI**) is strongly correlated with annual evapotranspiration (ET). Redrawn from Groeneveld *et al.* (2007).

consuming, and expensive to obtain accurate and spatially distributed estimates of vegetation groundwater-use from field measurements. Groeneveld and Baugh (2007) derived a new formulation of the standard NDVI (*NDVI**) and showed how this can be calibrated to quantify rates of evapotranspiration (ET_a; Fig. 19.8) using standard weather data from which to calculate (E_o). E_o is the grass reference *ET* calculated using the Penman-Monteith equation, as described in the FAO-56 method (Allen *et al.* 1998)).

The *NDVI** is functionally equivalent to the crop coefficient (K_c) commonly used in micrometeorology. The approach of Groeneveld and Baugh (2007) and Groeneveld *et al.* (2007) is particularly applicable to arid and semi-arid sites with a shallow water table. At such locations rainfall is low and usually erratic but water supply to roots is relatively constant. Consequently *ET* closely tracks ET_o, which varies as a function of solar radiation, wind speed, and vapour pressure deficit. By using summer peak season NDVI data across a fourteen-year study period, *NDVI** was calculated thus (Eq. 19.6):

$$NDVI* = (NDVI - NDVI_z)/NDVI_m - NDVI_z) \qquad (19.6)$$

where $NDVI_z$ and $NDVI_m$ are the NDVI values for zero vegetation cover and NDVI at saturation, respectively. In their study, Groeneveld and Baugh (2007) were able to disaggregate the influence of groundwater supply from that of recent rainfall.

Groeneveld *et al.* (2007) then applied this approach to three contrasting arid sites in the United States. A linear correlation between measured annual ET_a and mid-summer *NDVI** was obtained across the three sites, despite the sites having very different vegetation composition and structure (Fig. 19.8). However, the regression of ET_a/ET_o versus *NDVI** did not pass through the origin and therefore will introduce an offset error if *NDVI** is used to estimate ET_a. To overcome this, Groeneveld *et al.* (2007) transformed ET_a to ET_a* (Eq. 19.7):

$$ET_a* = (ET_a - rainfall)/(ET_o - rainfall) \qquad (19.7)$$

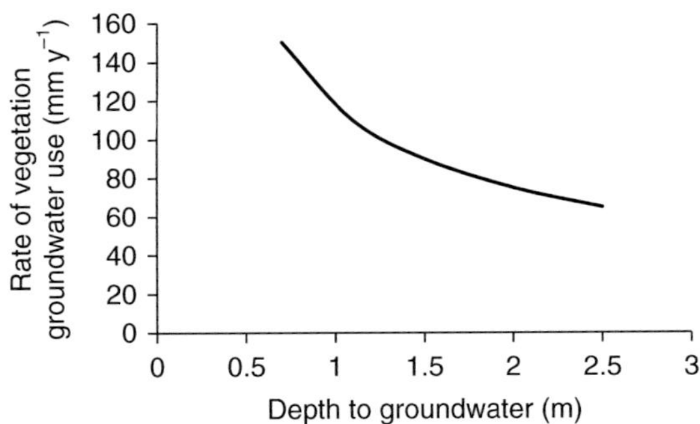

Figure 19.9. As depth to groundwater increased beneath an alkali scrub, groundwater-use declined. Redrawn from Groeneveld (2008).

The regression of *ET* versus NDVI** yielded a regression with a slope of 0.97 and an R^2 of 0.96 with a zero intercept. Thus *NDVI** is a reliable indicator of *ET**. By re-arranging the equation above and substituting *NDVI** for *ET**, they showed that:

$$ET_a \text{ (estimated)} = (ET_o - rainfall)NDVI^* + rainfall \qquad (19.8)$$

By deducting the contribution of annual rainfall to annual ET_a it is possible to calculate the amount of groundwater that is transpired by vegetation (ET_{gw}). Thus, $ET_{gw} = (ET_o - rainfall)\ NDVI^*$.

Groeneveld (2008) applied the methodology of Groeneveld *et al.* (2007), using mid-summer NDVI data, to estimate annual total *ET* of alkali scrub vegetation in Colorado. Annual groundwater-use was then estimated as the difference between annual rainfall and annual *ET* for each year. On-site estimates of groundwater-use were larger than those estimated using NDVI data and ET_o because the remote sensing method does not include surface evaporation of groundwater. Annual ET_{gw}^* was compared to measurements made by Cooper *et al.* (2006) at the same site agreed to within 20%. A strong relationship between depth-to-groundwater and *ET* was observed (Fig. 19.9).

A key question for managers of groundwater resources to answer is: How much water does a GDE use? It is apparent that application of sapflow technologies, eddy covariance technologies, coupled to remote sensing data and modelling, can provide reasonable estimates of annual groundwater-use. Table 19.3 summarises some recent studies of rates of groundwater-use using these methods.

19.7 Groundwater Recharge, Climate and Vegetation

A solid understanding of the controls of recharge of aquifers is required for sustainable management of aquifers. Since Roman times humans have examined relationships among climate and groundwater depth; more recently climate and vegetation

Table 19.3. A summary of some studies that estimated rates of groundwater-use by vegetation

Vegetation type	Location	Rate of GW use	Technique	Reference
Woodland Shrubland Grassland	San Pedro River, Arizona, USA	475–517 mm y^{-1} 340–421 mm y^{-1} 330–403 mm y^{-1}	Eddy covariance; MODIS eVI	Scott *et al.* 2008
Riparian forest: *Populus deltoides;* *Tamarix chinensis;* *Elaeagnus angustifolia*	Rio Grande, New Mexico, USA	0.1–18 mm d^{-1}	"White" method of daily fluctuations in GW depth	Martinet *et al.* 2009
Nevada saltbush; Alkali sacaton; Greasewood; Rabbitbrush; saltgrass	Owens Valley, California, USA	20–85% of annual vegetation water-use is sourced from groundwater	Water balance approach	McLendon *et al.* 2008
Riparian meadow: willows, rushes and sedges	Red Canyon Creek, Rocky Mountains, Wyoming, USA	Growing season average: 0.2–1.6 mm d^{-1}; maximum average 3.1 mm d^{-1}	"White" method of daily fluctuations in GW depth	Lautz 2008
Simulations of GDEs	N/a	1–5 mm d^{-1}	Simulation modelling of daily fluctuations in GW depth	Loheide *et al.* 2005
Riparian forest and *Eucalyptus* woodland	Daly River, NT, Australia	0–50% of daily ET in the dry season	Stable isotope analyses	Lamontagne *et al.* 2005
Alkali scrub vegetation	San Lui Valley, Colorado, USA	64–141 mm y^{-1} across 14 y study period	Remote sensing of NDVI and RS estimates of ET	Groeneveld 2008
Riparian forest	Sopron Hills, Hungary	2.4–6.8 mm d^{-1} during the growing season	"White" method of daily fluctuations in GW depth	Gribovszki *et al.* 2008
Riparian forest: *Populus* spp; *Morus* spp; *Salix* spp.	Arkansas River, Kansas, USA	3–5 mm d^{-1}	"White" method of daily fluctuations in GW depth	Butler *et al.* 2007

interactions have been examined (Kim and Jackson 2012). Generally recharge increases as the total annual rainfall and rainfall intensity increase but declines as potential evapotranspiration rates increase because of increased evaporation and transpiration of soil water. While such general trends are well known, only recently has the importance of vegetation in controlling recharge been accepted (Eamus *et al.* 2006). In a recent global analysis of recharge, Kim and Jackson (2012) identified the following major determinants of rates of recharge:

1. Total landscape rates of water receipt (rainfall plus irrigation plus lateral inflows) was the major determinant of rates of recharge;
2. Vegetation type was the second largest determinant of rates of recharge;
3. Rate of potential evapotranspiration was the third largest determinant of rates of recharge;
4. Saturated soil hydraulic conductivity was the fourth largest determinant.

The relative differences between vegetation types in rates of recharge were largest in dry regions and clayey soil and Kim and Jackson (2012) concluded that future land-use change is likely to significantly influence recharge rates. Global land-surface models rarely include consideration of the role of vegetation in recharge and this limits their application in ecohydrological assessments of groundwater dependent ecosystems.

Palaeohydrological evidence strongly supports the view that the great regional aquifers of the world (for example: The Great Artesian Basin, the High Plains Aquifer, the Nubian Plains Aquifer) store water that is several thousands of years old and that recharge mostly occurred in past millennia (Taylor *et al.* 2012) and very little, if any, significant recharge has occurred since then. Stable isotope analyses of groundwater suggest that recharge in these massive basins occurred before and during the Late Pleistocene (about 126,000–12,000 years ago) glaciation. Because recharge rates are currently very low, mining of this fossil groundwater requires sensitive management. Future increases in temperature and more frequent and extensive droughts (see Chapter 20) will be associated with decreased recharge, because of reduced rainfall, but also increased groundwater abstraction for irrigation and human consumptive use, both of which will decrease groundwater resources. Estimating the direct effects of future climate change on recharge is problematic and likely to be overshadowed by the impacts of land-use change on rates of recharge (Taylor *et al.* 2012).

19.8 References

Allen RG, LS Pereira, D Raes and M Smith, (1998). Crop evapotranspiration-Guidelines for computing crop water requirements–FAO Irrigation and drainage paper 56. FAO.
Barron OV, I Emelyanova, TG van Niel, D Pollock and G Hodgson, (2014). Mapping groundwater dependent ecosystems using remote sensing measures of vegetation and moisture dynamics. *Hydrological Processes* 28, 372–385.

Bauer P, G Thabeng, F Stauffer and W Kinzelbach, (2004). Estimation of the evapotranspiration rate from diurnal groundwater level fluctuations in the Okavango Delta, Botswana. *Journal of Hydrology* 288, 344–355.

Brown L, (2007). Water tables falling and rivers running dry: international situation. *International Journal of Environment* 3, 1–5.

Butler JJ, GJ Kluitenberg, DO Whittemore, SP Loheide, W Jin, MA Billinger and X Zhan, (2007). A field investigation of phreatophyte-induced fluctuations in the water table. *Water Resources Research* 43, W02404, DOI: 10.1029/2005WR004627

Caldwell MM, TE Dawson and JH Richards, (1998). Hydraulic lift: consequences of water efflux from the roots of plants. *Oecologia* 113, 151–161.

Canham CA, RH Froend and WD Stock, (2009). Water stress vulnerability of four *Banksia* species in contrasting ecohydrological habitats on the Gnangara Mound, Western Australia. *Plant, Cell and Environment* 32, 64–72.

Canham CA, RH Froend, WD Stock and M Davies, (2012). Dynamics of phreatophyte root growth relative to a seasonally fluctuating water table in a Mediterranean-type environment. *Oecologia* 170, 909–916.

Clifton CA and R Evans, (2001). Environmental water requirements to maintain groundwater dependent ecosystems. Environmental Flows Initiative Technical Report Number 2. Canberra: Commonwealth of Australia.

Cook PG, G Favreau, JC Dighton and S Tickell, (2003). Determining natural groundwater influx to a tropical river using radon, chlorofluorocarbons and ionic environmental tracers. *Journal of Hydrology* 277, 74–88.

Cook PG, TJ Hatton, D Pidsley, AL Herczeg, A Held, A O'Grady and D Eamus, (1998). Water balance of a tropical woodland ecosystem, Northern Australia: a combination of micro-meteorological, soil physical and groundwater chemical approaches. *Journal of Hydrology* 210, 161–177.

Cooper DJ, JS Sanderson, DI Stannard and DP Groeneveld, (2006). Effects of long-term water table drawdown on evapotranspiration and vegetation in an arid region phreatophyte community. *Journal of Hydrology* 325, 21–34.

Eamus D, J Cleverly, N Boulain, N Grant, R Faux and R Villalobos-Vega, (2013). Carbon and water fluxes in an arid-zone *Acacia* savanna woodland: An analyses of seasonal patterns and responses to rainfall events. *Agricultural and Forest Meteorology* 182–183, 225–238.

Eamus D, R Froend, R Loomes, G Hose and B Murray, (2006). A functional methodology for determining the groundwater regime needed to maintain the health of groundwater-dependent vegetation. *Australian Journal of Botany* 54, 97–114.

Eamus D, CM Macinnis-Ng, GC Hose, MJ Zeppel, DT Taylor and BR Murray, (2005). TURNER REVIEW No. 9. Ecosystem services: an ecophysiological examination. *Australian Journal of Botany* 53, 1–19.

Engel V, EG Jobbagy, M Stieglitz, M Williams, RB Jackson, (2005). Hydrological consequences of eucalyptus afforestation in the argentine pampas. *Water Resources Research* 41, W10409

Evans R, (2007). The Effects of Groundwater Pumping on Stream Flow in Australia. Technical Report, Canberra, Land & Water Australia.

Fan Y, Li H, and G Miguez-Macho, (2013). Global patterns of groundwater depth. *Science* 339, 940–943.

Froend R and P Drake, (2006). Defining phreatophyte response to reduced water availability: preliminary investigations on the use of xylem cavitation vulnerability in *Banksia* woodland species. *Australian Journal of Botany* 54, 173–179.

Froend R and B Sommer, (2010). Phreatophytic vegetation response to climatic and abstraction-induced groundwater drawdown: examples of long-term spatial and temporal variability in community response. *Ecological Engineering* 36, 1191–1200.

Giordano M, (2009). Global groundwater? Issues and solutions. *Annual Review of Environment and Resources* 34, 153–178.

Gleick PH and M Palaniappan, (2010). Peak water limits to freshwater withdrawal and use. *Proceedings of the National Academy of Sciences* 107, 11155–11162.

Gribovszki Z, P Kalicz, J Szilágyi and M Kucsara, (2008). Riparian zone evapotranspiration estimation from diurnal groundwater level fluctuations. *Journal of Hydrology* 349, 6–17.

Gribovszki Z, J Szilágyi and P Kalicz, (2010). Diurnal fluctuations in shallow groundwater levels and streamflow rates and their interpretation–a review. *Journal of Hydrology* 385, 371–383.

Groeneveld DP, (2008). Remotely-sensed groundwater evapotranspiration from alkali scrub affected by declining water table. *Journal of Hydrology* 358, 294–303.

Groeneveld DP and WM Baugh, (2007). Correcting satellite data to detect vegetation signal for eco-hydrologic analyses. *Journal of Hydrology* 344, 135–145.

Groeneveld DP, WM Baugh, JS Sanderson and DJ Cooper, (2007). Annual groundwater evapotranspiration mapped from single satellite scenes. *Journal of Hydrology* 344, 146–156.

Jin XM, ME Schaepman, JG Clevers, ZB Su and G Hu, (2011). Groundwater depth and vegetation in the Ejina area, China. *Arid Land Research and Management* 25, 194–199.

Kalma JD, TR McVicar and MF McCabe, (2008). Estimating land surface evaporation: A review of methods using remotely sensed surface temperature data. *Surveys in Geophysics* 29, 421–469.

Kim JH and RB Jackson, (2012). A global analyses of groundwater recharge for vegetation, climate and soils. *Vadose Zone Journal* 11, DOI: 10.2136/vzj2011.0021RA

Kolb TE, SC Hart and R Amundson, (1997). Box elder water sources and physiology at perennial and ephemeral stream sites in Arizona. *Tree Physiology* 17, 151–160.

Kray J, D Cooper and J Sanderson, (2012). Groundwater-use by native plants in response to changes in precipitation in an intermountain basin. *Journal of Arid Environments* 83, 25–34.

Lamontagne S, PG Cook, A O'Grady and D Eamus, (2005). Groundwater-use by vegetation in a tropical savanna riparian zone (Daly River, Australia). *Journal of Hydrology* 310, 280–293.

Lautz LK, (2008). Estimating groundwater evapotranspiration rates using diurnal water-table fluctuations in a semi-arid riparian zone. *Hydrogeology Journal* 16, 483–497.

Li F and T Lyons, (1999). Estimation of regional evapotranspiration through remote sensing. *Journal of Applied Meteorology* 38, 1644–1654.

Loheide II, SP, JJ Butler and SM Gorelick, (2005). Estimation of groundwater consumption by phreatophytes using diurnal water table fluctuations: A saturated-unsaturated flow assessment. *Water Resources Research* 41, W07030.

Lv J, XS Wang, Y Zhou, K Qian, L Wan, D Eamus and Z Tao, (2012). Groundwater-dependent distribution of vegetation in Hailiutu River catchment, a semi-arid region in China. *Ecohydrology* 6, 142–149.

Máguas C, K Rascher, A Martins-Loucao, P Carvalho, P Pinho, M Ramos, O Correia and C Werner, (2011). Responses of woody species to spatial and temporal ground water changes in coastal sand dune systems. *Biogeosciences Discussions* 8, 1591–1616.

Martinet MC, ER Vivoni, JR Cleverly, JR Thibault, JF Schuetz and CN Dahm, (2009). On groundwater fluctuations, evapotranspiration, and understory removal in riparian corridors. *Water Resources Research* 45 (5), DOI: 10.1029/2008WR007152.

McLendon T, PJ Hubbard and DW Martin, (2008). Partitioning the use of precipitation- and groundwater-derived moisture by vegetation in an arid ecosystem in California. *Journal of Arid Environments* 72, 986–1001.

Münch Z and J Conrad, (2007). Remote sensing and GIS based determination of groundwater dependent ecosystems in the Western Cape, South Africa. *Hydrogeology Journal* 15, 19–28.

Murray BR, GC Hose, D Eamus and D Licari, (2006). Valuation of groundwater-dependent ecosystems: a functional methodology incorporating ecosystem services. *Australian Journal of Botany* 54, 221–229.

Nagler PL, K Morino, K Didan, J Erker, J Osterberg, KR Hultine and EP Glenn, (2009). Wide-area estimates of saltcedar (*Tamarix* spp.) evapotranspiration on the lower Colorado River measured by heat balance and remote sensing methods. *Ecohydrology* 2, 18–33.

Nagler PL, RL Scott, C Westenburg, JR Cleverly, EP Glenn and AR Huete, (2005). Evapotranspiration on western US rivers estimated using the Enhanced Vegetation Index from MODIS and data from eddy covariance and Bowen ratio flux towers. *Remote Sensing of Environment* 97, 337–351.

O'Grady A, P Cook, P Howe and G Werren, (2006). Groundwater-use by dominant tree species in tropical remnant vegetation communities. *Australian Journal of Botany* 54, 155–171.

Phillips DL and Gregg JW (2003). Source partitioning using stable isotopes: coping with too many sources. *Oecologia* 136, 261–269.

Querejeta JI, H Estrada-Medina, MF Allen and JJ Jiménez-Osornio, (2007). Water source partitioning among trees growing on shallow karst soils in a seasonally dry tropical climate. *Oecologia* 152, 26–36.

Scott ML, PB Shafroth and GT Auble, (1999). Responses of riparian Cottonwoods to alluvial water table declines. *Environmental Management* 23, 347–358.

Scott RL, WL Cable, TE Huxman, PL Nagler, M Hernandez and DC Goodrich, (2008). Multiyear riparian evapotranspiration and groundwater-use for a semiarid watershed. *Journal of Arid Environments* 72, 1232–1246.

Seckler D, R Barker and U Amarasinghe, (1999). Water scarcity in the twenty-first century. *International Journal of Water Resources Development* 15, 29–42.

Shafroth PB, JC Stromberg and DT Patten, (2000). Woody riparian vegetation response to different alluvial water table regimes. *Western North American Naturalist* 60, 66–76.

Shiklomanov IA, (2008). World water resources: A new appraisal and assessment for the 21st century, United Nations Educational, Scientific and Cultural Organisation.

Stock WD, L Bourke and RH Froend, (2012). Dendroecological indicators of historical responses of pines to water and nutrient availability on a superficial aquifer in south-western Australia. *Forest Ecology and Management* 264, 108–114.

Taylor RG, B Scanlon, P Doll and M Rodell, (2013). Groundwater and climate change. *Nature Climate Change* 3, 322–329.

Wada Y, LPH and LPH Beek, CM van Kempen, JWTM Reckman, S Vasak and MFP Bierkens, (2010). Global depletion of groundwater resources. *Geophysical Research Letters* 37, L20402.

Whitley R and D Eamus, (2009). How much water does a woodland or plantation use: a review of some measurement methods, Canberra, Land & Water Australia.

Zencich SJ, RH Froend, JV Turner and V Gailitis, (2002). Influence of groundwater depth on the seasonal sources of water accessed by *Banksia* tree species on a shallow, sandy coastal aquifer. *Oecologia* 131, 8–19.

Zeppel MJ, CM Macinnis-Ng, IA Yunusa, RJ Whitley and D Eamus, (2008). Long term trends of stand transpiration in a remnant forest during wet and dry years. *Journal of Hydrology* 349, 200–213.

20

Global-Change Drought and Forest Mortality

20.1 Introduction

In the past century, atmospheric CO_2 concentrations have increased by approximately 110 ppm. During this time, average global surface temperatures have increased by almost 1°C. Over the past fifty years, annual rainfall across large areas of all continents has substantially declined periodically, often for several-to-many years consecutively, resulting in widespread and prolonged drought conditions. While drought is a recurrent feature of global weather patterns, the areal extent, duration, and intensity, of these droughts is unprecedented. Consequently, widespread forest (and grassland) mortality is occurring in response to these droughts. The loss of such large areas of vegetation has major negative impacts on the provision of ecosystem services, landscape albedo, carbon, water and energy balances, soil erosion and biodiversity.

Field assessments of mortality have been undertaken extensively in the United States and Europe, but less extensively in Asia and Australia, where population density is often low across extensive regions and scientific staff and resources tend to be concentrated in cities. Application of remote sensing to regional-scale mortality has been undertaken in the past decade and these studies have been invaluable in quantifying the timing and extent of regional die-back and mortality. In addition, models of the mechanisms of mortality are being developed and these are briefly discussed.

Vegetation mortality (forests and grasslands in particular) have experienced climate-change induced droughts that have persisted for two to seven years across all continents over the past fifty years. The combination of reduced water supply and increased vapour pressure deficit and temperature have resulted in excessive and unrepaired embolism, fatal canopy dehydration, and fatal depletions of carbon stores. Associated decreased resistance to biotic (fungal, insect) stress can also provide the final push into widespread mortality. Remote sensing has provided indices of the climatic drivers, the aerial extent, and the timing of these events.

The three big questions addressed in this case study are: (1) where is regional-scale vegetation mortality occurring; (2) what in regional climate systems is causing mortality; and (3) what mechanisms may link climate to vegetation mortality?

20.2 Global Change-Type Droughts

Drought can be defined as a period of time when rainfall is lower than the long-term average for a given site. Droughts can persist for periods of months-to-years and to more than a decade (Table 20.1). Droughts occur in arid, semi-arid, mesic, temperate and tropical regions and can have major impacts on agricultural production, native landscape productivity, river flow, and groundwater recharge. Consequently droughts negatively affect food supplies, human wellbeing, ecology, the provision of ecosystem services, biogeochemical cycling and local and regional energy, carbon, and water balances. In this case study we examine global-change-type droughts (Breshears *et al.* 2005) and focus on forest mortality, using a combination of field ecophysiological and laboratory plant physiological studies, remote sensing and modelling. However, examples of grassland and shrubland mortality following drought are also extensively documented (Scott *et al.* 2010, McAuliffe and Hamerlynck 2010).

Drought has been a recurrent feature of terrestrial landscapes for longer than Humans have walked across them. Medieval mega-droughts (regional in extent, decades in duration) occurred in North America, while China has experienced extreme and exceptional drought conditions repeatedly since the 1500s, with typical durations of two to five years (Shen *et al.* 2007). Analyses of tree rings and lake sediments both provide important data sets for quantifying both the extent and duration of droughts prior to the development of extensive and reliable instrument records.

During the twentieth century mean land and sea-surface temperatures have increased significantly (almost 1°C; Fig. 20.1). Simultaneously, although the saturated water vapour pressure of the atmosphere has increased because of this increased temperature, vapour pressure deficits (the difference between observed and saturated vapour pressures) have increased because the increase in saturation vapour pressure occurs at a faster rate than the increase in observed vapour pressure (Fig. 20.2). This is a poorly recognised aspect of recent droughts and has important implications for understanding links between climate and mortality.

Global-change type droughts are thus defined as droughts that are associated with above-average temperatures and, it is now increasingly realised, above average vapour pressure deficits (Eamus *et al.* 2013, Breshears *et al.* 2013).

During drought, when rainfall is reduced, rates of evapotranspiration from vegetated surfaces decline and the partitioning of incoming solar radiation between latent and sensible energy fluxes is skewed towards sensible energy fluxes. Thus drought is associated with not only reduced rainfall but also increased air temperatures, increased canopy temperature and increased vapour pressure deficits. This latter effect arises

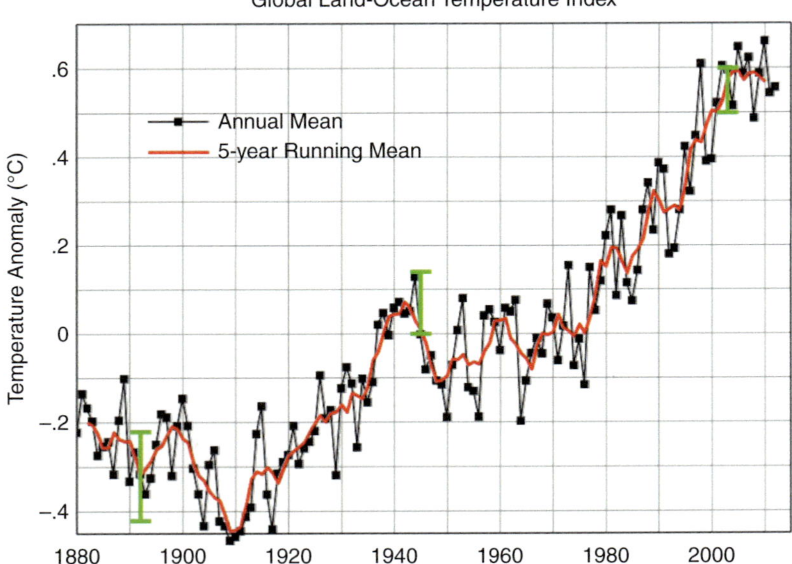

Figure 20.1. Trends in global surface temperature anomalies since 1880. Source of data: NASA.

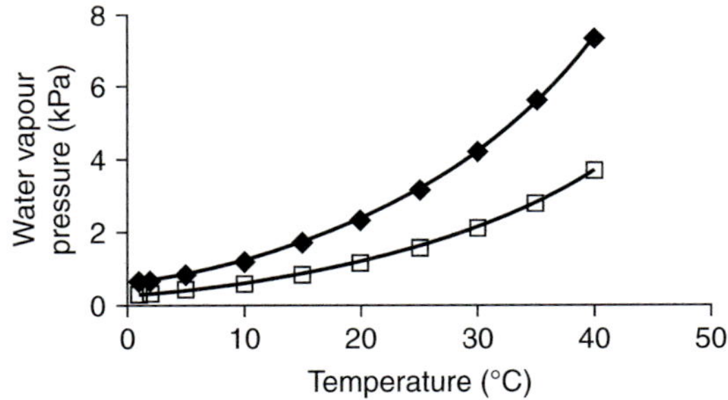

Figure 20.2. Saturated water vapour pressure (diamonds) increases faster than increases in observed water vapour pressure (squares) as air temperature increases. Consequently vapour pressure deficit (the difference between the two) increases with increased air temperature.

from a combination of increased temperature and reduced rates of evapotranspiration from vegetation and soils as soil water content declines.

In an analysis of rainfall, the Palmer Drought Severity Index (PDSI), streamflow and model-simulated soil moisture content, Dai (2011) showed trends in annual surface air temperature, rainfall and run-off for the period 1950–2008 (Fig. 20.3). Since 1950, surface temperatures have increased by 1–3°C and rainfall has declined over much of Africa, southern Europe, south and east Asia and eastern Australia. Note that the rainfall and run-off data are independent of each other but there is good agreement between the two, both spatially and temporally.

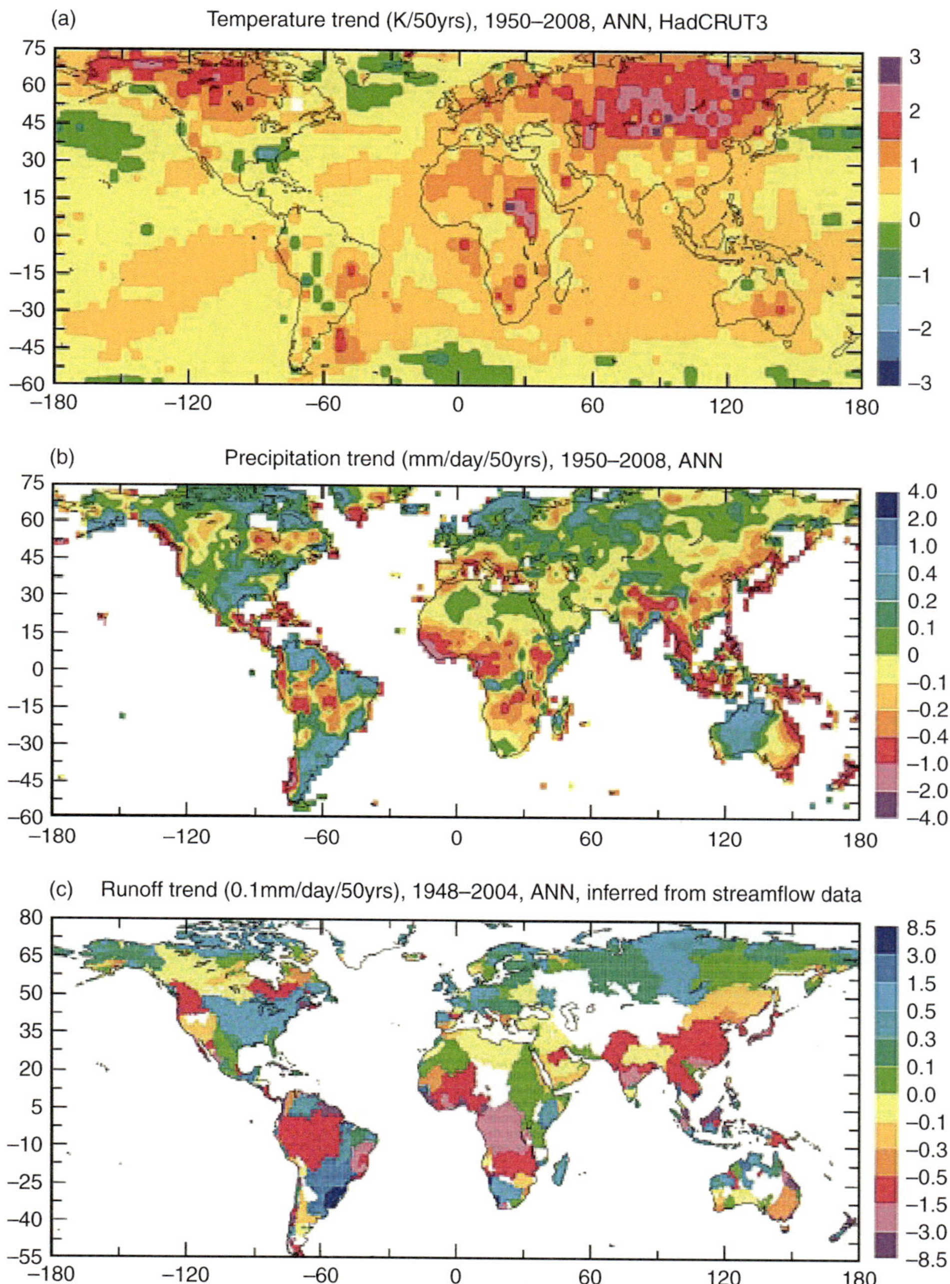

Figure 20.3. Trends in air temperature (1950–2008), rainfall (1950–2008) and run-off (1948–2004). Reproduced with permission from Dai (2011).

As would be expected, given the trends in temperature and rainfall, the PDSI also shows a trend of increasing aridity (a more negative PDSI) for the same period (Fig. 20.4). The PDSI is an index of drought that ranges from −4 (severe drought) to +4 (very wet conditions with much larger than average rainfall). It uses rainfall

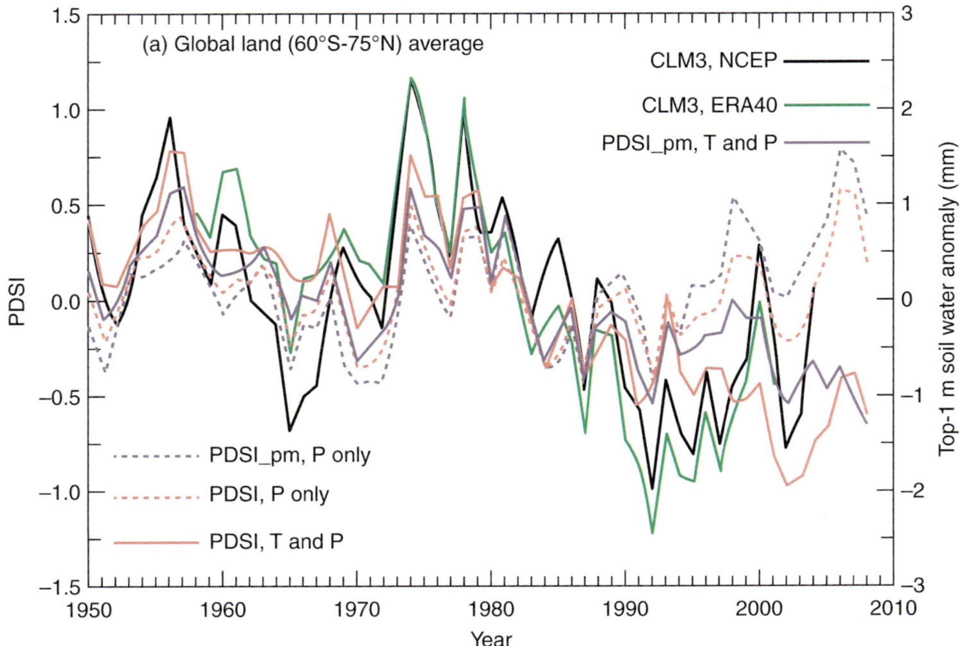

Figure 20.4. Global land surface averaged annual time series of soil moisture anomaly for the top 1 m of soil derived from the land surface model CLM3 and the Palmer drought severity index. Reproduced with permission, from Dai (2011).

and temperature data and works best on longer (several months or more) rather than short (days-to-weeks) time periods. Modelled soil water content also shows a decline for this period. Consequently, a combination of field observations and modelled outputs suggest an increase in aridity for large regions across the globe for the period 1950–2008.

20.3 Field Observations of Drought and Mortality

In a review of published literature for the period 1970–2009, Allen and colleagues identified eighty-eight examples of regional-scale forest mortality (Fig. 20.5). It is clear that drought induced forest mortality has occurred across all forested continents.

Table 20.1 summarises some of the examples of drought induced forest mortality in each continent. It is apparent that all types of forest are susceptible to drought-induced mortality. Asia and Australia are strongly influenced by the El Niño weather phenomenon (Chapter 1) and severe El Niño years of 1982/3 and 1997/8 are associated with extensive regional drought and increased forest mortality in these regions. The Mediterranean region of Europe is characterised by warm-to-hot dry summers and consequently this region has experienced a large frequency of drought-induced mortality in oak, fir, beech, and pine forests. In North America and Canada, approximately 20 million hectares of forests have experienced significant mortality. El Niño

Table 20.1. *A summary of some of the forests of each continent that have experienced drought-induced mortality*

Continent	Forest type	Dates
Africa	Savanna	1972–73; 1988–1992
North America and Canada	Upland temperate mixed forest	1990–2002
	Boreal forest	1990–1997
	Montane mixed coniferous forest	1983–2004
	Coniferous forest	2001–2004
Australia	Tropical savanna	1992–1996
	Tropical savanna	1990–2002
Russia	Boreal and temperate forest	2005–2008
	Montane mixed forest	1987–1988
Asia	Montane tropical forest	1976–1980
	Temperate montane mixed forest	2003–2008
South America	Tropical rainforest	1982–1985
	Tropical rainforest	2005
Europe	Temperate broadleaf forest	1980–1985
	Temperate mixed coniferous	1991–1997
	Temperate mixed coniferous and broadleaf forest	1985–1998 2003–2006
	Temperate broadleaf Mediterranean coniferous	2003–2008

Source: From Allen *et al.* (2010).

seasonal droughts also affect South and Central America and increased mortality is reported for the Amazon Basin, Costa Rica, Panama, and Patagonia. Even forests that are not normally water limited (for example tropical humid rainforest in the Amazon and Borneo) can show drought-induced mortality.

All of the studies reporting forest mortality in Figure 20.5 and Table 20.1 relied on field-based assessments of mortality. These field assessments often use total loss of canopy cover to identify mortality and such studies have revealed much species-level and topography-related variability in rates of mortality within a single catchment. Local variation in stand density, age-distribution, soil characteristics (especially depth and water holding capacity), and run-on/run-off characteristics, all influence local rates of mortality. Understanding the causes of forest mortality and the interplay of drought, temperature, and vapour pressure, remains a challenge and is discussed later in this chapter.

Forests are not the only ecosystems to suffer drought-induced mortality. The twenty-year trend of increased annual rainfall in the arid and semi-arid regions of the American southwest from mid-1970s to the late 1990s, were followed by a drought between 1998 and 2002. For example, the Sonoran and Mojave deserts in this region have experienced large, negative values of the six-monthly standardised precipitation index (SPI), indicating drought conditions for the period 1999–2003. This index compares rainfall at a site for a specified six-month period with the long-term average

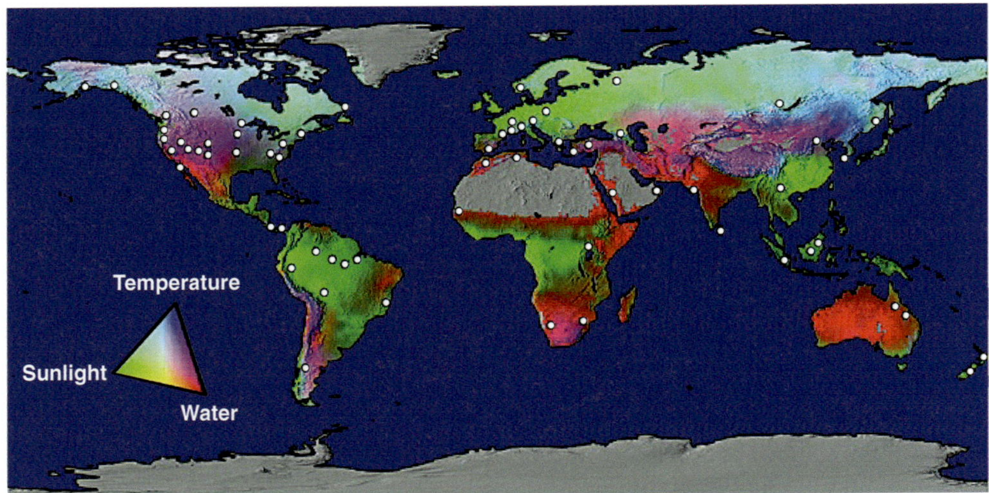

Figure 20.5. A global map indicating the environmental limits to net primary productivity (colours of the continents) and the approximate locations of documented forest mortality related to drought and climate stress. Reproduced with permission from Allen *et al.* (2010).

rainfall (at least thirty years of monthly data are required) for the same time of year. The six-month SPI is good at revealing seasonal patterns in medium-term trends in precipitation and is perhaps more sensitive than the Palmer Drought Severity Index. *Larrea tridentate* (creosote bush) and *Ambrosia dumosa* (burro-weed) and *Ambrosia deltoidea* (triangle bursage) are dominant in both deserts. *Ambrosia* spp. are small drought-deciduous shrubs while *L. tridentate* are longer-lived, larger, drought-tolerant, evergreen shrubs. Rates of mortality for both *Ambrosia* species were far larger (ranging from 12% to 100%) than for *L. tridentate* (ranged from 0 to 35%) across the two deserts (Fig. 20.6). Changes in competitive outcomes, species composition and ecosystem function are likely outcomes of these responses to drought in these deserts.

20.4 Remotely Sensed Observations of Drought and Mortality

Remote sensing provides a method by which changes in forest vigour can be assessed across the globe for the entire period covered by the satellite record. Thus temporal trends and geographic distributions can be assessed using a single protocol.

There are two principle themes in the use of remote sensing for assessing drought and its impacts on forests. The first is a quantification of the climatological degree of drought/non-drought (e.g. Chen *et al.* 1994) and the second is an assessment of the vegetation response (loss of canopy cover) to drought. Both have often used the Normalised Difference Vegetation Index (NDVI) as a principal input.

Land surface temperature is strongly correlated with canopy water content or soil moisture content and clearly during drought, canopy and soil water content decline and

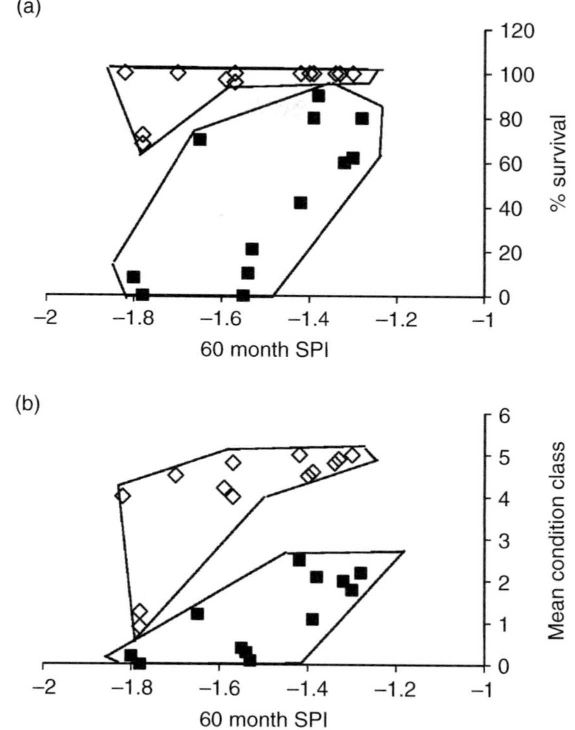

Figure 20.6. Changes in survival and condition class for three co-occurring desert shrubs in the Mojave and Sonoran deserts of the south-western United States. SPI is the standardised precipitation index (see text). Open diamonds are *Larrea tridentata* and closed squares are two *Ambrosia* species. Redrawn from McAuliffe and Hamerynck (2010).

surface temperatures increase (because of the lack of evaporative cooling and increased partitioning of solar radiation into sensible heat flux). Consequently, calculating the ratio of surface temperature to NDVI improves the correlation with canopy and soil water content. Thus Sandholt *et al.* (2002) developed the temperature-vegetation dryness index (TVDI; Eq. 20.1):

$$\mathrm{TVDI} = (T_s - T_{s,\min})/(a + b\mathrm{NDVI} - T_{s,\min}) \qquad (20.1)$$

where $T_{s,\min}$ is the minimum surface temperature observed across the landscape, T_s is the observed surface temperature for each pixel, NDVI is the normalised difference vegetation index for that pixel, and a and b are parameters defining the linear regression of $T_{s,\max}$ against NDVI ($T_{s,\max}$ = a + bNDVI). An area large enough to represent the entire range of surface moistures (from wet to dry) is required to get a reasonable estimate of these parameters. Whether this index is capable of application to monitor drought, which requires deep soil moisture rather than just the upper shallow soil profile (as measured in the TVDI) to be depleted, remains to be determined.

Most recently, Yao et al. (2010) developed an evaporative drought index (EDI) to quantify drought, as given in equation 20.2:

$$EDI = 1 - ET/PET \qquad (20.2)$$

where ET and PET are actual and potential rates of evapotranspiration, respectively.

Yao *et al.* (2010) used a simple evapotranspiration model for actual ET (Eq. 20.3):

$$ET = R_n(0.144 + 0.6495NDVI + 0.009T_{ad} - 0.0163DTAR) \qquad (20.3)$$

where R_n is net radiation, NDVI is normalised difference vegetation index from MODIS, T_{ad} is daytime average air temperature and DTAR is diurnal air temperature range.

PET was estimated from:

$$PET = 0.0023R_a(T_{mean} + 17.8)\sqrt{T_{max} - T_{min}} \qquad (20.4)$$

where R_a is extra-terrestrial solar incident radiation. Yao and co-workers obtained very good agreement between their EDI and on-ground assessments and eddy covariance data. The following scale for EDI was identified:

- Wet conditions: EDI < 0.2
- Normal conditions: EDI = 0.2 to 0.4
- Moderate drought: EDI = 0.4 to 0.6
- Severe drought: EDI = 0.6 to 0.8
- Extreme drought: EDI > 0.8

This EDI (Eq. 20.2) was shown to have a better spatial resolution (4 km) than the Palmer Drought Severity Index. Also it incorporates information about energy fluxes in response to soil moisture stress without the need for large numbers of difficult-to-get meteorological variables and it can be applied at regional and continental scales.

Kogan (1995) developed a vegetation condition index (VCI) also based on the NDVI (Eq. 20.5):

$$VCI = 100*(NDVI - NDVI_{min})/(NDVI_{max} - NDVI_{min}) \qquad (20.5)$$

where NDVI, $NDVI_{min}$ and $NDVI_{max}$ are the smoothed weekly NDVI values and the minimum and maximum NDVI values observed across several years for a specific site. Similarly, Chen and co-workers developed the anomaly vegetation index (AVI) by analysing annual dynamics in NDVI (Chen *et al.* 1994). These indices allow quantification of the response of vegetation through changes in canopy reflectance, itself a function of canopy cover and hence LAI. As tree mortality progresses, tree canopy cover declines (LAI declines) and during drought there is minimal growth of the understorey because of the lack of rain and so a strong signal in the VCI is apparent throughout the duration of the drought.

Hybrid NDVI-based indices of drought have also been developed, including Veg-DRI, an index that combines satellite-based observation of vegetation condition, climate data and biophysical characteristics at 1 km spatial scales (Brown *et al.* 2008). The satellite data include percentage annual seasonal greenness and the start-of-season anomalies, the climate data are the self-calibrated Palmer Drought Severity Index, and the standardised precipitation index and the biophysical data are land use, land cover type, soil available water capacity, and elevation. The VegDRI index is currently applied across the continental United States by the United States Geological Survey and the National Drought Mitigation Centre to produce maps every two weeks at fine resolution (1 km²) of the drought status of the United States.

Remote sensing can also be used to map the temporal trends and spatial extent of mortality events. In the south-western region of the United States are extensive (220,000 km²) piñon–juniper woodlands and forests. A multi-year drought started in 1998 in this region and massive tree dieback (crown thinning; not mortality) began in 2000 for piñon but not juniper. Widespread mortality (in piñon but not juniper) was recorded from 2002 onwards. Breshears and co-workers (Breshears *et al.* 2005) used climate (rainfall, temperature) and soil moisture data for the period 1990 to 2003 and NDVI data to document the extent of forest mortality (Fig. 20.7). As is often observed, drought dramatically reduced tree resistance to secondary stressors, including fungal and beetle infestations and it was the beetle infestation that was the proximal cause of tree death (see discussion on causes of mortality).

Widespread forest mortality alters the regional carbon balance (Potter *et al.* 2009, 2011). Huang *et al.* (2010) used woody photosynthetic vegetation cover (PVC) derived from multi-spectral satellite images (Landsat TM and Enhanced Thematic Mapper images) collected in the dry season, when only the evergreen tree component is present and the understorey herbaceous layer is absent. Time series PVC data were derived using an automated image processing system (see Asner *et al.* 2006 for a full description). By obtaining field-based allometric relationships between crown cover and above-ground biomass for each species, Huang and co-workers were able to accurately quantify the loss of living above-ground biomass (23.3 Mg ha^{-1}) across the 4100 km² study area. Since annual above-ground net primary productivity for these woodlands is about 1–1.5 Mg C ha^{-1} y^{-1}) such a loss represents more than a decade's worth of growth. It is also likely that this rate of loss exceeds that induced by wildfire and other disturbances.

20.4.1 Remote Sensing of Vegetation Function and Drought in High Northern Latitudes

Drought is, of course, associated with reduced rates of evapotranspiration (ET). In high northern latitudes where boreal forest (Chapter 14) and Arctic tundra are found, the warming trend of the twentieth century has had major impacts on hydrology, gas fluxes, loss of permafrost, earlier onset and lengthening of the growing

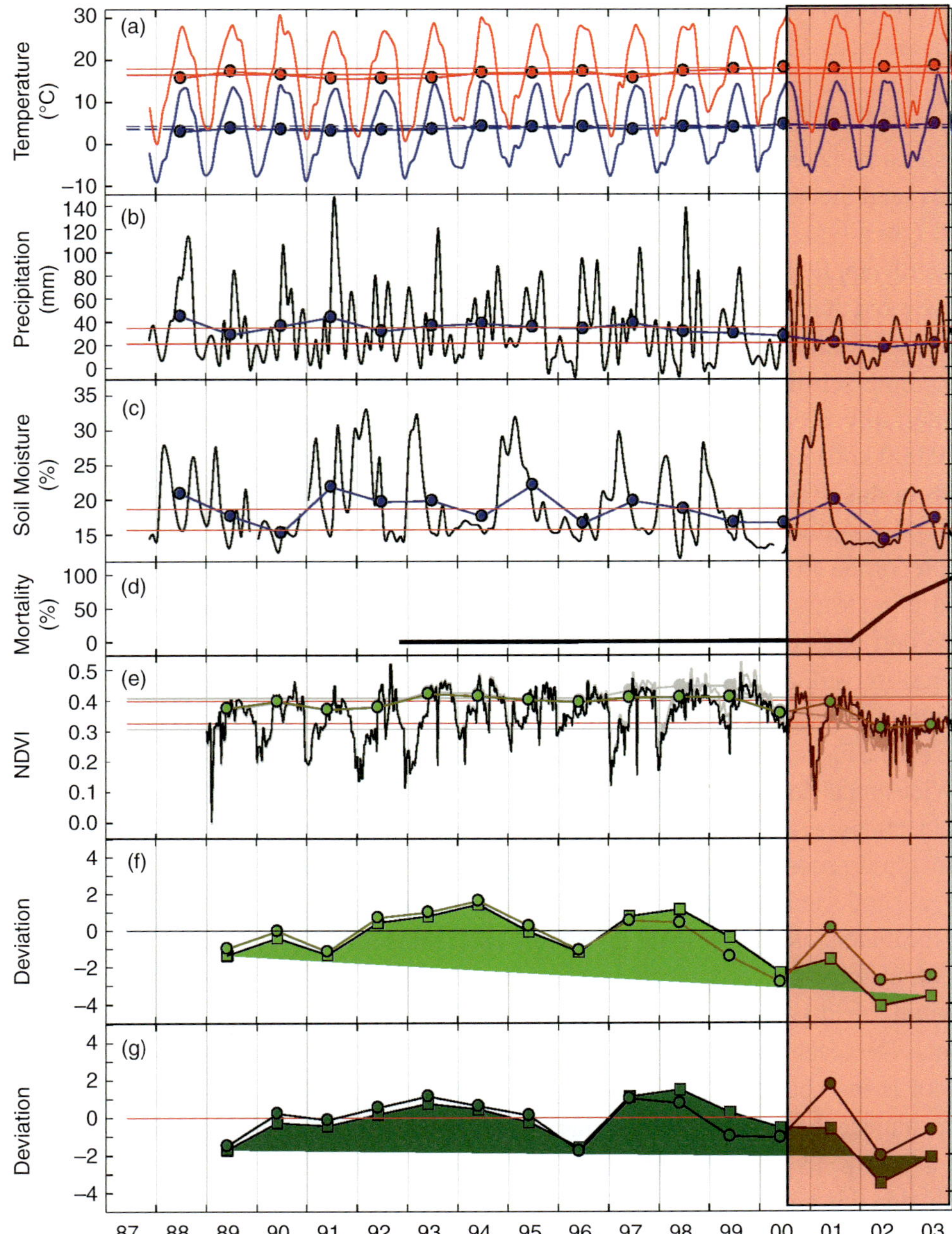

Figure 20.7. Drought induced mortality in a piñon-juniper forest in northern New Mexico. During the drought (2000 onwards) air temperatures were larger, rainfall much lower, rates of mortality larger and the normalized difference vegetation index (NDVI) much smaller, than in the pre-drought period (1990–1999). The last four years (red panel) were subject to drought. Reproduced from Breshears *et al.* (2005), with permission.

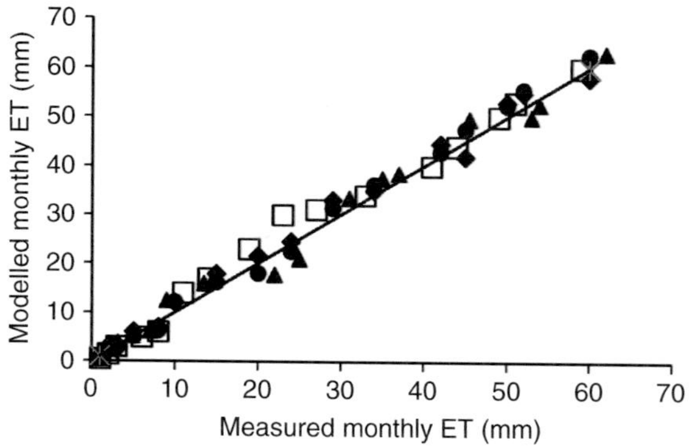

Figure 20.8. Modelled and measured monthly evapotranspiration (ET) for the Boreal region of the northern hemisphere. The three symbols represent three different eddy covariance tower sites. Redrawn from Zhang *et al.* (2009).

season and persistent droughts have been recorded in the late twentieth and early twenty-first century (Zhang *et al.* 2008). Vegetation cover is highly variable in density, structure, and function, both spatially and temporally, and there are few long-term intensive field-based measures of landscape structure and function. Consequently, polar orbiting satellites provide a particularly important tool for investigating these regions.

Zhang *et al.* (2009) developed an ET algorithm that used the Penman-Monteith (PM) equation and a fractional vegetation cover (F_c) model based on NDVI (eq. 20.6):

$$F_c = (NDVI - NDVI_{min})/(NDVI_{max} - NDVI_{min}) \tag{20.6}$$

where $NDVI_{min}$ and $NDVI_{max}$ are the NDVI values for bare soil and dense green vegetation, respectively. In the PM equation surface and aerodynamic resistances are required and these can be estimated using the Jarvis-Stewart formulations (Chapter 2) combined with NDVI estimates of canopy cover. F_c is used to partition available incoming energy into canopy and soil components and these were used in the inverted PM equation to estimate surface conductance. Good agreement between surface conductance derived from eddy covariance data, NDVI values for the tower footprint, and the empirical relationship for surface conductance derived from remote sensing, was obtained for the four dominant vegetation types. Consequently strong correlations of calculated ET *versus* tower data were also obtained (Fig. 20.8).

While small positive trends in both rainfall and ET were apparent across the pan-Arctic region, the standardised summer vegetation moisture stress index revealed drought to have occurred in 1989–1991, 1995, 1998, 2001 and 2003, especially in eastern Alaska, Yukon, southern and central Canada and eastern Siberia and northern Mongolia. Thus boreal forests and grasslands in these regions showed widespread drying trends for the 1983–2005-period in agreement with tree-ring analyses of boreal forests in Canada. Intensification (or speeding up) of the hydrological cycle was apparent, as predicted from atmospheric warming and a lengthening of the growing season.

20.5 Mechanisms that May Explain Tree Mortality

Defining "death" in trees is problematic. Trees are complex, multi-compartment organisms exhibiting internal feedbacks and homeostasis. A complete understanding of the proximal and distal causes of mortality during drought remains elusive. However, significant steps have been made in our understanding.

Death can be defined as the irretrievable loss of function of the tree as an integrated unit. This does not mean the loss of all functions of all parts of the tree simultaneously. It means the loss of integration of function across the various compartments (e.g. xylem, phloem, leaves, roots). Having defined death, the question arises, how does drought induce forest mortality?

There are two main hypotheses proposed to explain tree mortality in response to drought. The first is hydraulic failure; the second is carbon starvation (McDowell *et al.* 2008). This (probably false) dichotomy is exemplified in Figure 20.9. However, it is likely that there is no single causal mechanism. Rather death occurs in response to a systematic failure of integration across the sub-compartments of a tree (see later in this chapter; Anderegg *et al.* 2012a, b).

20.5.1 Hydraulic Failure

To maintain adequate tissue hydration plants must prevent rates of canopy water loss that exceed rates of root water uptake. Rates of canopy water loss are determined by

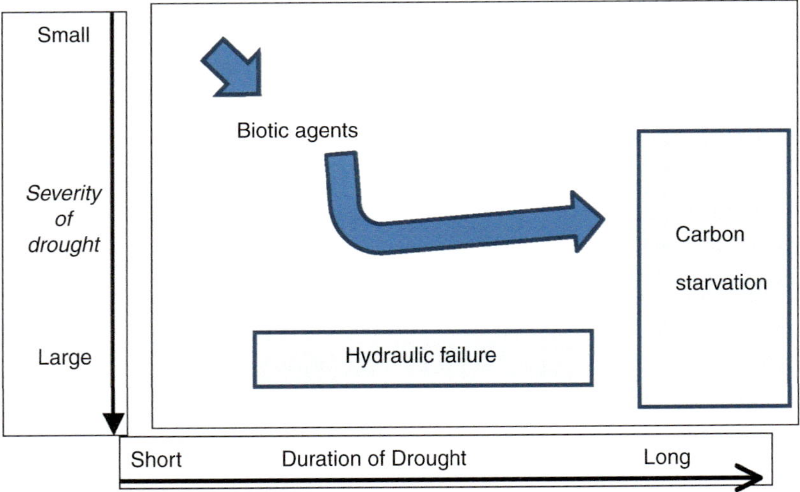

Figure 20.9. A schematic representation of the hydraulic failure and C starvation hypothesis of tree mortality. Based on McDowell *et al.* (2008).

soil moisture content, canopy leaf area, leaf stomatal conductance, vapour pressure deficit, solar radiation, and wind speed (Chapter 3). Rates of root water uptake and transport to the canopy are determined by root distribution, soil water content, and the hydraulic conductance of the transport pathway (principally radial root conductance and xylem conductance in wet soils). This can be summarised in a simple way thus (Eq. 20.7):

$$E = K_l \, (\psi_s - \psi_l - h\rho_w g) \tag{20.7}$$

where E is the rate of canopy water use, K_l is leaf specific hydraulic conductance of the soil-plant continuum, $\psi_s - \psi_l$ is the difference between the soil and leaf water potential and $h\rho_w g$ is the gravitational pull on a column of water of density ρ_w and height h (i.e. the height of the water column in the xylem from root to canopy). If K_l is constant, as the rate of canopy water use increases the difference in water potential between leaf and root increases; that is, the tension in the xylem increases. At some critical value, the water column can snap (Chapter 3) and xylem embolism occurs. The degree of xylem embolism increases as soil water potential declines and different species differ in their xylem vulnerability curves. In addition to hydraulic failure occurring in xylem, soils also exhibit hydraulic failure, whereby the hydraulic connection between root surface and soil water is broken. Both examples of hydraulic failure result in a reduced capacity to supply water to the canopy.

There is a critical xylem water potential at which 50 percent or 88 percent or 100 percent loss of hydraulic conductance occurs. Plant canopies have several strategies available to avoid xylem water potentials inducing these levels of embolism, including reducing stomatal conductance in the short-to-medium (hours-to-weeks) term, reducing canopy leaf area in the medium-to-long term (weeks-to-months), increasing root growth as the upper soil profile dries, and increasing xylem hydraulic conductance. Structural adjustments can be summarised mathematically as in equation 20.8 (Whitehead and Jarvis 1981):

$$E = [(k_s A_s)/(h \, \rho A_l)] \, (\psi_s - \psi_l - h \, \rho_w g) \tag{20.8}$$

where k_s is sapwood specific hydraulic conductivity, A_s is sapwood area, A_l is leaf area and all other terms have been previously defined. Thus, structural adjustments (in canopy leaf area, sapwood area, hydraulic conductivity) can occur in a way such that the gradient in water potential can be kept within constraints to avoid run-away xylem embolism. Such adjustments have been demonstrated in *Pinus*, for example, to maintain a constant stomatal conductance in xeric and mesic environments (Addington *et al.* 2006). For the piñon-juniper forest in northern New Mexico, juniper shows a much lower rate of mortality during drought than piñon (25% compared to 80%) and has more cavitation resistant xylem, a lower leaf area to sapwood area (Huber value; Chapter 3), a lower leaf area to root area ratio, a larger gradient in water potential between root and shoot, and, is shorter. All of these responses are consistent with the adjustments expected to increase drought tolerance implicit in equation 20.8 (McDowell *et al.* 2008).

The hydraulic failure hypothesis postulates that during drought, the gradient in water potential between root and canopy exceeds the critical limit and xylem embolism exceeds any repair processes. Consequently the canopy dehydrates. Growth ceases, phloem transport and hence export of C to roots ceases, extensive leaf mortality and leaf fall occurs, and, respiration rates increase because of increased canopy temperatures (because of the absence of cooling effects of transpiration).

20.5.2 *Hydraulic Failure and Isohydric* Versus *Anisohydric Stomatal Behaviour*

Isohydric species regulate their leaf water potential within relatively small limits through a tight regulation of stomatal conductance (g_s) (Chapter 3). As soil water content declines, g_s declines, thereby limiting transpiration and limiting the decline in leaf water potential as soil water potential declines. In contrast, anisohydric species keep their stomata open as soil water potential declines and consequently leaf water potential declines as soil water potential declines. Anisohydric species may be more common in xeric than mesic environments and consequently have xylem that is more resistant to embolism, although this is not a universally observed rule.

Piñon is isohydric and therefore experiences higher (closer to zero) leaf water potentials than juniper, which is anisohydric. However, piñon is also much more sensitive to xylem embolism (Fig. 20.10). Thus, the isohydric behaviour of piñon, coupled to the high degree of sensitivity to embolism, means that stomata remain closed for much longer during drought in piñon than juniper. Conversely, juniper maintains open stomata for much longer and its leaf water potential declines to much lower values than that seen in adjacent piñon trees. The consequences of

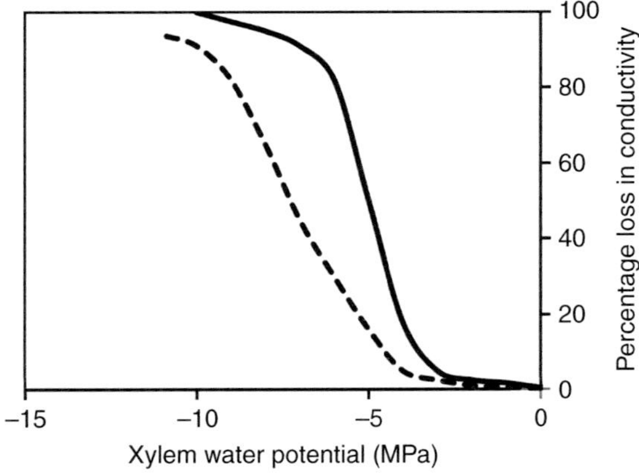

Figure 20.10. Xylem embolism sensitivity curve for stem segments of piñon (solid black line) and juniper (dashed line) showing how juniper (anisohydric) is more resistant to embolism than piñon (isohydric). Redrawn from McDowell *et al.* (2008).

this difference are that piñon trees do not maintain a positive carbon (C) balance once stomata are shut relatively early into a drought. Is this the cause of mortality in piñon? Do droughted trees with closed stomata starve? These issues are now discussed.

20.5.3 Carbon Starvation

The C starvation hypothesis suggests that after prolonged periods of stomatal closure during drought, C reserves are depleted and death occurs, often because the tree cannot fend-off insect and fungal attacks. In support of this hypothesis is the universal observation that the rate of photosynthesis in water stressed plants declines with time. For example, photosynthesis in piñon trees declines to zero when leaf water potentials decline to −2.3 MPa. Given that about 50 percent of current photosynthetic C gain is respired, then a negative C balance is reached when leaf water potential is about −1.0 MPa. In contrast, juniper trees reach a negative C balance when leaf water potentials are about −3.0 MPa, indicating that C starvation is likely to occur much earlier into a drought for piñon trees than for juniper trees. Juniper leaf water potentials can descend to −7.0 MPa during drought, but only about 60% embolism occurs, whereas piñon has a water potential during drought of about −2.5 MPa with very little embolism (Fig. 20.10) but has closed stomata.

Drought, as previously noted, is almost always associated with increased leaf and air temperatures because of increased partitioning of incoming solar radiation into sensible heat flux. This increased temperature also exacerbates C starvation because: (a) temperatures are likely to become supra-optimal for photosynthesis but not respiration, thereby pushing the balance of C gain and C loss towards C loss; and, (b) increased temperatures cause VPD to increase and this tends to reduce stomatal aperture, thereby confounding the reduction in C gain associated with drought.

During the early stage of drought, growth is inhibited earlier than photosynthesis and carbohydrate reserves can increase during these early stages. Enzymes involved in starch degradation and protein synthesis decline in activity while enzymes involved in starch synthesis have increased activity. Stores of carbohydrates decline only after a prolonged drought, not in the early stages. Reductions in phloem transport reduce carbohydrate supplies to roots thereby limiting their ability to explore new depths of soil for water. Furthermore, refilling of embolised xylem is an energy-expensive process (Chapter 3) and the depletion of carbohydrate stores in stems and reduced translocation in the phloem results in a reduced capacity to rebound from the embolism induced by drought.

Changes in carbohydrate status of plants during drought are rarely simple, with some organs showing accumulation of some, but not all, carbohydrates while other organs may show a depletion of some, but not all, carbohydrates. This can be observed in Table 20.2 where, for example, soluble sugar content of leaves of droughted *E. globulus* increased during drought, but leaf starch contents declined during drought.

Table 20.2. Changes in carbohydrate content of three species subject to a long-term drought

		Leaf		Stem		Root		Whole plant		Whole plant
		Starch	*Soluble sugars*	*Starch*	*Soluble sugars*	*Starch*	*Soluble sugars*	*Starch*	*Soluble sugars*	*TNC*
E. globulus	C	27.6	68.1	3.8	28.5	7.9	38.2	14.8	45.6	60.4
	D	**4.3**	**123**	2.1	26.6	**2.8**	**30.6**	**2.9**	**62.5**	65.4
E. smithii	C	55.3	70.6	2.6	26.7	13.5	44.6	20.1	41.7	61.9
	D	**3.7**	**103.3**	2.8	28.3	**3**	**27.4**	**2.9**	51	53.9
P. radiata	C	22.4	54.8	11.6	33.1	31	35.1	19	35.3	54.3
	D	**5.4**	**33.2**	**1.5**	**21**	**2.2**	**47.4**	**2.8**	**30.8**	**33.7**

Abbreviations: C = control (well-watered); D = Drought applied; TNC = total non-structural carbohydrates. Values in bold are significantly different from the respective control (well-watered) value. Units are mg $(gDW)^{-1}$.

Source: From Mitchell *et al.* (2013).

Similarly the roots of *Pinus radiata* accumulated soluble sugars during drought, but stems and leaves were depleted in both starch and soluble sugars (Table 20.2).

In this study, Mitchell *et al.* (2013) compared the responses of three tree species to drought. The two eucalypt species were fast growing and exhibited high rates of water-use. In contrast *Pinus radiata* was slower growing with a smaller rate of water-use. The eucalypts therefore depleted soil moisture more quickly (as evidenced by a more rapid decline in pre-dawn leaf water potential) than the pine. Furthermore, the eucalypts suffered total loss of whole-plant hydraulic conductance in less than one hundred days of drought. In contrast, the pine maintained a measurable whole-plant conductance (albeit much reduced relative to controls) for more than 125 days. Patterns of change in total plant carbohydrate contents differed between the pine and the two eucalypts such that at the end of the drought the eucalypts had depleted starch levels but increased soluble sugar levels, and consequently total non-structural carbohydrate contents for the eucalypts did not differ between drought and well-watered plants. In contrast, the pine exhibited depletions in both starch and soluble sugar contents and consequently whole-plant total non-structural carbohydrate contents for the pine were reduced in the drought plants relative to the well-watered controls (Table 20.2). The cause of this difference in behaviour between the two eucalypts and the pine was the significantly lower leaf water potential at the turgor-loss point in the eucalypts than the pine, allowing the eucalypts to photosynthesise for longer. Hydraulic regulation of plant water status was coordinated with the leaf water potential at the turgor-loss point in all three species but the lower leaf water potential at the turgor-loss point and the maintenance of open stomata in the eucalypts for longer resulted in earlier mortality (90–130 days for the eucalypts but 215 days for the pine), and a more rapid decline in whole-plant conductance, but no loss of total non-structural carbohydrates.

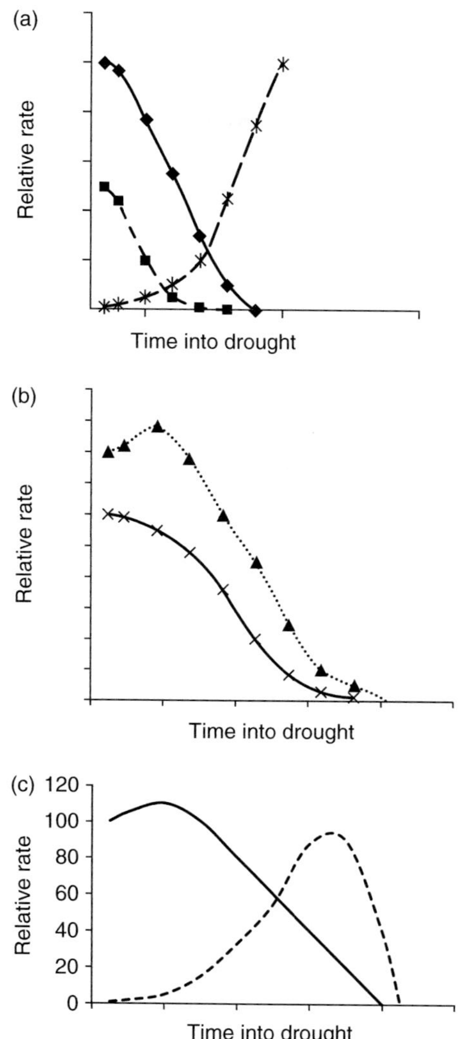

Figure 20.11. Trajectories for key processes and their responses to drought. The upper panels show changes in canopy photosynthesis (diamonds), percentage loss in hydraulic conductivity (stars, dashed line), growth rate (squares) and rates of phloem transport (cross, solid line) and non-structural carbohydrate contents (triangles, dotted line). The lower panel shows changes in the capacity for defence against biotic stress (solid line) and the population density of biotic organisms in the region. Redrawn from McDowell *et al.* (2011).

It is clear that the hydraulic failure hypothesis and the carbon starvation hypothesis are not mutually independent or exclusive. Rather they act synergistically to cause "multiple organ failure" and consequent death (McDowell *et al.* 2011). Some of the known and hypothesised trajectories for key processes and their responses to drought are presented in Figure 20.11.

Not only does the non-structural carbohydrate content of foliage and stems decline during drought in many (but not all) species, there is increasing evidence that the impact of drought on surviving trees can be very long-lasting. Thus Galiano *et al.* (2011) demonstrate for Scots pine that depleted carbon reserves were maintained four years after a major drought, and that this was the result of a long-lasting reduction in leaf area per tree induced by drought. They also showed that mortality of individual trees was correlated with significantly lower rates of carbon storage in those trees.

20.6　Global Convergence in Vulnerability of Forests to Drought

Predicting how forests are likely to respond to future global-change type droughts requires a quantitative understanding of the mechanisms underlying forest responses to drought (Choat *et al.* 2012). The idea of a "hydraulic safety margin" may provide a relatively simple, relatively quick, way to compare, contrast, and quantify, the response of multiple tree species to drought. In the context of forest mortality the hydraulic safety margin is defined as the difference between the minimum xylem water potential experienced by a species at a site (ψ_{min}) and the water potential at which either 50 percent or 80 percent of xylem hydraulic conductance is lost (ψ_{50}, ψ_{80} respectively) by that species. (An alternate definition is briefly discussed in Chapter 3). The values of ψ_{50} and ψ_{80} are derived from xylem vulnerability curves (Chapter 3). While large between-species differences in both ψ_{50} (or ψ_{80}) and ψ_{min} are observed in the literature, it is apparent that there is global convergence to two linear relationships between ψ_{min} and ψ_{50}, one for angiosperms and one for gymnosperms (Fig. 20.12). These two regressions were derived from a data-set of 223 species across sites that differ in mean annual rainfall from 200 to 4500 mm. Three conclusions are apparent from the regressions in Figure 20.12. First, resistance to embolism, as measured by ψ_{50}, is tightly correlated with the level of drought stress, as measured by ψ_{min}. Second, the safety margin of gymnosperms is larger than that of angiosperms. Third, 70 percent of all species sampled operated at narrow safety margins of less than 1.0 MPa, and this small margin was apparent in both xeric and mesic species so that the size of the safety margin was independent of annual rainfall.

An alternate method to quantifying sensitivity to drought has been used by Brodribb and Cochard (2009) and Urli *et al.* (2013). These studies used the minimum recoverable water potential (ψ_{rec}; Urli *et al.* 2013) or the maximum survivable water stress (Brodribb and Cochard 2009) for a range of ecophysiologically meaningful variables, including predawn leaf water potential, rate of whole-plant water-use, leaf-scale transpiration, g_s, and, rate of photosynthesis. In a study of five angiosperm species, Urli *et al.* (2013) found a linear correlation between individual values of ψ_{rec} for each of the five ecophysiologically meaningful variables and cavitation resistance, as measured through P_{50} (the water potential at which 50% loss of xylem conductance was observed) for all five angiosperm species. Significantly, in all five species the water potential that caused an 88 percent loss of xylem conductance was associated with non-recovery from drought. This study specifically selected five angiosperm species that differ significantly in their drought tolerance and specifically used a range of

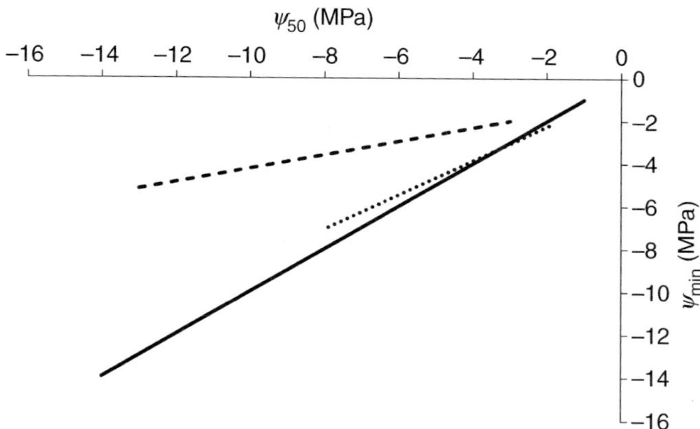

Figure 20.12. The minimum xylem water potential (ψ_{min}) declines linearly with increasing resistance to embolism as inferred from the xylem water potential that results in 50 percent loss of hydraulic conductance (ψ_{50}) for both angiosperms (dotted line) and gymnosperms (dashed line) but gymnosperms maintain a larger safety margin (the vertical distance between the regression and the 1:1). Redrawn from Choat *et al.* (2012).

ecophysiologically-meaningful variables. Consequently, the authors conclude that for angiosperms, a global convergence for the upper limit of drought tolerance is the water potential that causes 88 percent loss of xylem conductance (Urli *et al.* 2013). Importantly the five ecophysiologically meaningful variables (rate of photosynthesis, leaf- and whole-plant scale water-use, g_s, pre-dawn leaf water potential), when taken together, were reflective of a loss of integration of function across multiple plant compartments (leaves, roots, xylem). This point is discussed further later in this chapter. The minimum recoverable water potential for the five angiosperm species examined by Urli *et al.* (2013) was very close to P_{88} values for each of the five species. This is in contrast to the results of the four gymnosperm species comparison of Brodribb and Cochard (2009) who found that the minimum recoverable water potential was close to P_{50} for each species. These differences between angiosperms and gymnosperms reflect differences in xylem structure between the two taxa. Gymnosperm xylem is almost entirely composed of tracheids and these have both a transport and structural role. In contrast the xylem of angiosperms contains vessels and thick-walled fibres. The former are longer and wider than tracheids and it is the fibres that provide structural support, not the tracheids.

20.7 Modelling the Interactions Amongst Drought and Increased Temperatures and VPD

The soil-plant-atmosphere model of Williams *et al.* (2001) has been successfully applied to describe water, carbon, and energy fluxes in a range of boreal, tropical, and temperate ecosystems (Whitley *et al.* 2011, Zeppel *et al.* 2008). Most recently

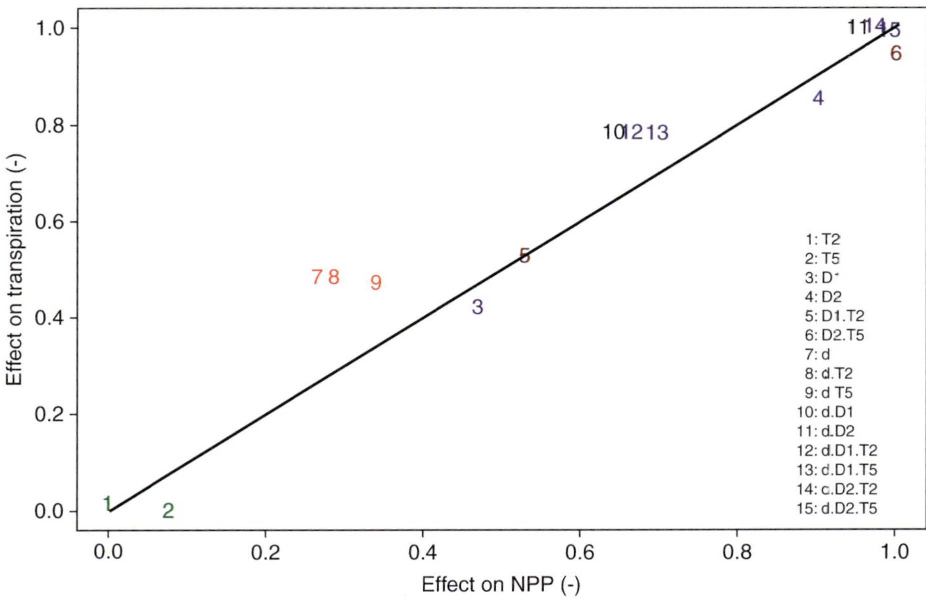

Figure 20.13. Normalised changes in net primary productivity (NPP) and transpiration resulting from increased in temperature, drought and vapour pressure deficit, applied as single factors or in combination. d = drought; *D* = vapour pressure deficit; T = temperature. Simulations of climates with increased temperature applied as a single factor or increased temperature and drought, have a smaller impact on transpiration and NPP than increased *D* applied alone or increased *D* and drought applied together. From Eamus *et al.* (2013). The impact of a treatment is larger as the normalised value increases towards a value of one.

the SPA model was parameterised for a range of climate scenarios in which the interaction of drought with increased temperature and increased vapour pressure deficit (*D*) were examined, singly and in combination (Eamus *et al.* 2013). The behaviour of net primary productivity (NPP), stomatal conductance, and soil water content, were examined across a two-year simulation period. The unequivocal finding was that it was drought plus increased *D* that had a far larger negative impact on NPP than drought plus increased temperature (Fig. 20.13; Table 20.3). Thus drought, applied as a three-month absence of summer rain, resulted in an approximately seven-fold increase in the number of days when NPP was zero (Table 20.3) and an 18 percent decline in total annual NPP. Increasing air temperature (without drought) had little impact on either NPP or the number of days when NPP was zero but increasing *D* alone did have a large negative impact on the number of days when NPP was zero and a large negative impact on total NPP. The interaction of drought with increased *D* was highly significant, and the reduction in NPP was far larger than when drought occurred with increased temperature alone (Table 20.3). The reasons for this differential effect of temperature and *D* was because increased *D* increased the evaporative demand of the atmosphere more than increased temperature and the frequency of stomatal closure occurring in response to limiting soil moisture content was dramatically increased (Eamus *et al.* 2013).

Table 20.3. The influence of drought, increased air temperature, and increased vapour pressure deficit, and their interactions on NPP, of an Australian, temperate eucalypt woodland

Simulated conditions	Meteorological variables adjusted	Number of days across the 2 years for which NPP = zero	Two-year sum of NPP (gCO_2 m^{-2} y^{-1})
1) Control	None; two-year measured field meteorological data used as inputs.	13	7700
2) Drought (defined as an absence of rainfall for the three summer months)	Rainfall excluded for the summer (December 1 year 1 to February 28 year 2, inclusive), all other days identical to the control simulation (1).	88	6300
3) Increased air temperature, D unchanged	Air temperature was increased by 5°C above the control data for the two year simulation. All other input data remained unchanged.	14	7200
4) Increased D, air temperature unchanged	D increased by 1 kPa above the observed (control) data. All other input data remained unchanged.	197	5000
5) Small increase in air temperature and D	Air temperature increased by 2°C and D increased by 1 kPa throughout the two year simulation. All other input data remained unchanged.	175	4100
6) Larger increase in air temperature and D	Air temperature increased by 5°C and D increased by 2.5 kPa throughout the two year simulation. All other input data unchanged.	284	2000
7) Drought plus increased D, temperature unchanged	3 month summer drought (as per (2) above) and D increased by 1 kPa.	246	4279
8) Drought plus increased air temperature, D unchanged	3 month summer drought (as per (2) above) and temperature increased by 2°C and D unchanged.	87	6250
9) Drought plus increased D, small increase in air temperature	3 month summer drought (as per (2) above) and D increased by 1.0 kPa with temperature increased by 2°C.	251	4200
10) Drought plus increased D, larger increase in air temperature	3 month summer drought (as per (2) above) and D increased by 1.0 kPa with temperature increased by 5°C.	453	4100

Source: From Eamus *et al.* (2013).

20.8 An Integrated View of Mortality

Whilst the C starvation hypothesis is attractive and makes intuitive sense, it is not universally accepted as a mechanism for mortality in trees during drought (Anderegg *et al.* 2012a). Rather it is suggested that it is the loss of integration of function across multiple compartments that leads to, and defines, death in trees (Anderegg *et al.* 2012b). In an examination of non-structural carbohydrates (NSC) of roots of aspen (*Populus tremuloides*) in the western United States, Anderegg *et al.* (2012a) could find no depletion of NSC at the limit of its distribution where NSC reserves were assumed to be smallest, during a period of documented aspen die-back. Furthermore, in experimentally-induced droughts (in glasshouse and field locations) no depletion of NSC was observed in leaves, branches, stems, or roots. Finally, in a detailed field assessment of trees that were experiencing widespread drought, and where 68 percent of trees that were sampled subsequently died, no depletion of NSC was observed in any tissue. In contrast, evidence of hydraulic failure was observed in field measurements of hydraulic conductivity of roots and branches and leaf water potentials.

Anderegg *et al.* (2012b) propose that drought leads to death because the gradual dehydration of all tree tissues results in:

1. Decreased nutrient uptake by roots
2. Decreased export of carbohydrates to roots
3. Decreased xylem-phloem communication
4. Decreased turgor and hence decreased cell growth and stomatal conductance
5. Decreased C uptake
6. Increased biotic stresses (fungal, bacterial and insect attack)
7. Reduced xylem repair and regrowth during and after drought, leading to "cavitation fatigue" whereby sensitivity to a second drought is larger than the sensitivity observed to the first drought.
8. Loss of integration through long-distance transport of hormones because of changes in rates of xylem and phloem transport.

Thus Anderegg *et al.* (2012b) conclude that transport failure at the whole-tree level is equivalent to multiple organ failure in animals, which invariably leads to death and transport failure arises because of an accumulation, over time, of the water debt arising from drought. An interaction across carbon, water, energy, and nutrient cycles results in whole-tree-scale loss of integration and subsequent death.

20.9 Ecological Modelling of Mortality at Landscape Scales

Manion was possibly one of the earliest scientists to attempt to link climate, environmental stress (for example, acid rain), biotic stresses (for example insect or fungal attack), and forest health (Manion 1981). Manion talked about predisposing, inciting, and contributing, factors in the transition of a forest from healthy to unhealthy and eventually to mortality. Predisposing factors include old age and species-specific niche requirements (e.g. shade tolerant or intolerant) and are chronic stresses that reduce resistance to inciting factors that arrive after the onset of predisposing factors. Inciting factors (for example drought) and contributing factors (e.g. insect attack or fungal disease) are, perhaps, interchangeable categories.

Under a changing climate, with warmer temperatures, larger vapour pressure deficits, and reduced rainfall, the climate envelope at a site is altered (Wang *et al.* 2012; Fig. 20.14). This can reduce plant function and viability and hence increase the rate of mortality. Countering this are 'stabilising factors"; for example increased atmospheric CO_2 concentrations can, in some locations, act to counter the effect of climate change by stimulating growth (through a CO_2 fertilisation effect) and increasing water-use-efficiency, thereby increasing drought resistance.

Whilst representing the key processes of photosynthesis, stomatal conductance, energy exchange, nutrient cycling, growth, and transpiration, mechanistically in a number of soil-plant-atmosphere models and land-surface models is now routine, capturing the process of mortality is not adequately represented because we do not have a complete mechanistic understanding of the causes of mortality of vegetation. However, one approach has been to base the causes of mortality on starvation, when respiration exceeds GPP for sufficient period of time (see Eamus *et al.* 2013 for application of a SPA model where NPP = zero was taken as a measure of the causes of mortality). Thus mortality (M) is given by equation 20.9 (Wang *et al.* 2012):

$$\mu - \begin{cases} 0 & \dfrac{GPP + C_s}{C_{Demand}} \geq 1 \\ a & \dfrac{GPP + C_s}{C_{Demand}} < 1 \end{cases} \tag{20.9}$$

where a is a constant value ranging from 0 to 1, GPP is gross primary productivity, C_s is stored non-structural carbohydrate, and C_{demand} is the total demand for carbon for respiration and other metabolic processes (Guneralp and Gertner 2007). Some models have a mortality function that does not range from 0 to 1 but from a threshold value greater than zero.

Alternatively some models have mortality arising from xylem cavitation and the loss of hydraulic conductivity (Martinez-Vilalta *et al.* 2002). Dynamic global vegetation models (DGVMs) are models that represent biogeochemical and hydrological cycling within a model that also includes vegetation dynamics at seasonal, annual, and multi-decadal, time-scales. However, current DGVMs such as ORCHIDEE and CABLE are not able to represent drought-induced mortality well, although some attempts are leading to improvements (e.g. in LPJ; Sitch *et al.* 2003).

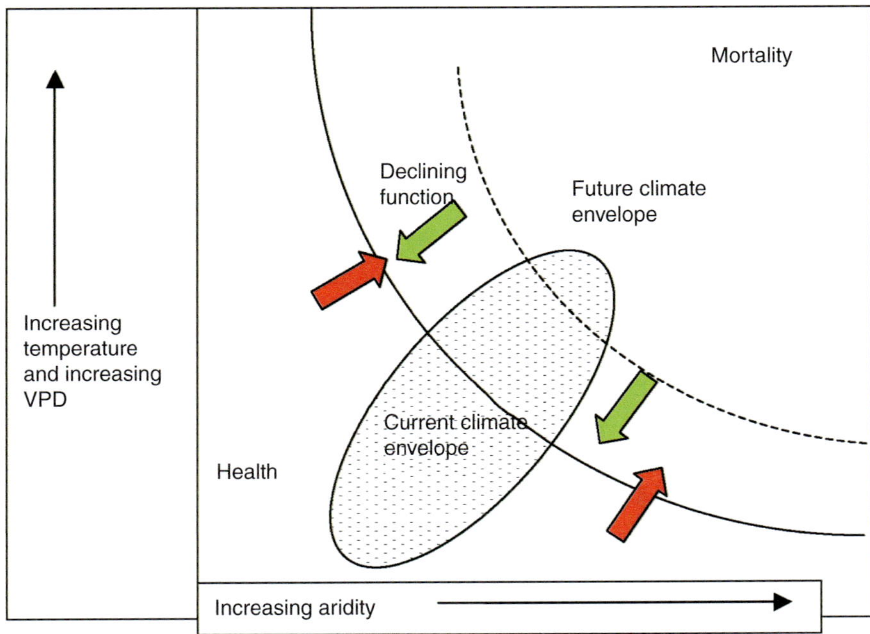

Figure 20.14. A conceptual representation of the influence of a change in climate envelope arising from changes in temperature and vapour pressure deficit (VPD) and increasing aridity. Red arrows represent the effects of inciting and predisposing factors while green arrows represent the counter-effect of stabilising factors. Modified from Wang *et al.* (2012).

20.10 Summary

This chapter has demonstrated that widespread forest mortality has been extensively documented on all forested continents over the past fifty years and that the extent, severity, and frequency, of these climate-change-type droughts might be increasing. The decline in water supplies and associated increased vapour pressure deficit and temperature result in the loss of hydraulic conductivity and consequent canopy dehydration, along with, in some cases, apparently fatal depletions of carbon stores within the plant. Insect and fungal attacks are generally secondary agents associated with forest mortality. Remote sensing of regional landscapes has provided indices of the climatic characteristics of drought as well as measurement of the timing and aerial extent of the loss of vegetation function.

20.11 References

Allen CD, AK Macalady, H Chenschouni, D Bachelet, N McDowell and M Vennetier, (2010). A global overview of drought and heat-induced tree mortality reveals emerging climate change risks for forests. *Forest Ecology and Management* 259, 660–684.
Anderegg WRL, JA Berry and CB Field, (2012a). Linking definitions, mechanisms, and modelling of drought-induced tree death. *Trends in Plant Science* 17, 693–700.

Anderegg WRL, JA Berry, DB Smith, JS Sperry, LDL Anderegg and CB Field, (2012b). The roles of hydraulic and carbon stress in a widespread climate-induced forest die-off. *Proceedings of the National Academy of Science* 109, 233–237.

Asner GP, EN Broadbent, PJC Oliveira, M Keller, DE Knapp and JNM Silva, (2006). Condition and fate of logged forests in the Brazilian Amazon. *Proceedings of the National Academy of Sciences* 103, 12947–12950.

Breshears DD, NS Cobb, PM Rich, KP Price, CD Allen and RG Balice, (2005). Regional vegetation die-off in response to global-change type drought. *Proceedings of the National Academy of Sciences* 95, 15144–15148.

Breshears DD, HD Adams, D Eamus, NG McDowell, DJ Law, RE Will, AP Williams and CB Zou, (2013). The critical amplifying role of increasing atmospheric moisture demand on tree mortality and associated regional die-off. *Frontiers in Plant Science* 4, 266. DOI 10.3389/fpls.2013.00266

Brodribb TJ and H Cochard, (2009). Hydraulic failure defined the recovery and point of death in water-stressed conifers. *Plant Physiology* 149, 575–584.

Brown JF, BD Wardlow, T Tadesse, MJ Hayes and BC Reed, (2008). The Vegetation Drought Response Index (VegDRI): A new integrated approach for monitoring drought stress in vegetation. *Remote Sensing* 45, 16–46.

Chen W, Q Xiao and Y Sheng, (1994). Application of the anomaly vegetation index to monitoring heavy drought in 1992. *Remote Sensing of the Environment* 9, 106–112.

Choat B, S Jansen, TJ Brodribb, H Cochard, S Delzon, R Bhaskar, SJ Bucci, (2012). Global convergence in the vulnerability of forests to drought. *Nature* 491, 752–755.

Dai A, (2011). Drought under global warming: a review. *Wiley Interdisciplinary Reviews: Climate Change,* 2, 45–65.

Eamus D, N Boulain, J Cleverly and DD Breshears, (2013). Global change-type drought-induced tree mortality: vapour pressure deficit is more important than temperature *per se* in causing decline of tree health. *Ecology and Evolution* 3, 2711–2729.

Fisher RA, M Williams, AL da Costa, Y Malhi, RF da Costa, S Almeida and PW Meir, (2007). The response of an Eastern Amazonian rainforest to drought stress: Results and modelling analyses from a through-fall exclusion experiment. *Global Change Biology* 13, 1–8.

Galiano L, J Martínez-Vilalta and F Lloret, (2011). Carbon reserves and canopy defoliation determine the recovery of Scots pine 4 yr after a drought episode. *New Phytologist* 190, 750–759.

Güneralp B and G Gertner, (2007). Feedback loop dominance analysis of two tree mortality models: relationship between structure and behavior. *Tree Physiology* 27(2): 269–280.

Huang C-Y, GP Asner, NN Barger, JC Neff, ML Floyd, (2010). Regional aboveground live carbon losses due to drought-induced tree dieback in piñon-juniper ecosystems *Remote Sensing of Environment* 114, 1471–1479.

Huete AR, K Didan, YE Shimabukuro, P Ratana, SR Saleska, LR Hutyra, W Yang, RR Nemani and R Myneni, (2006). Amazon rainforests green-up with sunlight in dry season. *Geophysical Research Letters* 33 doi:10.1029/2005GL025583.

Kogan FN, (1995). Application of vegetation index and brightness temperature for drought detection. *Advances in Space Research* 15, 91–100.

Manion PD, (1981). *Tree disease concepts.* Prentice Hall, Englewood Cliffs, USA.

Martínez-Vilalta J, J Piñol and K Beven, (2002). A hydraulic model to predict drought-induced mortality in woody plants: an application to climate change in the Mediterranean. *Ecological Modelling* 155, 127–147.

McAuliffe JR and EP Hamerlynck, (2010). Perennial plant mortality in the Sonoran and Mojave deserts in response to severe multi-year drought. *Journal of Arid Environments* 74, 885–896.

McDowell NG, WT Pockman and CD Allen, (2008). Mechanisms of plant survival and mortality during drought: why do some plants survive while others succumb to drought? *New Phytologist* 178, 719–739.

McDowell NG, DJ Beerling, DD Breshears, RA Fisher, KF Raffa and M Stitt, (2011). The interdependence of mechanisms underlying climate-driven vegetation mortality. *Trends in Ecology and Evolution* 26, 523–532.

Mitchell PJ, AP O'Grady, DT Tissue, DA White, ML Ottenschlaeger and EA Pinkard, (2013). Drought response strategies define relative contributions of hydraulic dysfunction and carbohydrate depletion during tree mortality. *New Phytologist* 197, 862–872.

Potter C, S Klooster, C Hiatt, V Genovese and JC Castilla-Rubio, (2011). Changes in the carbon cycle of Amazon ecosystems during the 2010 drought. *Environmental Research Letters* 6, doi:10.1088/1748–9326/6/3/034024.

Potter C, S Klooster, A Huete, V Genovese, M Bustamante, LG Ferreira, de Oliveira RC Junior and R Zepp, (2009). Terrestrial carbon sinks in the Brazilian Amazon and Cerrado region predicted from MODIS satellite data and ecosystem modelling. *Biogeosciences Discussion* 6, 947–969.

Sandholt I, K Rasmussenand and J Andersen, (2002). A simple interpretation of the surface temperature/vegetation index space for assessment of surface moisture status. *Remote Sensing of Environment* 79, 213–224.

Scott RL, ET Hamerynck, GD Jenerette, MS Moran and GA Barron-Gafford, (2010). Carbon dioxide exchange in a semi-desert grassland through drought-induced vegetation change. *Journal of Geophysical Research* 115 GO3026.

Shen C, W-C Wang, Z Hao and W Gong, (2007). Exceptional drought events over eastern China during the last five centuries. *Climatic Change* 85, 453–471.

Sitch S, B Smith, IC Prentice, A Arneth, A Bondeau, W Cramer, JO Kaplan, S Levis, W Lucht, MT Sykes, K Thonicke and S Venevsky, (2003). Evaluation of ecosystem dynamics, plant geography and terrestrial carbon cycling in the LPJ dynamic global vegetation model. *Global Change Biology* 9, 161–185.

Urli M, AJ Porte, H Cochard, Y Guengant, R Burlett and S Delson, (2013). Xylem embolism threshold for catastrophic hydraulic failure in angiosperm trees. *Tree Physiology* 33, 672–683.

Wang W, C Peng, DD Kneeshaw, GR Larocque and Z Luo, (2012). Drought induced tree mortality: ecological consequences, causes and modelling. *Environmental Reviews* 20, 109–121.

Whitehead D and PG Jarvis, (1981). Coniferous forests and plantations. In *Water Deficits and Plant Growth, Vol. VI.* TT Kozlowski (Ed), Academic Press, New York, pp. 49–52.

Whitley RJ, CMO Macinnis-Ng, LB Hutley, J Berringer, M Zeppel, M Williams, D Taylor and D Eamus, (2011). Is productivity of mesic savannas light-limited or water-limited? Results of a simulation study. *Global Change Biology* 17, 3130–3149.

Williams M, BE Law, PM Anthoni and MH Unsworth, (2001). Use of a simulation model and ecosystem flux data to examine carbon–water interactions in ponderosa pine. *Tree Physiology* 21, 287–298.

Yao Y, S Liang, Q Qin and K Wang, (2010). Monitoring drought over the conterminous United States using MODIS and NCEP reanalysis-2 data. *Journal of Applied Meteorology and Climatology* 49, 1665–1680.

Zeng N, A Mariotti and P Wetzel, (2005). Terrestrial mechanisms of interannual CO_2 variability. *Global Biogeochemical Cycles* 19, GB1016, doi:10.1029/2004GB002273.

Zeppel MJ, C Macinnis-Ng, A Palmer, D Taylor, R Whitley, S Fuentes, I Yunusa, M Williams and D Eamus, (2008). An analysis of the sensitivity of sap flux to soil and plant variables assessed for an Australian woodland using a soil-plant-atmosphere model. *Functional Plant Biology* 35, 509–520.

Zhang XY, YQ Wang, XC Zhang, W Guo and SL Gong, (2008). Carbonaceous aerosol composition over various regions of China during 2006. *Journal of Geophysical Research* 113, D14; DOI: 10.1029/2007JD009525

Zhang K, JS Kimball, Q Mu, LA Jones, SJ Goetz and SW Running, (2009). Satellite based analysis of northern ET trends and associated changes in the regional water balance from 1983 to 2005. *Journal of Hydrology* 379, 92–110.

Index

Printed in the United Kingdom by TJ Clays Ltd.